HANDBOOK OF PHARMACEUTICAL ANALYSIS BY HPLC

This is Volume 6 of

SEPARATION SCIENCE AND TECHNOLOGY

A reference series edited by Satinder Ahuja

HANDBOOK OF PHARMACEUTICAL ANALYSIS BY HPLC

Edited by

Satinder Ahuja

Ahuja Consulting
Calabash, North Carolina

Michael W. Dong

Purdue Pharma
Ardsley, New York

ELSEVIER
ACADEMIC
PRESS

Amsterdam – Boston – Heidelberg – London – New York – Oxford – Paris – San Diego – San Francisco – Singapore – Sydney – Tokyo

ELSEVIER B.V.
Radarweg 29
1043 NX
Amsterdam
The Netherlands

ELSEVIER Inc.
525 B Street, Suite
1900 San Diego
CA 92101-4495
USA

ELSEVIER Ltd
The Boulevard
Langford Lane
Kidlington
Oxford OX5 1GB
UK

ELSEVIER Ltd
84 Theobalds Road
London WC1X 8RR
UK

Notice
No responsibility is assumed by the Publisher for any injury and/or damage to persons or property as a matter of products liability, negligence or otherwise, or from any use or operation of any methods, products, instructions or ideas contained in the material herein. Because of rapid advances in the medical sciences, in particular, independent verification of diagnoses and drug dosages should be made.

First edition 2005

ISBN: 0-12-088547-6

♾ The paper used in this publication meets the requirements of ANSI/NISO Z39.48-1992 (Permanence of Paper).

Printed in the United Kingdom.

CONTENTS

2. Key Concepts of HPLC in Pharmaceutical Analysis

RAPHAEL M. ORNAF AND MICHAEL W. DONG

3. HPLC Instrumentation in Pharmaceutical Analysis: Status, Advances, and Trends

MICHAEL W. DONG

4. HPLC Columns for Pharmaceutical Analysis

UWE D. NEUE, BONNIE A. ALDEN, PAMELA C. IRANETA, ALBERTO MÉNDEZ, ERIC S. GRUMBACH, KIMVAN TRAN, AND DIANE M. DIEHL

5. Sample Preparation for HPLC Analysis of Drug Products

CLARK K. CHOI AND MICHAEL W. DONG

6. HPLC Method Development

HENRIK T. RASMUSSEN, WEIYONG LI, DIRK REDLICH, AND M. ILIAS JIMIDAR

7. Validation of HPLC Methods in Pharmaceutical Analysis

ASHLEY S. LISTER

10. Regulatory Considerations in HPLC Analysis

CHRISTINE F. RICHARDSON AND FRITZ ERNI

11. HPLC System Calibration for GMP Compliance

MICHAEL W. DONG

12. System Validation: Pre-Installation, IQ, OQ, and PQ

WILHAD M. REUTER

13. Assay and Stability Testing

ANISSA W. WONG AND ARUNA DATLA

14. Impurity Evaluations

TODD CECIL AND ERIC SHEININ

15. HPLC in Dissolution Testing

QINGXI WANG AND VIVIAN GRAY

16. Cleaning Validation using HPLC for Analysis

ANDREW PLASZ

17. LC/MS Application in High-Throughput ADME Screen

RON KONG

18. Chiral Separations

Y. VANDER HEYDEN, D. MANGELINGS, N. MATTHIJS, AND C. PERRIN

19. Applications of LC/MS in Pharmaceutical Analysis

LEON Z. ZHOU

PREFACE

High-pressure liquid chromatography is frequently called high-performance liquid chromatography (both are abbreviated HPLC or, simply, LC) because it offers improved performance over classical liquid chromatography. HPLC is the premier analytical technique in pharmaceutical analysis, which is predominantly used in the pharmaceutical industry for a large variety of samples. It is the method of choice for checking the purity of new drug candidates, monitoring changes or scale-ups of synthetic procedures, evaluating new formulations, and scrutinizing quality control/assurance of final drug products. To support each new drug application or commercial product, tens of thousands of HPLC tests are conducted by a host of dedicated scientists to assure the potency and quality of the new drug product.

Presently there is no definitive text in HPLC that specifically addresses the needs of the busy pharmaceutical scientist on the pivotal subject of pharmaceutical analysis. This handbook strives to offer a complete yet concise reference guide for utilizing the versatility of HPLC in drug development and quality control. The *Handbook of Pharmaceutical Analysis by HPLC* can be broadly classified into six major sections:

1. Overview, theory, instrumentation, and columns (Chapters 1–4)
2. HPLC methods and practices, including sample preparation and ion chromatography (Chapters 5–9)
3. Regulatory aspects of ICH guidelines, instrumental calibration, and validation (Chapters 10–12).
4. HPLC applications: assays, impurity evaluation, dissolution testing, cleaning validation, high-throughput screening, and chiral separations (Chapters 13–18).

5. Post-chromatographic and tandem techniques: LC/MS, LC/NMR, and chromatographic data handling (Chapters 19–21)
6. New developments in HPLC (Chapter 22).

Each of the 22 chapters (see table of contents), written by selected experts in their respective fields, provides the reader with an in-depth understanding of HPLC theory, hardware, methodologies, regulations, applications, and new developments.

The main focus of this book is on small drug molecules and pharmaceutical dosage forms. This handbook provides practical guidelines using case studies on sample preparation, column or instrument selection, and summaries of "best practices" in method development and validation, as well as "tricks of the trade" in HPLC operation. It captures the essence of major pharmaceutical applications (assays, stability testing, impurity testing, dissolution testing, cleaning validation, high-throughput screening, and chiral separations) in addition to nuances in interpreting ICH guidelines and instrument qualification. The book also highlights novel approaches in HPLC and the latest developments in hyphenated techniques, such as LC-NMR or LC-MS, and in data handling.

We would like to thank the authors for their contributions, which have enabled us to put together a unique handbook that provides the reader with an in-depth understanding of HPLC theory, hardware, methodologies, regulations, and applications. Their excellent contributions will serve as a definitive reference source for laboratory analysts, researchers, managers, and executives in industry, academe, and government agencies, who are engaged in various phases of analytical research and development or quality control.

Satinder (Sut) Ahuja
Michael Dong

CONTRIBUTORS

Numbers in parentheses indicate the page on which the authors' contributions begin.

Satinder Ahuja (1, 611) Ahuja Consulting, 1061 Rutledge Ct, Calabash, NC 28467 USA

Bonnie A. Alden (77) Waters Corp., 34 Maple St, Milford, MA 01757 USA

Todd Cecil (359) U.S. Pharmacopeia, 12601 Twinbrook Parkway, Rockville, MD 20852 USA

Clark K. Choi (123) Purdue Pharma, 444 Saw Mill River Rd, Ardsley, NY 10502 USA

Aruna Datla (335) ALZA Corporation, 1015 Joaquin Rd, Mountain View, CA 94043 USA

Diane M. Diehl (77, 611) Waters Corp., 34 Maple St, Milford, MA 01757 USA

Michael W. Dong (19, 47, 123, 255, 291) Purdue Pharma, 444 Saw Mill River Rd, Ardsley, NY 10502 USA

Fritz Erni (273) Novartis Pharma AG, Lichtstrasse 25, CH-4056, Basel, Switzerland

Vivian Gray (379) V.A. Gray Consulting Inc., 9 Yorkridge Trail, Hockessin, DE 19707 USA

Eric S. Grumbach (77) Waters Corp., 34 Maple St, Milford, MA 01757 USA

Pamela C. Iraneta (77) Waters Corp., 34 Maple St, Milford, MA 01757 USA

M. Ilias Jimidar (145) Johnson & Johnson Pharmaceutical Research and Development a division of Janssen Pharmaceutica n.v., Turnhoutseweg 30 B-2340, Beerse, Belgium

Ron Kong (413) Synaptic Pharmaceutical Corporation, 215 College Rd, Paramus, NJ 07652 USA

Weiyong Li (145) Johnson & Johnson Pharmaceutical Research and Development, 1000 Rt 202, POB 300, Raritan, NJ 08869 USA

Ashley S. Lister (191) Analytical Sciences Worldwide Research & Development, Purdue Pharma, 444 Saw Mill River Rd, Ardsley, NY 10502 USA

D. Mangelings (447) Vrije Universiteit Brussel, Faculty of Medicine and Pharmacy Dept. Pharmaceutical and Biomedical Analysis, Laarbeeklaan 103 B-1090, Brussels, Belgium

N. Matthijs (447) Vrije Universiteit Brussel, Faculty of Medicine and Pharmacy Dept. Pharmaceutical and Biomedical Analysis, Laarbeeklaan 103 B-1090, Brussels, Belgium

Robert P. Mazzarese (581) Waters Corp., 34 Maple St, Milford, MA 01757 USA

Alberto Méndez (77) Waters Corp., 34 Maple St, Milford, MA 01757 USA

Uwe D. Neue (77) Waters Corp., 34 Maple St, Milford, MA 01757 USA

Raphael M. Ornaf (19) Purdue Pharma, 444 Saw Mill River Rd, Ardsley, NY 10502 USA

C. Perrin (447) Laboratoire de chimie analytique-UMR 5625, Faculté de pharmacie, 15 avenue, Charles Flahault. 34093, Montpellier, France

Andrew Plasz (401) Abbott Laboratories Pharmaceutical Analytical Research and Development, 1401 Sheridan Rd, North Chicago, IL 60064 USA

Christopher Pohl (219) Dionex Corp., 1228 Titan Way, PO Box 3603, Sunnyvale, CA 94088 USA

Henrik T. Rasmussen (145) Johnson & Johnson Pharmaceutical Research and Development, 1000 Rt 202, POB 300, Raritan, NJ 08869 USA

Dirk Redlich (145) Johnson & Johnson Pharmaceutical Research and Development, 1000 Rt 202, POB 300, Raritan, NJ 08869 USA

Wilhad M. Reuter (303) PerkinElmer LAS, 710 Bridgeport Ave., Shelton, CT, 06484 USA

Christine F. Richardson (273) Novartis Ringaskiddy Limited, Ringaskiddy, Co. Cork, Ireland

Eric Sheinin (359) U.S. Pharmacopeia, 12601 Twinbrook Parkway, Rockville, MD 20852 USA

KimVan Tran (77) Waters Corp., 34 Maple St, Milford, MA 01757 USA

Y. Vander Heyden (447) Vrije Universiteit Brussel, Faculty of Medicine and Pharmacy Dept. Pharmaceutical and Biomedical Analysis, Laarbeeklaan 103 B-1090, Brussels, Belgium

Qingxi Wang (379) Pharmaceutical Research & Development, Merck Research Laboratories, West Point, PA 19486 USA

Mieng-Hua Wann (569) Purdue Pharma, 444 Saw Mill River Rd, Ardsley, NY 10502 USA

Anissa W. Wong (335) ALZA Corporation, 1010 Joaquin Rd, Mountain View, CA 94043 USA

Leon Z. Zhou (499) Analytical Sciences, Purdue Pharma, 444 Saw Mill River Rd, Ardsley, NY 10502 USA

1

OVERVIEW: HANDBOOK OF PHARMACEUTICAL ANALYSIS BY HPLC

SATINDER AHUJA

Ahuja Consulting, Calabash, NC

I. INTRODUCTION

A variety of methods are available for analyzing pharmaceutical compounds; however, high-pressure liquid chromatography is currently the method of choice for the analysis of these compounds. High-pressure

Handbook of Pharmaceutical Analysis by HPLC,
S. Ahuja and M.W. Dong, editors.

liquid chromatography is at times called high-performance liquid chromatography because it offers high performance over ambient pressure or low-pressure liquid chromatography. The use of the term high performance can be debated; however, its use in the context described above is acceptable. The terms high-pressure or high-performance liquid chromatography have been used interchangeably in this book. Coincidentally, they both have the same abbreviation, i.e., HPLC. More commonly, this technique is simply called liquid chromatography or LC.

HPLC is used in the pharmaceutical industry for a wide variety of samples. It is the method of choice for checking the purity of new drug candidates, monitoring changes or the scale up of synthetic procedures, in-process testing for developing new formulations, and quality control/assurance of final drug products. A large number of books have been published on HPLC and are referenced throughout this text; a select list of books that may be found useful by the interested reader is provided at the end of this chapter (1–15).

This handbook provides busy pharmaceutical scientists with a complete yet concise reference guide for utilizing the versatility of HPLC in new drug development and quality control. It can be broadly divided into six major sections:

1. Overview, theory, and instrumentation (Chapters 1–3)
2. HPLC methods and practices (Chapters 4–9)
3. Regulatory aspects (Chapters 10–12)
4. Applications (Chapters 13–18)
5. Post-chromatographic techniques (Chapters 19–21)
6. New developments in HPLC (Chapter 22)

All of the chapters have been written by selected experts in their respective fields, and will undoubtedly provide the readers with an in-depth understanding of HPLC theory, hardware, methodologies, regulations, and applications.

Pharmaceuticals, for the purpose of this book, means chemical compounds that are used in pharmaceutical production. This can comprise the active ingredient, which is also called active pharmaceutical ingredient (API) or drug substance or drug product; and the inert pharmaceutical ingredients (excipients) that are used to formulate a drug product in the form of tablets, capsules, ointments, creams, lotions, parenterals, inhalers, and a variety of drug delivery systems.

The primary object of this book is to provide the HPLC practitioner with a handy guide to the use of HPLC for analyzing pharmaceutical compounds of interest. This means familiarizing the practitioner with the theory, instrumentation, regulations, and numerous applications of HPLC. This handbook provides practical guidelines using case studies on sample preparation, column or instrument selection, and summaries of "best practices" in method development and validation, as well as "tricks of the trade" in HPLC operation. It describes the essence of major pharmaceutical applications of HPLC (assays, stability testing, impurity testing, dissolution testing, cleaning validation, high-throughput screening, and chiral separations) in addition to nuances in interpreting International Conference on Harmonisation (ICH) guidelines and instrument qualification. This book also highlights the latest developments in hyphenated techniques, such as LC-NMR or LC-MS, and in data handling. Finally, it discusses new and upcoming developments in HPLC.

The contents of this book have been highlighted below to provide the reader with an overview of pharmaceutical analysis by HPLC.

II. KEY CONCEPTS IN HPLC

A concise view of HPLC concepts essential to the performance of successful pharmaceutical analysis is provided in Chapter 2 for the benefit of laboratory analysts and pharmaceutical professionals. The goal is to present the basic theory and terminology to the beginner, and, at the same time, delineate the more advanced concepts in gradient elution and high-pH separations for the more experienced practitioners. The important concepts in HPLC that are covered include retention, selectivity, efficiency, and resolution. This chapter discusses their relationship to key column and mobile phase parameters such as particle size, column length and diameter, mobile phase strength, pH, and flow rate. The impact of various other concepts important in chromatographic analysis such as peak capacity, gradient time, void volume, and limit of quantification are also discussed.

III. HPLC INSTRUMENTATION IN PHARMACEUTICAL ANALYSIS

An overview of HPLC instrumentation, operating principles, and recent advances or trends that are pertinent to pharmaceutical analysis is provided in Chapter 3 for the novice and the more experienced analyst. Modern liquid chromatographs have excellent performance and reliability because of the decades of refinements driven by technical advances and competition between manufacturers in a two billion-dollar-plus equipment market. References to HPLC textbooks, reference books, review articles, and training software have been provided in this chapter. Rather than summarizing the current literature, the goal is to provide the reader with a concise overview of HPLC instrumentation, operating principles, and recent advances or trends that lead to better analytical performance. Two often-neglected system parameters—dwell volume and instrumental bandwidth—are discussed in more detail because of their impact on fast LC and small-bore LC applications.

IV. HPLC COLUMNS IN PHARMACEUTICAL ANALYSIS

An accurate choice of column configuration and column chemistry is very important for carrying out pharmaceutical analysis by HPLC successfully (see Chapter 4). In the section on column configuration, it is emphasized that column performance should be seen from the standpoint of the ratio of column length to particle size. The overall performance characteristics of a column remain constant when this ratio is kept constant: the same maximum resolving power and the same shortest analysis time can be achieved. Other subjects discussed are elevated temperature, column diameter, high-speed analysis, and monolithic columns. In the section on column chemistry, the silica-based stationary phases are described in great detail, including new developments such as water-wettable stationary phases and packings with embedded polar groups. An overview of commonly applied surface chemistries of packings used in reversed-phase applications is included. A separate discussion is dedicated to hybrid packings and zirconia-based packings. Because the long-term reproducibility of an assay is of utmost importance in the pharmaceutical industry, the brief review of the reproducibility of stationary phases should be of great interest to the readers. As many modern chemical entities are polar compounds, hydrophilic interaction chromatography may be a suitable technique for these compounds. Finally, a synopsis of what is currently understood about column selectivity in reversed-phase HPLC is provided.

V. SAMPLE PREPARATION

Sample preparation (SP) is generally not given adequate attention in discussions of pharmaceutical analysis even though its proper execution is of paramount importance in achieving fast and accurate quantification (see Chapter 5). Non-robust SP procedures, poor techniques, or incomplete extraction are the major causes of out-of-trend and out-of-specification results. The common SP techniques have been reviewed with a strong focus on tablets or capsules, as they are the primary products of the pharmaceutical industry. Detailed descriptions of SP methods for assays and impurity testing are provided with selected case studies of single- and multi-component products.

Sample preparation refers to a family of solid/liquid handling techniques to extract or to enrich analytes from sample matrices into the final analyte solution. While SP techniques are well documented, few references address the specific requirements for drug product preparations, which tend to use the simple "dilute and shoot" approach. More elaborate SP is often needed for complex sample matrices (e.g., lotions and creams). Many newer SP technologies such as solid-phase extraction

(SPE), supercritical fluid extraction (SFE), pressurized fluid extraction or accelerated solvent extraction (ASE), and robotics are topics of numerous research papers, symposia and commercial promotion, and are adequately referenced in this chapter. Both novices and experienced analysts can learn the usefulness of sample preparation techniques for the solid dosage forms. The "best practices" are described in the dilute and shoot approach, and the "tricks of the trade" in grinding, mixing, sonication, dilution, and filtration of drug products. Selected case studies of sample preparation for assays and impurity testing are used to illustrate the strategies, trade-offs, and potential pitfalls encountered during method development.

VI. METHOD DEVELOPMENT

The method development of pharmaceuticals by HPLC begins with the generation of an orthogonal array of methods suitable for the separation of early drug substance lots, synthetic intermediates and starting materials, excipients (where known), and products from forced decomposition studies (see Chapter 6). From the array, two orthogonal methods are selected to support subsequent stages of pharmaceutical development. One of these methods is optimized to separate all known related substances and excipients, using chromatographic optimization software such as DryLab. The other method is used to analyze pivotal lots of drug substance and drug product formulations to assure that the primary method continues to be viable for the separation of all relevant components.

Following finalization of drug substance synthetic routes and drug product formulation, the focus shifts to the development of robust and transferable methods for post-approval support at quality control units. It is important to remember during the final stage of method development that achievement of separation conditions is only one of the necessary parameters for successful method implementation. Extensive studies to measure robustness and quantitative method performance are conducted to assure that the method performs as intended in quality control laboratories. It should be emphasized that successful method development requires extensive cooperation between the development laboratory and the receiving quality control laboratories.

VII. METHOD VALIDATION

Method validation is discussed in Chapter 7 as it relates to the HPLC methods of analysis. Validation is a process required by law, and the concept is described by regulatory agencies in the guidance documents. The analyst performing method validation is responsible for interpreting the guidance into acceptable practices. Establishing a clear, well-planned validation program that incorporates the concepts discussed here is critical to success. Method validation is not a one-time process that can be ignored once completed; it is a constant, evolving process. A well-executed and documented validation serves as evidence to regulatory agencies that the method in question is compliant and performs as intended. Periodic reviews should occur to evaluate validation data for compliance with the current regulatory requirements. To be successful, applicants need to remain up to date with the current thinking of regulatory agencies and anticipate the changes to occur in the regulations. Validation programs that are able to incorporate these changes quickly and efficiently save both time and money for the applicant and help streamline the process to an approvable application.

VIII. ION CHROMATOGRAPHY

Ion chromatography has become an essential tool for the pharmaceutical analytical chemist. The high sensitivity of the technique, coupled with the wide dynamic operating range made possible with modern high-capacity stationary phases, makes it ideal for the analysis of ions in pharmaceutical applications (see Chapter 8). The combination of gradients and suppressed conductivity detection provides a powerful screening tool for the analysis of ions in drug substances and in pharmaceutical formulations, thus offering the basis for analysis of counter ions, additives, and manufacturing by-products. Ion chromatography has been expanded on a number of fronts, now covering an extremely wide variety of analytes, and has seen significant improvements in stationary phase design, chromatographic performance, detection sensitivity, suppressor design, and electrolytic generation of eluents. With all of these improvements, ion chromatography is much more than simply ion-exchange chromatography by another name. Ion chromatography is used in HPLC for separation of inorganic anions, small organic acids, and simple amines.

This chapter reviews the underlying principles of ion chromatography and its application in pharmaceutical analysis. It provides an overview of eluent systems, applications of gradients, electrolytic eluent generation, suppressors, and stationary phases. Applications of ion chromatography to the confirmation of counter ions, active ingredient analysis, competitive analysis and development work are discussed.

IX. HOW TO BE MORE SUCCESSFUL WITH HPLC ANALYSIS

Chapter 9, by summarizing a series of standard operating procedures representing the best practices of experienced HPLC analysts, describes how we can be more successful in HPLC operation. It offers brief guidelines on maintenance and troubleshooting and the means for enhancing HPLC assay precision. Many years of hands-on experience is required to be truly proficient in HPLC operation. Most analysts agree that the best way to learn HPLC is to be an apprentice to an expert chromatographer. Barring that, it may be desirable to use various reference books, user's manuals, training videos, hands-on training courses, and computer-based training software. This chapter focuses on the practical aspects of HPLC including the "best practices" or operating guides for mobile phase preparation and HPLC system operation. Guidelines on maintenance and troubleshooting are highlighted as well as ways to improve HPLC assay precision. The goal is to offer the analyst a concise overview and a guide on how to be more successful in HPLC analysis.

X. REGULATORY PRACTICES

To guarantee the quality, clinical performance, efficacy, and safety of a pharmaceutical product, specifications are fixed and approved by the competent regulatory authorities of each country in which the drug is marketed. Analytical testing ensures that these specifications are met by confirming the identity, purity, and potency of drug substances and drug products prior to their release for commercial use (see Chapter 10). To assure reliable bioavailability, drug products' performance testing is needed. This performance testing comes in the form of either dissolution or disintegration tests.

HPLC has become the most important separation technique for analyzing drug substance and drug product. In today's laboratory, testing is for the most part conducted by HPLC, and most wet chemistry methods have become obsolete. For drug substances, HPLC testing is focused on assay and impurity testing. For drug products, HPLC is also used for dissolution testing (an indicator of bioavailability) and content uniformity (used to ensure that the content of the drug substance is uniform throughout a batch). The technology surrounding HPLC instruments themselves and the software used to run them have advanced to such a degree that in many labs analyses are run overnight and in some cases 24 hours a day unattended. This chapter reviews the regulatory guidelines and requirements for the successful use of HPLC in drug substance and drug product testing in development and routine quality control.

XI. HPLC SYSTEM CALIBRATION

Calibration, in current Good Manufacturing Practices (GMP) terminology, refers to instrument "qualification" or performance verification of HPLC. In most pharmaceutical laboratories, calibration is performed every 6–12 months. If the results are acceptable according to some predetermined criteria, a sticker is placed on the instrument to indicate its readiness for GMP work. This periodic calibration, coupled with initial system qualification (Installation Qualification and Operation Qualification or IQ/OQ) and daily system suitability testing, is part of an overall system validation program to ensure data validity and regulatory compliance (see Chapter 11). A company-wide system calibration program can also facilitate method transfer by minimizing system-to-system variability among different testing sites. This chapter reviews the principles and strategies used for HPLC system calibration that includes the pump, the detector, the autosampler, and the column oven. A case study is used to illustrate the development of the calibration procedures for all system modules and the rationale of setting up acceptance criteria that balance productivity and compliance.

XII. SYSTEM VALIDATION

Chapter 12 defines the terms, responsibilities, requirements, and recommended procedures involved in pre-installation: Installation Qualification (IQ), Operational Qualification (OQ) and Performance Qualification (PQ); all of these are part of a typical HPLC system validation process. As the Food and Drug Administration (FDA) does not publish a definitive reference or "cookbook" for these procedures, the suggestions listed are only recommendations. The information in this chapter should be viewed as guidelines and not as inflexible mandates. The important thing is to take reasonable, verifiable action, armed with well-thought-out SOPs, protocols, security, and documentation procedures to verify that an HPLC system used in the development or production of pharmaceutical products is performing both in accordance with the Functional Specifications set forth by the vendor and the User Requirement Specification (URS) set forth by the user. Remember that the only thing worse than having no validation protocol is to have one that is poorly designed and/or one that is not followed.

There are various stages to the validation process and all have to be thought out as part of an overall *time line*. The validation process starts with preliminary planning, and selecting/training those qualified in defining and designing the various procedures, experiments, and pertinent documents. This leads to the drafting of specific SOPs and IQ/OQ/PQ protocols, then moves on to carrying out the pre-installation procedures, and finally culminates with implementing the IQ, OQ, and PQ procedures, in that order. One must also have plans/procedures for continued maintenance and re-calibration. All of this is ultimately the responsibility of the user, but especially for the IQ and OQ the vendor can be of invaluable help, both in supplying the needed documentation and in implementing the various tasks.

XIII. ASSAY AND STABILITY TESTING

The general concept of drug stability and the factors affecting drug substance and drug product stability are covered in Chapter 13. This chapter also discusses the guidelines from the ICH for stability testing on drug substances and products and the requirements at different phases of the drug development process in the pharmaceutical industry. The role of HPLC in stability testing along with the chromatographic techniques and the procedures involved in developing a stability-indicating method are also described. Case studies of challenging HPLC method development on dual drug systems in solid dosage and surfactants with different molecular weights in liquid formulation are presented. Stability testing is an integral part of pharmaceutical product development and is ongoing throughout the entire drug development process. Product integrity and shelf life are based on stability testing. The pharmaceutical industry is challenged by the frequently changing regulatory requirements. The need for improvement in analytical techniques poses challenges and opportunities in pharmaceutical stability testing. A wealth of specific information is available in the FDA stability guidance and in ICH stability guidelines. HPLC and other analytical techniques play important roles in stability testing. The precision, ease of use, and ruggedness of HPLC methods are preferred by far over other separation and quantification techniques. A continuous effort to learn updated technology and to remain informed of constantly changing regulations is the only way to succeed in the pharmaceutical industry.

XIV. IMPURITY EVALUATIONS

Impurities in drug substances and drug products continue to be the source of great concern, discussion, debate, and research (see Chapter 14). These concerns and debates typically center on the potential safety risks associated with impurities resulting from contamination. However, most of the work being performed in the pharmaceutical industry with respect to impurities is focused on the isolation, identification, and quantification of impurities that could potentially be found as part of the manufacturing process or through chemical decomposition. On the surface it appears that there is a divergence between the perceived need for public safety and the industry focus. However, it should be appreciated that a thorough understanding of all known potential impurities (also known as the impurity profile) helps find unknown impurities and contaminants.

The most effective means to increasing public safety is to create and maintain a public specification, e.g., those given in the United States Pharmacopoeia (USP). In theory, a public specification defines the boundaries of safety and efficacy of a given drug substance and product. The boundaries are generally set by the manufacturer and approved by the regulatory agency (e.g., the United States Food and Drug Administration or the Ministry of Health in Japan). These boundaries are created by evaluating critical parameters of a drug substance/product, with clinical data collected on human subjects. The first lot of material that is tested in the clinic is often called the pivotal lot. Once the pivotal lot is manufactured, it is evaluated by a battery of tests to ensure that the material is well characterized. Through this testing, a manufacturer may qualify impurities as nontoxic, and can create goalposts to evaluate additional lots, and include this information in the submission to the appropriate regulatory agency. The tests and procedures used to evaluate and characterize the pivotal lot and the acceptance criteria created to define the goalposts are the basis of a public specification. A public specification such as those found in the USP or the British Pharmacopoeia (BP) provides a benchmark of expected impurities, both toxic and nontoxic, that are typically found in the manufacture of a drug substance. With this information and a standardized approach, an analyst can ensure that additional lots of a material are equivalent to the pivotal lot. More important in terms of public safety, these public specifications aid the industry, government, and the affected public look for new impurities and contaminants. Therefore, it is vital that the pharmaceutical industry regularly update these publications for the continued safety of the patient.

Impurity testing is pivotal in pharmaceutical development for establishing drug safety and quality. In this chapter, an overview of impurity evaluations of drug substances and products by HPLC is presented from both the laboratory and regulatory standpoints. Concepts from the development of impurity profiles to the final establishment of public specifications are described. Useful strategies in the identification and quantification of impurities and degradation products are summarized with practical examples to illustrate impurity method development.

XV. DISSOLUTION TESTING

HPLC instrumentation and column technologies have progressed to the point that pharmaceutical researchers have tremendous flexibility in selecting an optimal method for any dissolution testing method during drug development and routine quality control tests (see Chapter 15). As formulation development tends toward increasing complexity, HPLC methodology will be embraced even more in dissolution testing, as it can help solve unique analytical challenges posed either by the media or by dosage form characteristics. Equipped with a variety of detection methods, HPLC can be used to analyze dissolution samples that otherwise cannot be analyzed by UV methods either because of significant interferences or poor UV absorbance. Automation of dissolution systems with on-line HPLC analysis has been accomplished by the development of fast chromatography, which is being increasingly used in the pharmaceutical industry to improve throughput and productivity.

XVI. CLEANING VALIDATION

The HPLC method used for validation of cleaning of the manufacturing equipment has to be linear; this usually requires additional evaluations to extend the lower end of the linear range used for the potency assay (see Chapter 16). In many laboratories, analysts realize that they will have to evaluate cleaning samples and push the linear range as low as possible in the initial HPLC method development for the API potency assay. Many times during the development cycle, toxicology or clinical data indicate that the cleaning limit must be lowered. It is important for the analyst to know the limits of existing methodology to ascertain the possibility of accommodating lower cleaning limits. Poor chromatographic response by many of today's molecules above 220–240 nm means that many cleaning assays are forced into the lower end of the UV range. This presents additional challenges for the analyst in both linearity and sample analysis because of solvent and impurity interferences at the low wavelengths.

XVII. HIGH-THROUGHPUT SCREENING

There is a significant increase in demand for new drug candidates that exhibit improved efficacy, selectivity, and safety. In response to this demand, drug discovery strategy has dramatically changed during the last two decades. Automated and parallel organic synthesis can generate a large number of compounds that can be screened for biological activity and potency against the appropriate targets. These high-throughput screens (HTS) usually identify many hits with potent in vitro pharmacological activity (see Chapter 17). The need for a high-throughput ADME (absorption, distribution, metabolism, and elimination) screen is driven by the necessity for not only acceleration of the discovery process identifying new leads rapidly, but also for the improvement of the success rate of clinical development. Therefore, ADME screens for desirable drug-like properties have been developed to help identify promising new chemical entities (NCEs) among libraries in order to optimize the ADME properties and minimize the potential of toxic liabilities. These ADME analyses generate a significant amount of data that can be utilized to categorize and rank NCEs according to their aqueous solubility, chemical stability, intestinal permeability, hepatic metabolism, etc. The results from ADME screens can be utilized by medicinal chemists to design and synthesize NCEs with more favorable ADME properties in the lead optimization phase.

This chapter reviews some of the applications of high-throughput liquid chromatography–mass spectrometry (LC/MS) techniques in ADME screens in the lead optimization phase of drug discovery. Several routinely used techniques that are utilized in performing these ADME assays are described, with a strong focus on either the instrumental methods or experimental designs that significantly improve the throughput. Selected case studies on both instrumental methods and experimental setups are discussed to illustrate the strategies, trade-offs, and any potential pitfalls of these various techniques. The preparation of biological samples is also presented in great detail for the LC/MS quantitative bioanalysis.

HPLC coupled with mass spectrometry is widely used in drug discovery and development. Mass spectrometric applications are being used in qualitative analysis, quantitative analysis, and preparative HPLC analysis where the mass spectrometer serves as a detector to trigger fraction collection. This chapter presents applications of LC/MS tandem mass spectrometry (LC/MS and LC/MS/MS) in in vitro ADME screens and bioanalysis of in vivo biological samples for evaluation of hits as well as the lead optimization phase in drug discovery. An effort has been made to report the most recent advances in LC/MS software and hardware, analytical instrumentation, and automated sample handling and processing techniques where a significant increase in analysis throughput has been reported.

XVIII. CHIRAL SEPARATIONS

Chiral separation of drug molecules and of their precursors, in the case of synthesis of enantiomerically pure drugs, is one of the important application areas of HPLC in pharmaceutical analysis. Besides HPLC, capillary electrophoresis (CE) is another technique of choice for chiral separations. Chapter 18 provides an overview of the different modes (e.g., direct and indirect ones) of obtaining a chiral separation in HPLC and CE. The direct approaches, i.e., those where the compound of interest is not derivatized prior to separation, are discussed in more detail since they are currently the most frequently used techniques. These approaches require the use of the so-called chiral selectors to enable enantioselective recognition and enantiomeric separation. Many different molecules have been used as chiral selectors, both in HPLC and CE. They can be classified into three different groups, based on their structure: (i) low-molecular-weight selectors, (ii) macrocyclic selectors, and (iii) macromolecular selectors. Finally, some practical guidelines about screening conditions to test the enantioselectivity of a given compound on a limited number of chromatographic systems, and about method optimization when an initial separation is already obtained, are given briefly. The possibilities potentially applicable to induce the separation of enantiomers of interest are numerous. Not only can different techniques be applied, but also within each technique large numbers of chiral selectors might be used. This assures that almost any chiral separation can be achieved with at least one of the dozens of commercially available chiral selectors for the different techniques. Difficulties, however, originate from the selection of the appropriate chiral selector for a given separation. Since chiral recognition mechanisms are complex, the prediction of a suitable chiral selector is very difficult. Frequently, the identification of suitable selectors for a specific pair of enantiomers requires considerable experimentation and might therefore be highly demanding with respect to time, material, and labor. There is a clear need for fast method development strategies to avoid to some extent trial-and-error approaches, and recently some attempts in this context have been made.

Among the different chiral stationary phases (CSPs) used in chiral HPLC, the polysaccharide CSPs have proved to be very useful and might be recommended as first-choice columns when one wants to develop chiral separations. These CSPs know many applications, both in normal-phase and reversed-phase mode. The chiral selectors of these CSPs have a broad enantiorecognition range (80% of marketed drugs), which makes them excellent candidates for developing both screening strategies and individually optimized separations. Moreover, the number of different types of stationary phases can be limited to three or four (Chiralcel OD-H and OJ, Chiralpak AD, and occasionally, AS) since both reversed-phase and normal-phase versions exist, while the latter can even be used for chiral SFC.

XIX. LC-MS

The power and versatility of LC/MS and LC/MS/MS are discussed in Chapter 19 on the basis of successful practices and how an individual investigator can follow these practices to obtain useful information during different stages of the drug development process. The primary focus is on various applications of LC and electrospray mass spectrometry for pharmaceutical analysis. The goal is to provide some ideas and experimental designs for the solution of qualitative problems including drug impurity profiling, HPLC method development, deficit in drug mass balance, drug discoloration, and drug counterfeiting. Many examples are presented to illustrate how LC/MS and LC/MS/MS can be used to provide answers to these real-world problems.

LC/MS has emerged as one of the most important techniques in pharmaceutical analysis. It is used to ensure the quality and safety of pharmaceutical products. Mass spectrometry can detect molecules in the positive and negative ionization modes, in which molecules are transformed, primarily, into protonated or deprotonated pseudo-molecular ions. The obvious benefit of using LC/MS is the opportunity it affords for the rapid identification of unknown compounds by generating compound-specific information such as molecular weight and elemental composition. Other benefits can also be realized, such as increased gain in confidence and decreased time required for HPLC method development. Using a "generic HPLC/MS method," data can be collected in a consistent manner throughout the development life cycle of a pharmaceutical product.

However, LC/MS is not a panacea. In mass spectrometry, molecules must exist as ionic species in order to be detected. The ionization efficiency may vary over many orders of magnitude, depending on the structure of the compounds studied. For non-polar and neutral drug compounds, an ESI or APCI interface may not be sensitive enough to allow meaningful structure determination. Although assignment of various signals in an ESI/MS mass spectrum is usually straightforward, artifacts can arise from electrochemical reactions during the electrospray ionization process. Thermally induced degradation is also a common phenomenon with both ESI and APCI interfaces because of the heat input that has been used to aid in the evaporation of solvents. The presence of artifacts will obviously complicate accurate determination of the molecular weight for trace-level drug impurities. A thorough knowledge of the sample relating to its formulation, packaging materials, synthetic route, chemical activity of drug compound, etc. is beneficial in the identification of unknown impurities.

XX. LC-NMR

Hyphenated analytical techniques such as LC-MS, which combines liquid chromatography and mass spectrometry, are well-developed laboratory tools that are widely used in the pharmaceutical industry. For some compounds, mass spectrometry alone is insufficient for complete structural elucidation of unknown compounds; nuclear magnetic resonance spectroscopy (NMR) can help elucidate the structure of these compounds (see Chapter 20). Traditionally, NMR experiments are performed on more or less pure samples, in which the signals of a single component dominate. Therefore, the structural analysis of individual components of complex mixtures is normally time-consuming and less cost-effective. The combination of chromatographic separation techniques with NMR spectroscopy offers advantage for the on-line separation and structural elucidation of unknown compounds. Mixtures such as crude reaction mixtures in drug discovery can be analyzed without prior separation. Experiments combining an HPLC with NMR on the study of mixtures were introduced to the scientific community in the early 1980s. However, LC-NMR was not widely practical because of its low sensitivity, approximately six orders in magnitude inferior to that of MS. Another challenge is from the measurement of proton signals in mobile phase. Recent developments in higher magnetic field strength and electronics that have improved the sensitivity of probes, together with advanced solvent suppression techniques, have made LC-NMR measurement practical.

During the last decade, LC-NMR has been fully commercialized, and its applications in both academia and industry have been growing rapidly. One example is LC-NMR-MS, which is a more powerful hyphenated technique combining LC-MS and LC-NMR to provide complementary information about structures simultaneously and offer a method for more accurate and rapid structural analysis. In this chapter, some routine subjects such as the theory and design of flow probes have been avoided, as there are many papers published in the 1990s that have covered these topics comprehensively. The emphasis is therefore on describing experimental design, practical applications, and recent developments in technology. With all the applications to date, LC–NMR spectroscopy is still a relatively insensitive technique because of the poor mass sensitivity of the NMR detection system. To this end, several other hyphenated NMR techniques have been developed to enhance the sensitivity of the technique. LC-SPE-NMR dramatically increases the sensitivity up to a factor of four by utilizing a solid-phase extraction (SPE) device after the LC column. Capillary LC-NMR also significantly lowers the detection limit to low nanogram range through the integration of capillary LC with NMR detection. Other breakthroughs, such as cryo-LC-probe technology, combine the advantages of sample flow and enhanced sensitivity from a cryogenically cooled NMR probe. Most of the latest developments in LC-NMR technology have been covered in this chapter.

In general, LC-NMR is an efficient analytical technique for the identification of components in pharmaceutical mixtures. It is especially efficient when it is combined with MS; the two on-line detectors are complementary in providing unequivocal structural identification for both expected components and unknown substances. However, the relatively low NMR sensitivity is still the major drawback that hampers the utility of LC-NMR technique. Hopefully, the newly developed cryogenic LC-NMR probes coupled with recent interface enhancements and higher magnetic field strengths will further enhance the utility of this technique in solving more challenging problems.

XXI. CHROMATOGRAPHY DATA PROCESSING

Chromatographic techniques represent one of the most significant sources of analytical data found in today's pharmaceutical laboratories. All of the chromatographic techniques produce data that must be acquired, interpreted, quantified, compared, reported, and finally archived (see Chapter 21). Whether the analysis is qualitative or quantitative in nature, the data must somehow be interpreted and reported so that meaningful decisions can be made. It may be as simple as a qualitative decision that indicates whether a reaction has reached completion successfully, or it may be a series of quantitative analyses that help determine if a batch or lot of product meets its specifications and may now be released. This chapter also examines the evolution and perhaps the revolution that has taken place within the chromatography data system (CDS) marketplace.

XXII. NEW DEVELOPMENTS IN HPLC

A few glimpses have been provided into the new developments in HPLC that will impact on the future progress in this field (see Chapter 22). Many of these advances revolve around reducing not only HPLC cycle times by improving the columns and the instrumentation but also improving sample preparation techniques. Under Simplifying Sample Preparation, the latest trends in reducing sample preparation time, such as direct sample infusion/injection and on-line SPE, are discussed. In the section on Column Technologies, newer trends in stationary phases such as metal oxide-based phases and hybrids may offer novel selectivities as compared to C18-based phases. The section on Improvements in Detectors provides some observations on how new detectors are increasing the versatility of HPLC. In the last section, on Improvements in HPLC Throughput, monolithic columns, small particles packed in short columns, high-temperature LC, ultra high-pressure LC, and parallel injection techniques are covered. All of these approaches focus on developing both columns and instruments capable of achieving the types of separations that theory predicts. It is expected that new developments on lab-on-chip will eventually produce more functional systems for small molecules. All in all, the future of HPLC is exciting, and time will unravel newer advances that will make the work of the chromatographer even faster and more productive.

REFERENCES

1. Ahuja, S. and Alsante, K., *Handbook of Isolation and Characterization of Impurities in Pharmaceuticals*, Academic Press, New York, 2003.

2. Ahuja, S., *Chromatography and Separation Science,* Academic Press, New York, 2003.
3. Ahuja, S. and Scypinski, S., *Handbook of Modern Pharmaceutical Analysis,* Academic Press, New York, 2001.
4. Ahuja, S., *Chiral Separations by Chromatography,* Oxford Press, New York, 2000.
5. Ahuja, S., *Impurities Evaluation of Pharmaceuticals,* Marcel Dekker, New York, 1998.
6. Meyer, V. R., *Practical HPLC,* Wiley, New York, 1998.
7. Katz, E., Eksteen, R., Schoenmakers, P. and Miller, N., *Handbook of HPLC,* Marcel Dekker, New York, 1998.
8. Adamovics, J. A., *Chromatographic Analysis of Pharmaceuticals,* Marcel Dekker, New York, 1997.
9. Lunn, G. and Schmuff, N. R., *HPLC Methods for Pharmaceutical Analysis,* Wiley, New York, 1997.
10. Giddings, J., *Unified Separation Science,* Wiley, New York, 1991
11. Ahuja, S., *Selectivity and Detectability Optimizations in HPLC,* Wiley, New York, 1989.
12. Ahuja, S., *Trace and Ultratrace Analysis by HPLC,* Wiley, New York, 1989.
13. Snyder, L., Glajch, J. and Kirkland, J., *Practical HPLC Method Development,* Wiley, New York, 1988.
14. Ahuja, S., *Ultratrace Analysis of Pharmaceuticals and Other Compounds of Interest,* Wiley, New York, 1986.
15. Snyder, L. and Kirkland, J., *Introduction to Modern Liquid Chromatogaphy,* Wiley, New York, 1979.

2

KEY CONCEPTS OF HPLC IN PHARMACEUTICAL ANALYSIS

RAPHAEL M. ORNAF AND MICHAEL W. DONG

Purdue Pharma, Ardsley, NY

Handbook of Pharmaceutical Analysis by HPLC,
S. Ahuja and M.W. Dong, editors.

ABSTRACT

This chapter provides the reader with an overview of high-performance liquid chromatography (HPLC) concepts focusing on reversed-phase HPLC and its application in pharmaceutical analysis. Basic terminology and fundamental concepts of retention, selectivity, efficiency, and resolution are described. Their relationships to column and mobile phase parameters are delineated. The resolution and the van Deemter equations are discussed to illustrate their utilities for enhancing assay speed and resolution. Separation of acidic and basic analytes using ion suppression, ion-pairing, and high-pH mobile phase techniques are described. Key gradient parameters leading to higher peak capacity and throughput are discussed.

I. INTRODUCTION

A. Scope

The objective of this chapter is to provide laboratory analysts and pharmaceutical professionals with a concise overview of high-performance liquid chromatography (HPLC) concepts essential to pharmaceutical analysis. Our approach is to present the basic theory and terminology to the beginner while presenting the more advanced concepts in gradient elution and high-pH separations to the more experienced practitioners. The reader is referred to the numerous HPLC textbooks,[1–6] training courses,[7–9] journals, and Internet resources for a more formal and detailed treatment of HPLC theory.

This chapter is divided into the following sections:

- General axioms in HPLC
- Fundamental concepts of retention, selectivity, efficiency, and peak resolution
- The resolution equation
- The van Deemter equation
- Mobile phase parameters (solvent strength, pH, ion-pairing reagent, and temperature)
- Gradient parameters (flow rate, gradient time, peak capacity, and dwell volume)
- Limit of quantitation (LOQ)

Note that our primary focus is on reversed-phase HPLC (RPLC) since it is the predominant mode for pharmaceutical analysis. Many of these concepts, however, are applicable to other modes of HPLC such as ion-exchange, adsorption, and gel-permeation chromatography.

B. What Is HPLC?

HPLC is a physical separation technique conducted in the liquid phase in which a sample is separated into its constituent components (or analytes) by distributing between the mobile phase (a flowing liquid) and a stationary phase (sorbents packed inside a column). An online detector monitors the concentration of each separated component in the column effluent and generates a chromatogram. HPLC is the most widely used analytical technique for the quantitative analysis of pharmaceuticals, biomolecules, polymers, and other organic compounds.

C. Some HPLC Axioms

The goal of HPLC analysis is to separate the analyte(s) from the other components in the sample in order to obtain accurate quantitation for each analyte. Towards this goal, several common-sense axioms are worth mentioning since they are at times overlooked even by experienced practitioners. These axioms are:

1. *The sample must be soluble*: "If it's not in solution, it cannot be analyzed by HPLC." Although this may seem obvious, solubility issues complicate real assays of low-solubility drugs and controlled-release formulations. Many situations encountered in pharmaceutical analysis, such as low recovery, lack of mass balance, and out-of-specification results, might stem from solubility problems in a sample preparation step, rather than the HPLC analysis itself.
2. *For separation, analytes must be retained and have differential migration in the column*: Separation of components cannot occur without retention and interaction with the stationary phase. In addition, analytes must show differential retention vs. other components.
3. *The mobile phase controls HPLC separation*: While the HPLC stationary phase provides retention and influences the separation mechanism, it is the mobile phase which controls the overall separation. HPLC method development efforts focus on finding a set of mobile phase conditions that provide adequate separation of the analyte peak(s) from other components in the sample.
4. *The final analyte solution should be prepared in the mobile phase*: The final analyte solution, if possible, should be dissolved in the mobile phase or a solvent "weaker" in strength than the starting mobile phase. Many chromatographic anomalies such as splitting peaks or fronting peaks are caused by injecting an analyte dissolved in a strong solvent. If a strong solvent must be used because of solubility considerations, a smaller injection volume (5–10 μL) should be considered.

II. FUNDAMENTAL CONCEPTS

First, we will explore the three fundamental factors in HPLC: retention, selectivity, and efficiency. These three factors ultimately control the separation (resolution) of the analyte(s). We will then discuss the van Deemter equation and demonstrate how the particle diameter of the packing material and flow rate affect column efficiencies.

A. Retention

Figure 1 shows a typical chromatogram, which includes a time axis, an injection point, and an analyte peak. The time between the sample injection point and the analyte reaching a detector is called the retention time (t_R). The retention time of an unretained component (often marked by the first baseline disturbance caused by the elution of the sample solvent) is termed void time (t_0). Void time is related to the column void volume (V_0), which is an important parameter that will be elaborated later.

The peak has both width (W_b) and height (h). The subscript b denotes that the width was measured at the base line. Sometimes the width halfway up the peak ($W_{1/2}$) or at 5% of peak height ($W_{0.05}$) is used to meet the method or compendial requirements.

The height or the area of a peak is proportional to the concentration or the amount of that particular component in the sample. Either attribute can be used to perform quantitative calculations. The peak area is most commonly used since it provides a more accurate quantitative measurement.

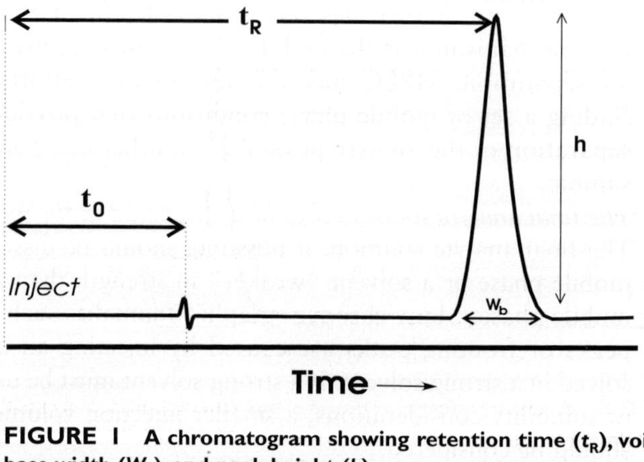

FIGURE 1 A chromatogram showing retention time (t_R), void time (T_0), peak at base width (W_b), and peak height (h).

B. Capacity Factor (k')

While retention time is used for peak identification, it is dependent on the flow rate, the column dimension, and other parameters. A more fundamental term that measures the degree of retention of the analyte is the capacity factor or retention factor (k'), calculated by normalizing the net retention time (t_R', retention time minus the void time) by the void time. The capacity factor measures how many times the analyte is retained relative to an unretained component.[10]

$$\text{Capacity factor, } k' = \frac{t_R - t_0}{t_0}$$

By rewriting the equation, we get

$$t_R = t_0(1 + k') = t_0 + t_0 k'$$

which indicates that retention time is proportional to k'.

A k' value of zero means that the component is not retained and elutes with the solvent front. A k' value of 1 means that the component is slightly retained by the column while a k' value of 20 means that the component is highly retained and spends much time interacting with the stationary phase. In most assays, analytes elute with k' between 1 and 20 so that they have sufficient opportunity to interact with the stationary phase resulting in differential migration. Peaks eluting at high k' (>20) are problematic because of long run times and poor sensitivity as a result of excessive peak broadening in the column. Figure 2 shows how k' is measured by dividing t_R' by t_0.

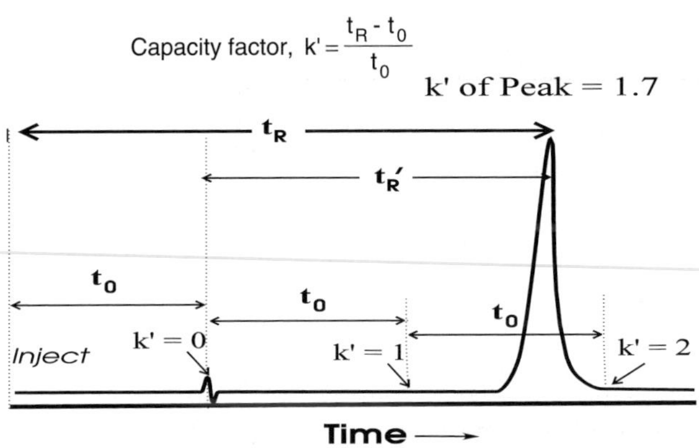

FIGURE 2 A chromatogram showing the calculation for capacity factor (k') which is equal to t_R'/t_0. k' is an important parameter defining the retention of the analyte. Typical k' values for isocratic analyses are 1 to 20.

Liquid chromatography is a thermodynamic method of separation, where each component in the sample is distributed between the mobile phase and the stationary phase.

$$X_m \leftrightarrow X_s$$

where X_m and X_s are solutes in the mobile phase and stationary phase, respectively, and the distribution is governed by the partition coefficient, K

$$\text{Partition coefficient, } K = \frac{[X_s]}{[X_m]}$$

In RPLC (using a hydrophobic stationary phase such as a C18 phase and a hydrophilic mobile phase such as a mixture of methanol and water), the partitioning is somewhat analogous to a two-phase liquid–liquid extraction process occurring in a separatory funnel between a non-polar (i.e., hexane) and a polar solvent (i.e., water). In RPLC, the retention time of the analyte is strongly influenced by the strength or polarity of the mobile phase (i.e., the % of organic solvent content). Table 1 and Figure 3 shows typical retention data of ethylparaben vs. the % of organic solvent content in the aqueous mobile phase. Note that methanol is a strong solvent and ethylparaben shows very little retention or low k' at 80% to 100% methanol. By lowering the methanol content in the mobile phase, retention time of ethylparaben increases exponentially from less than 2 min at >80% methanol to 70 min at 20% methanol. Figure 3 also shows that—as is typical in RPLC—log k' is inversely proportional to % organic solvent content in the mobile phase. We will discuss the use of other even stronger organic solvents such as acetonitrile (ACN) and tetrahydrofuran (THF) in RPLC later.

TABLE 1 Retention Data of Ethyl Paraben

% MeOH in mobile phase	Retention time t_R (min)	Capacity factor k'	Log k'
100	1.3	0.18	−0.74
90	1.3	0.18	−0.74
80	1.5	0.36	−0.44
70	1.9	0.73	−0.14
60	2.7	1.45	0.16
50	4.7	3.27	0.51
40	10.4	8.45	0.93
30	25.4	22.09	1.34
20	70.4	63.00	1.80

Column: Water Symmetry C18, 5-μm, 150×3.9 mm, 1 mL/min, 30°C.

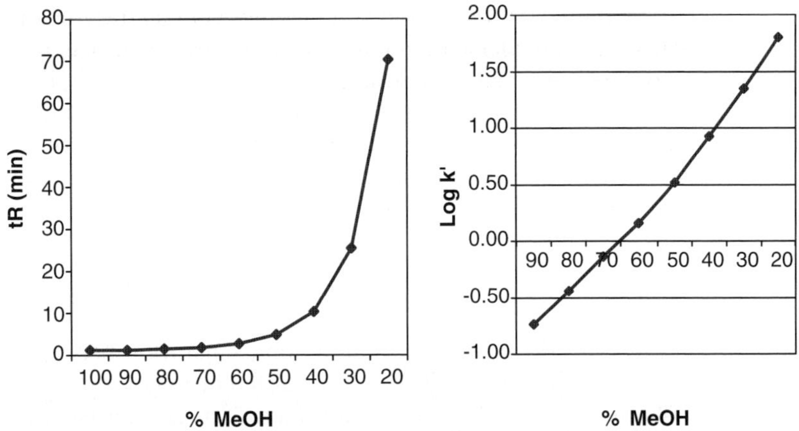

FIGURE 3 Retention data plots for ethylparaben vs. % methanol of the mobile phase. Note that log *k*′ is inversely proportional to % MeOH. The proportionality is linear over a short range. LC conditions were: Column: Water Symmetry C18, 5-μm, 150 × 3.9 mm, 1 mL/min, 30°C.

1. Void Volume (V_0)

The concept of column void volume (V_0) is important for several reasons. Void volume is the volume of the empty column minus the volume occupied by the solid packing materials. It is the liquid holdup volume of the column that each analyte must elute from. Note that the void volume is equal to the void time multiplied by the flow rate (F).

$$V_0 = t_0\,F$$

Typically, V_0 is equal to ~60–70% of the volume of the empty column with 30–40% of the volume occupied by the porous solid stationary phase.

$$\text{Void volume} \quad V_0 = 0.65\,\pi\,r^2\,L$$

where r is the inner radius of the column and L is the length of the column.

Note that V_0 is proportional to the square of the inner radius of the column. It is important to have a rough idea of the void volume of the column since it often dictates the operating flow-rate range, sample-loading capacity, and mass sensitivity (the minimum detectable amount) of the assay. For instance, a typical analytical column (150 mm × 4.6 mm i.d.) has a V_0 of about 1.5 mL and is operated at ~1.0 mL/min. In contrast, by reducing the inner diameter to 2.0 mm, a typical LC/MS column (150 mm × 2.0 mm i.d.) has a V_0 of about 0.3 mL and is operated at

~0.2 mL/min. Column void volumes also control the volumes of the eluting peaks. Smaller column void volumes lead to smaller peak volumes and, therefore, higher analyte concentration. As a result, if the same mass of analyte is injected, small-diameter columns lead to higher sensitivity. However, extracolumn band-broadening effects are more pronounced in small-diameter columns as discussed in the instrumentation chapter of this book.

C. Selectivity (α)

As mentioned earlier as an axiom, separation between two components is only possible if they have different migration rates through the column. Selectivity or separation factor (α) is a measure of differential retention of two analytes. It is defined as the ratio of the capacity factors (k') of two peaks as shown in Figure 4. Selectivity must be >1.0 for peak separation.

Selectivity is dependent on the nature of the stationary phase (i.e., C18, C8, phenyl, cyano, etc.)[6] and the mobile phase composition. The selectivity effects of the mobile phase can be skillfully exploited by experienced chromatographers to enhance separation of key analytes in the sample.

D. Column Efficiency (N)

As readily observed in most chromatograms, peaks tend to be Gaussian in shape and broaden with time, where W_b becomes larger with longer t_R. This is caused by band-broadening effects inside the column, and is fundamental to all chromatographic processes.[1,6,9] The term, plate number (N), is a quantitative measure of the efficiency of the column, and is related to the ratio of the retention time and the standard deviation of

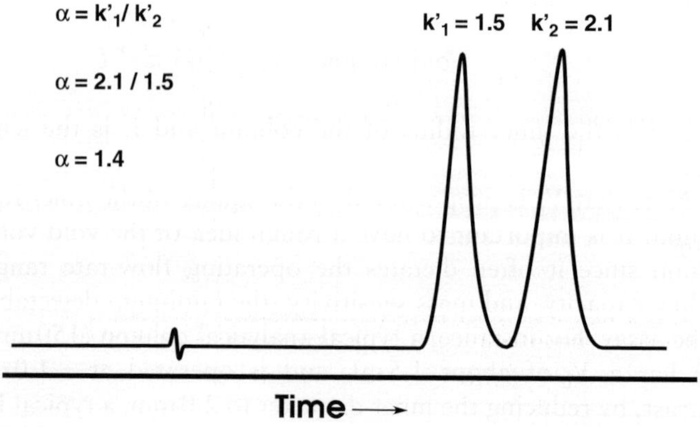

FIGURE 4 A chromatogram of two peaks with a selectivity factor (α) of 1.4.

the peak width (σ). Since W_b is equal to 4σ, the following equation has been derived:

$$\text{Number of theoretical plates, } N = \left(\frac{t_R}{\sigma}\right)^2 = 16\left(\frac{t_R}{w_b}\right)^2$$

Figure 5 shows a chromatogram with a peak width of 10 units and a t_R of 135 units. The column efficiency (N) can therefore be calculated as follows:

$$N = 16\left(\frac{135}{10}\right)^2 = 2916$$

Since it is difficult to measure σ or W_b manually, a relationship using width at half height or $W_{1/2}$ is often used to calculate N as described in the United States Pharmacopoeia (USP).

$$N = 5.54\left(\frac{t_R}{w_{1/2}}\right)^2$$

I. Height Equivalent of a Theoretical Plate or Plate Height (HETP or *H*)

The concept of a plate is traditionally derived from the industrial distillation process using a distillation column consisting of individual plates where the condensing liquid is in equilibrium with the rising vapor. Thus, a longer distillation column would have more "plates" or equilibration steps. Similarly in chromatography, the height equivalent of a theoretical plate (HETP or H) is equal to the length of the column (L) divided by the number of theoretical plates (N) even though there are no discrete plates inside the HPLC column.

$$\text{Height Equivalent of a Theoretical Plate, HETP } (H) = L/N$$

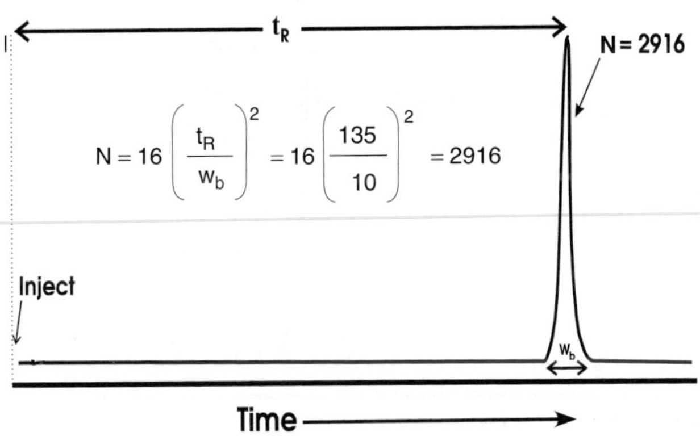

FIGURE 5 **A chromatogram showing a peak from a column with plate count of 2916.**

As will be shown later, HETP is determined by particle diameter of the packing material (d_p) and to a lesser extent by the flow rate (F) and column temperature (T). For a well-packed column, the plate height is roughly equal to $2.5d_p$. Therefore, a column packed with 5-μm particles would have a plate height of 12.5 μm. A typical 15 cm long column packed with 5-μm materials should have $N = L/H = 150{,}000\,\mu m/12.5\,\mu m$ or 12,000 plates. Similarly, a typical 15 cm long column packed with 3-μm materials should have $N = L/H = 150{,}000\,\mu m/7.5\,\mu m$ or 20,000 plates. For this reason, columns packed with smaller particles are usually more efficient.

E. Resolution (R_s)

The goal of most HPLC analysis is the separation of one or more analytes from other components in the sample in order to obtain quantitative information for each analyte. Resolution (R_s) is the degree of separation of two adjacent analyte peaks, and is defined as the difference in retention times of the two peaks divided by the average peak width (Figure 6). As peak widths of adjacent peaks tend to be similar, the average peak width can be equal to the width of one of the two peaks.

$$\text{Resolution, } R_s = \frac{t_{R1} - t_{R2}}{((w_{b1} + w_{b2})/2)} = \frac{\Delta t_R}{w_b}$$

In Figure 6, if $\Delta t = 23$ units and $W_b = 14$ units, then $R_s = 23/14 = 1.5$.

Figure 7 provides graphic representations of two peaks with various resolutions from 0.6 to 2.0. Whereas an R_s of 0 means complete coelution

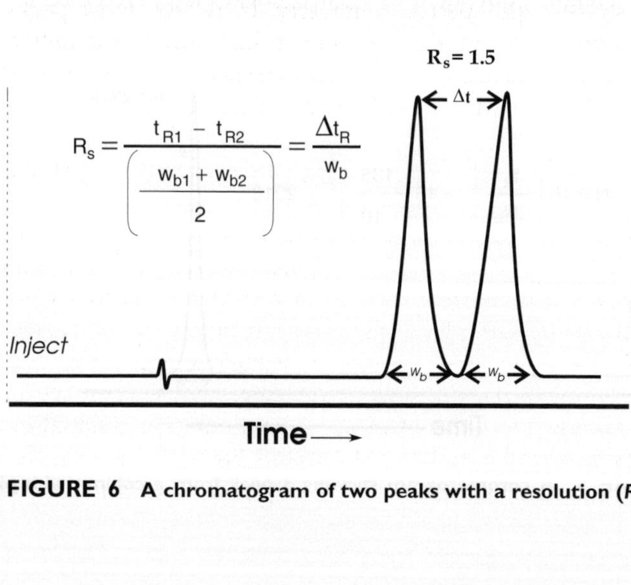

FIGURE 6 A chromatogram of two peaks with a resolution (R_s) of 1.5.

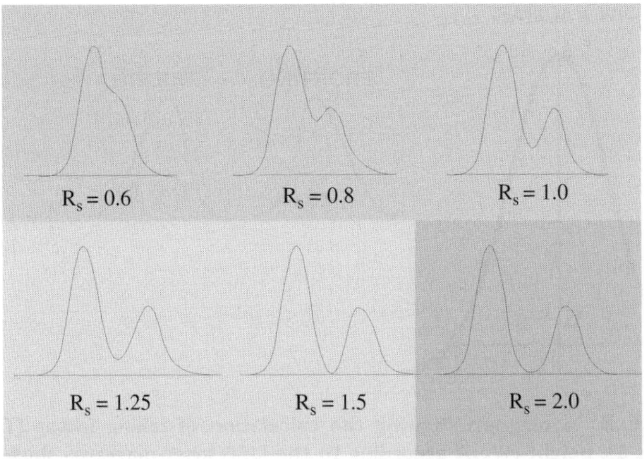

FIGURE 7 A schematic diagram showing two closely eluting peaks at various resolution values from 0.6 to 2.5. Figure reprinted with permission from Reference 7.

or no separation, a shoulder can be discernible at an R_s of 0.6. An R_s of 1 indicates that partial separation is achieved and an R_s of 1.5 indicates that baseline separation is achieved. The goal of most pharmaceutical HPLC analysis is to achieve baseline separation for all key analytes. Most often, a target value of $R_s > 2.0$ is desirable because such a condition leads to more robust separation and quantitation.

F. Tailing Factor (T_f)

Under ideal conditions, chromatographic peaks will have Gaussian peak shapes with perfect symmetry. In reality, most peaks are either slightly fronting or tailing. As shown in Figure 8, the tailing factor (T_f) as defined by the USP[11] is a measure of peak asymmetry. In this calculation, peak width at 5% peak height ($W_{0.05}$) is used.

$$\text{Tailing factor } (T_f) = W_{0.05}/2f = AC/2AB \qquad \text{(see Figure 8)}$$

Tailing factors for most peaks should fall between 0.9 and 1.4, with a value of 1.0 indicating a perfectly symmetrical peak. Peak tailing is typically caused by adsorption or other strong interactions of the analyte with the stationary phase while peak fronting can be caused by column overloading, chemical reaction or isomerization during the chromatographic process. For instance, many basic analytes (amines) display some peak tailing due to the strong interaction with residual acidic silanol groups in silica-based bonded phases.[6] Figure 9 shows a chromatogram with three components. Uracil is very polar and elutes with the solvent

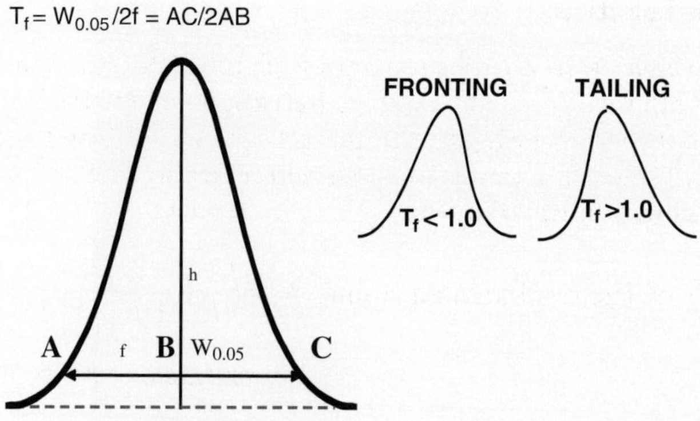

FIGURE 8 A diagram showing the calculation of tailing factor (T_f) from peak width at 5% height ($W_{0.05}$) according to the USP. Inset diagrams show fronting and tailing peaks.

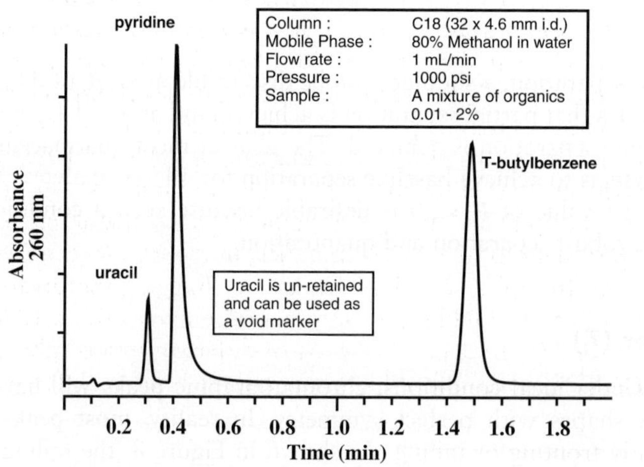

FIGURE 9 An HPLC chromatogram of three components with LC conditions shown in the inset. Note that the basic pyridine peak is tailing while the t-butylbenzene peak (neutral) is symmetrical.

front with a $k' = 0$. It is often used as a void volume marker for the measurement of t_0. Pyridine is a base and is ionized in a neutral mobile phase. Pyridine exhibits considerable peak tailing on this column due to strong interaction with residual silanol groups. *T*-butylbenzene is a neutral (non-ionizable) and fairly hydrophobic molecule, which elutes much later with excellent peak symmetry. One of the most important advances in HPLC was the development of high-purity silica with low silanophilic activity. Most modern HPLC columns are packed with these silica materials and display significant reduction of peak tailing for basic analytes.[6]

G. The Resolution Equation

The effectiveness of the separation (R_s) in HPLC analysis is dependent on both thermodynamic factors (retention and selectivity) and kinetics factors (peak width and column efficiency).[10,12] The relationship of resolution to other parameters can be expressed somewhat quantitatively in the resolution equation:

$$\text{The resolution equation} \quad R_s = \left(\frac{k'}{k'+1}\right)\left(\frac{\alpha-1}{\alpha}\right)\left(\frac{\sqrt{N}}{4}\right)$$

Retention Selectivity Efficiency

Scrutiny of the resolution equation indicates that R_s is controlled by three relatively independent terms: retention, selectivity, and efficiency (Figure 10). To maximize R_s, k' should be relatively large. However, a value of k' over 10 will approach a point of diminishing returns as the retention term of $k'/(1+k')$ approaches unity. No separation is possible if $k'=0$, since R_s must equal zero if k' is zero in the resolution equation.

Selectivity (α), the ratio of k' for two closely eluting peaks, is close to 1.0, and typically between 1.01 and 1.50. Selectivity is maximized by using column or mobile phase conditions that resolve all critical analytes. Figure 11 illustrates how resolution can be achieved by exploiting the selectivity effect of different organic solvent modifiers. Note that a small change of selectivity can have a major effect on resolution as resolution is proportional to ($\alpha-1$). For instance, in Figure 11, R_s is increased significantly from 0.8 to 2.5 (~3 fold) by changing the organic modifier from ACN to MeOH because α is increased from 1.07 to 1.19 (as ($\alpha-1$) goes from 0.07 to 0.19). Columns of different bonded phases (i.e., C8, phenyl, CN, polar-imbedded)[6] can also provide a different selectivity

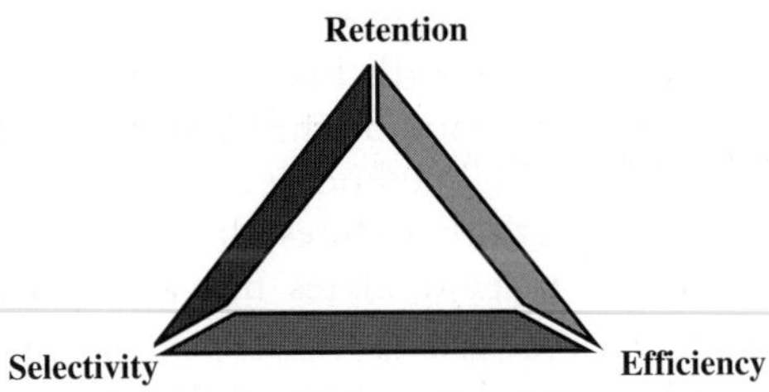

$$R_s = \left(\frac{k'}{k'+1}\right)\left(\frac{\alpha-1}{\alpha}\right)\left(\frac{\sqrt{N}}{4}\right)$$

Retention Selectivity Efficiency

FIGURE 10 **The resolution equation, which is governed by three factors: retention, selectivity, and efficiency.**

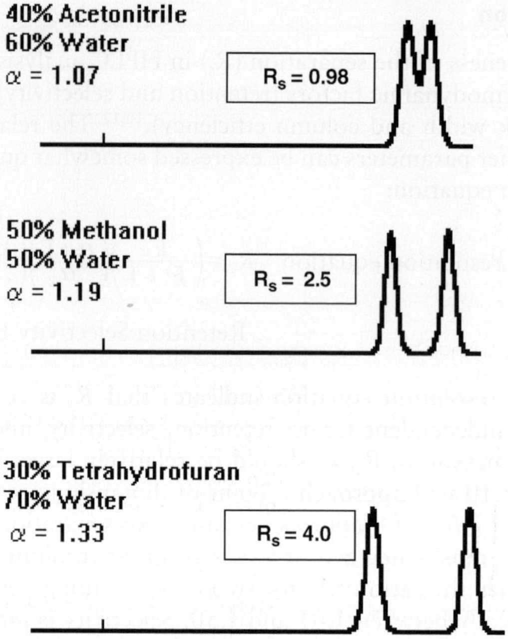

FIGURE 11 **Three chromatograms illustrating mobile phase selectivity using different organic solvent modifiers in reversed-phase chromatography. Note that small changes of selectivity can have a significant impact on resolution. THF is a stronger solvent and yields the best resolution in this instance. Figure reprinted with permission from Reference 7.**

effect as illustrated in the separation of four antidepressants in Figure 12. Finally, the plate count of the column (N) should be maximized by using a longer column or using columns packed with smaller particles. However, increasing N is not an efficient way to achieve resolution since R_s is proportional to the square root of N. For instance, doubling N by increasing the column length by 2 would increase analysis time by 2, but only increase resolution by the $\sqrt{2}$ or by 41%.

In summary, during method development, the general strategy is to find a solvent strength that elutes all solutes between a k' of 1 and 20 and to separate all key analytes by varying the organic solvent and other mobile phase modifiers. If this is unsuccessful, a different bonded phase might be tried. For simple separation, it is more productive to vary selectivity by changing the mobile phase than bonded phase. Efforts to increase N by increasing column length is "expensive" in terms of analysis time. High column efficiency, however, is needed for complex mixtures such as formulations based on natural products or impurity testing of multi-ingredient drugs. Often gradient analysis is required for analysis of such complex mixtures.

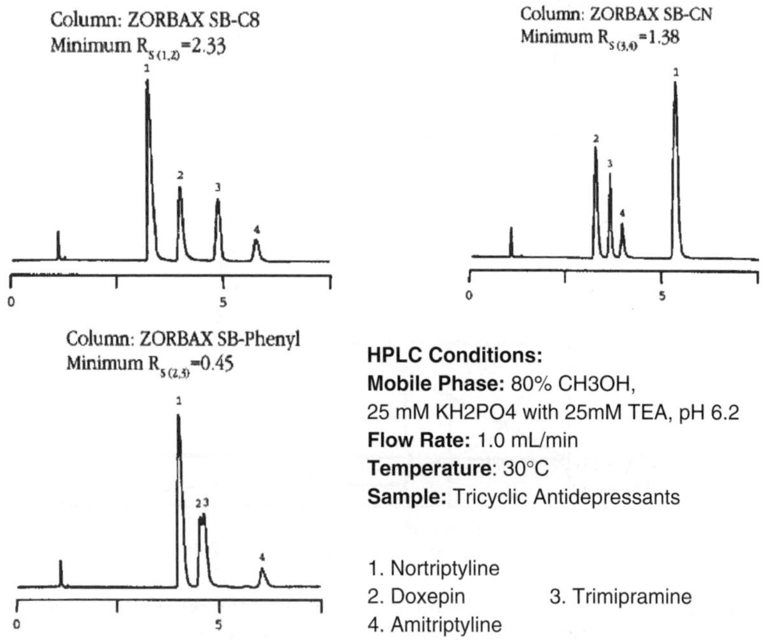

FIGURE 12 Three HPLC chromatograms of four antidepressant drugs illustrating column selectivity effects using three bonded phases (C8, phenyl, and cyano). Note changes of peak elution order. HPLC conditions and peak identification are shown in the inset. Figure reprinted with permission from Agilent Technologies.

H. The van Deemter Equation

The phenomenon of band broadening in the gas chromatographic process was first studied by van Deemter in the 1950s resulting in the formulation of the van Deemter equation, correlating HETP or plate height with linear flow velocity (V).[13] Figure 13 shows how the van Deemter curve is a composite curve stemming from three separate terms (A, B/V and CV) which are in turn controlled by factors such as particle size (d_p), and diffusion coefficients (D_m), where

$$\text{HETP} = A + B/V + CV$$

The A term represents "eddy diffusion or multi-path effect" and is proportional to (d_p).
The B term represents "longitudinal diffusion" and is proportional to (D_m).
The C term represents "resistance to mass transfer" and is proportional to (d_p^2/D_m).

While there are some disagreements among experts on the best quantitative relationship between HETP and flow rate, the van Deemter

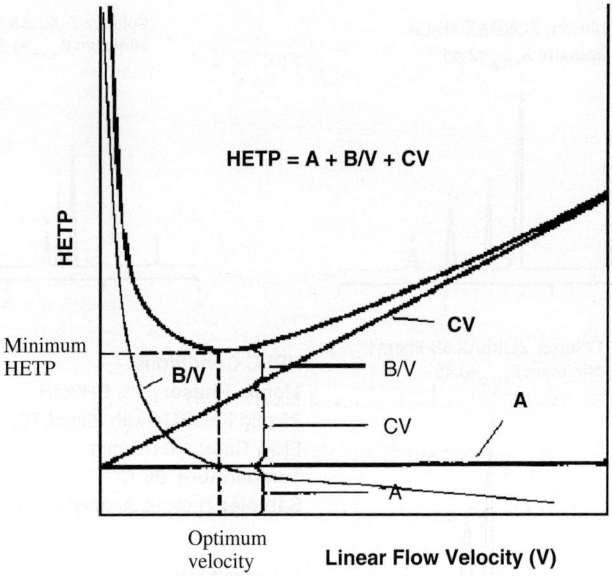

FIGURE 13 van Deemter curve showing the relationship of HETP vs. average linear velocity. The van Deemter curve has a classical shape and is a composite plot of *A*, *B* and *C* terms (plotted below to show their contributions). This diagram is adapted from Reference 9.

equation is the best known and appears to explain the band-broadening concept in HPLC well,[6] although it was developed for gas chromatography.

Figure 14 shows the experimental van Deemter curves for three columns packed with 10-, 5- and 3-μm particles.

- Plots show that small d_p yields lower HETP (or a small particle column has more efficiency per unit length) since the *A* term is proportional to d_p.
- Small-particle columns have less efficiency loss at high flow rates since the *C* term is proportional to d_p^2.

For these reasons, smaller-particle columns are particularly well suited for fast LC and high-throughput screening applications.[14,15]

A more vigorous treatment similar to the van Deemter equation but developed specifically for HPLC is the Knox equation,[16] which uses a number of reduced parameters where h is the reduced plate height (h/d_p) and v is reduced velocity ($V\ d_p/D_m$).

$$\text{Reduced plate height, } h = Av^{1/3} + B/v + C/v$$

The reader is referred to several key references for a more detailed treatment of the band-broadening theory, which is fundamental to the chromatographic process.[1,5,6]

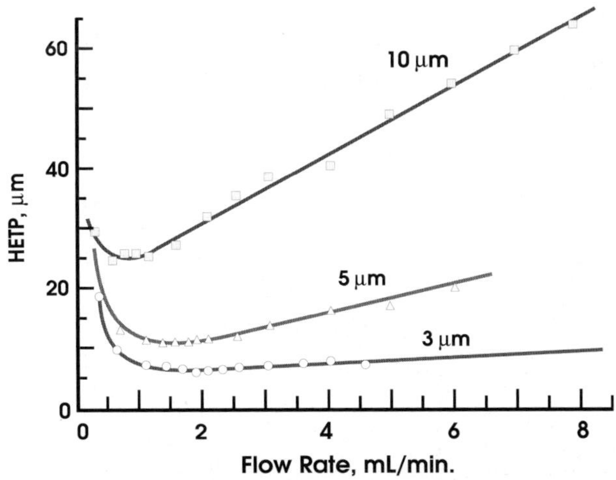

FIGURE 14 van Deemter curves of three columns packed with 10-, 5-, and 3-µm particles.

III. MOBILE PHASE PARAMETERS

A. Organic Solvent Strength and Selectivity

We have discussed previously that the solvent strength or % organic solvent content in the mobile phase controls the retention time of the analyte (Figure 3) and that different organic solvents (MeOH, ACN, THF) can have a dramatic effect on selectivity. Figure 15 graphically depicts the linear relationships of log k' vs. % organic solvent content for the three common RPLC organic solvents. Note that THF is stronger than ACN, which in turn is stronger than MeOH in RPLC. Consequently, a much lower % THF is needed (vs.% MeOH) to attain the same k' for a given solute. While THF yields different selectivity in RPLC, its use is not very popular (except in gel-permeation chromatography) due to its toxicity and its tendency to form peroxide (unless a preservative is added). Another ether, methyl *t*-butyl ether (MTBE), is often used instead of THF. MTBE, however, must be used at low levels or with a co-solvent due to its low miscibility with water.

B. Buffers

Many drugs have either acidic or basic functional groups and can exist in solutions in ionized or non-ionized forms. The ionic state and degree of ionization greatly affect their chromatographic retention in RPLC. Typically, the ionic form does not partition well into the hydrophobic stationary phase and therefore has significantly lower k' than the neutral,

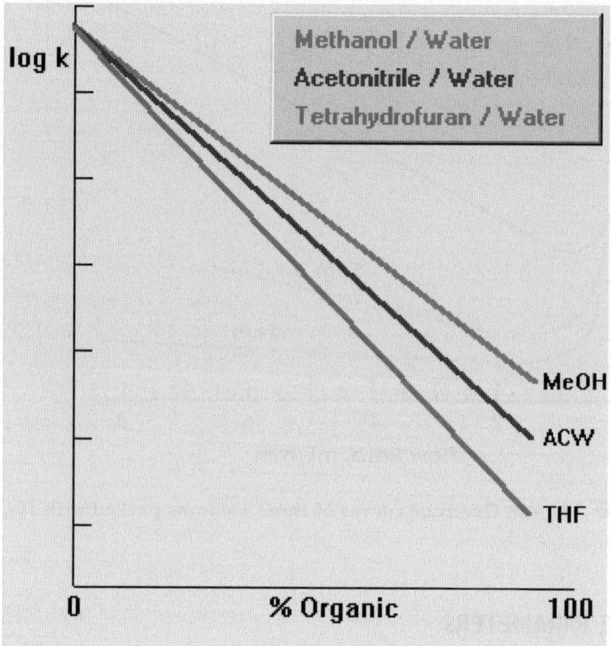

FIGURE 15 Plots of log k' vs. % organic solvent modifier for methanol, acetonitrile, and tetrahydrofuran. Reprinted with permission from Reference 7.

TABLE 2 Common HPLC Buffers and their Respective pK_a and UV Cutoffs

Buffer	pK_a	UV cutoff (nm)
Trifluoroacetic acid[1]	0.3	210
Phosphate	2.1, 7.2, 12.3	190
Citrate	3.1, 4.7, 5.4	225
Formate[1]	3.8	200
Acetate[1]	4.8	205
Carbonate[1]	6.4, 10.3	200
Tris(hydroxymethyl) aminomethane	8.3	210
Ammonia[1]	9.2	200
Borate	9.2	190
Diethylamine	10.5	235

[1]Volatile buffer systems that are MS compatible.

un-ionized form. Buffers are commonly used to control the pH of the mobile phase for the separation of acidic or basic analytes. Table 2 summarizes the common buffers for HPLC and their respective pK_a and UV cutoffs. Readers are referred to the chapter on HPLC operation on buffer use. Volatile acids and their ammonium salts are used for the development of mass spectrometer (MS) compatible HPLC methods. Since a buffer is only effective within ±1 pH unit from its pK_a, judicious selection of the proper buffer within its buffering range is paramount.

C. Acidic Mobile Phase

A mobile phase at acidic pH of 2.5–3 is a good starting point for most pharmaceutical applications because the low pH suppresses the ionization of most acidic analytes resulting in their higher retention.[2,3] Common acids used for mobile-phase preparations are phosphoric acid, formic acid, and acetic acid. Low pH also minimizes the interaction of basic analytes with surface silanols on the silica packing (because silanols do not ionize at acidic pH). Also, the lifetime of most silica-based columns is excellent in the pH range of 2–8. However, basic analytes are ionized at low pH and might not be retained unless ion-pairing reagents are used.

D. Ion-pairing Reagents

Ion-pairing reagents are detergent-like molecules added to the mobile phase to provide additional retention or selectivity for the analytes with opposite charge. Long-chain alkyl sulfonates are commonly used for the separation of water-soluble basic analytes as shown in Figure 16 in the analysis of water-soluble vitamins (WSV). Hexanesulfonate binds with

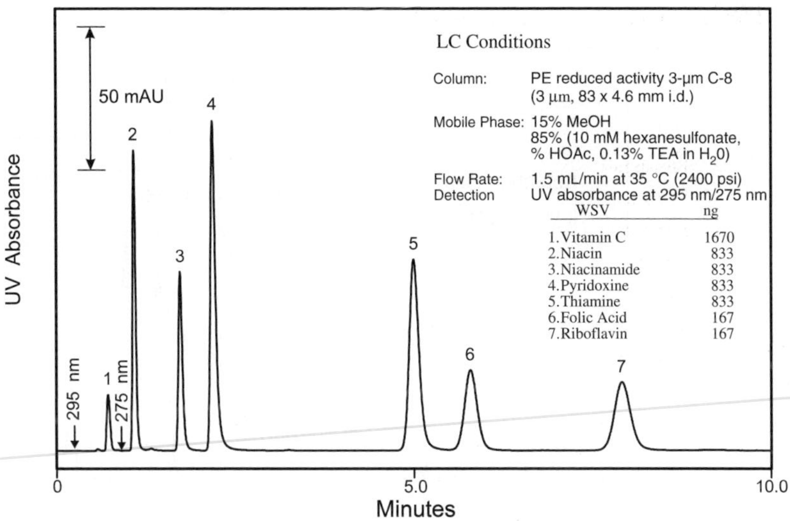

FIGURE 16 HPLC chromatogram of water-soluble vitamins using ion-pair chromatography. LC conditions and peak identification are shown in the inset. The retention times of basic analytes (pyridoxine and thiamine) are strongly dependent of the concentration of ion-pairing reagent (1-hexanesulfonate) in the mobile phase. Reprinted with permission from Reference 17.

basic analytes such as pyridoxine and thiamine to form ion-pairs. These neutral ion-pairs are now more hydrophobic than their unassociated protonated amines resulting in higher retention times. An ion-pairing reagent provides an additional parameter to facilitate resolution of basic analytes from other components in the sample.[17,18] Note that amine modifiers such as triethylamine (TEA) are often added in the mobile phase to reduce peak tailing caused by the strong interaction of basic analytes with acidic surface silanols. For acidic analytes, ion-pairing reagents such as *tetra*-alkylammonium salts are often used.

E. High pH Mobile Phase

The advent of silica-based columns stable under high-pH conditions (e.g., pH 1–12) offers an important alternate approach for the separation of basic analytes.[6,19] Figure 17 shows the basis of this approach in the separation of amitriptyline and nortriptyline. Note that at low pH, both analytes are ionized and co-elute with the solvent front at t_0. At pH 8, both analytes are partially ionized and are well separated from each other with nortriptyline eluting first due to its higher pK_a (and therefore more ionized at pH 6). At pH 10, both analytes exist as unprotonated bases and are well retained and resolved. The application of this approach is increasing for assay and impurity testing of many water-soluble basic drugs.[19] The advantages of high-pH separation as opposed to ion-pairing are mass-spectrometry compatibility, better sensitivity and unique selectivity.

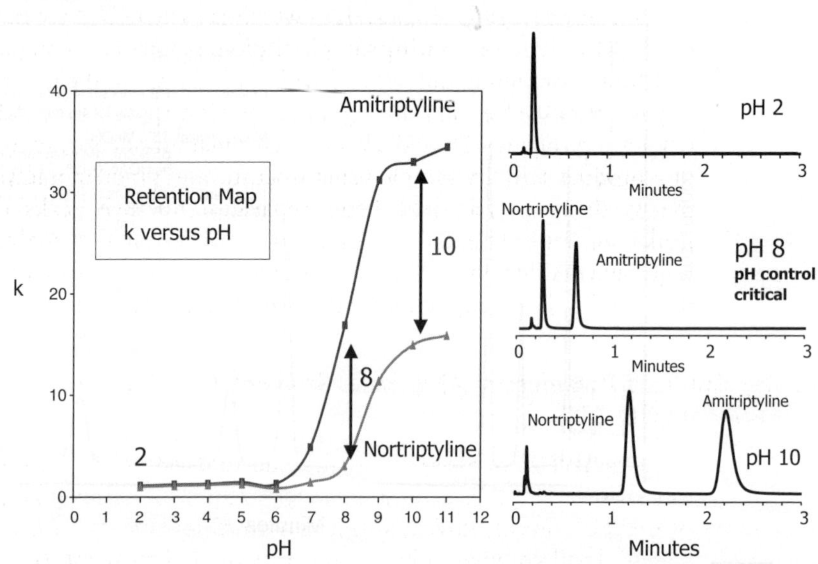

FIGURE 17 Retention map and chromatograms of two basic antidepressants using mobile phases at various pH illustrating the importance of pH in the separation of basic analytes. Figure reprinted with permission from Waters Corporation.

F. Other Parameters (Flow rate [*F*], Column Temperature [*T*])

For isocratic analysis, flow rate has no impact on k' or α, since flow has the same effect on t_R and t_0. Both operating pressure and analysis time are inversely proportional to flow rate. Flow rate also has a significant effect on N, as efficiency is reduced (with H increasing) at higher flow rate due the higher resistance to mass transfer (the van Deemter C term).

Higher column temperatures (T) lower the viscosity of the mobile phase (and thus the pressure) and increase the diffusion coefficient of the analytes (increase N). The effects of these operating parameters are discussed further in the next section on gradient analysis.

IV. ISOCRATIC VS. GRADIENT ANALYSIS

Traditionally, most pharmaceutical assays are isocratic analysis employing the same mobile phase throughout the elution of the sample. Isocratic analyses are particularly common in quality control applications since they use simpler HPLC equipment and premixed mobile phases. Notable disadvantages of isocratic analysis are limited peak capacity (the maximum number of peaks that can be accommodated in the chromatogram), and problems with samples containing analytes of diverse polarities. Also, late eluters (such as dimers) are particularly difficult to quantitate in isocratic analysis due to excessive band broadening with long retention times.

In contrast, gradient analysis in which the strength of the mobile phase is increased with time during sample elution, is suited for complex samples and those containing analytes of wide polarities. Gradient chromatography is amenable for high-throughput screening applications and for impurity testing. Figures 18 and 19 show chromatograms for impurity testing of a product stability sample using isocratic and gradient analyses.[19] Note that gradient analysis yields better separation for early peaks and sharper peaks for late eluters. The disadvantages of gradient analysis are the requirements for more complex instrumentation and greater skills in method development, and difficulties in method transfer.[2]

A. Key Gradient Parameters (Flow rate, Gradient time [t_G], Peak Capacity [*P*])

Gradient methods are more difficult to develop because the optimization of several additional parameters is required—starting and ending % of organic solvent, flow rate (F), and gradient time (t_G). Also, the HPLC system dwell volumes, which can vary from 0.5 to 10 mL (see instrumentalation chapter) can make method transfer problematic. Unlike isocratic analysis, flow rate has a dramatic influence on retention and selectivity in gradient analysis. Peaks have similar width in gradient elution since they

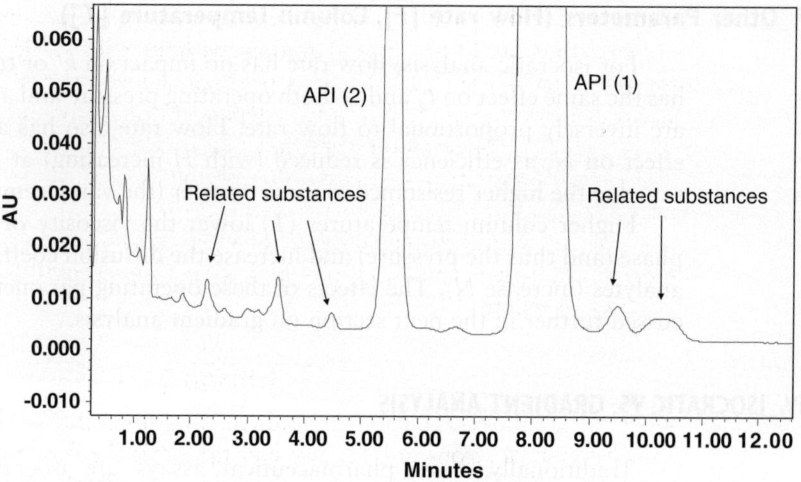

FIGURE 18 **HPLC separation of a stability sample under isocratic conditions showing low levels of various related substances (API: active pharmaceutical ingredient). Reprinted with permission from Reference 19.**

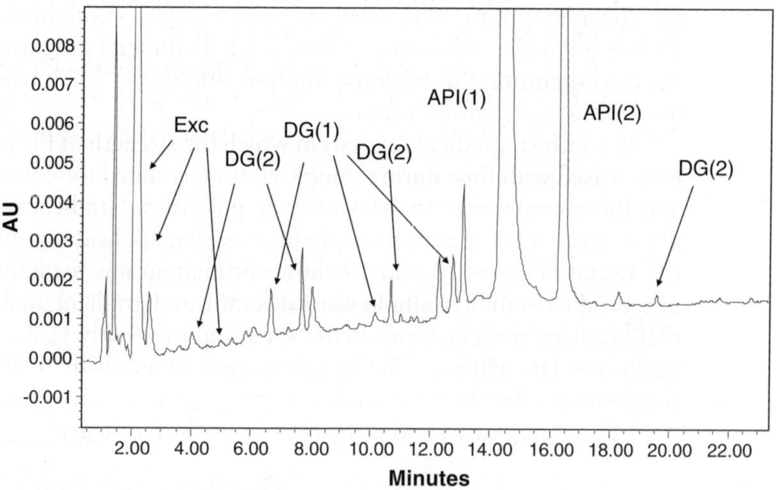

FIGURE 19 **HPLC separation of the same stability sample shown in Figure 18 under gradient conditions showing better resolution and increased sensitivity of trace impurities, degradants (DG), and excipients (Exc). Reprinted with permission from Reference 19.**

are forced to elute with the increasing by stronger mobile phase composition. Therefore, column efficiency cannot be measured under gradient conditions, as late eluting peaks would yield higher plate counts than early peaks in the same column. Instead, the concept of peak capacity (P) is introduced and typically used.[20,21] Peak capacity is the maximum number of peaks that can fit in a chromatogram with an R_s of 1. Figure 20 shows the parameters used to calculated peak capacities in gradient analysis, where P approximates (t_G/w_b) and is dependent on factors including t_G, N, d_p, L, and F.[20–22] Due to the narrower peak widths in gradient analysis, a higher peak capacity is possible (100–200) as oposed to the much lower peak capacities for isocratic analysis (50–100).

The concept of the capacity factor k' is also considerably more complex in gradient chromatography and is best represented by an average k' or k^* [18] where

$$k^* = \frac{t_G F}{1.15 S \Delta\phi V_0}$$

where k^* is the average k' under gradient conditions, $\Delta\phi$ the change in volume fraction of strong solvent, S a constant, F is the flow rate (mL/min), t_G the gradient time (min), and V_0 the column void volume (mL).

Note that in gradient analyses operated at higher flow rates, more mobile phase is pumped through the column per unit time. As a result, in terms of the total volume of mobile phase used, it is equivalent to operating at a lower flow rate and a longer gradient time. Figure 21 shows comparative chromatograms of two gradient HPLC peptide maps illustrating the dramatic reduction of analysis time by reducing t_G, increasing F, and lowering V_0 by reducing the column length.[22]

Figure 22 shows a high-throughput application in which high flow rate, a short column with very small particles packings, short t_G, and

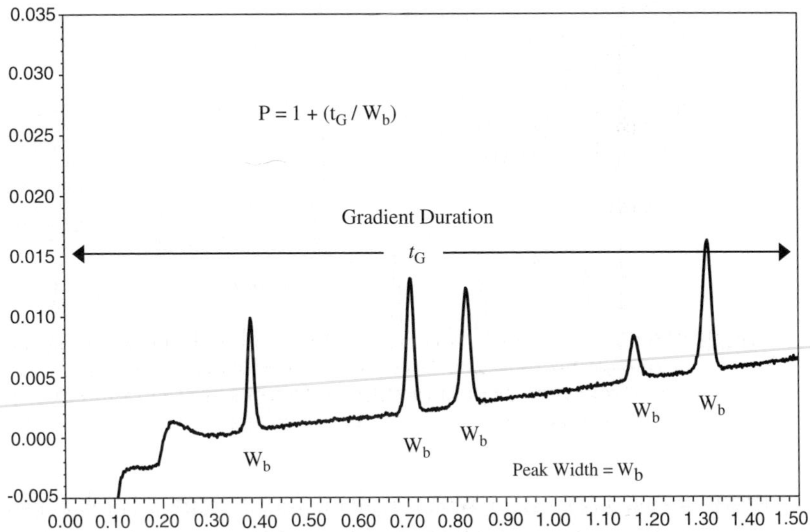

FIGURE 20 **Chromatogram illustrating the concept of peak capacity (*P*) which is the maximum number of peaks that can be accommodated in a chromatogram.**

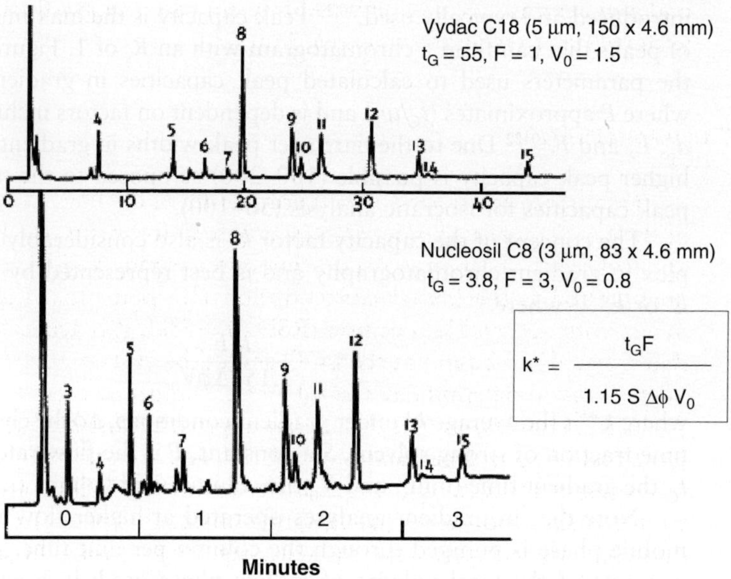

FIGURE 21 **Two HPLC gradient chromatograms (tryptic maps of lysozyme) illustrating the dramatic effect of flow rate (*F*), gradient time (t_G), and void volume (V_0) on analysis time. Figure reprinted with permission from Reference 22.**

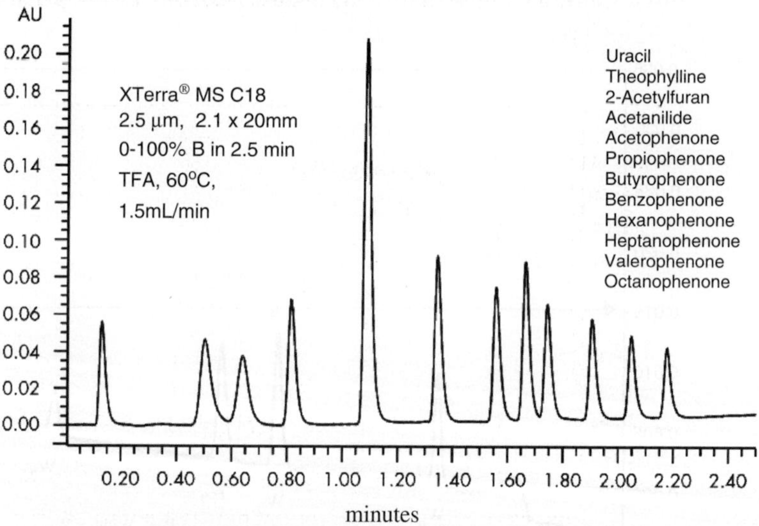

FIGURE 22 **An example of high-throughput screening utilizing all the concepts of reducing column length (*L*), particle size (d_p), column diameter (d_c), gradient time (t_G), increasing flow (*F*), and column temperature (*T*) to achieve very fast gradient separations. LC conditions and peak identification are shown in the inset. Figure reprinted with permission from Waters Corporation.**

high temperature are all combined to yield very fast gradient separation with excellent resolution in under 3 min.[20]

V. LIMIT OF QUANTITATION

Finally, we will discuss the determination of the limit of quantitation or LOQ. Establishing an LOQ is required for many pharmaceutical applications such as impurity testing and cleaning validation. Limit of quantitation is the lowest concentration (%, ppm) that can be determined with acceptable precision (RSD of ~5%). It is generally accepted that a signal/noise ratio at the LOQ should be at least 10.[23,24] There are many ways to determine the LOQ and a number of them are discussed in the ICH guidelines,[23] including an approach illustrated in Figure 23. Note that LOQ is dependent on V_0, k', N, and volume of injection (V_{inj}). However, in practice, higher sensitivity or lower LOQ is often achieved by increasing the signal (using larger injection volume) or switching detection wavelength, and by reducing noise (using high-sensitivity detectors).

VI. GLOSSARY OF HPLC TERMS

- Column—a tube that contains packed sorbents. Typical HPLC columns are stainless-steel tubes packed with silica-based bonded phases.

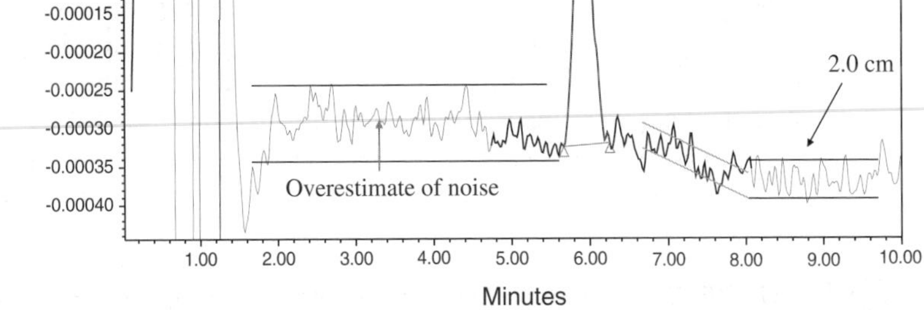

FIGURE 23 HPLC chromatogram illustrating the calculation of LOQ at a signal-to-noise ratio of 10. Note that the peak-to-peak noise is often overestimated by the analyst as shown.

- Mobile phase—a liquid that carries the sample through the column. Typical mobile phases in reversed phase LC are a mixture of water with acetonitrile or methanol.
- Retention—the tendency of a solute to partitioned into the stationary phase of the column.
- Resolution (R_s)—the degree of separation between two peaks as defined by the difference in their retention time divided by the average peak width.
- Efficiency or plate count (N)—an assessment of column performance. N should be fairly constant for a particular column and can be calculated from the retention time and the peak widths.
- Selectivity (α)—the ratio of retention (k') of two adjacent peaks.
- Sample capacity—the maximum mass of sample that can be loaded on the column without destroying peak resolution.
- Capacity factor (k')—a measure of solute retention obtained by dividing the net retention time by the void time.
- Peak Capacity—the maximum number of peaks that can be resolved in a chromatogram with a resolution of one.

VII. SUMMARY AND CONCLUSION

This chapter provides an overview of essential concepts in HPLC including retention, selectivity, efficiency, and resolution as well as their relationships with key column and mobile phase parameters such as particle size, column length and diameter, mobile phase strength, pH, and flow rate. The significance of several concepts important in pharmaceutical analysis such as peak capacity, gradient time, void volume, and limit of quantitation are discussed.

ACKNOWLEDGMENTS

The authors acknowledge helpful ideas and suggestions from C. Davidson, C. Choi, and Lane Gehrlein from Purdue Pharma. This manuscript has been presented in parts as short courses on "Advanced HPLC in Pharmaceutical Analysis" at Eastern Analytical Symposium 2001–2002, HPLC 2002, and Pittcon 2003. Numerous comments and suggestions from the attendees were incorporated into this manuscript.

REFERENCES

1. Snyder, L. R. and Kirkland, J. J., *Introduction to Modern Liquid Chromatography*, John Wiley & Sons, New York, 1979.
2. Snyder, L. R., Kirkland, J. J. and Glajch, J. L., *Practical HPLC Method Development*, 2nd Edition, Wiley-Interscience, New York, 1997.

3. Meyer, V. R., *Practical HPLC,* Wiley, New York, 1998.
4. Adamovics, J. A., *Chromatographic Analysis of Pharmaceuticals*, 2nd Edition, Marcel Dekker, New York, 1997.
5. Katz, E., Eksteen, R., Schoenmakers P. and Miller, N., *Handbook of HPLC,* Marcel Dekker, New York, 1998.
6. Neue, U. D., *HPLC Columns: Theory, Technology, and Practice,* Wiley-VCH, New York, 1997.
7. Introduction to HPLC, CLC-10 and HPLC Method Development, CLC-20, (Computer-based Instruction), Savant, Fullerton, CA, http://www.savant4training.com/savant2.htm.
8. LC Resources, Training Courses, Walnut Creek, CA, http://www.lcresources.com/training/trprac.html.
9. Modern Liquid Chromatography Short Course, presented by J. Dorsey at Purdue Pharm, 1991.
10. Rizzi, A., Retention and selectivity, In: Katz, E., Eksteen, R., Schoenmakers, P. and Miller, N., Eds., *Handbook of HPLC*, Marcel Dekker, New York, 1998.
11. United States Pharmacopeia 26—National Formulatory 21, The United States Pharmacopeial Convention, 2003.
12. Tijssen, R., The mechanism and importance of zone-spreading, In: Katz, E., Eksteen, R., Schoenmakers, P. and Miller, N., Eds., *Handbook of HPLC*, Marcel Dekker, New York, 1998.
13. Van Deemter, J. J., Zuidenweg, F. J. and Klinkenberg, A., *Chem. Eng. Sci.,* 5:271, 1956.
14. Dong, M. W., *Today's Chemist at Work*, 9(2):46–45, 2000.
15. Dong, M. W. and Gant, J. R., *LC.GC*, 2(4):294, 1984.
16. Knox, J. H., *J. Chromatogr. Sci.*, (15):352, 1977.
17. Dong, M. W. and Pace, J. L., *LC.GC,* 14(9):794, 1996.
18. Dong, M. W., Lepore, J. and Tarumoto, T., *J. Chromatogr. Sci.*, 442:81, 1988.
19. Dong, M. W., Miller, G. and Paul, R., *J. Chromatogr. Sci.*, 987:283, 2003.
20. Neue, U. D. and Mazzeo, J. R., *J. Sep. Sci.*, 24:921, 2001.
21. Snyder, L. R. and Stadalius, M. A., In: Horvath, Cs. Ed., *High-performance Liquid Chromatography—Advances and Perspectives,* Vol. 4, Academic Press, New York, pp. 195–309, 1986.
22. Dong, M. W., Tryptic mapping by reversed-phase liquid chromatography, In: Brown, P. Ed., *Advances in Chromatography*, Vol. 32, Marcel Dekker, New York, pp. 21–51, 1992.
23. ICH Guidelines for Method Validation: Q2B Validation of Analytical Procedures: Methodology, 1996, http://www.ich.org/pdfICH/Q2B.pdf.
24. Dong, M. W., *Today's Chemist at Work*, 9(8):28, 2000.

3 HPLC INSTRUMENTATION IN PHARMACEUTICAL ANALYSIS: STATUS, ADVANCES, AND TRENDS

MICHAEL W. DONG

Purdue Pharma, Ardsley, NY

Handbook of Pharmaceutical Analysis by HPLC,
S. Ahuja and M.W. Dong, editors.

ABSTRACT

This chapter presents an overview of current trends in high-pressure liquid chromatography (HPLC) instrumentation focusing on recent advances and features relevant to pharmaceutical analysis. Operating principles of HPLC modules (pump, detectors, autosampler) are discussed with future trends.

I. INTRODUCTION

A. Scope

High-pressure liquid chromatography (HPLC) is the most widely used analytical technique. Today's liquid chromatographs have excellent performance and reliability—the result of decades of refinements driven by technical advances and competition between manufacturers in a two-billion-dollar-plus equipment market. Literature on HPLC equipment abounds, including textbooks,[1–3] reference books,[4,5] review articles,[6,7] and training software.[8] Rather than summarizing the current literature, the goal of this chapter is to provide the reader with a concise overview of HPLC instrumentation, operating principles and recent advances or trends that lead to better assay performance. Two often-neglected system parameters, dwell volume and instrumental bandwidth, are discussed in more detail because of their impacts on Fast liquid chromatography (Fast LC) and small-bore LC applications.[9,10]

This book's focus is on analytical scale HPLC, so preparative scale LC and micro LC are not discussed in detail here. In-depth treatments of these techniques are found elsewhere.[11,12] LC/MS, data handling and practical HPLC operation are discussed in later chapters.

B. HPLC Systems and Modules

A typical HPLC system consists of several modules: a pump, an injector, a detector, and a data-handling device (Figure 1). The simplest

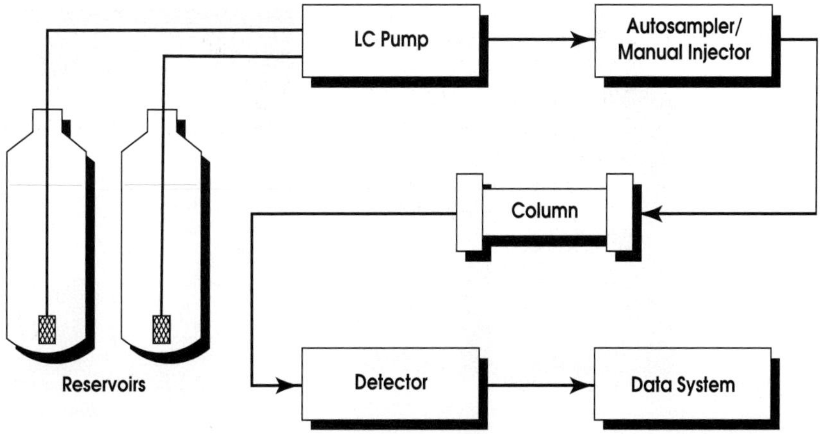

FIGURE 1 A schematic of an HPLC system showing the column and major modules.

system comprises an isocratic pump, a manual injector, a UV detector and a strip-chart recorder. Such bare-bone configurations are rare in modern pharmaceutical laboratories, which tend to have high-end instruments for better precision, productivity and regulatory compliance. A typical HPLC system for pharmaceutical analysis is likely to consist of a multi-solvent pump, an autosampler, an on-line degasser, a column oven and a UV/Vis and/or a photodiode array detector—all connected to, and controlled by, a data-handling workstation.

C. Modular Vs. Integrated Systems

HPLC instrumentation is classified as modular or integrated systems. In a modular system, separate modules are stacked, plumbed and connected to function as a unit. In an integrated system, modules are built inside a single housing and often share a common controller board. These built-in modules cannot function outside the system. Solvent lines and electrical wires are internal, giving the system a "cleaner" appearance. Figure 2 shows examples of modular and integrated HPLC systems. Modular systems are supposedly more serviceable since internal components are more easily accessible and the malfunctioning module can be "swapped." Integrated systems, however, may have better module integration. For instance, the autosampler can be designed to inject samples right at the beginning of a pump stroke yielding better performance in retention time precision. In practice, the demarcations of these two types are not significant since well-designed modules linked by high-speed electrical connections often look and function as a single well-integrated unit.

FIGURE 2 Examples of modular and integrated HPLC systems. Upper row (l to r) modular systems: Agilent 1100 Series, Jasco LC-2000 Series; Lower row (l to r) integrated systems: Waters Alliance, Shimadzu LC-2010.

II. HPLC SOLVENT DELIVERY SYSTEMS

A modern HPLC solvent delivery system consists of one or more pumps, solvent reservoirs and a degassing system. HPLC pumps can be categorized in several ways: by flow range, by driving mechanism or by blending methods. A typical analytical pump has flow range of 0.001–10 mL/min, which handles comfortably the flow rates required by most pharmaceutical assays (e.g., 0.5–3 mL/min). Preparative pumps with flow ranges of 30 mL/min up to liters/min are discussed elsewhere.[11] Most pumps use a reciprocating driving mechanism. The exceptions are screw-driven syringe pumps for capillary LC applications.[12,13] Another way to classify pumping systems is by the way solvent blending is achieved—under either low-pressure or high-pressure mixing conditions.

Today's HPLC pumps have sophisticated designs honed by decades of incremental improvements. They are also more reliable and easier to maintain than their predecessors. Short seal life and check valve malfunctioning seen in the early models are no longer problems. In this section, the principle of pump operation is described with an emphasis on advances leading to higher reliability and performance. A discussion of system dwell volume is included.

A. Operating Principles

Most HPLC pumps use a reciprocating piston design as shown in Figure 3. Here, a motorized cam (or a direct screw-drive system) drives a piston back and forth to deliver solvent through a set of inlet and outlet check valves. All wettable components are made from inert materials; examples include stainless-steel pump heads, ruby balls and sapphire seats in check valves, sapphire pistons and fluorocarbon pump seals. Since liquid is only delivered during the inward stroke of the piston, the resulting sinusoidal liquid stream is usually passed through a pulse dampener to reduce pulsation. Another approach is to use a dual-piston in-parallel pump design where a single motor drives two pistons in separate pump heads (Figure 4). Since the pistons are 180° out of phase, the combined output results in a steadier flow pattern. The digital stepper motor that drives the cam is controlled by a microprocessor

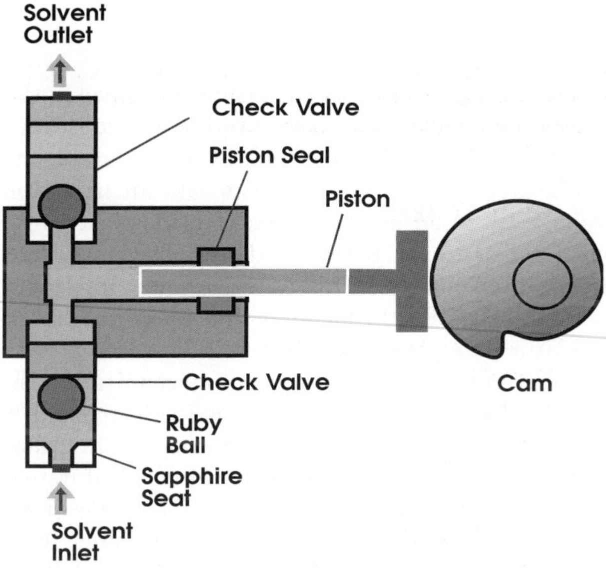

FIGURE 3 **A schematic of a reciprocating single-piston pump.**

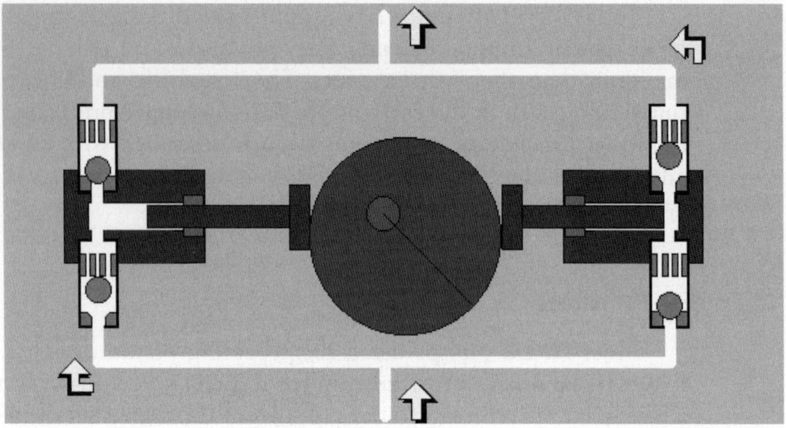

FIGURE 4 A schematic of a dual-piston in-parallel pump design. Diagram from HPLC equipment, CLC-30, reprinted with permission from Savant, Fullerton, CA.

that coordinates the piston speed with other components such as the proportioning valve and the pressure monitor.

Syringe pumps driven by screw mechanisms were popular in the 1960s because of their inherent precision and pulseless flow characteristics. Their disadvantages are a higher manufacturing cost and the problems associated with syringe refill cycles. Syringe pumps are currently used in specialized systems for microbore and capillary LC.[13]

B. Recent Advances

The simple reciprocating pump design developed in the 1960s was perfected through numerous innovations. The life of the piston seal was improved by better designs, such as more durable spring-loaded seals, a self-aligning piston mechanism and an irrigation system to eliminate any salt build-up (piston seal wash). Most pumps also have front panel access to the pump heads for easier maintenance. Pump performance often suffers at low flow rates. For instance, as a typical piston size is about 100 μL, blending and pulse-free flow are difficult to achieve at a flow rate of <0.2 mL/min or at <2 pump strokes per min. High-end pumps use sophisticated mechanisms such as variable stroke length (20–100 μL), micro-pistons (5–30 μL) and hybrid dual-piston in-series designs to improve performance. In a dual piston in-series design (Figure 5), two pistons (often of different sizes) are independently driven by separate motors. The pre-piston is synchronized with the secondary piston to provide smoother flow and higher compositional accuracy. Variations of this design are used in many

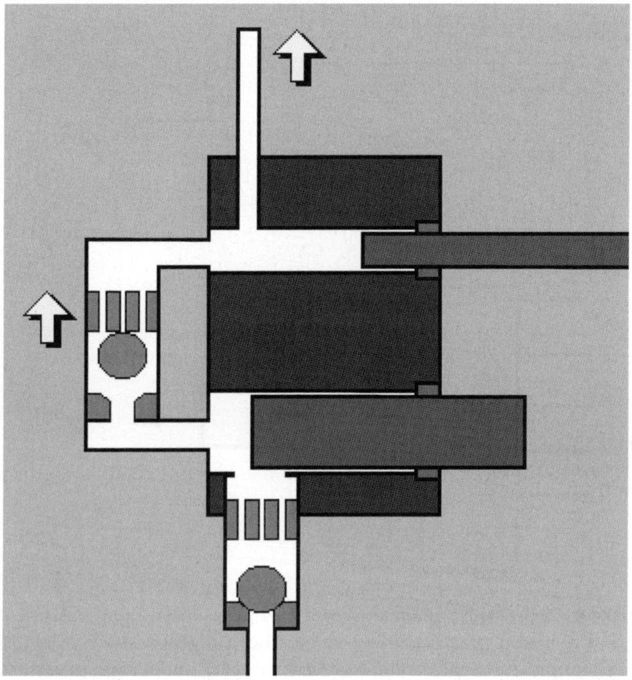

FIGURE 5 A schematic of a dual-piston in-series pump design. Diagram from HPLC equipment, CLC-30, reprinted with permission from Savant, Fullerton, CA.

pumps such as Waters Alliance 2695, Agilent Series 1100 Quatenary and Shimadzu LC10-AD.

C. High-pressure Vs. Low-pressure Mixing

Pumps can be categorized by the ways in which solvent blending is achieved, as shown in Figure 6. The low-pressure mixing design was developed in the late 1970s for multi-solvent pumps. Here a single pump draws pre-composed mobile phases with each piston stroke configured by a 4-port proportioning valve (a Teflon block with four solenoid valves attached to different solvent reservoirs) (Figure 6). Examples are Waters Alliance 2695, Agilent 1100 Series Quaternary and PerkinElmer Series 200 pumps. In contrast, high-pressure mixing systems form gradients by using two or more pumps to mix solvents at high pressures. A separate controller unit is often needed to change the flow rates of each pump to generate gradient profiles. A major disadvantage of the high-pressure mixing system is the cost of purchasing and maintaining several pumps in a multi-solvent system. An important

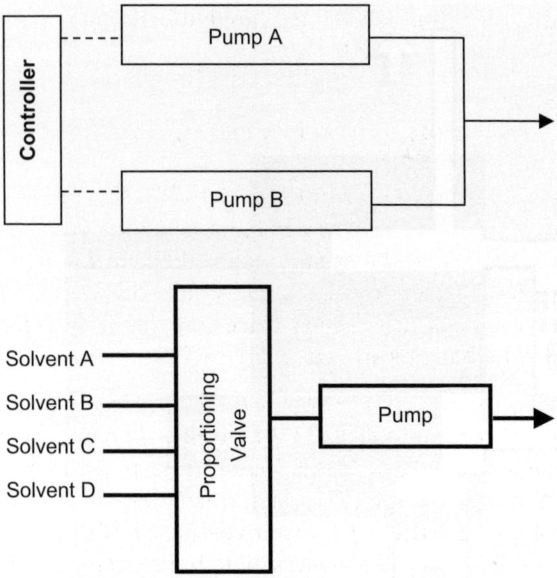

FIGURE 6 Schematic diagrams of a low-pressure mixing system using a single pump and a 4-port proportioning valve, and a high-pressure mixing system using multiple pumps and a controller to blend solvents under high pressure. Many newer pumps can also act as controllers for other pumps.

advantage is its lower dwell volumes. Examples of high-pressure mixing pumps (mostly binary) are Agilent 1100 Series Binary, Jasco 2000 Binary and Shimadzu LC10-AD.

D. Dwell Volume

System dwell volume, also known as gradient delay volume, is an important but often overlooked attribute of an HPLC system. The dwell volume is the liquid hold-up volume of the HPLC system from the point of solvent mixing to the head of the column. In a low-pressure mixing system, this volume includes the internal piston cavity, the pulse dampener, the priming valve and the mixer, in addition to the fluidic volumes in the autosampler and all connection tubing before the column. The typical dwell volume of a modern low-pressure mixing system is about 1–2 mL but can be as high as 5–7 mL in older systems. High-pressure mixing systems have inherently lower dwell volumes because the point of mixing is external to the pumps. Note that dwell volume is inconsequential in isocratic analyses but is important in gradient analyses because it translates into gradient delay time.

System dwell volume can be measured as follows:

- First, disconnect the column and replace it with a zero-dead-volume union.
- Next, place 0.5% acetone in water in reservoir A and water in reservoir B.
- Set a gradient program from 0% to 100% A in 10 min and a flow rate of 1.0 mL/min. Set the detection wavelength to 254 nm.
- Start the solvent gradient and record the detector signal.
- Draw a tangent line from the mid-point (50% absorbance value) of the resulting absorbance trace and let it intersect with the extended baseline as shown in Figure 7.

The time of the intersection point (which marks the gradient onset time) multiplied by the flow rate is the dwell volume of the system. For instance, the intersection point was found to be at 1 min for the Waters Alliance 2695 system in Figure 7, which translates into 1 min × 1 mL/min or a dwell volume of 1 mL. This dwell volume of 1000 μL is considered excellent for a low-pressure mixing system, but represents ~5 min of gradient delay time at 200 μL/min (for narrowbore columns) or ~20 min at 50 μL/min (microbore columns). This is wasted time from the gradient start in the pump to the actual onset of the gradients at the head of the column. Larger system dwell volumes are responsible for longer column equilibration and sample injection cycles. For these reasons, HPLCs for

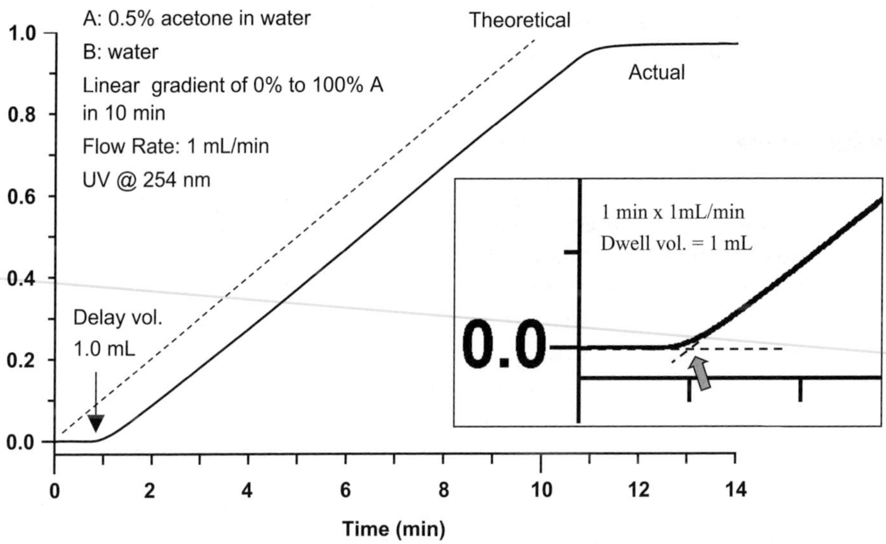

FIGURE 7 The absorbance gradient trace used to measure the system dwell volume of a Waters Alliance system. Inset shows the intersection point marking the gradient onset.

high-throughput screening using 1–2 min ballistic gradients utilize mostly high-pressure mixing pumps.

The above experiment can also be used to evaluate the mixing efficiency of the pump by directly connecting it to the detector and bypassing the autosampler. Here, a smooth gradient curve indicates a well-mixed system while a noisy line signifies inadequate mixing. As discussed earlier, it is more difficult to design a pump that performs gradients well at <0.2 mL/min. Factors such as piston volumes, pulsation, dwell volume and mixer efficiency all play critical roles. Thus, specially designed pumps are required for micro LC unless flow splitting is used.

E. Specifications

Manufacturers publish their product's performance characteristics as specifications, which are often used by the customer for comparison during the selection process. Table 1 shows the specifications of an Agilent 1100 Series Quaternary Pump, which is quite representative of other high-end analytical pumps. Note pulsation is particularly detrimental to the performance of flow-sensitive detectors (e.g., mass spectrometer, refractive index detector). Differences in dwell volumes and composition accuracy between HPLC systems might cause problems during method transfers.

F. Accessories

1. Solvent Degasser

Solvent degassing is needed because gas bubbles in the pump head can cause malfunctions (air-lock) or flow errors (low flow). Dissolved air often

TABLE 1 Specifications of Agilent 1100 Series Quaternary Pump

Parameter	Specification
Flow precision	<0.3% RSD (typically <0.15% RSD)
Flow range	Setpoints from 0.001 to 10.0 mL/min, in 0.001 mL/min increments
Pressure	
Operating range	0–40 MPa (0–400 bar, 0–5880 psi) up to 5 mL/min 0–20 MPa (0–200 bar, 0–2950 psi) up to 10 mL/min
Pulsation	<2% amplitude (typically <1%)
Compressibility compensation	User selectable, based on mobile phase compressibility
Gradient delay volume	800–1100 μL dependent on back pressure
Composition precision	<0.20% SD, at 0.2 and 1 mL/min

Data extracted from published specifications by Agilent Technologies in 2003.

out-gases during solvent blending due to heat generated from mixing (e.g., methanol/water). Low-pressure mixing systems are particularly prone to this out-gassing problem, causing blending errors. Degassing by stirring or ultra-sonication under vacuum is inadequate since solvents re-gas after 3–4 h.[14] Helium sparging is inconvenient and expensive in most countries. Its use is eclipsed by on-line vacuum degassers which are convenient and effective. Figure 8 shows a schematic of a vacuum degasser in which solvents are passed through semi-porous polymer membranes in an evacuated chamber where the smaller gaseous molecules (dissolved oxygen and nitrogen) are eliminated. Today, vacuum degassers are found in most HPLC systems. Their reliability has improved in recent times with a concomitant reduction of membrane volumes (from >10 to <1 mL).

2. Column Oven

In reversed-phase HPLC, column temperature is a strong determinant of retention time and also affects column selectivity.[2] A column oven is therefore required for most automated pharmaceutical assays to improve retention time precision, typically at temperatures of 30–50°C. Temperatures >60°C are atypical due to concerns about thermal degradation of the analytes and column lifetimes. Exceptions are found in high-throughput screening where higher temperatures are used to increase flow and efficiency. Ambient or sub-ambient operation is sometimes found in chiral separations to enhance selectivity. Column ovens

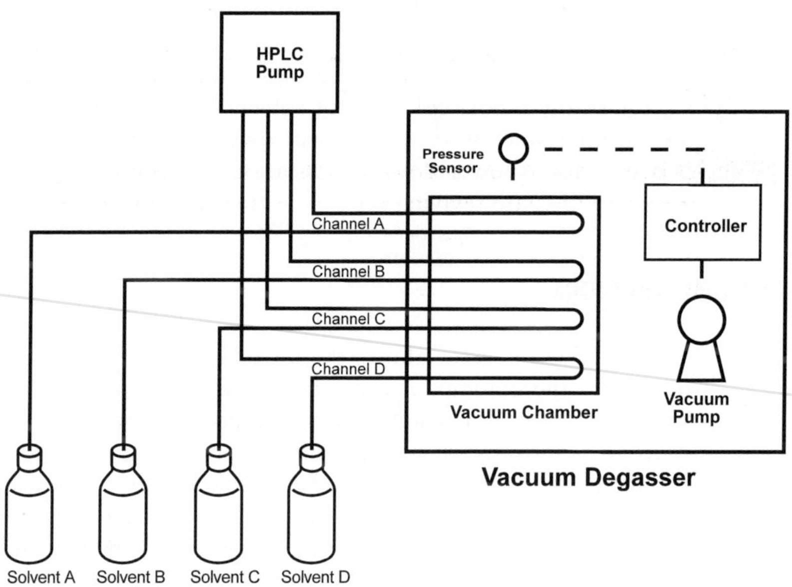

FIGURE 8 **A schematic of a 4-solvent on-line vacuum degasser. The preferred device for solvent degassing.**

operate either by circulating heated air or by direct contact (clam-shell type). Solvent pre-heating is achieved by passing the mobile phase through a long coiled tube embedded onto the heating element before the column. This pre-heating coil is often bypassed unless precise temperature control is needed because it causes additional extra column band-broadening. Trends in column oven design are moving towards wider temperature ranges (e.g., 4–80°C) with the use of a Peltier cooling device.

3. Column Selector Valve

Column selector valves can be added on as accessories to allow column switching for multi-dimensional chromatography (to increase the resolution of very complex samples such as in proteomics) or for automatic column selection (up to six columns) to facilitate methods development using different columns.

G. Trends

Multi-solvent pumps (i.e., quaternary pumps) have become standard gear in the pharmaceutical laboratory for their flexibility in handling gradients and column flushing and for their convenience in blending multiple solvents in method development. Modern pumps have better precision, reliability and longer seal life than their predecessors. This is made possible by innovative designs such as dual-piston in-series, variable stroke mechanism, micro-piston and piston seal wash. Manufacturers are vigilant in adopting new materials for enhanced performance, such as Titanium or PEEK pump heads, ceramic or Zirconia pistons and polyethylene/fluorocarbon composite piston seals. Trends in the use of small-bore columns for solvent savings and MS compatibility are pushing systems towards lower dwell volumes (<1 mL) and better performance at lower flows. Applications in high-throughput screening and micro LC are reviving high-pressure mixing systems.

III. MANUAL INJECTORS

An HPLC injector allows the introduction of a precise sample volume onto the column. A typical manual injector consists of a 6-port valve with a rotor, a sample loop and a needle port (Figure 9). A sample solution is introduced into the sample loop using a 22-gauge blunt tip syringe in the LOAD position. The sample is then injected into the column by switching the valve to INJECT. The typical external sample loop size ranges from 6 μL to 2 mL. For many years, the Rheodyne 7125 injector was the industry-standard. In the early 1990s, it was replaced by the Rheodyne 7725 injector, which injects samples without momentary flow disruptions.[15]

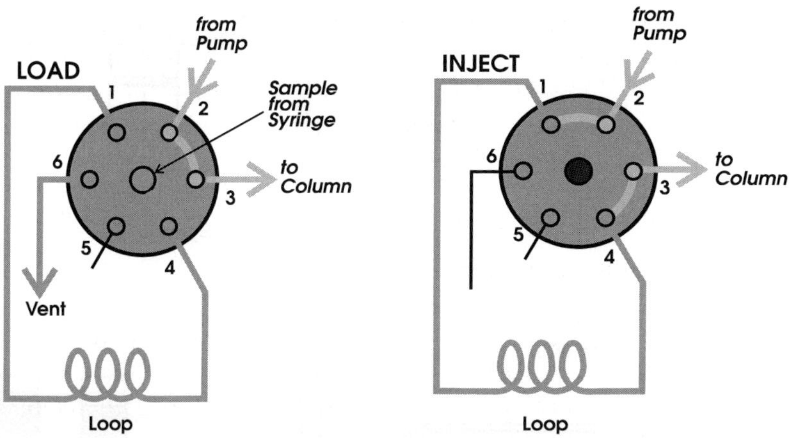

FIGURE 9 A schematic of a Rheodyne 7125 injector valve during the LOAD and INJECT cycle.

An injector valve operates in two modes—the fixed-loop mode or the partial-loop mode. In the fixed-loop mode, a sample is overfilled into the loop at 2–4 times the loop volume and the entire loop content is injected. In the partial-loop fill mode, a variable sample aliquot, measured precisely by a syringe at <50% of the loop volume, is injected. Note that the sample slug is introduced into the end of the sample loop and is back flushed onto the column to minimize band dispersion by the sample loop (Figure 9). Due to the emphasis on productivity, manual injectors are seldom used in the pharmaceutical laboratory except for preparative applications.

IV. AUTOSAMPLERS

A. Operating Principles

An autosampler allows the automatic injection of samples from a set of vials. Most laboratories use autosamplers to reduce labor cost and to increase productivity and precision. There are two common types of autosamplers: the push-loop and the integrated-loop designs.[16] Figure 10 shows a schematic diagram of the push-loop autosampler. It consists of a motorized 6-port valve assembly and a moving sample needle mounted on an *x*–*y*–*z* platform that allows random access to sample vials in trays or to 96-well plates. In some autosamplers, sample vials are brought to the injector by robotic arms. The sample needle is connected to a sampling syringe driven by a precise stepper motor. The sampling syringe can also draw flush solvents (depending on the positions of a set of solenoid valves). The sampling needle normally parks above a flush port. A typical

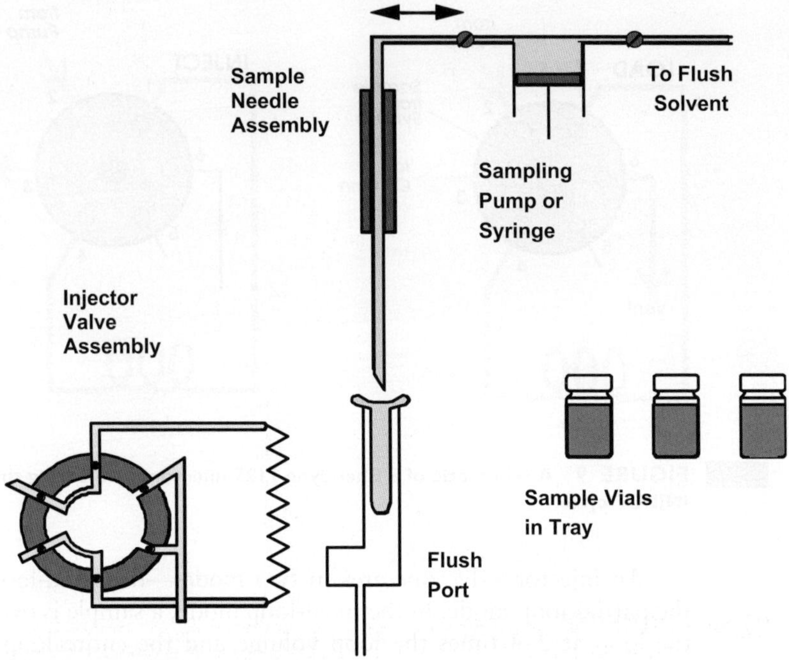

FIGURE 10 **A schematic of a push-loop autosampler which mimics the operation of a manual injection.**

injection sequence mimics the operation of a manual injection. First, the needle is cleaned out by pushing the flush solvent into the flush drain port to waste. Then, the needle moves to the sample vial, withdraws a sample aliquot, moves and enters into the injector needle port and delivers the sample aliquot. The motor then turns the injection valve to INJECT and starts the sample analysis. Small air bubbles are often used to segment the sample aliquot from the flush solvent inside the sampling needle. The advantages of this design are its relative simplicity and the flexibility of the sampling volume. Examples are Waters Alliance 2790, Shimadzu SIL-HT, PerkinElmer Series 200 and Agilent 1100 Series micro.

In recent years, the integrated-loop autosampler has become popular. Figure 11 shows a typical injection sequence of this type of autosampler. In the flush position, the needle is fitted in a high-pressure seal and flush solvent purges the sample loop. In the sampling position, the needle moves to the sample vial and withdraws an aliquot. In the inject position, the needle moves back to the high-pressure seal, the valve switches to INJECT to move the entire sample aliquot into the column. The strong points of this design are its better precision and carryover characteristics because the sample aliquot is contained in the swept portion of the loop. "Carryover" is the term used to describe how much of the previous

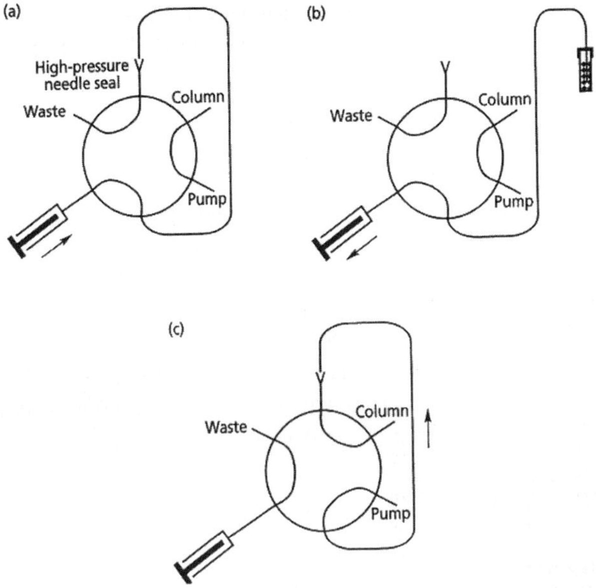

FIGURE 11 A schematic of an integrated-loop autosampler under various operational cycles: (a) flush, (b) sample fill, (c) inject. Reprinted with permission from Reference 16.

sample is "carried over" to the next sample.[16] The weak point of this design is its mechanical complexity. Also, the reliability of the high-pressure seal requires annual service maintenance. Examples are Waters Alliance 2695 and Agilent 1100 Series standard autosampler. Note that instead of using an injection valve, the fluidic path can also be controlled by a series of solenoid valves as in the case of Waters Alliance 2695.

B. Trends

Important characteristics of an autosampler are its programmability (random vial access, variable volume injections and number of injections per vial) and minimum sample volume requirements. The most important performance criteria are its sampling precision and carryover. Autosamplers have undergone tremendous improvements in recent years. Precision of 0.2% relative standard deviation (RSD) as well as very low carryover (<0.05%) is now routinely achievable. Faster turnaround time (up to 3 injections/min) and 96-well microplate compatibility are desirable for high-throughput screening and bioanalytical applications. Other autosamplers have additional liquid handling capability for dilution, standard addition and derivatization. Many have options for Peltier cooling of samples. One innovative design is the Waters 2690D dissolution interface which allows the autosampler to perform double-duty as a fraction collector to collect aliquots from dissolution baths at specified time points.

V. DETECTORS

An HPLC detector is often a modified spectrophotometer equipped with a small flow cell, which monitors the concentration (or mass) of eluting analytes.[17] Common detectors in the pharmaceutical laboratory are listed in Table 2 with their respective attributes and sensitivity levels. A recent survey found that 85% of pharmaceutical applications use absorbance detectors such as UV/Vis or photodiode array detectors (PDA).[18] These two detectors are covered in more detail in this section.

VI. UV/VIS ABSORBANCE DETECTORS

A. Operating Principles

The UV/Vis absorbance detector monitors the absorption of UV or visible lights by analytes in the HPLC eluent. They are prevalent because most pharmaceuticals have absorbance in this region. A typical UV/Vis detector consists of a deuterium source, and a monochromator (a moveable grating controlled by stepper motor to select wavelength through an exit slit) to focus the light through a small flow cell (Figure 12). A dual-beam optical

TABLE 2 Common HPLC Detectors and Their Attributes

Detector	Analyte/attributes	Sensitivity
UV/Vis absorbance	Specific: compounds with UV chromophores	ng
Diode array	Specific: same as UV/Vis detectors also provides UV spectra	ng
Fluorescence	Specific: compounds with native fluorescence or with fluorescent tag	fg–pg
Refractive index	Universal: polymers, sugars, triglycerides, organic acids, excipients	0.1–10 μg
Evaporative light-scattering	Universal: non-volatile or semi-volatile compounds, compatible to gradients	Low ng
Electrochemical	Specific: electro-active compounds	pg
Conductivity	Specific: anions and cations, organic acids, surfactants	ng ppm–ppb
Radioactivity	Specific, radioactive-labeled compounds	Low levels
Mass spectrometry	Both universal and specific, Definitive identification	fg–pg–ng

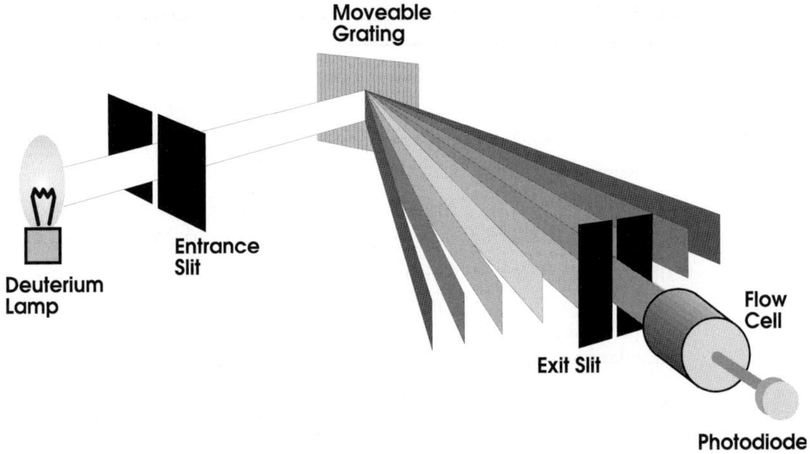

FIGURE 12 A schematic of an UV-Vis absorbance detector.

bench is typical for reducing drift. Two photodiodes are used to measure light intensities of sample and reference beams. The principle for an absorbance detection is Beer's Law[1] where

$$\text{Absorbance } (A) = \text{molar absorptivity } (\varepsilon) \times \text{pathlength } (b) \times \text{concentration } (c)$$

Most UV absorption bands correspond to transitions of electrons from $\pi \rightarrow \pi^*$, $n \rightarrow \pi^*$, or $n \rightarrow \sigma^*$ molecular orbitals. Besides aromatic compounds, organic functional groups such as carbonyl, carboxylic, amido, azo, nitro, nitroso and ketone groups have absorbance in the UV region.[1,19]

B. Specifications

Important performance characteristics of UV/Vis detectors are sensitivity, linearity and band dispersion. These are controlled by the design of the optics and the flow cell—more specifically by spectral bandpass, stray light characteristics and the volume and pathlength of the flow cell. Specifications of the Waters 2487 detector are shown in Table 3 to illustrate performance levels of modern detectors.

Sensitivity is the most important specification of the detector and is specified by baseline noise. The dual-wavelength detection feature is useful for the simultaneous monitoring of active ingredients and preservatives in drug products. Note that baseline noise can be substantially higher in the dual-wavelength mode (see Table 3) because this feature is achieved by toggling the monochromator between the two wavelengths. While a wavelength range of 190–700 nm is typical, the sensitivity performance above

TABLE 3 Specifications of Waters 2487 Dual Absorbance Detector

Wavelength range	190–700 nm
Bandwidth	5 nm
Wavelength accuracy	±1 nm
Wavelength repeatability	±0.1 nm
Linearity[1]	5% at 2.5 AU propylparaben, 257 nm
Baseline noise, single wavelength[1]	<± 0.35×10^{-5}AU, dry cell, 254 nm
Baseline noise, dual wavelength[1]	<± 2.5×10^{-5} AU
Drift	1×10^{-5} AU/h
Flow cell design	TaperSlit™
Standard analytical cell	Path length 10 mm Cell volume 10 μL Pressure limit 1000 psi
Optional flow cells	Semi-preparative, inert (titanium), high pressure for LC/MS, microbore

[1] ASTM E1657-94.
Data extracted from published specifications by Waters Corporation in 2003.

400 nm is lower due to the lower energy of the Deuterium source in the visible region. Some detectors have a secondary tungsten source to augment the deuterium source and extend the wavelength range to 190–1000 nm.

Spectral bandwidth is defined as the width in nm of the selected wavelength region and is related to the optical slit width of the spectrometer. A spectral bandwidth of 5–8 nm is typical in HPLC detectors. Increasing the bandwidth by widening the slit width improves detection sensitivity at the expense of linearity. Wavelength accuracy is an important parameter for instrument calibration.[20] Wider linearity and lower baseline noise are critical for achieving acceptable limits of quantitation (LOQs) for International Conference on Harmonisation (ICH) impurity testing. Flow cell design is critical for increasing sensitivity since signals are proportional to cell pathlengths. Increasing pathlengths with a larger flow cell, however, is often detrimental to system band dispersion and leads to extra-column band-broadening. Advanced flow cell designs such as tapered cells or flow cells equipped with focusing lenses as cell windows are often used to reduce gradient baseline shifts stemming from refractive index changes in the mobile phase. Most manufacturers offer a number of optional flow cells for applications such as semi-micro, microbore, semi-prep or LC/MS.

C. Trends

Sensitivity performance has improved greatly in the last two decades. The benchmark noise level of $\pm1\times10^{-5}$ AU/cm, thought at one time to be the physical limit of UV/Vis detection imposed by short-term source fluctuations, thermal flow noise and electronic noise, is now surpassed by many detectors. Extending the linear dynamic range to >2 AU is possible by lowering stray lights in the optical bench. The typical lifetime of the deuterium source is now 1000–2000 hs. Many modern detectors have dual- or multiple-wavelength detection and stop-flow scanning features. Most detectors have front panel access to self-aligned sources and flow cells for easy maintenance. Others have self-validation features such as power-up diagnostics, leak sensors, time logs for lamps, built-in holmium oxide filters for wavelength calibration or filter-wheels for linearity verification.[17]

VII. PHOTODIODE ARRAY (PDA) DETECTORS

A PDA detector provides UV spectra of eluting peaks in addition to monitoring the absorbance of the HPLC eluent like the UV/Vis absorbance detector. It is the preferred detector for testing impurities and for method development. PDA facilitates peak identification during methods development and peak purity evaluation during method validation. Detector sensitivity was an issue in earlier models but has improved significantly (more than ten-fold) in recent years.[21]

A. Operating Principles

Figure 13 shows the schematic of a PDA detector where the entire spectrum (or a selected portion) from the deuterium source passes through the flow cell and is dispersed onto a diode array element that measures the intensity of light at each wavelength. Most PDAs use a charge-coupled diode array with 512 to 1024 diodes, capable of spectral resolution of ~1 nm. Sophisticated spectral evaluation software allows the convenient display of both chromatographic and spectral data along three axes (absorbance vs. wavelength vs. retention time).

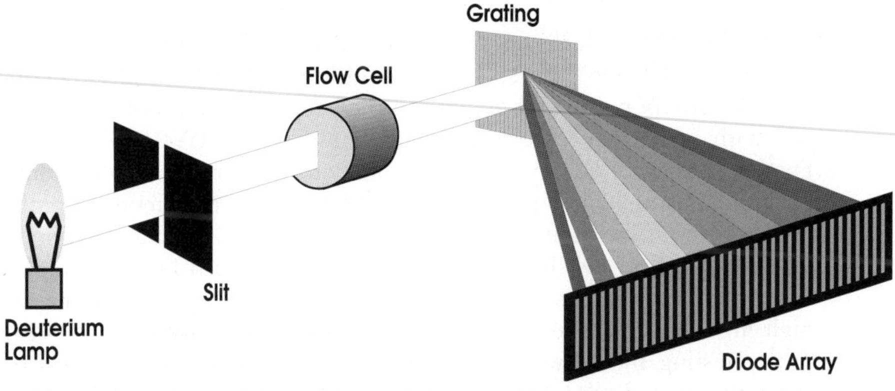

FIGURE 13 A schematic of a photodiode array detector.

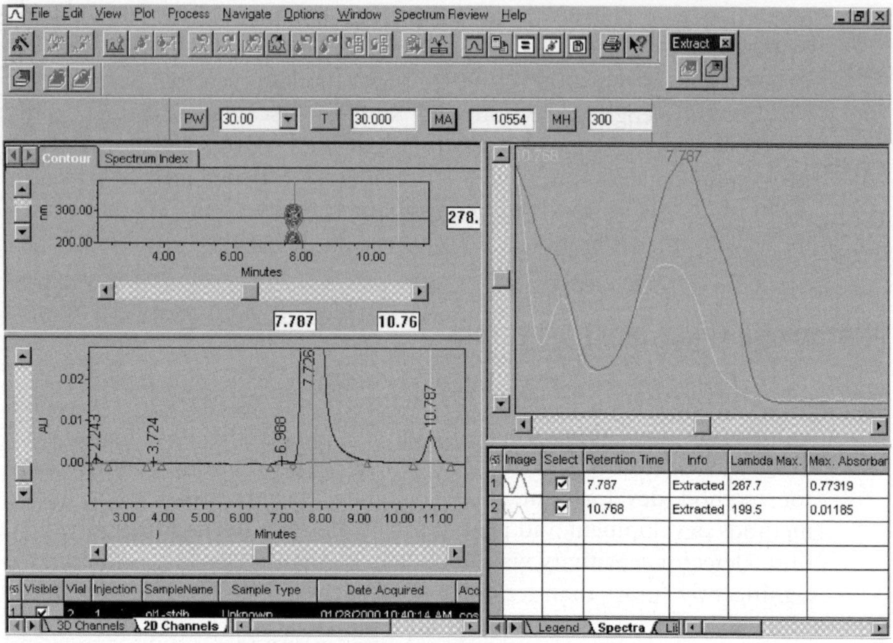

FIGURE 14 **Waters Millennium screen showing contour map, chromatogram at 278 nm and spectra of an active ingredient and a degradant of an impurity test sample.**

Figurc 14 shows a Waters Millennium screen with multiple windows displaying a contour map, a chromatogram at a specified wavelength and the spectrum of the active ingredient peak in the test sample. Most software also allow automated spectral annotations of λ_{max}, peak matching, library searches and peak purity evaluation.[17,19–21]

B. Trends

Earlier PDAs suffered from poor detector sensitivity (noise of $\pm1\times10^{-4}$ AU) and low spectral resolution (~5 nm). Modern PDAs have sensitivity performance close to the benchmark level of $\pm1\times10^{-5}$ AU with performance equivalent to that of many UV/Vis detectors. Further sensitivity enhancements are likely to stem from advanced flow cell design using light-pipe (fiber-optics) technology to extend the path-length without increasing noise or chromatographic band dispersion.[22] This type of flow cell is especially important in micro LC for samples of limited availability such as in proteomics.[13] Linearity performance at high absorbance (up to 2.0 AU) is also important in pharmaceutical purity testing by allowing the injection of a higher concentration of the active ingredient to enhance the detection of trace impurities. One interesting innovation is the development of a programmable slit feature so that the user can choose between high detector sensitivity (wider slit) or higher spectral resolution (narrower slit).[23]

VIII. OTHER DETECTORS

A. Fluorescence (Fl)

A fluorescence detector monitors the emitted fluorescent light of the HPLC eluent. It is selective and extremely sensitive (pg to fg) to highly fluorescent compounds. Its application is limited since few pharmaceutical ingredients have strong innate fluorescence. A fluorescence detector can be a regular fluorescence spectrophotometer fitted with a small flow cell. However, most detectors are built specifically for HPLC. An fluorescence detector consists of a xenon source, an excitation monochromator, an emission monochromator, a square flow cell and a photomultiplier for amplifying the emitted photons. The xenon source can be a high-power continuous source (150 W) or a pulsed source (<20 W). The pulsed source is becoming popular because it requires less power, has more energy in the far UV region, and allows detection modes such as phosphorescence, chemiluminescence and bioluminescence. All high-end units have double monochromators for wavelength programmability. Filters are used in lower-cost units. Sensitivity specification is often quoted by the signal/noise ratio of the Raman band of water which ranges from ~100 in older models to >300 in modern units. Sensitivity is often enhanced by widening the optical slits (e.g., up to 20 nm).[24]

B. Refractive Index (RI)

An RI detector measures the difference in RI between the sample cell containing the eluting analyte and the reference cell (containing pure eluent). It offers universal detection but has lower sensitivity (0.01–0.1 μg) than UV/Vis absorbance detection and is more prone to temperature and flow changes. RI detection is used for components of low chromophoric activities such as sugars, triglycerides, organic acids, pharmaceutical excipients and polymers. It is the standard detector for polymer characterization in gel permeation chromatography (GPC).[25] Modern RI detectors are mostly differential deflection type with a wide RI range of 1.00–1.75 RIU (refractive index unit). They have thermostated flow-cell assemblies and allow unattended operation via auto-purging the reference flow cell. Their sensitivity, baseline stability and reliability have improved significantly in recent years. Their major disadvantages are their low sensitivity and their incompatibility with gradient elution.

C. Evaporative Light Scattering Detector (ELSD)

An ELSD reduces the HPLC eluent into a particle stream and measures the scattered radiation. It offers universal detection for non-volatiles or semi-volatiles and has higher sensitivity than the RI detector (low ng) in addition to being compatible with gradient analysis.[17,26] It is routinely used in combinatorial screening. Response factors are less variable than those of other detectors. An ELSD consists of a nebulizer equipped with a constant temperature drift tube where a counter-current of heated air or nitrogen reduces the HPLC eluent into a fine stream of analyte particles. A laser or a polychromatic beam intersects the particle stream and the scattered radiation is amplified by a photomultiplier. Manufacturers include Alltech, Polymer Laboratories, Shimadzu, Waters, Sedere and ESA.

D. Electrochemical Detector (ECD)

An ECD measures the current generated by electroactive analytes in the HPLC eluent between electrodes in the flow cell. It offers sensitive detection (pg-levels) of catecholamines, neurotransmitters, sugars, glycoproteins, and compounds containing phenolic, hydroxyl, amino, diazo or nitro functional groups.[1,27] The detector can be of amperometric, pulsed-amperometric or coulometric type with the electrodes made from vitreous or glassy carbon, silver, gold or platinum, operated in the oxidative or reductive mode. Manufacturers include BSA, ESA and Shimadzu. A research-grade coulometric array detector with multiple electrodes operated with increasing voltage is also available.[28]

E. Conductivity

A conductivity detector measures the electrical conductivity of the HPLC eluent stream and is amenable to low-level determination (ppm–ppb levels) of ionic components such as anions, metals, organic acids and surfactants.[17,29] It is the primary detection mode for ion chromatography.[30] Manufacturers include Dionex, Alltech, Shimadzu and Waters.

F. Radioactivity

A radioactivity detector is used to measure radioactivity in the HPLC eluent using a flow cell. The detection principle is based on liquid scintillation technology to detect phosphors caused by the radiation, though a solid-state scintillator is often used around the flow cell.[17,31] This detector is very specific and can be extremely sensitive. It is often used for conducting experiments using tritium or C-14 radiolabeled compounds in toxicological, metabolism or degradation studies. Manufacturers include Packard and EG&G Berthrod.

G. Mass Spectrometry (MS)

LC/MS is the ultimate analytical technique, which combines the versatility of HPLC with the identification power of MS.[32] The weak link in LC/MS has always been the interface which connects the liquid stream at atmospheric pressure to the high vacuum present inside the mass spectrometer. The development of several atmospheric pressure interfaces, electrospray and atmospheric pressure chemical ionization (APCI), has contributed to the tremendous success and popularity of LC/MS and LC/MS/MS in bioresearch, drug discovery, combinatorial analysis and pharmacokinetic assays. This topic is covered in more depth in a later chapter.

IX. DATA HANDLING

Chromatographic data handling devices range from a strip chart recorder, an integrator, a PC-based workstation to a client–server network system. Most PC-based data handling workstations also incorporate full instrumental control of the HPLC system including those from other manufacturers. In recent years, major pharmaceutical companies have installed centralized data network systems to ensure data security and integrity. Data handling systems are described in a later chapter.

X. INSTRUMENTAL BANDWIDTH (IBW)

Chromatography is a band-broadening process in which the sample aliquot broadens and differentiates into analyte bands as they travel through the HPLC column. This innate broadening process is reduced by using tightly packed columns filled with small particles.[2] Broadening also occurs outside the column (extra-column band-broadening) due to the parabolic flow profiles in connection tubing (Poiseuille flow) and other devices in the fluidic path such as the injector valve and detector flow cell. This extra-column band-broadening or dispersion in the HPLC system is detrimental to column resolution and must be controlled by using small-diameter connection tubing, a low-dispersion injector and a small detector flow cell. This instrumental broadening effect is particularly damaging to applications of small-diameter or short columns which produce smaller peak volumes.[9,10]

With the trend toward using Fast LC and small-bore columns (3- and 2-mm i.d.) in pharmaceutical analysis, increasing demand is placed on reducing the system dispersion or IBW. Figure 15 shows the deleterious effect of IBW on the performance of a Fast LC column showing that

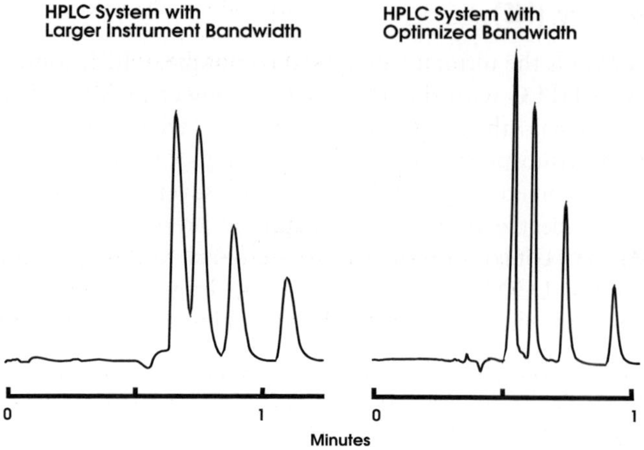

FIGURE 15 Comparative chromatograms showing the deleterious effect of instrumental bandwidth to the performance of a Fast LC separation of four paraben anti-microbials. HPLC conditions: column: C18 (33×4.6 mm, 3 μm), mobile phase: 50% acetonitrile/water, 2 mL/min, detection at 254 nm.

a system with larger IBW can significantly reduce the efficiency and resolution of the column. This broadening effect can be expressed quantitatively by the additive relationship of variance where the total variance of the observed peak is equal to the sum of the true column peak variance plus the variance of the IBW.

$$\sigma^2_{\text{total}} = \sigma^2_{\text{column}} + \sigma^2_{\text{IBW}}$$

where the variance of the IBW is equal to the sum of the variances stemming from the injector, the detector, and the connection tubing.

$$\sigma^2_{\text{IBW}} = \sigma^2_{\text{injector}} + \sigma^2_{\text{detector}} + \sigma^2_{\text{connection tubing}}$$

This IBW of the HPLC system can be measured by replacing the column with a zero dead-volume union and recording the peak base width of a small-volume injection. For instance, by injecting 2 μL of a 0.5% caffeine or uracil solution at 0.5 mL/min and detection at 254 nm on a Waters Alliance system and collecting data at 10 points/s, the resulting trace shown in Figure 16 was obtained. The base width (4 sigma) of this peak of ~60 μL is equal to the IBW. Figure 17 shows a chart plotting column efficiency of a 100-mm long 3-μm column vs. peak capacity (k') on an HPLC system with a 60 μL IBW. Note that since peaks widen with high retention or k', the effect of extra-column band-broadening is lessened in late eluting peaks. Also note that the loss of efficiency is more significant

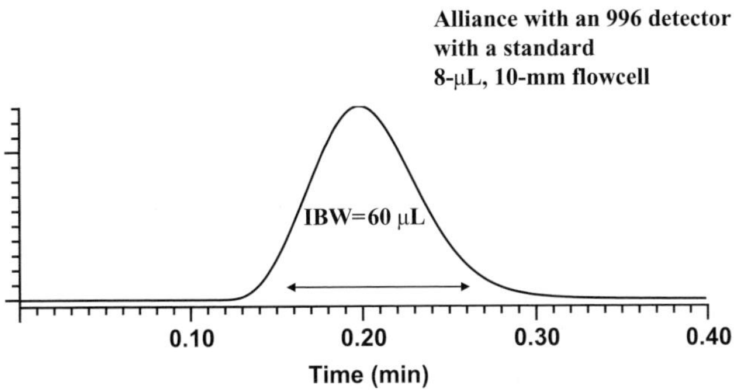

FIGURE 16 The chromatogram of an injection of a caffeine solution without the column showing the instrumental bandwidth of a Waters Alliance HPLC system with a 966 PDA detector with a standard flow cell.

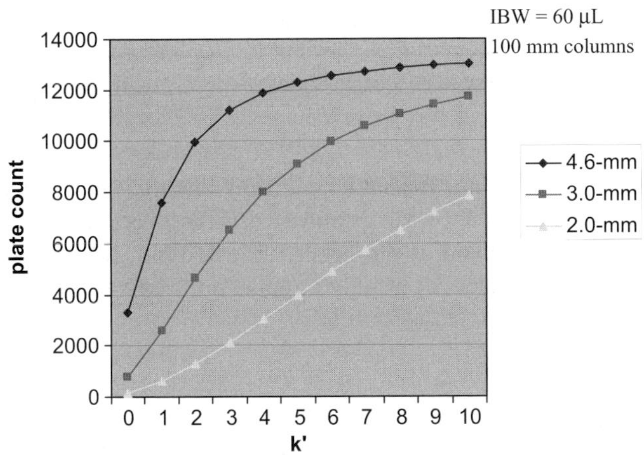

FIGURE 17 Chart showing the efficiency loss at various k' of 100-mm long 3-μm columns of various diameters from an HPLC with IBW of 60 μL. Note that while column efficiency is mostly retained for the 4.6 mm at high k', substantial losses are experienced with that of smaller i.d. columns.

when smaller-diameter columns are used since the peak volumes are smaller.

The IBW of a standard system can be reduced to 30–40 μL by using shorter lengths of 0.005–0.007″ i.d. tubing and a semi-micro flow cell (2–3 μL).[10] Further reduction might involve a low dispersion micro-injector or a redesign of the autosampler. Table 4 summarizes the typical IBW and other instrumental requirements of various column types from conventional (4.6 mm), Fast LC, minibore (3 mm), narrowbore (2 mm) and microbore (1 mm) to micro LC (<0.5 mm) columns. Note that the dispersion

TABLE 4 Summary of Typical Instrumental Requirements for Different Column Types

Column type	Peak vol. $k'=2$ (μL)	IBW (μL)	Detector flowcell (μL)	Tubing i.d. (in)	Injector	Inject vol. (μL)[1]	Pump	System dwell vol. (μL)
Conventional	200	100	8–10	0.009	Standard	100	Existing	2000
Fast LC	60	30	6	0.007	Standard	20	Existing	1000
Minibore	100	40	6	0.007	Standard	50	Existing	1000
Narrow bore	40	20	4	0.006	Standard	20	Modern existing	400
Microbore	10	5	2	0.005	Micro	5	h-p mixing micro-pump	100
Capillary	3	2	On-line capillary, LightPipe	0.002 or none	Special	<1	Split flow, Syringe pump	20

[1]Max inj. vol. recommended for isocratic assays.

contribution of the injector or the injection volume is not important in gradient analysis because of the gradient focusing effect in which the analyte band is re-concentrated at the head of the column. IBW requirements are therefore less stringent for gradient applications and only post-column dispersion is critical.

XI. INSTRUMENTATION TRENDS IN PHARMACEUTICAL ANALYSIS

In summary, HPLC is a mature analytical technique based on well-developed instrumentation with a definitive trend towards higher performance, better reliability, more features and easier maintenance. The pharmaceutical laboratory is using more information-rich detectors such as diode array and mass spectrometer, or is placing multiple detectors in series to generate more data per run. Client/Server data networks are installed with centralized data archiving. HPLCs can be fully controlled and data can be conveniently accessed from multiple locations including the laboratory, the office, and at home via web links. High-throughput screening in drug discovery is driving HPLC towards higher productivity,[33] while proteomics applications are pushing micro LC towards higher sensitivity and resolution. Major productivity or resolution gain might stem from further instrument development in diverse areas such as multi-dimensional chromatography,[34] ultra-high-pressure LC,[35] parallel analysis[36] and lab-on-a-chip.[37]

XII. MANUFACTURERS

Major manufacturers of HPLC instruments include Waters, Agilent (HP), and Shimadzu, PerkinElmer, Thermo, Beckman, Varian, Hitachi, Jasco, Dionex, Gilson and Isco. The Internet addresses of these companies can be found in the reference section. Many pharmaceutical laboratories tend to purchase HPLC from a single vendor (or two) in an effort to reduce costs in equipment validation and service. Data handling software considerations rather than product performance specifications dominate in the product-selection process. For instance, once a client–server data handling system is selected, HPLCs from the same manufacturer are typically purchased. Using systems from various vendors with different control and data handling software becomes prohibitively expensive as the cost in software training and validation escalates.

XIII. SUMMARY AND CONCLUSION

This chapter provides the novice and the more experienced analyst with an overview of HPLC instrumentation, operating principles and recent advances or trends that are pertinent to pharmaceutical analysis.

ACKNOWLEDGMENTS

The author acknowledges helpful ideas and suggestions from R. Ornaf, D. Barker, C. Davidson, J. Bonilla, C. Choi and G. Miller from Purdue Pharma. This chapter has been presented in parts in short courses on "Advanced HPLC in Pharmaceutical Analysis" at Eastern Analytical Symposium 2001–2002, HPLC 2002, and Pittcon 2003. Numerous comments and suggestions from the attendees of these short courses were incorporated during the development of this chapter.

REFERENCES

1. Skoog, D. A., Holler, F. J. and Nieman T. A., *Principles of Instrumental Analysis*, 5th ed., Brooks Cole, Pacific Grove, CA, 1997.
2. Snyder, L. R. and Kirkland, J. J., *Introduction to Modern Liquid Chromatography*, John Wiley & Sons, New York, 1979.
3. Meyer, V. R., *Practical HPLC*, Wiley, New York, 1998.
4. Adamovics, J. A., *Chromatographic Analysis of Pharmaceuticals,* 2nd ed., Marcel Dekker, New York, 1997.
5. Miller, J. M. and Crowther, J. B., *Analytical Chemistry in a GMP Environment: A Practical Guide,* Jossey-Bass, New York, 2000.
6. Katz, E., Eksteen, R., Schoenmakers P. and Miller, N., *Handbook of HPLC*, Marcel Dekker, New York, 1998.

7. LaCourse, W. R., *Anal. Chem.*, 74:2813, 2002.
8. HPLC equipment, CLC-30, (Computer-based Instruction), Savant, Fullerton, CA.
9. Dong, M. W., *Today's Chemist at Work*, 9(2):46–51, 2000.
10. Neue, U. D., *HPLC Columns: Theory, Technology, and Practice*, Wiley-VCH, New York, 1997.
11. Guiochon, G., Goshen, S. and Katti, A., *Fundamentals of Preparative and Non-linear Chromatography*, Academic Press, Boston, 1994.
12. Grimes, R., Serwe M. and Chervet, J. P., *LC.GC,* 15(10):960, 1997.
13. Waters CapLC System, Waters Corp., Milford, MA, April 2000, 72000133en.
14. Dong, M. W., *Today's Chemist Work*, 9(8):28–32, 2000.
15. Sample Injector for HPLC 7725/7725i/9725/9725i, Product Bulletin 209, Rheodyne, Rohnert Park, CA, http://www.rheodyne.com/lit/docs/2320617A.pdf
16. Dolan, J., *LC.GC*, 19(4):389, 2001.
17. Scott, R. P. W., *Chromatographic Detectors. Design, Function, and Operation*, Marcel Dekker, New York, 1996.
18. Market Analysis and Perspectives, Laboratory Analysis Instrument Industry, 1998–2002, Strategic Directions International, Inc., 5th ed., July, 1998.
19. Owen, T., *Fundamentals of Modern UV-Vis Spectroscopy: Primer*, Agilent Technologies, Waldronn, Germany.
20. Dong, M. W., Paul, R. and Roos, D., *Today's Chemist Work*, 10(2):42–48, 2001.
21. Dong, M. W., Passalacqua, P. V. and Choudhury, D., *J. Liq. Chromatogr.* 13(11): 2135–60, 1990.
22. DeLand, P., *Amer. Lab.*, 19, Nov, 1997.
23. Specifications, Agilent 1100 Series UV-Visible Detectors, Agilent Technologies, 5980–1532e, 2001.
24. Specifications, Agilent 1100 Series Fluorescence Detector, Agilent Technologies, 5968–9105e, 1999.
25. Reuter, W. M., Dong, M. W. and McConville, J. *Amer. Lab.* 23(5):45–58, 1991.
26. Hsu, B. H., Orton, E., Tang, S.-Y. and Carlton, R. A., *J. Chromatogr.* B, 725(1):103, 1999.
27. Erickson, B. E., *Anal. Chem.*, 71(9):353A, 2000.
28. Acworth, I. N. and Bowers, M., Progress in HPLC-HPCE, Chapter 6. In *Coulometric Electrode Array Detectors for HPLC,* Acworth, I. N., Naoi, M., Parvez, S. and Parvez, H., Eds., VSP Publications, The Netherlands, pp. 3–50, 1997.
29. Tsuyama, Y., Uchida, T. and Goto, T. *J. Chromatogr.*, 596:181, 1992.
30. Fritz, J. S. *Anal. Chem.* 59:335A–344A, 1987.
31. Devol, T. A., Duffey, J. M. and Paulenova, A. J. *J. Radioanal. Nucl. Chem.*, 249(2): 295, 2001.
32. Wehr, T., *LC.GC*, 18(4):406, 2000.
33. Wehr, T., *LC.GC*, 20(1):40, 2002.
34. Wehr, T., *LC.GC*, 20(10):954, 2000.
35. Jorgenson, J. E., *Anal. Chem.*, 69:983, 1997.
36. Wu, J. T., *Rapid Commun. Mass Spectr.*, 15(2):73, 2001.
37. McEnery, M., Tan, A., Glennon, J. D., Aldermen, J., Patterson, J., O'Mathuna, S. C., *Analyst*, 125(1):25, 2000.

INTERNET RESOURCES

http://www.waters.com
http://www.chem.agilent.com
http://www1.shimadzu.com/products/lab/hplc.html
http://instruments.perkinelmer.com/ai/chrom/index.asp

http://www.thermo.com
http://www.beckman.com
http://www.varianinc.com
http://www.hitachi-hhta.com
http://www.jascoinc.com
http://www.dionex.com
http://www.gilson.com
http://www.esainc.com/
http://www.bioanalytical.com/
http://www.rheodyne.com
http://www.savant4training.com
http://www.isco.com
http://www.sedere.com/
http://www.polymerlabs.com/
http://www.alltechweb.com
http://www.berthold.com.au/bioanalytical.html
http://www.microlc.com

4 HPLC COLUMNS FOR PHARMACEUTICAL ANALYSIS

UWE D. NEUE, BONNIE A. ALDEN, PAMELA C. IRANETA, ALBERTO MÉNDEZ, ERIC S. GRUMBACH, KIMVAN TRAN, AND DIANE M. DIEHL

Waters Corporation, Milford, MA

ABSTRACT

This chapter deals with the properties of high-pressure liquid chromatography columns. It is divided into two sections: column physics and column chemistry. In the section on column physics, we discuss the properties that influence column performance, such as particle size, column length and column diameter, together with the effect of instrumentation on the quality of a separation. In the section on column chemistry, we examine in depth the surfaces of modern packings, as well as the newer developments such as zirconia-based packings, hybrid packings or monoliths. We have also included a short section on

Handbook of Pharmaceutical Analysis by HPLC,
S. Ahuja and M.W. Dong, editors.

hydrophilic interaction chromatography, a technique for the analysis of polar compounds that is drawing interest again in the pharmaceutical industry. Finally, we review what is currently understood about the selectivity of reversed-phase columns.

I. INTRODUCTION

Since its creation around 1973, modern high-pressure liquid chromatography (HPLC) has played a dominant role in the analysis of pharmaceuticals. It is used in many different applications: for example, in content uniformity assays and stability-indicating methods, for the purity profiles of drug substances, or in the analysis of drug metabolism in animals and humans. The heart of all of these assays is the HPLC column. In this chapter, we will describe the fundamental properties of HPLC columns as well as how these properties influence column performance and separation characteristics in pharmaceutical assays.

The chapter is divided into two major subsections: column physics and column chemistry. In column physics, we examine the influence of column dimensions and particle size on the performance of a column. Special emphasis is placed on fast analysis, which has become an important issue in the pharmaceutical industry. In the section on column chemistry, we focus primarily on modern stationary phases based on high-purity silica, but we do not ignore other important developments such as hybrid packings or zirconia-based stationary phases. Similarly, the primary emphasis is on reversed-phase chromatography, but we also cover hydrophilic interaction chromatography, an important technique for polar compounds. We end with a section on column selectivity in reversed-phase chromatography to guide the user in finding columns that are similar or dissimilar.

Additional background literature can be found in textbooks on HPLC or specifically on HPLC columns.[1–5] For further information on surface derivatization and column chemistry in general, Reference 6 or 7 is recommended.

II. COLUMN PHYSICS

How fast a separation can be performed depends primarily on the choices of column length and particle size. A secondary factor is the viscosity of the mobile phase under the operating conditions. All three factors together determine the backpressure of the column, which in turn is limited by the capabilities of the HPLC instrument. How good a separation is depends on particle size and column length. Longer columns have a higher plate count. For the same column length, smaller particles give narrower peaks. Let us examine the relationship of column performance and analysis time in more detail. Instead of using complex equations, we will use simple diagrams to explain the concepts.

A. Column Performance Determined by Particle Diameter and Column Length

First, we look at isocratic separations. Let us assume that the analysis can be accomplished within a retention factor of 10. We also suppose that the analysis is carried out with a typical reversed-phase solvent and a sample with a typical molecular weight of a pharmaceutical entity. In order to manipulate the analysis time, we will keep the mobile phase composition the same and vary the flow rate. The maximum backpressure that we will be able to apply is 25 MPa (250 bar, 4000 psi). In Figure 1, we have plotted the plate count as a function of the analysis time for a 5 μm 15-cm column. We see that the column plate count is low at short analysis times and reaches a maximum at an analysis time of about 1 h. A further increase in analysis time is not useful, since the column plate count declines again. This is the point where longitudinal diffusion limits the column performance. The graph also stops at an analysis time of just under 5 min. This is the point when the maximum allowable pressure drop has been reached.

In the next graphs we will use the square root of the plate count as a measure of column performance. After all, resolution is proportional

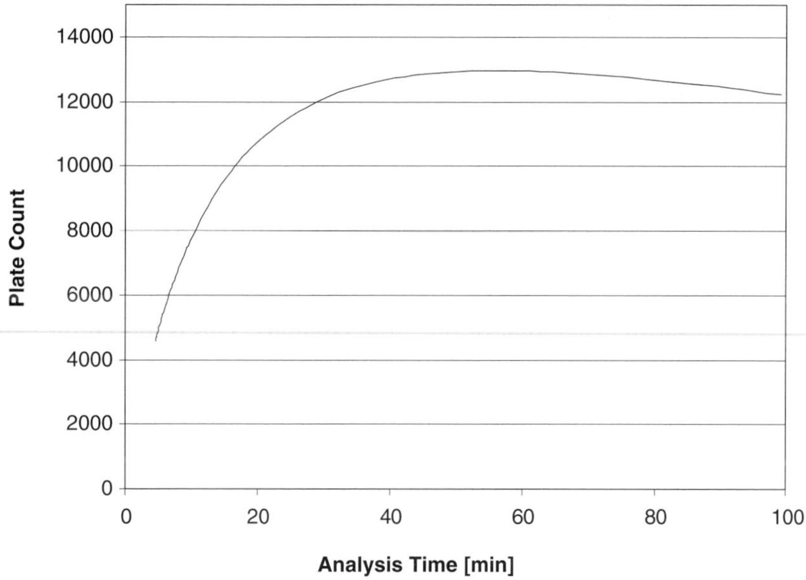

FIGURE 1 Plate count vs. analysis time for a 5-μm 15-cm column.

to this value. Therefore we use the expression "resolving power" for the square root of the plate count. We will also stop the graphs at the point where the column performance begins to decline again. Now let us examine the relationship between resolving power and analysis time for different columns.

Let us first investigate the performance of columns packed with 5-μm particles (Figure 2a). The resolving power is plotted vs. the analysis time for columns with a length of 5, 10, 15 cm, etc. up to 30 cm. We see clearly that the resolving power that is achievable increases with the column length. We also observe that one needs to operate the column at a rather long analysis time to achieve the maximum performance. Under our standard conditions, it takes just over 100 min to accomplish the analysis with maximum resolving power with a 30 cm 5-μm column. Conversely, if we want to get the analysis done in 20 min, the 30-cm column is barely capable of doing this since we will be operating close to the pressure limit that we have imposed. On the other hand, a 25-cm column or even a 20-cm column achieves nearly the same resolving power in 20 min, but at a much lower backpressure. The seemingly trivial change from a 30-cm column to a 25-cm column reduces the backpressure by 44% for the same analysis time. Half of this reduction is due to the shortening in column length, and half of it stems from the fact that a slower flow rate is needed with the shorter column to get to the same analysis time.

If we do not need to care about resolving power, for example, if we are doing a dissolution test and we have only one ingredient that we need to measure, the analysis can be accomplished on a 5 cm 5-μm column in about half a minute. The resolving power is only about 1/12 of what can be accomplished with a 30 cm 5-μm column. On the other hand, if we still want a reasonable resolving power, say a value of 50, we can do this with either a 5-cm column or a 10-cm column at a run time of about 3 min, but the backpressure of the 5-cm column is 4 times lower than the backpressure of the 10-cm column.

Now let us look at the performance characteristics of 3-μm columns shown in Figure 2b. The scales of this figure are identical to the scales in Figure 2a. The same column lengths are shown as in Figure 2a. The first impression is that the span of the graphs for the 3-μm columns is much shorter than that of the 5-μm columns. This is indeed the case: the maximum resolving power is always attained at a shorter analysis time. At the same time, the pressure limit is reached at a longer analysis time, i.e., a smaller flow rate, compared to the 5-μm columns. The combination of both effects limits the analysis time span that can be covered with a particular column length. As an example, a 30 cm 3-μm column can be used for an analysis of about 60 min in our example. A faster analysis is prohibited by the backpressure, and a longer analysis is not worthwhile since one will lose performance. We also see that there is less overlap in terms of resolving power between the different columns. Thus every one

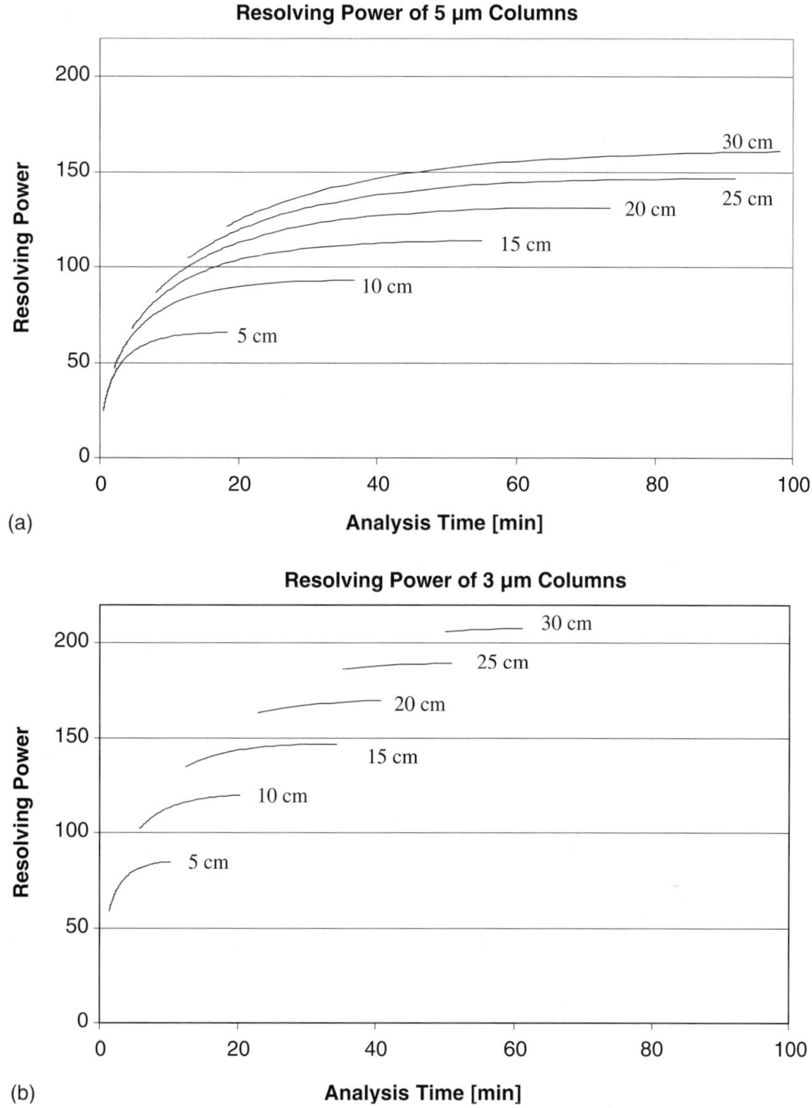

FIGURE 2 Comparison of the resolving power of 5 µm (a) and 3 µm columns (b) of various lengths.

of the selected column lengths has its own clear range of analysis time and performance. Finally, for a particular column length, the resolving power of a 3-µm column is higher than that of a 5-µm column. Every practitioner of HPLC knows this from experience.

An interesting aspect of column design is shown in Figure 3. Here we plot the characteristics of columns with a constant ratio of column

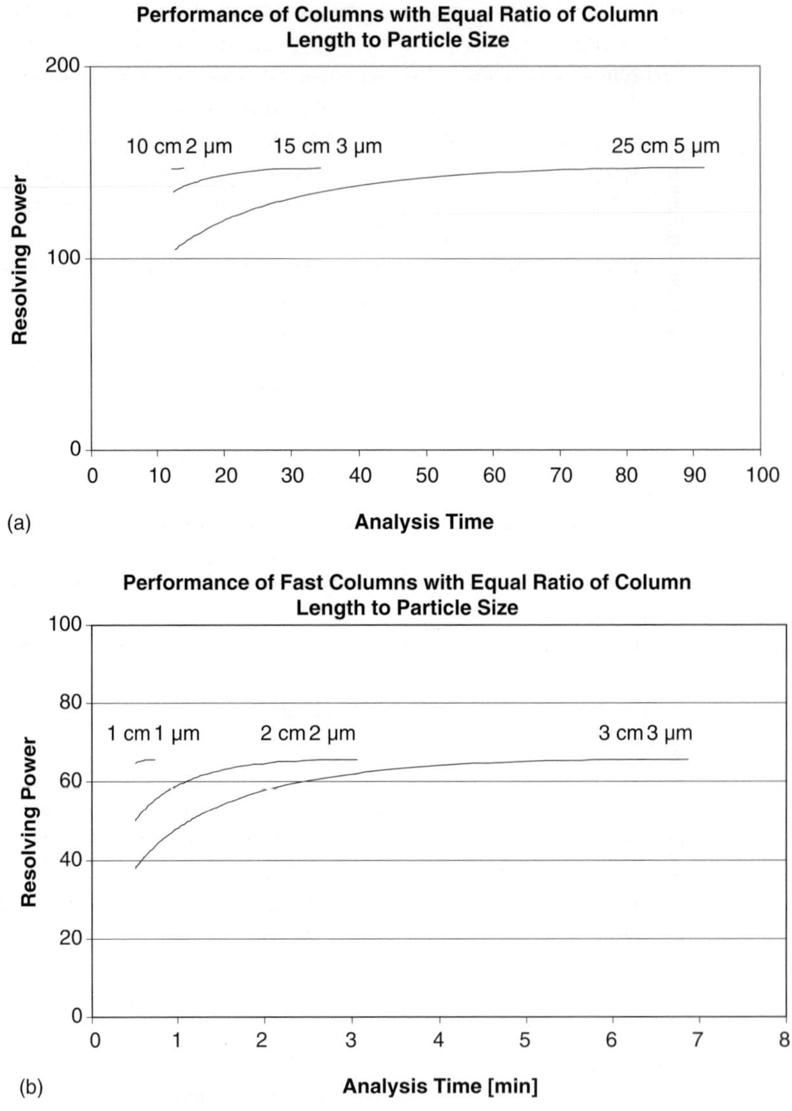

FIGURE 3 **Column performance at constant ratio of column length to particle size: (a) long columns and (b) short columns.**

length to particle size. In Figure 3a we see the properties of three columns with a rather high performance: a 25 cm 5-µm, 15 cm 3-µm, and a 10 cm 2-µm column. The maximum resolving power of all three columns is identical. In addition, the shortest analysis time that can be reached with these three columns is identical. This is due to the fact that all three columns reach the same column dead time at the same backpressure. One needs to realize that the same column dead time is achieved using a slower flow rate with the shorter columns. Overall, if one scales the column length proportionally to the particle size, the same maximum performance is reached, and the same shortest analysis time is possible. As we decrease the particle size, the velocity at which the optimum performance is occurring slowly moves toward the pressure limit, as shown in Figure 3a for the 10 cm 2-μm column. The optimum performance of a 5 cm 1-μm column would be outside the pressure limit that we had imposed on our study.

Now let us look at short fast columns that are suitable for rapid analysis. In Figure 3b, we are comparing the performance of a 3-cm column packed with 3-μm particles, a 2-cm column packed with 2-μm particles and a 1-cm column packed with 1-μm particles. As above, the columns have the same maximum performance and the same speed capability. The resolving power of all columns is at 66, and the shortest analysis time that can be achieved with any one of these columns is about half a minute with the standard pressure limitation. As above, the larger particle size achieves the maximum separation power at a longer analysis time than the smaller particle size: about 7 min are needed for the 3-μm particles, 3 min for the 2-μm particles and less than 1 min for the 1-μm particles. Despite the fact that these columns are rather short, we should not underestimate their separation power: the resolving power of 66 translates to about 4400 theoretical plates. If we want to get a good separation power in a 2- to 3-min time frame, either a 3 cm 3-μm or a 2 cm 2-μm column is suitable.

Until now, we have discussed isocratic separations only. In principle, the same thought processes apply to gradient separations as well. However, we have another tool available in gradient separations to manipulate the resolving power of the gradient: we can change the gradient volume. If we are interested in a 5-min gradient separation, we can run a gradient from 0% to 100% organic solvent at several different flow rates. At a slow flow rate, we operate the column close to its optimum plate count. At fast flow rate, we expand the gradient volume, and an expanded gradient should give a better separation. At the same time, we will lose theoretical plates because of the high flow rates. What is the best approach?

First, we need to establish how we measure the quality of a gradient separation. Theoretically, the best-way to do this is to calculate the peak capacity.[8–10] Peak capacity describes the number of peaks with a fixed resolution that can be resolved in a gradient. It is also something that can easily be calculated and measured for a particular gradient:

$$P = 1 + \frac{t_g}{w} \tag{1}$$

where P is the peak capacity, t_g is the gradient run time, and w is the peak width measured for a peak in the gradient. In general, most peaks in a gradient have about the same width, and the peak capacity of a gradient can often be measured using one or two typical peaks in the chromatogram.

The peak capacity of a gradient depends on the gradient operating conditions. This is discussed in detail in Reference 9. Here we will just give the relevant equation for gradients over a broad range of solvent composition without going into further detail:

$$P = 1 + \frac{\sqrt{N}}{4} \frac{B\Delta c}{B\Delta c t_0/t_g + 1} \tag{2}$$

where N is the column plate count (which varies with the flow rate), B is the slope of the relationship between the logarithm of the retention factor and the solvent composition, and Δc is the difference in solvent composition from the beginning to the end of the gradient. In order to keep things simple, we will always apply the same gradient—from 0% organic to 100% organic—and the value of Δc is always 1. The column dead time t_0 is a function of the flow rate. Thus the peak capacity will vary for two reasons, when we change the flow rate: the plate count will change, and the gradient span (the ratio of gradient time to column dead time) will change. The interested reader is referred to Reference 9 for a more complete explanation of the use of this equation and its implications.

Let us now compare columns of different lengths packed with different particle sizes as a function of the gradient run time. We also use a fixed column diameter, since we want to work with different flow rates. This allows us to see how the expansion and contraction of the gradient with different flow rates affects the peak capacity. In all the following studies, we will use a column diameter of 4.6 mm. As seen above in our isocratic examples, the highest pressure will be 25 MPa (250 bar, 4000 psi).

Let us first take a standard 15 cm 5-μm column and run a gradient at 1 mL/min over different gradient run times. In Figure 4a, we see the results of this experiment. The peak capacity is plotted against the gradient duration. At very short gradient run times, the peak capacity is low. It increases steadily with increasing gradient run time. Above a 1-h gradient duration, the gain in resolution is small, barely worth the long time waiting. For this column, the best gradient run time seems to be between 30 min and 1 h. This combines good resolution with a reasonable analysis time.

In today's world, we really do not want to wait for such a long time to complete an analysis. Let us assume that we want to finish the analysis in under 10 minutes. A gradient run time of 8 min may be about right to accomplish this. The resolution at a flow rate of 1 mL/min does not look good. However, we can change the flow rate. In Figure 4b, we see what is happening when we change the flow rate for an 8-min gradient run. The previously selected flow rate of 1 mL/min is clearly not the best choice. At 1 mL/min, the peak capacity for the 8-min gradient is only about 80. At 4 mL/min, the peak capacity is around 120–50% better

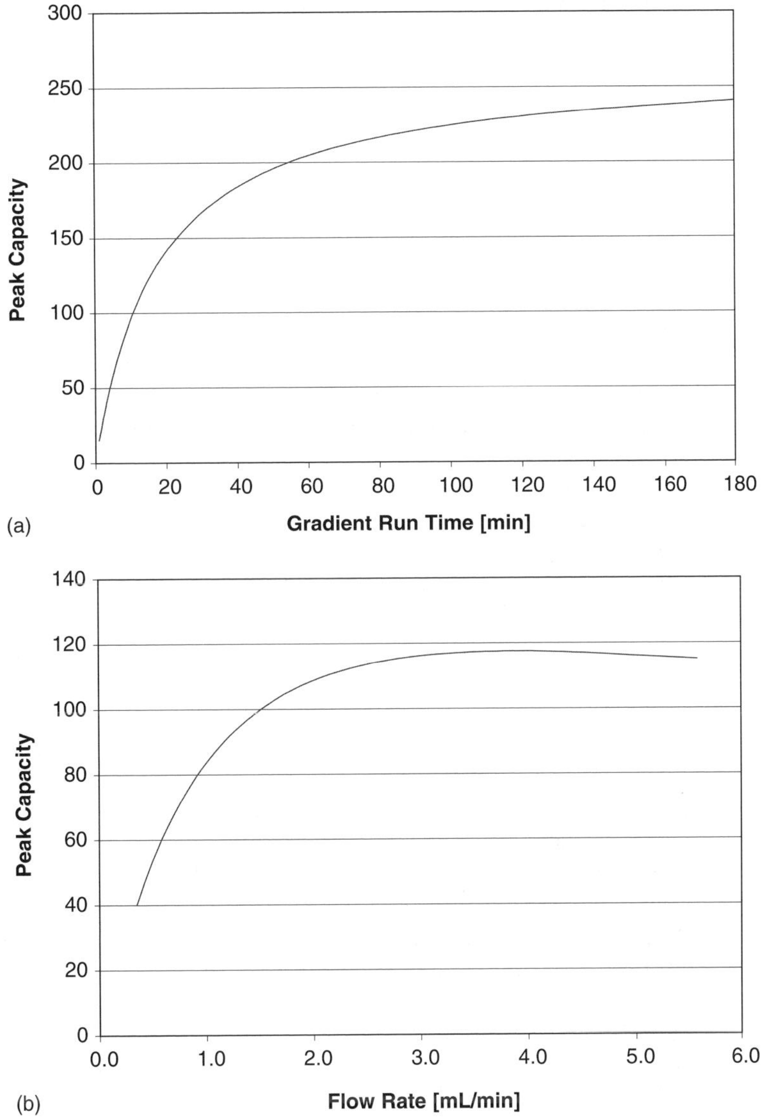

FIGURE 4 **Peak capacity in gradient separations. (a) Peak capacity as a function of gradient duration at constant flow rate. (b) Peak capacity as a function of flow rate at constant gradient run time.**

than at the slow flow rate. The column backpressure is also still tolerable, far away from the limit that the instrument imposes on us. We conclude that for fast gradients we should be operating the columns at fast flow rates. The graph also shows that the optimum is rather flat: there is not much difference in peak capacity, if the gradient is run at 3 or at

5 mL/min. This allows for some flexibility in the actual operating conditions. From a practical standpoint, we want to choose conditions that are not too close to the pressure limit of the instrumentation.

It is possible to combine the two variables—gradient run time and flow rate—together with the resulting peak capacity in a three-dimensional (3-D) graph. This gives us a complete view of the separation capabilities of a column under gradient conditions. In addition, it shows us the entire set of compromises that we had explored in the two examples shown in Figure 4. The resulting 3-D plot for our 15 cm × 4.6 mm 5-μm column is shown in Figure 5. The *x*-axis represents the flow rate, the *y*-axis is the gradient duration. The vertical *z*-axis shows the resulting gradient peak capacity. Both the flow-rate axis and the gradient-duration axis are on a logarithmic scale. The graph itself contains lines of equal height that allow us to see conditions of equal peak capacity. As above, the highest flow rate is limited to just under 6 mL/min.

For very long gradient run times, for example 6 h, we can reach a peak capacity of around 250 with this column. The flow rate for this gradient would be around 0.6 mL/min. For more realistic conditions, for example a 45-min gradient, a peak capacity of 194 can be achieved at a flow rate of 1.4 mL/min. Even for a 45-min gradient, the flow rate is

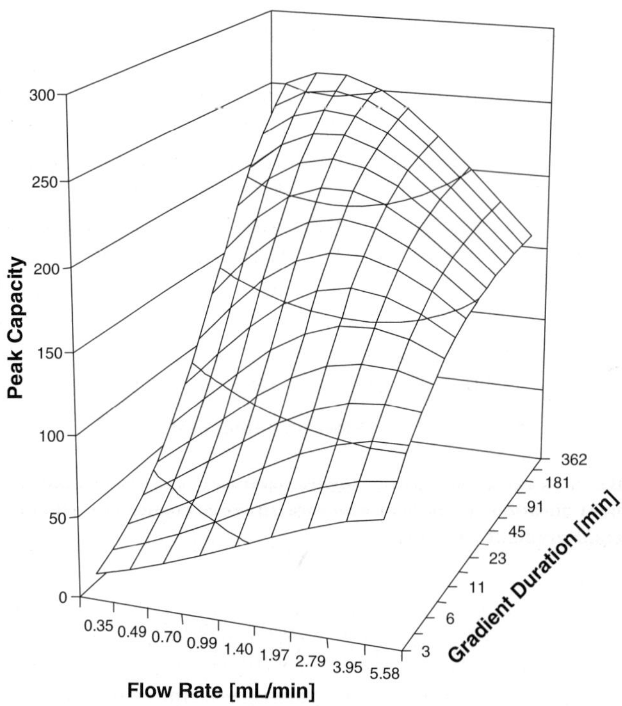

FIGURE 5 **3-D plot of peak capacity as a function of flow rate and gradient duration.**

higher than what we would have selected intuitively. We had seen above that the flow rate increases to 4 mL/min for an 8-min gradient. At a still shorter gradient duration, the maximum peak capacity is reached at flow rates outside the pressure range that is available. In addition, the peak capacity continues to drop with decreasing gradient run time.

We can create similar graphs for other columns. If we were to select a 25-cm column packed with 5-μm particles, the maximum flow rate that can be used would be lower, just over 3 mL/min. For the long, 6-h gradient, the peak capacity of this column would approach 320. The optimum flow rate is somewhat higher compared to the 15-cm column, around 0.8 mL/min. For a 45-min gradient, we can still reach a peak capacity of 220, at 1.6 mL/min. This flow rate is marginally higher than the flow rate used for the 15-cm column to achieve the optimal peak capacity for this gradient. The maximum of the peak capacity shifts outside our pressure capability for a 16-min gradient. For the 8-min gradient discussed above, the 25-cm column reaches a peak capacity of 119 at the pressure limit, the same as we were able to achieve with the 15-cm column at a lower pressure.

What will happen if we select a short 5 cm 5-μm column? We expect that this column will not perform well for the long gradients. Indeed, the best we can do is a peak capacity of about 160 for the 6-h gradient at 0.4 mL/min. While this is low, it is only about half of the peak capacity of the 25-cm column. For shorter gradients, the peak capacity barely changes for the 5-cm column: it is still around 150 for a 2-h gradient. For the 45-min run time used as an example for the other 5-μm columns, the maximum peak capacity is 135 at a flow rate of just under 1 mL/min. For the 8-min gradient, the 5-cm column starts to become competitive: the peak capacity of 96 does not quite reach the value of the 15-cm column, which was 120, but the pressure is much lower. There are two reasons for this. The column is shorter, and the best flow rate is lower (2 mL/min) than for the 15-cm column (4 mL/min). We are also far away from the pressure limit: this short column can be used for up to about 15 mL/min.

For longer gradients, columns behave as expected. Longer columns give more resolution. But for short gradients, there are some surprises. These are pointed out for a really fast 4-min gradient in Figure 6. The three 5-μm columns all have about the same performance at a flow rate of 3 mL/min. While this is the optimum for the 5-cm column, it is the pressure limit of the 25-cm column. The 15 cm 5-μm column is still capable of achieving a higher peak capacity, but higher flow rates are needed. At 3 mL/min, it performs as well as the 5-cm column and slightly better than the 25-cm column. The 15-cm and 25 cm 3-μm columns reach their pressure limit long before the optimum performance is reached. The 25 cm 3-μm column cannot even get close to the performance of the 5 cm 5-μm column. The 15 cm 3-μm column is doing OK, but it also is limited by the backpressure limitation. However, it is outperformed under all

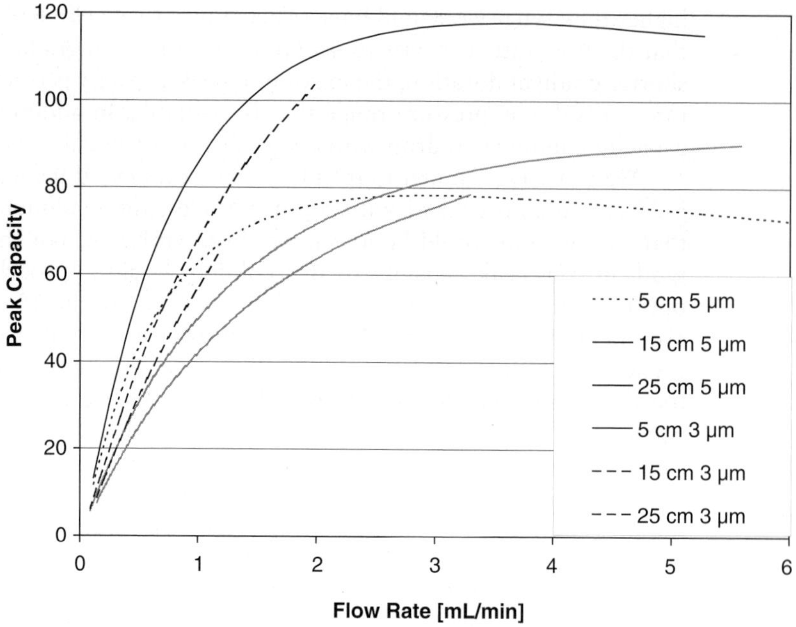

FIGURE 6 Plot of peak capacity vs. flow rate for 5-µm columns and 3-µm columns of various lengths.

conditions by the short 5 cm 3-µm column, which reaches optimum performance somewhere between 3 and 4 mL/min. We learn from this example that one should use short columns for short gradient analyses.

Earlier in this chapter, we discussed the column scaling rules for isocratic chromatography. Let us now see if and how these scaling rules apply to gradient chromatography. We learned that columns with a constant ratio of column length to particle size have the same separation power, but that a smaller particle size leads to faster separations. Let us compare a 5 cm 5-µm, 3 cm 3-µm and a 2 cm 2-µm column now for gradient separations. This comparison is shown in Figure 7 in the form of the 3-D plots introduced earlier. The x-axis on the front shows the flow rate on a logarithmic scale. The largest value is limited by the column backpressure. It decreases with decreasing particle size and column length, but—as shown before for isocratic chromatography—the shortest retention time for the unretained peak is the same for all three columns. In terms of the retention time of an unretained peak, the x-axis on all three graphs are identical. The y-axis represents the gradient duration, also on a logarithmic scale. It ranges from 1 to 64 min. The vertical z-axis shows the peak capacity.

If we look at the rear of the graphs, at long analysis times, we see that the maximum in the resolving power shifts towards faster flow

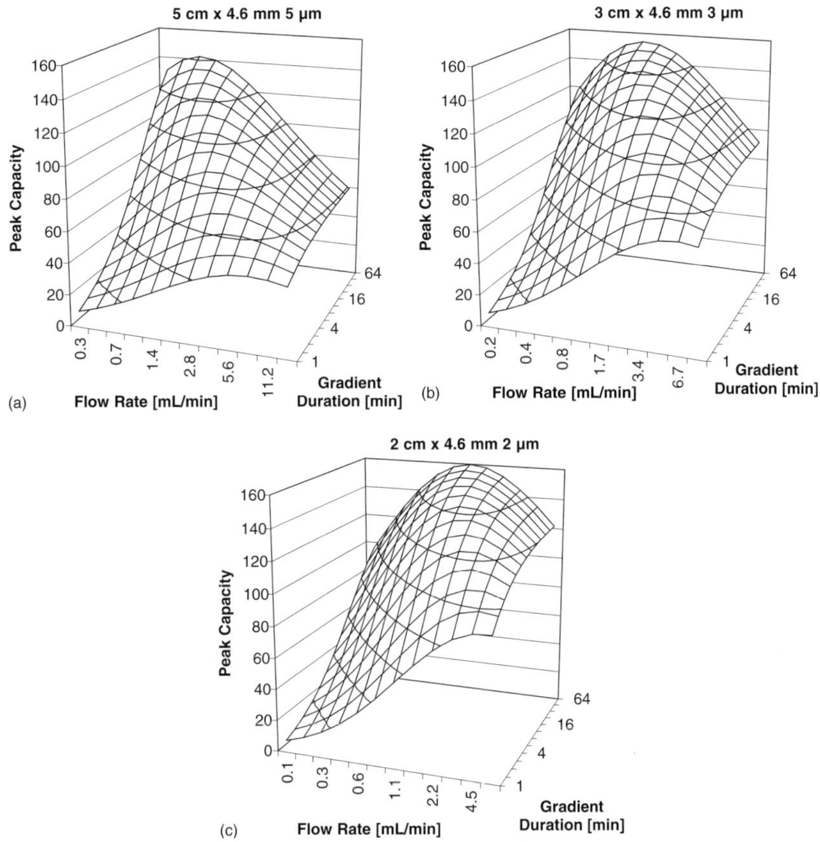

FIGURE 7 Peak capacity plots for columns of equal ratio of column length to particle size: (a) 5 μm 5 cm, (b) 3 μm 3 cm, (c) 2 μm 2 cm.

rates with smaller particle size. In addition, the maximum peak capacity is identical for all three columns. This is in complete agreement with our experience for isocratic chromatography. For very fast chromatography, the maximum of the curve shifts towards the right with decreasing particle size. This means that the optimum peak capacity is obtained at a faster column dead time with the smaller particles. More importantly, the peak capacity increases with smaller particles packed into shorter columns. For a 1-min gradient, the peak capacity drops from over 90 to just over 45 going from the 2-μm column to the 5-μm column. If we used a still smaller particle size, for example a 1 cm 1-μm column, the maximum peak capacity for a 1-min gradient would be outside the pressure range available with the current instrumentation. Short columns packed with small particles give the best results. While the details are a bit more complicated for a gradient separation, the overall rules are similar to our experience for isocratic chromatography shown earlier.

It should be mentioned that high-speed analysis is of major interest in today's pharmaceutical analysis. The latest trends in high-throughput screening push the sample turnaround time to its limits. The resolving power of the MS detector has in the past put less emphasis on the quality of the chromatographic separation, but this trend is slowly reversing due to the recognition that ion-suppression effects can falsify analytical results. This is especially true in the analysis of biological samples[11,12] such as plasma and urine. To avoid these pitfalls, it is always better to maximize the chromatographic resolving power as outlined here, even in conjunction with MS-detection. In Section II. D we will cover the combination of high speed with high resolution in more detail.

B. The Effect of Elevated Temperature

In today's world, elevated temperature is often considered to be a tool in method development and optimization. The increase in temperature decreases the viscosity of the solvent and reduces the backpressure. It also diminishes retention. For a moderate increase in temperature, the changes are rather small. For a variation in temperature by 10°C, the retention time decreases by about 20% in isocratic chromatography. The column backpressure also decreases by 10% to 20% for the same increase in temperature. The change in retention needs to be compensated for with a lower concentration of organic solvent. If acetonitrile is the solvent, a smaller fraction of acetonitrile is needed, which increases the viscosity and the backpressure. Thus for most practical applications, an increase in temperature does not result in an advantage if acetonitrile is the organic solvent. The story is completely different for methanol. Methanol/water mixtures have a viscosity maximum of around 50%. If the mobile phase contains less than 50% methanol, the viscosity decreases if a smaller percentage of methanol is used. Under these circumstances, an increase in temperature has the dual effect of reducing the pressure due to the increased temperature and due to the lower water content of the mobile phase required for equal retention. Of course, for compounds eluting in a solvent strength higher than about 60% methanol, an increase in temperature is counterproductive.

Another interesting consequence arises from the fact that diffusion and viscosity both depend on temperature. As the viscosity decreases, the diffusivity increases. As a consequence of this, the column performance as measured by the plate count is exclusively determined by the column backpressure. While this is not 100% accurate, it is at least a good rule of thumb. To get about the same column performance, you should increase the flow rate when you increase the temperature such that the column backpressure is about the same. Then, the benefit of the increased temperature is directly the reduction in analysis time.

C. The Choice of Column Diameter: Sensitivity, Detector Compatibility, Bandspreading

Until now, we have changed column length and particle size, but not the column diameter. In principle, column performance is independent of column diameter, until one gets into really small columns. Only when the ratio of column diameter to particle diameter becomes less than 30, are new effects observed due to changes in the structure of the packed bed.[1] For our purposes, we can stick with the simple rule that the column diameter has no effect on column performance. Of course, we also assume that the flow rate is changed in proportion to the column cross-section. Under these circumstances, the compromises between performance and run time outlined above remain the same. However, there is one area where smaller diameter columns may have advantages: if we are working with an application where the amount of sample is limited. If we can inject the available amount of sample onto a smaller i.d. column, the dilution will decrease and the sensitivity will therefore increase. For example, the same amount of sample injected onto a 4.6-mm column and a 2.1-mm column will increase the signal nearly five-fold.

However, those who have worked with smaller diameter columns have often experienced lower performance and other difficulties. This is primarily due to extra-column effects: the bandspreading in the injector, detector and tubing, or the gradient delay volume of the instrument. Troubleshooting guidelines for sorting out the causes of these difficulties are available in Reference 1. With proper care, 2-mm columns can be run on a standard modern HPLC instrument with few difficulties. Smaller i.d. columns require special instrumentation.

If the analysis is carried out under isocratic conditions, the bandspreading in the injector, detector and connection tubing all play a role, as do the injection volume and the injection solvent. In principle, one should inject the sample in a solvent composition of a lower elution strength than the mobile phase. This enriches the sample at the column top and reduces pre-column bandspreading. The connection tubing between injector and column and detector should be minimized. Often, the best solution is to use a smaller i.d. tubing for the smaller i.d. column. The detector volume will also contribute to bandspreading, but a reduction in the detector volume will result in a loss in sensitivity, and one needs to think through carefully what the gains and the losses are. If this is a gradient analysis, the concerns about the bandspreading between injector and column vanish. However, this assumes that the sample can be dissolved in the starting solvent composition of the gradient. If the sample is dissolved in an organic solvent, for example DMSO, strange effects can occur, including a breakthrough of a large fraction of the sample as an unretained peak. If the concern about a precolumn bandspreading is reduced (for example in gradient analysis), the column can be connected directly to the detector, and postcolumn bandspreading can thus be minimized. In low-pressure gradient instruments, the delay between the point where the gradient is mixed to the column inlet can be avoided by using a delayed injection, a feature that is available with modern equipment. If high-pressure gradient mixing with a dual pump is used, the gradient delay can be minimized without difficulty by just using appropriate plumbing. With modern instrumentation, this is not an issue any more, but with an older instrument these difficulties are insurmountable.

D. High-speed Analysis

Today, more analyses need to be performed in less time with the same manpower and instrumentation, and therefore it is of significant interest to minimize the analysis time. This is true in the quality control (QC) department as well as in drug discovery or development, although the target time frames and the available tools may be different. In some cases, a 5-min analysis may be a great breakthrough, in other cases investigators are trying to break the 1-min barrier. The LC/MS analysis of biological samples, especially, is pushing the sample throughput to less than one minute per analysis.[11] A review of the state of the art in high-speed LC/MS analysis of small molecules of interest in the pharmaceutical industry at the time of this writing is given in Reference 12.

In Figure 3 above we have shown that good resolving power is still achievable with a 2-cm long 2-μm column. The shortest analysis time at our pressure limit is half a minute, but at a reasonable pressure of about 15 MPa (150 atm, 2200 psi), an analysis time of less than 1 min can be achieved at a reasonable retention factor. The column plate count is around 3600 plates, if everything is working out right. The unretained peak will elute after 6 s. This means that the width of this peak is about 0.4 s. Thus the sampling rate of the instrument needs to be quite high to properly represent such a fast peak. The typical rule of thumb is that one needs a sampling rate that is high enough to collect about 20 to 40 data points per peak. Therefore the minimum sampling rate for such a peak would be about 1 data point every 20 ms. Of course, some compromises are possible. After all, the unretained peak is not what we need to quantitate. The peak of interest elutes often after a retention factor of 2, with a peak width of 1.2 s, and we are in a range that is achievable with comfort.

Similar considerations apply to very rapid gradient separations. We have shown above that a 1-min gradient with a 2 cm 2-μm column can achieve a peak capacity of about 90. This means that the peak width of all the peaks will be 2/3 of a second. Inappropriate settings of the data acquisition rate would reduce the performance of the separation. For ultra-rapid gradient analysis with short, high-performance columns, fast detectors and an appropriate setting of the data acquisition rate are mandatory.

An example of a fast, well-executed separation is shown in Figure 8. A 2-mm i.d. 3 cm 3.5-μm column was used in a rapid, 1-min gradient. In order to maximize the peak capacity, a high flow rate was used: 2 mL/min. To reduce the column backpressure, the separation was carried out at 60°C. A fast detector sampling rate was employed to keep up with the high speed of the separation. The injection was coordinated with the arrival of the gradient at the top of the column in order to eliminate the

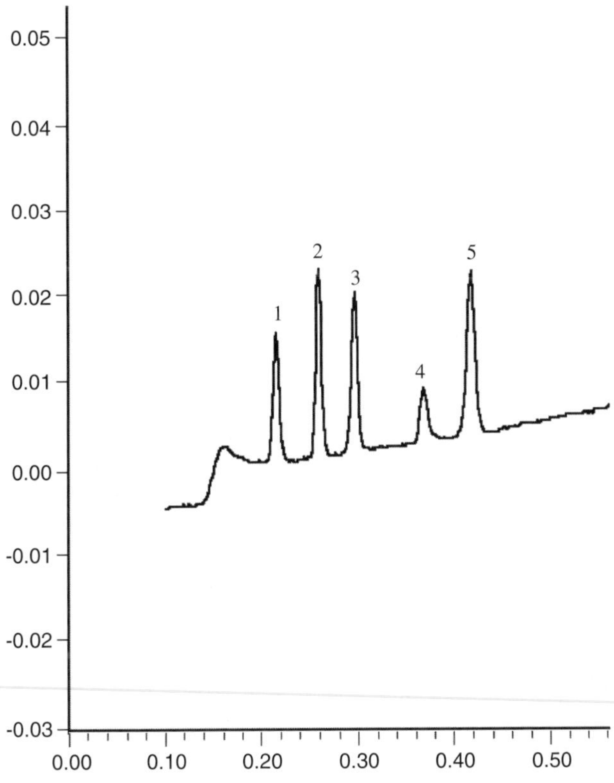

FIGURE 8 **Rapid gradient separation of five compounds. Column: XTerra® MS C_{18}, 2.1 × 30 mm, 3.5 μm. Conditions: gradient: 5 to 95% B over 1 min, A: H_2O with 3% acetonitrile, B: acetonitrile, both A and B containing 10 mM ammonium acetate buffer, pH 5. Flow rate: 2 mL/min. Temperature: 60°C. Detection: 210 nm with a 2.8 μL flow cell. Analytes: (1) lidocaine, (2) prednisolone, (3) naproxen, (4) amitriptyline, (5) ibuprofen. Chromatogram courtesy of Yung-Fong Cheng, Waters Corporation.**

influence of the gradient delay volume of the instrument. While the gradient separation is executed on the column, the pump is preparing already the reequilibration cycle for the next injection. With such a scenario, injection cycle times of not much more than 1 min are possible. In order to preserve the high-quality separation that was achieved in the column, a detector cell with a small cell volume, 2.8 μL, was used. The column was connected to the detector cell with a minimal amount of tubing. This also reduces the bandspreading. To heat up the mobile phase, we made use of the length of tubing needed to bridge the distance between the injector and the column. As mentioned above, the length of precolumn tubing in a gradient separation does not contribute to a broadening of the analyte peaks. As can be seen in this example, high-quality separations can be executed in less than a minute, if attention is paid to experimental details.

Even for more moderate speeds, short columns with small particles should be considered.[13,14] The rules remain similar to the ones discussed above. For optimal gradient separations in the 3- to 5-min time frame, a flow rate of roughly 3 to 5 mL/min (depending on the diffusion coefficient of the analytes and the details of the gradient execution) is a good starting point for a 4.6 mm i.d. column of 2 or 3-cm length packed with 3 or 3.5-μm particles. An example of such a separation is shown in Figure 9. This excellent gradient separation of standards with a broad polarity range was carried out in 4 min on a 2 cm 3.5-μm column. The column backpressure was 10 MPa (100 atm, 1500 psi). This demonstrates that high-quality separations can be executed on these short columns at moderate pressure.

The change to shorter columns packed with smaller particles is of interest for many applications in the QC department in the pharmaceutical industry. Often, the separations are executed using 25 or 15-cm columns in a 15- to 30-min time frame. In this chapter, we have demonstrated that the same separation power is available today with short columns packed with small particles. The use of short columns reduces the analysis time to a few minutes. Even faster separations, under 1 min, are possible today, but great attention needs to be paid to experimental details.

E. Monolithic Columns

In the mid 1990s, the first reports on the performance of monolithic columns created much excitement in the scientific community. High-performance separations at low backpressure and a short analysis time were the promise. Nevertheless, it was several years before the first commercial products became available,[15] and made it possible to obtain a proper judgement on the ability and the limitations of the technology.

What are monoliths? While packed beds are made from individual particles, monoliths form a continuous porous structure throughout the

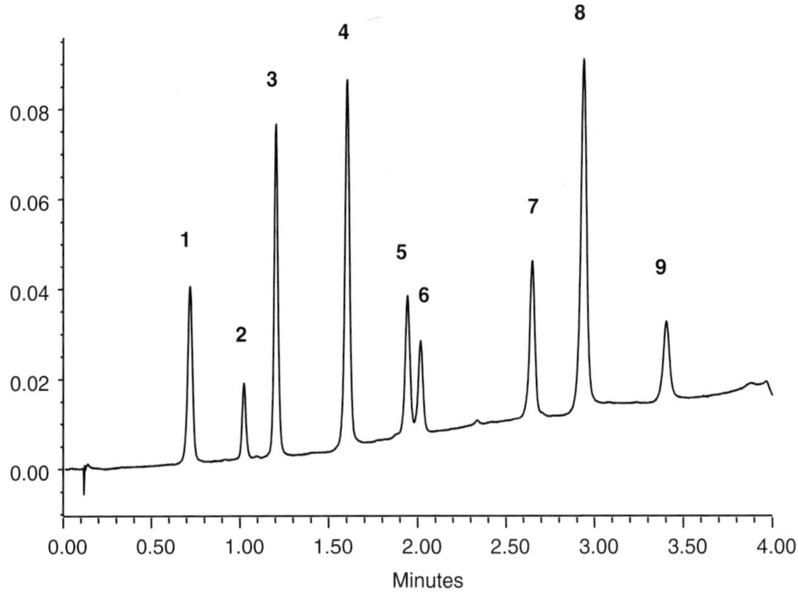

FIGURE 9 Moderately rapid gradient separation. Column: XTerra® MS C_{18} IS, 4.6× 20 mm 3.5 μm. Gradient: 0 to 100% B over 4 min, A: 0.1% formic acid in water, B: 0.1% formic acid in acetonitrile. Flow rate: 3.0 mL/min. Temperature: 30°C. Detection: UV at 254 nm. Instrument: Alliance™ 2795 with 996 photodiode array detector. Compounds: (1) acetanilide, (2) triamcinolone, (3) hydrocortisone, (4) 2-amino-7-chloro-5-oxo-5H-[1]benzopyrano[2,3-*b*]pyridine-3-carbonitrile, (5) 6*a*-methyl-17*a*-hydroxyprogesterone, (6) 3-aminofluoranthene, (7) 2-bromofluorene, (8) perylene, (9) naphtho(2,3-*a*)pyrene.

chromatography column.[15,16] Scanning electron micrographs of a monolith at appropriate scales are shown in Figure 10. In the center, one can clearly see the continuous coral-like primary structure of the monolith. Such a structure has a higher interstitial porosity than is possible to achieve with a packed bed of particles. The consequence of this feature is a lower backpressure for a given performance. The performance of the structure is determined by the length of the diffusion path from a flow channel into the center of the porous structure. As one can see, this distance is in the order of 3 μm. Thus it becomes intuitively clear that the performance of these monoliths is equivalent to the performance of a 3 to 5 μm packed bed. Van-Deemter plots in Reference 17 show an A-term between 7.4 and 9.8 μm for a monolith. Since the standard A-term in a well-packed column is about 2× the particle diameter, such an A-term would be observed with columns packed with 3.7 to 4.9-μm particles. At the same time we see that the holes in the structure, the macropores, have a dimension of about 2 μm. Such a distance is similar to the interstitial space between 8-μm particles. From this, one would

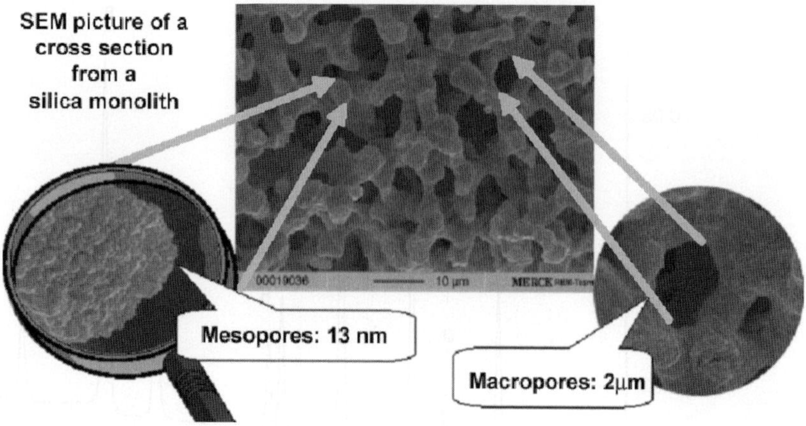

FIGURE 10 **Internal structure of a monolith. Courtesy of Karin Cabrera, Merck KGaA.**

expect that the pressure drop is comparable to the pressure drop of a packed bed of 8 to 10-μm particles. This is indeed the case: in Reference 17 it is shown that the pressure drop is about 3.5 times lower that the pressure drop of a 5-μm column. Combining these two facts, a well-designed monolith provides the performance of a bed of 3.5 to 5-μm particles with the backpressure of a 9 μm packed bed column.

The secondary structure, the mesopores, is similar to the internal structure of standard HPLC particles. This secondary structure provides the surface for retention. The standard pore size is in the order of 13 nm, resulting in a specific surface area of about 300 m^2/g. Due to the lower ratio of retentive structure to interstitial space, the retentivity of monoliths and the preparative loadability tends to be significantly lower than the retentivity and loadability of packed beds of 10-nm particles. Since the monolithic columns described here are made from silica, they can be derivatized in the same way and with the same technology as silica-based particles. Also, the useful pH range is the same as for silica-based particles.

The promise of monolith is the achievement of a higher performance at a lower backpressure than a packed bed. While this is true in principle, current implementations are limited by the fact that the external wall to the structure is made from PEEK. At the time of this writing, the commercially available monoliths can only be used up to a pressure of 20 MPa (200 atm, 3000 psi), while packed bed steel columns can be used up to double this pressure and higher. Also, the preparation of the monolith appears to be cumbersome. At the current time, the silica-based monoliths are available only with an internal diameter of 4.6 mm. The speed is thus also limited by the flow rate achievable by the HPLC instrument. At the same time, the detector of choice today is the mass spectrometer, which can tolerate only much

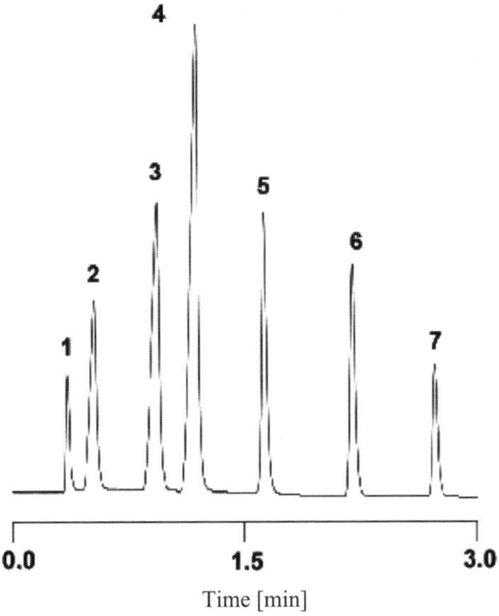

FIGURE 11 Example of a fast gradient separation using a monolith column. Column: Chromolith Performance $RP_{18}e$, 100 mm × 4.6 mm. Mobile phase: Gradient: A: acetonitrile, B: 20 mM phosphate pH 4.5. Flow rate 4 mL/min. Chromatogram courtesy of Karin Cabrera, Merck KGaA.

lower flow rates than are required in order to take advantage of the benefits of a monolith. In summary, there are significant differences between the column dimensions that are optimal for the user and what the existing monolith technology can offer. Therefore it appears that the technology is still in its infancy, now nearly a decade after the first reports. However, the future may still tease the full benefits out of the monolithic structures.

An example of a high-speed application is shown in Figure 11. Here a rapid gradient separation of a few acids and esters is shown. A rapid gradient is executed at a high flow rate, 4 mL/min, on a 100 mm × 4.6 mm Chromolith column. Seven compounds are separated in less than 3 min.

III. COLUMN CHEMISTRY

The column chemistries predominantly used today originated in the early 1990s. In this time frame, high-purity silicas became commercially available for the first time. Also, bonded phases with embedded polar groups were conceived first in the same time frame. Due to the beneficial features of a high-purity silica, stationary phases based on this technology are primarily used in today's pharmaceutical industry. At the same time, the use of phases with embedded polar groups continues to grow rapidly due to the additional advantages that such phases provide. The introduction of pH-stable phases based on inorganic–organic hybrids has expanded the mobile phase range that can be exploited by chromatographers. This too has had its impact on the methods that are used in the industry today. In this section, we will discuss the different options available.

A. Silica and Its Derivatization

Most HPLC separations in today's pharmaceutical industry are carried out using reversed-phase chromatography. This is largely due to the convenience and speed of column equilibration compared to normal-phase chromatography, and the broad range of samples that can be chromatographed conveniently. Before we go into a discussion of the use of modern packings, it will be worth our while to briefly review the surface chemistries that are commonly employed in reversed-phase chromatography. For additional information on bonding chemistry, Reference 6 or 7 is recommended.

Silica contains surface silanols, about 8 μmol/m^2 to be precise. They are derivatized with silanization agents in the presence of an appropriate scavenger of the side product(s). The most commonly used reagent is a monochloro silane:

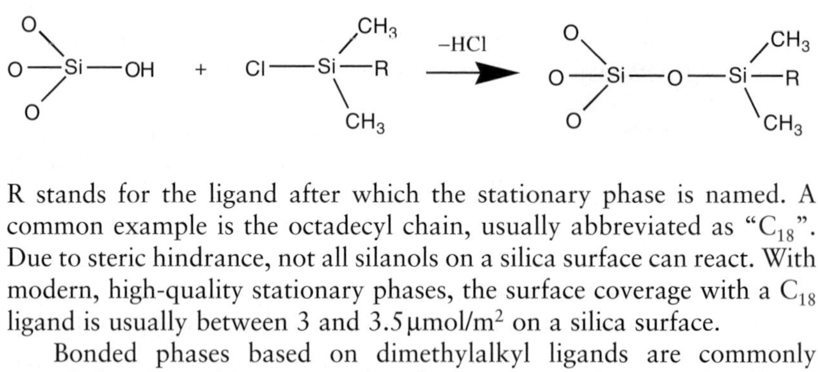

R stands for the ligand after which the stationary phase is named. A common example is the octadecyl chain, usually abbreviated as "C_{18}". Due to steric hindrance, not all silanols on a silica surface can react. With modern, high-quality stationary phases, the surface coverage with a C_{18} ligand is usually between 3 and 3.5 μmol/m^2 on a silica surface.

Bonded phases based on dimethylalkyl ligands are commonly regarded to be stable at pH-values higher than 2, but one should not regard this as a completely fixed number. The quality of the silica, the ligand density, and the ligands themselves all play a role. In addition, one should not consider a phase to be completely unstable at pH 1.9 and infinitely stable at pH 2.1. There is, therefore some flexibility around these rules. Instead of a monofunctional silane, di- and trifunctional (di- and trichloro-) silanes are used sometimes. Their advantage is an improved ligand stability at acidic pH, due to the fact that the common bond to the surface is bidentate. In one-for-one comparisons, it has been shown that the hydrolytic stability of a bonded phase based on a di- or trifunctional silane at low pH is five times higher than that of a monofunctional bonded phase. The structure of the common attachment of difunctional and trifunctional ligands is shown below:

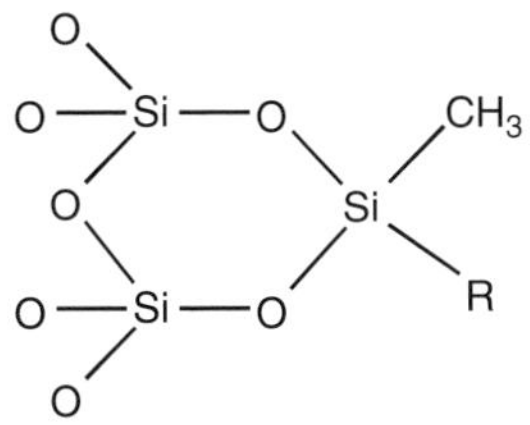

difunctional silane

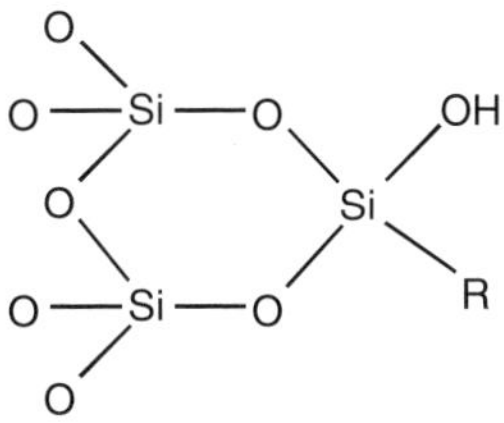

trifunctional silane

Another approach to improve the stability of bonded phases at acidic pH is the use of sterically hindered silanes,[18] which use, for example, isopropyl groups as side chains, instead of methyl groups. Due to the somewhat lower surface coverage, this technique leaves a larger amount of silanols on the surface than a high-quality bonding with a dimethylsilane, but this is balanced against the definite improvement in hydrolytic stability at acidic pH.

Silica-based bonded phases are commonly regarded to be stable to pH 8. Once again, this number is more a rule than a fixed value. In the alkaline pH, the silica itself dissolves. Efficient techniques to completely cover the surface can improve this situation. The most effective approach though to improve the stability in the alkaline pH is the creation of hybrid phases, which do not use silica as the carrier of the bonded phase (see below).

Figure 12 shows the structure of some of the most commonly used stationary phases in reversed-phase chromatography. The side groups X_1 and X_2 are, for the most part, either methyl groups, or an additional attachment to the surface, or silanol groups. As has been shown above, no more than two bonds to the silica surface can be created. Therefore, all bondings with trifunctional silanes create at least one additional silanol group for every ligand that is attached to the surface. Some of the residual silanols, as well as some of the newly created silanols can be removed by "endcapping," which is an exhaustive reaction with trimethylsilyl reagents. But even an exclusive reaction with a trimethylsilane provides only a surface coverage of about 4 μmol/m². Thus only half of the 8 μmol/m² silanol groups on a fully hydroxylated silica can be removed via this endcapping reaction.

The top two stationary phases in Figure 12 are the C_{18} and C_8 phases, which are the most frequently used phases in reversed-phase LC. Below that, three phases with an embedded polar group are shown: the carbamate phases (e.g., SymmetryShield), amide phases (e.g., Discovery RPAmide C_{16}) and urea phases (such as Prism or Spectrum). It should

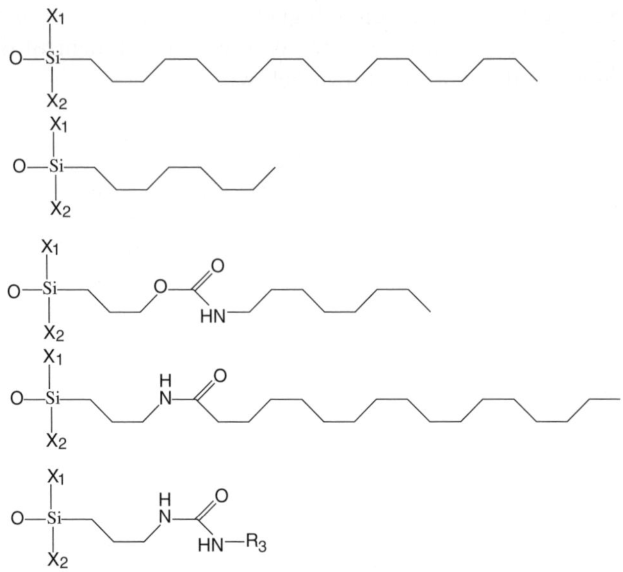

FIGURE 12 **Structure of common bonded phases.**

be pointed out that amide phases can be created via two synthetic routes. In the older approach, an aminosilane is reacted first with the silica surface, and the aminogroup is derivatized in a second step with a long-chain activated acid, such as an acid chloride. Multiple reactions on a surface are never complete, and amide packings prepared via this approach always contain residual amino-groups on the surface. Just as silanols create tailing of basic compounds, residual amines create tailing and irreproducible retention for acidic analytes. People have recognized this problem, and some of the newer amide phases are now apparently prepared in a single-step surface reaction, i.e., the ligand is assembled first.

In Figure 13, we show some of the less frequently used reversed-phase ligands. On top you see one of the possible phenyl ligands, i.e., a propylphenyl bonded phase. Other possibilities exist as well, from direct attachment of the phenyl group to the silicon atom, to branched chains, or longer chains, such as a hexylphenyl ligand. Below the phenyl ligand, a cyano bonded phase is shown. Cyano phases were very popular in the early 1980s, but they fell into disgrace due to mechanical stability problems. Recently, fluorinated phases have achieved an increased level of popularity. Most aliphatic perfluorinated phases are of the type shown here, but a branched version exists as well. The currently commercially available PFP-phases—PFP stands for pentafluorophenyl—are all of the same type and contain a 3-carbon spacer between the silicon atom and the phenyl group.

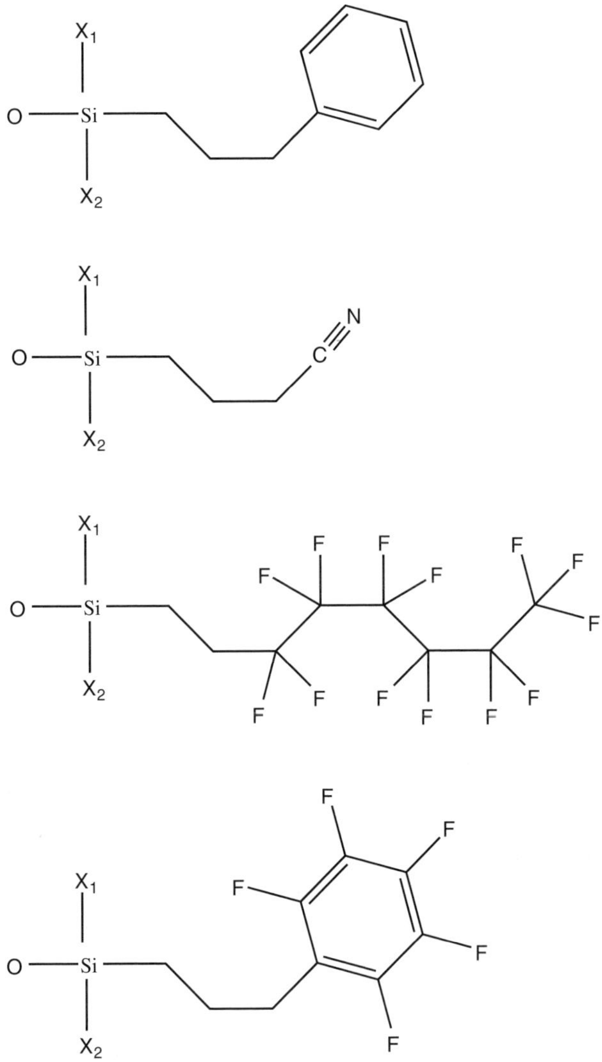

FIGURE 13 Structure of less commonly used bonded phases.

Besides the packings mentioned above, a range of other stationary phases exists as well, but they are rarely used in pharmaceutical analysis. Even with this more limited selection of ligands, packings with quite different selectivities can be created, as in shown in Section III. F. The diversity stems from the choices in the properties of the silica, such as pore volume, pore size, surface area and purity as well as differences in ligand density. Nevertheless, general statements about different phases can be derived, and these are covered in the next section.

B. Properties of Modern Silica-based Stationary Phases

What is the reason for the overwhelming acceptance of stationary phases based on high-purity silicas in the pharmaceutical industry? The answer is simple: superior peak shapes for analytes with basic functional groups, which has been a problem with older phases. The older, low-purity silicas contain metal ions buried in the matrix of the silica. These contaminants acidify the surface silanols, and the consequence is a strong and non-uniform interaction with basic analytes. This in turn results in tailing peaks, which is an impediment for accurate peak integration and peak resolution. Of course, adding appropriate additives, such as amine modifiers, to the mobile phase can solve these difficulties. But this is an unnecessary and undesired complication in methods development. Therefore, silicas that are free from this complication are much preferred.

The silanol induced peak tailing is also a function of the pH of the mobile phase. It is much less pronounced at acidic pH than at neutral pH. Therefore many of the older HPLC methods use acidified mobile phases. However, pH is an important and very valuable tool in methods development. The selectivity of a separation of ionizable compounds is best adjusted by a manipulation of the pH value.[1,19] The retention factor of the non-ionized form of an analyte is often by a factor of 30 larger than the one of the ionized form, and it can be adjusted to any value in between by careful control of the mobile phase pH. This control must include a good buffering capacity of the buffer to avoid random fluctuations of retention times.

Table 1 gives the tailing factor for the basic analyte amitriptyline at neutral pH on several commercially available packings. One can clearly see the difference between the older packings and the newer packings based on high-purity silicas. It is unquestionable today that surface silanols on a packing participate in the retention of basic analytes, and

TABLE 1 Peak Tailing for Amitriptyline for Various Packings at pH 7.0

Packing[1]	USP tailing factor
μBondapak C_{18}	4.60
Nova-Pak C_{18}	4.47
Chromolith $RP_{18}e$	4.92
Atlantis dC_{18}	4.71
Symmetry C_{18}	1.85
XTerra MS C_{18}	1.49
Luna $C_{18}(2)$	1.35
SymmetryShield C_{18}	1.14
XTerra RP_{18}	1.12

[1]Column dimensions: 4.6 mm × 150 mm, except for Chromolith $RP_{18}e$ 4.6 mm × 100 mm.

that the activity of such silanols depends strongly on the purity of the silica. Even among the newer packings based on high-purity silicas, differences can be observed. Much better results are obtained with phases with an embedded polar group such as SymmetryShield RP_{18}. Packings of this nature have been designed with a neutral polar functional group close to the surface of the silica. Useful polar groups for this purpose are amide, carbamate, urea or sulfonamide groups—all groups with strong hydrogen bonding properties. The true nature of the mechanism by which the polar functional group suppresses the undesirable interaction with silanol groups has not yet been demonstrated. Several possible mechanisms have been proposed:[20] a competition of the polar functional group in the ligand with the analyte, an interaction of analytes with the polar functional group instead of an interaction with surface silanols, or a strong binding of water by the polar functional group, thus suppressing any interaction of the analyte with the underlying structure. At this point in time, the first mechanism can be excluded for steric reasons. However, a clear distinction between the latter two principles is not yet possible. The selectivity pattern of packings with embedded polar groups suggests a specific interaction of the polar functional groups via hydrogen bonding with suitable analytes such as phenols or sulfonamides. On the other hand, a study of the retention as a function of the chain length of the ligand indicates that the methylene groups below the polar functional group do not participate in the retention and do not contribute to the hydrophobicity of the packing. This observation supports the third mechanism.

Another approach to a reduction in silanol activity has been the use of mixed silanes.[21] This was first practised in the design of the YMC Basic packing, but a recent and more rigorous application of this idea is the Luna $C_{18}{}^2$ packing. The surface is derivatized with a mixture of C_{18} and C_8 silanes. This improves the surface coverage that can be achieved, and reduces the access to surface silanols. The result of this procedure is an improvement in the peak shape of basic analytes (see Table 1).

In today's world, many analytes are very polar. Therefore, one would desire to use mobile phases with 100% water to maximize retention. Unfortunately, this is not possible with modern purely hydrophobic C_{18} packings. The wetting angle of water on such hydrophobic surfaces is unfavorable. If these packings are used in a fully aqueous mobile phase, the mobile phase is pushed out of the pores when the pressure on the packing is released. This results in a complete loss of retention, until the packing is rewetted preferentially with an organic solvent such as methanol or acetonitrile. This phenomenon does not happen with stationary phases with an embedded polar group. Therefore these phases can be used under such circumstances, i.e., with fully aqueous mobile phases. Recently, phases have emerged that claim to be using polar endcapping procedures to improve the wettability of a C_{18} packing in such mobile phases as well. However, the nature of the "polar endcapping" groups is not disclosed, and in cases that were tested a polar endcapping could not be found. Thus one would believe that at least many, if not all of the "polar endcapped" phases are simply un-endcapped phases. On a modern high-purity silica, the lack of endcapping is not as detrimental as it was with older, low-purity silicas.

A newer approach that maximizes retentivity and water wettability simultaneously is the technique used in the design of the Atlantis dC_{18} packing. The dewetting of the bonded phase surface is a function of the wetting angle and the pore size. The wetting angle in turn can be manipulated by the surface coverage with the hydrophobic C_{18} ligand. Silanol activity, on the other hand, can be constrained via endcapping without detrimental consequences to the wetting angle, if done properly. The result of a careful examination of the combination of these parameters resulted in the Atlantis dC_{18} packing. The key feature is a low C_{18} surface coverage, around 1.5 μmol/m^2, with a difunctional C_{18}, combined with a complete endcapping. This material is compatible with 100% aqueous mobile phases, exhibits higher retention than packings with embedded polar groups, and gives good peak shapes for analytes that interact strongly with surface silanols.

C. Modern Developments

1. Inorganic–Organic Hybrids

Additional improvements in silanol activity have been made with inorganic/organic hybrid packings. Such packings are not made from silica, but from an appropriate mixture of tetraethoxysilane and organofunctional silanes. Using this technique, a significant molar fraction of organic groups can be incorporated into the matrix of the packing. These groups can serve multiple purposes. They can improve the hydrolytic stability of the packing compared to silica-based packings, especially in the alkaline pH range. They also allow the incorporation of new functions into the matrix. In all cases of technical relevance, the advantageous mechanical properties of silica, its hardness and pressure stability, are preserved. Together, this technology permits the design of pressure-stable packings with novel combinations of surface functions or improved chemical properties. For example, the incorporation of methyl groups into the matrix results in a significant improvement in the hydrolytic stability of a packing compared to silica-based packings. This expands the useful pH range into the alkaline pH, to values above pH 11, which in turn permits the application of alkaline buffers. Most commonly used is ammonium bicarbonate, in the pH range of 10 to 9. With this buffer, the ionization of basic analytes can be suppressed. This results in much higher retention for such analytes, often improved by a factor of 30 compared to acidic pH. Even more important from a practical standpoint is the change in selectivity obtained by a change in pH. An example of this is shown in Figure 14. In this graph, the retention time under alkaline conditions is plotted vs. the retention under acidic conditions using gradient elution with the same gradient. The retention of the neutral compounds does not change with the pH, and they fall on the line in the center of the graph. Acidic compounds are more retained under acidic conditions, and are found below the center line. Basic compounds are more retained at alkaline pH, and are therefore seen above the centerline. One can also discriminate more specific patterns: the change in retention is more pronounced for tertiary amines than for secondary amines, as indicated by the lines marked with t and s. The fundamental message of this graph is that the retention patterns change quite drastically when the pH is changed from acidic to alkaline. This has important consequences in method development: pH is the most powerful tool in methods development, if ionizable compounds need to be separated.[19]

Additional improvements in the design of hybrid packings have been made recently.[22] If an ethyl-bridged silane is used in the preparation of the hybrid matrix, a still greater improvement in hydrolytic stability will be achieved. In one-for-one comparisons with silica-based C_{18} packings, it has been shown that hydrolytic stability increases immensely, permitting the use of bonded phases based on the ethyl-bridged hybrid packing at alkaline pH values. At the same time, the mechanical strength of the ethyl-bridged hybrid packing is identical to the best silica-based packings. Also, mass transfer is as good as with silica-based packings. As a

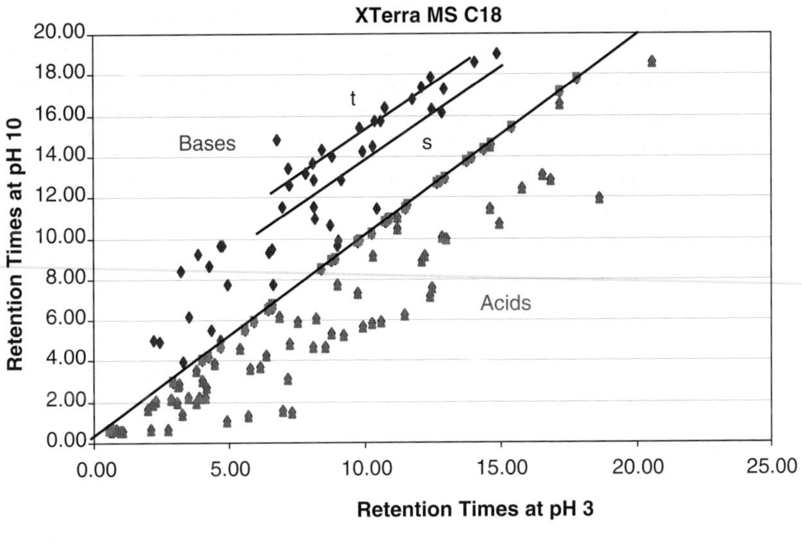

FIGURE 14 Difference in the retention pattern at acidic pH and alkaline pH.

consequence, a compromise-free highly pH stable hybrid packing is now available.

Due to the use of the organofunctional silane (i.e., a silane with an organic group permanently attached to the silicone) in the synthesis of the hybrid packings, the activity of surface silanols is much reduced compared to silica-based packings. There are two factors that combine to produce this effect. One factor is simply the much smaller population of surface silanols due to the presence of the organic groups in the packing. The second factor is the change in the silanol pK due to neighboring methyl groups or ethylene bridges. A comparison of the pK of surface silanols measured for silica and for the methyl hybrid packing shows a change from around 7 for a high-purity silica to values above 9 for a methyl-bridged hybrid packing.[23,24] This shift of the silanol pK into the alkaline pH reduces the influence of ionized silanols, which are believed to be responsible for the tailing of basic analytes. Thus hybrid packings exhibit much improved peak shapes compared to most silica-based packings, even under identical operating conditions around a neutral pH.

2. Zirconia-based Packings

In recent years, zirconia was developed as a base material for HPLC.[25] Its key advantage is its hydrolytic stability. Zirconia is stable over the entire pH range, from pH <1 to pH >13. It can be functionalized by coating and crosslinking polymers. With this technique, reversed-phase packings can be created. However, as with silica, it has been difficult to completely eliminate the influence of the underlying carrier packing. Mixed mode interactions arise, which are different from silica-based packings. In order to take full advantage of zirconia-based packings, it is necessary to understand the Lewis acid-base interactions that emerge, in the same way as it was necessary to understand silanol interactions on the older silica-based packings. On the other hand, as it is true with silica, the knowledgeable user can take advantage of the interactions with the zirconia matrix to tune the selectivity of the separation.

The stability of zirconia and zirconia-based packings also allows the use of these columns at an elevated temperature. The key advantage of this is the reduction in analysis time due to the reduction in the viscosity of the mobile phase. In addition, the much broader temperature range also permits a fine-tuning of the selectivity of a separation compared with silica-based packings.

A range of zirconia-based packings is available at the time of this writing, including anion and cation exchangers and reversed-phase packings. Among reversed-phase packings, there is polybutadiene-, polystyrene- and carbon-coated zirconia, and Diamond-bond C_{18}, a zirconia-based packing that is first coated with carbon, which is then further derivatized with C_{18}-phenyl groups. Significant selectivity differences can be found between the carbon-coated and the polybutadiene-coated zirconia. The same is true when carbon-coated zirconia and standard reversed-phase packings are compared. This is shown in Figure 15. The investigators measured the retention of 22 non-ionic solutes on several different packings and plotted the logarithm of the retention factors obtained on each packing against each other packing. The standard deviation of such a plot is a measure of the selectivity difference between the different chromatographic conditions. This standard deviation is plotted for each column pair in Figure 15. The designations CN, phenyl and ODS mark silica-based packings, and PRP stands for a styrene-divinylbenzene-based packing. The carbon-coated zirconia packing, C-ZrO_2, shows the largest selectivity differences compared to any of the other packings.

D. HILIC—A Complementary Technique to Reversed-phase Chromatography

In recent years, the interest in very polar compounds has increased in the pharmaceutical industry. On one hand, this has led to the development of reversed-phase columns that are compatible with fully aqueous mobile phases (see above). On the other hand, this has also revived the interest in

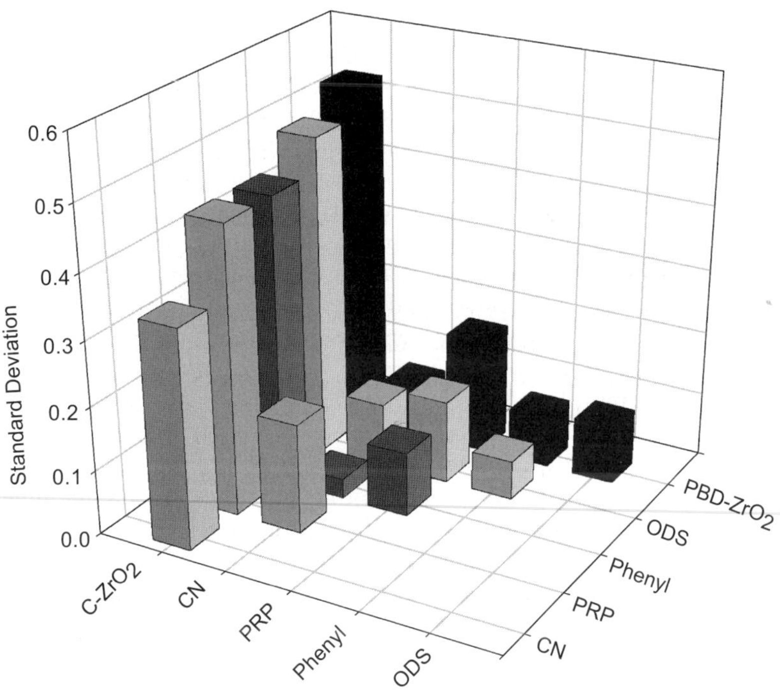

FIGURE 15 Selectivity differences between different packings. Plotted is the standard deviation of graphs of the logarithm of the retention factors for the different columns. Courtesy of Clayton McNeff, Zirchrom Separations.

alternative techniques such as hydrophilic interaction chromatography, or HILIC. HILIC is a technique that employs very polar stationary phases with reversed-phase mobile phases. The retention mechanism is predominantly a partitioning between a mobile phase with a low water content and a surface layer of water enriched on the surface of the packing.[1] The more polar the molecule, the stronger is the retention. Classically, this technique was employed for the separation of sugars on propylamino bonded phases. Today, two packings are predominantly used for this application. One is underivatized silica, the other is a silica derivatized with a zwitterionic bonded phase, commercially available as ZIC-HILIC. The structure of the zwitterionic ligand is shown below:

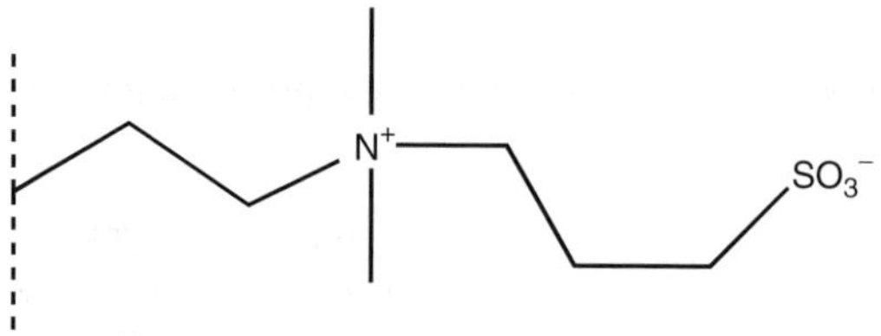

Due to the dual charge, it binds water rather strongly, and due to the charge balance, it does not exhibit ion-exchange properties.

Silica also binds water strongly and can be used for HILIC application. In addition, silanols are weak ion-exchange groups. Therefore retention of basic compounds happens via a combination of hydrophilic interaction chromatography and ion exchange.[1] It needs to be pointed out that basic compounds do not yield tailing peaks on silica. This is quite different from our experience in reversed-phase chromatography. The tailing in reversed-phase chromatography stems from steric hindrance of the access to silanols.[26,27] Once they are freely accessible—as is the case for underivatized silica—no peak distortions are encountered.

An example of a separation carried out in reversed-phase and in HILIC is shown in Figure 16. The analytes are rather polar, morphine and morphine 3-β-glucuronide. On the reversed-phase column, the glucuronide elutes first, since it is more polar than the parent compound. On the HILIC column, it elutes last. In addition, a gradient is used in the HILIC separation to elute both compounds in the same time frame. This demonstrates clearly the difference in retention between HILIC and reversed-phase chromatography.

Another important feature of HILIC is the improved sensitivity with electrospray mass spectrometry. At least one order of magnitude, but often more, can be gained compared with reversed-phase separations. This is significant for the analysis of parent drugs and metabolites in plasma or urine samples. Together with the ability to retain very polar metabolites, this feature makes HILIC now a very attractive technique in pharmacokinetics. In addition, HILIC can be combined conveniently

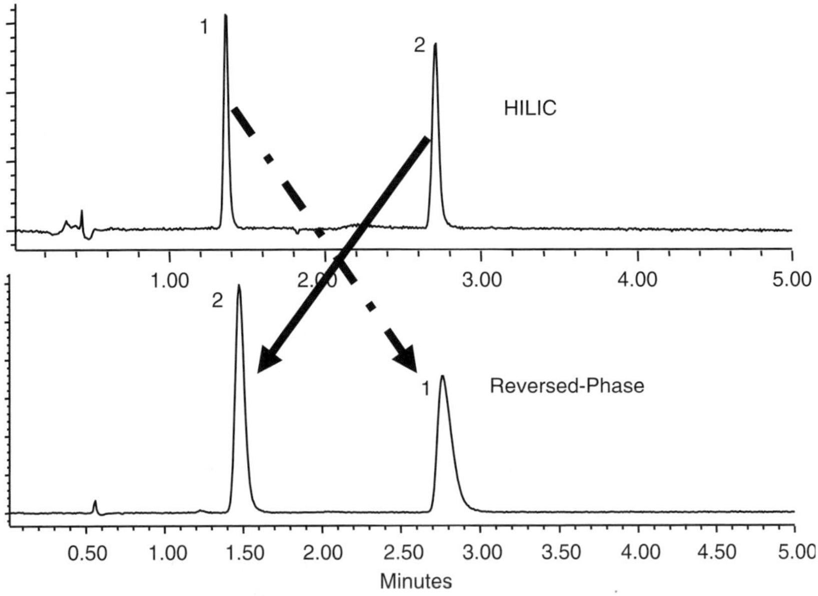

FIGURE 16 **Comparison of a HILIC separation (top) and a reversed-phase separation (bottom). Peak 1: morphine, peak 2: morphine 3-β-glucuronide. Top: column: Atlantis™ HILIC Silica, 4.6 × 50 mm, 3.0 µm; gradient from 90% to 50% acetonitrile with 10 mM ammonium formate buffer, pH 3.0; flow rate 2.0 mL/min. Bottom: Atlantis™ dC_{18}, 4.6 × 50 mm, 3.0 µm; mobile phase: 2% acetonitrile with 10 mM ammonium formate buffer, pH 3.0; flow rate 1.4 mL/min.**

with reversed-phase sample preparation: a plasma sample is loaded on a reversed-phase solid-phase extraction cartridge or well in a 96-well plate; proteins, salts, and sugars are washed off with water, and the analytes are reextracted with an organic solvent, such as a mixture of isopropanol and acetonitrile. The extract can then be injected directly onto a HILIC column for a final LC/MS analysis of the drug and metabolites.

E. Reproducibility of Stationary Phases

In the pharmaceutical industry, chromatographic QC methods are often used with no changes over long time frames. The basic reason for this is the amount of effort necessary for a complete revalidation of a method on a new column brand. Because of these constraints, chromatographic columns should be chosen from reliable and reputable manufacturers that can sustain a reproducible production of the columns over such extended time periods. Until recently, the capability of column manufacturers to reproduce the packing materials was not known, and users needed to rely on their intuition and trust the column manufacturers.

However, this has changed with a few recent publications, and it is worth our while to review the published results.

Kele and Guiochon[28–32,17] and Kele et al.[33] have carried out a consistent evaluation of several modern HPLC packings. The principle of the study was laid out in Reference 28. Columns from different manufacturers were then studied in the remainder of the publications.[29–33,17] Columns from the same batch of packing material were tested together with columns from different batches, using several chromatographic tests published in the literature. This permitted an assessment of the reproducibility of the test methods, the reproducibility from column to column of the same batch of packing, as well as the reproducibility between different batches of packing materials.

Typically, the reproducibility of results on the same column was excellent, with variations in retention time varying between 0.01% and 0.1%. The reproducibility from column to column was a function of the uniformity of the column's internal diameter and the consistency of the packing conditions. It varied from 0.2% to 0.5%, depending on the column. The reproducibility of the packing materials themselves was somewhat more difficult to assess, since different column manufacturers contributed different samples to the study. Waters provided 10 consecutive batches of the Symmetry C_{18} packing, with 8 out of these 10 being based on different batches of silica as well. Table 2 shows some selected results. The data presented include neutral hydrophobic compounds with and without polar functional groups, and basic hydrophobic compounds. As you can see, the retention factors as well as the relative retention factors within the same batch of packing could be reproduced to better than 0.3%. Some of the relative retention values had relative standard deviations under 0.07%. The batch-to-batch reproducibility values were between 0.85% and 1.93% for the retention factors, and 0.14% and 1.33% for the relative retention values. It can be seen that the reproducibility from batch to batch was quite excellent, exceeding the reproducibility typically achievable in the routine laboratory, even for difficult combinations of sample pairs, such as a strong base and a neutral compound.

Neue et al.[34] compared the results achieved in the quality control department at Waters for three columns: Symmetry C_{18}, a high-purity silica from the mid-1990s, Nova-Pak C_{18}, a spherical silica from the early 1980s, and μBondapak C_{18}, a packing introduced in 1973. Selected data are shown in Table 3. In this table, not only are the reproducibility data of the three types of packing materials shown, but also the difference between the time prior to 1985 and the time in the late 1990s. One can see an improvement in the reproducibility from the older packing materials to the newer packings. Also, for the older packings, one can see a difference between the results obtained prior to 1985 and the newer results. This improvement was due to a concerted effort by the manufacturer in the mid- and late 1980s to improve the control

TABLE 2 Kele and Guiochon Test Result for Symmetry C_{18} Using the Waters Test

1. Reproducibility from column to column using the same batch of packing

RSDs of the retention factors

Propranolol	Butylparabene	Dipropyl-phthalate	Naphthalene	Acenaphthene	Amitriptyline
0.160%	0.282%	0.252%	0.244%	0.269%	0.228%

RSDs of the retention relative to acenaphthene

Propranolol	Butylparabene	Dipropylphthalate	Naphthalene	Amitriptyline
0.239%	0.053%	0.068%	0.040%	0.296%

2. Reproducibility from batch to batch of symmetry C_{18} packing

RSDs of the retention factors

Propranolol	Butylparabene	Dipropyl-phthalate	Naphthalene	Acenaphthene	Amitriptyline
1.330%	0.828%	1.043%	1.146%	1.247%	1.857%

RSDs of the retention relative to acenaphthene

Propranolol	Butylparabene	Dipropylphthalate	Naphthalene	Amitriptyline
0.847%	0.870%	0.439%	0.136%	1.325%

TABLE 3

	α Base/acenaphthene[1] (%)		α Toluene/acenaphthene[1] (%)	
	< 1986	1991–1998	<1986	1991–1998
μBondapak C_{18}	13	7.3	1.9	1.4
Nova-Pak C_{18}	7	3.3	1.5	1.3
Symmetry C_{18}		2.2		0.4

[1]The base is doxepin for μBondapak C_{18} and Nova-Pak C_{18} and amitriptyline for Symmetry C_{18}.

over the manufacturing process with the goal of providing a more reproducible product to the pharmaceutical customer.

F. Column Selectivity in Reversed-phase HPLC

The adjustment of the selectivity of a separation has always been of significant interest for the chromatographer. Much can be done with the manipulation of the mobile phase. Replacing methanol with acetonitrile or THF results in many selectivity changes. For ionizable compounds, the manipulation of pH is important (see above). These selectivity effects are familiar to most chromatographers, but the selectivity effects caused by reversed-phase columns are less known and less understood.

Several investigators have studied column selectivity. Early studies were published by Engelhardt and Jungheim[35] and later by Tanaka and co-workers.[36] The Tanaka procedures were then also employed by Euerby and Petersson[37] to compare a larger set of commercially available columns. We used our test designed for the measurement of batch-to-batch reproducibility[34] to compare the properties of different commercially available columns.[38] The interpretation was later refined,[39] and forms the basis for the assessment of the properties of commercially available reversed-phase columns described here. Snyder and co-workers[40–45] established a new procedure by first testing a large set of analytes with the target of sorting out relevant parameters that govern the interaction between analytes and columns. From this data set, he established a column test in a single mobile phase and at two pH values with a small set of representative compounds. This test then measures the hydrophobicity, steric selectivity, hydrogen-bond donor and hydrogen-bond acceptor properties, and ion-exchange properties at acidic and neutral pH. At the time of this writing, over 150 commercially available reversed-phase columns have been tested with this method, and their characteristics have been established.

Unfortunately, the different column characterization methods correlate only superficially with each other.[39] The reason for these difficulties is not completely understood yet, but there are indications that they are primarily related to a stronger influence of the mobile phase than had been anticipated by the investigators (see the brief discussion on the subject in Reference 42). It now appears that many selectivity effects are the product of a combination of the mobile phase solvent and the stationary phase rather than the stationary phase alone. Other influences are the specific properties of analytes that are ignored in some tests.[39] Rutan has published a critical comparison of methods for the characterization of reversed-phase selectivity.[46] The article covers linear solvation energy relationships, the strengths and weaknesses of principle component analysis, a discussion of the key factor method, some selected results of the "test solute method" (a term used for our method or the Engelhardt/Jungheim test), and Snyder's iterative subtraction method. Some correlations between the different methods were found that make sense, but other associations were rather unexpected. The interested reader is encouraged to read this publication.

Despite these fundamental difficulties, the character of columns can still be described in a general way. Any of the methods can tell which packings are more hydrophobically retentive and can discriminate between high and low silanol activity. However, the relative position of closely related packings may change depending on the method, and we should be careful not to overstress such details. For example, silanol activity is a function of the pH of the mobile phase, and this function is related to, among other things, metal impurities in the matrix of the silica.[47,6,7]

In the stationary phase test discussed here, we use naphthalene and acenaphthene as our hydrophobic reference compounds, propranolol and amitriptyline as the compounds with basic functional groups, and dipropylphthalate and butylparaben for the determination of other polar selectivities of a packing. The mobile phase is a pH 7.00 phosphate buffer (35.0%) mixed with methanol (65.0%). The details of the test procedure are discussed in References 34, 38 and 39.

The hydrophobicity of a packing is best described by the retention factor of a neutral hydrophobic compound, such as naphthalene or acenaphthene. More sophisticated methods are possible, but they are of little concern for the practitioner. The natural logarithms of the retention factors of both compounds correlate linearly for all 91 commercial packings tested.[39] Thus we can use either one of these compounds as reference compounds for the hydrophobicity of a packing. We chose acenaphthene as our hydrophobic reference compound.

For the silanol interaction S, we found that the retention of amitriptyline corrected for the hydrophobic contribution to its retention is most suitable:[39]

$$S = \ln (k_{\text{amitriptyline}}) - 0.7124 \times \ln (k_{\text{acenaphthene}}) + 1.9748 \tag{3}$$

We can now plot the silanol activity of packings vs. their hydrophobic retention and obtain a *selectivity chart* (Figure 17). Both axes are on a logarithmic scale. The x-axis covers a 400-fold difference between the largest and the smallest value, and the y-axis accommodates a 90-fold range. Packings with closely related properties are located close to each other, and packings far away from each other are clearly different. In Figure 17a, the different types of packing materials are coded to give an overview of the different types of packings. In Figure 17b, all data points are numbered for identification of the different columns. The scales of Figure 17b are identical to Figure 17a. Figure 17c is an expansion of Figure 17b, and Figure 17d gives us a closer view of the indicated section of Figure 17c.

In Figure 17a, C_8 and C_{18} packings are marked by squares. The C_{18} packings are the black squares, and the C_8 packings are the gray squares. Open circles identify phenyl packings, and open diamonds signalize cyano packings. The stars are the tags for the fluorinated packings. All packings with an embedded polar group are identified by an open triangle, independent on the chain length or the nature of the bonded group. Packings that either do not fit into these categories or where the bonded phase type is not clear are shown with small closed diamonds.

Let us now look at the different categories. On the left side, we find packings with a low hydrophobicity, such as cyano packings. On the

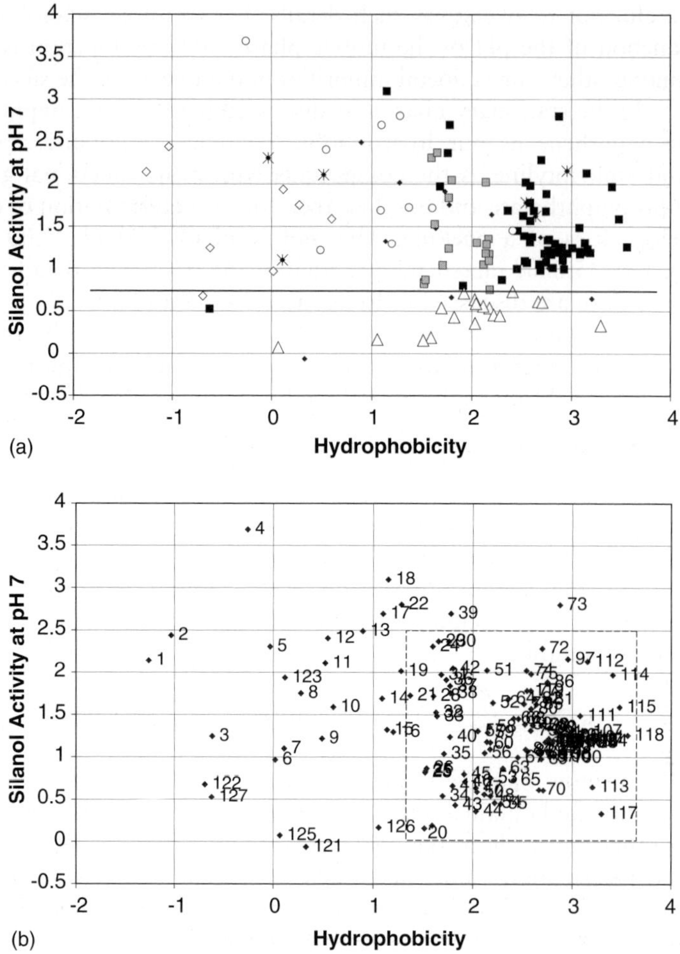

FIGURE 17 **Selectivity chart. Figure 17a is a summary chart. Figure 17c is an expansion of Figure 17b, and Figure 17d is an expansion of Figure 17c, as indicated. Designations in (a): black squares: C_{18}, gray squares: C_8, circles: phenyl, diamonds: cyano, stars: fluorinated packings, triangles: packings with embedded polar groups, small back diamonds: uncategorized packings. Designation in (b)–(d): 1 Nova-Pak CN HP, 2 Waters Spherisorb CN RP, 3 Hypersil CPS CN, 4 Waters Spherisorb Phenyl, 5 Keystone Fluofix 120N, 6 YMC-Pack CN, 7 Ultra PFP, 8 Zorbax SB-CN, 9 Hypersil BDS Phenyl, 10 Inertsil 3 CN, 11 Fluophase RP, 12 Hypersil Phenyl, 13 Zorbax SB-Aq, 14 YMC-Pack Ph, 15 YMC Basic, 16 Ultra Phenyl, 17 Inertsil Ph3, 18 Platinum EPS C_{18}, 19 Synergi Polar-RP, 20 XTerra RP_8, 21 Nova-Pak Phenyl, 22 Zorbax SB-Phenyl, 24 Zorbax Rx C_8, 25 XTerra MS C_8, 26 Prodigy C_8, 28 Zorbax Eclipse XBD Phenyl, 29 Zorbax SB C_8, 30 μBondapak C_{18}, 31 YMC J'Sphere L80, 32 Supelcosil LC DB-C_8, 33 ZirChrom PBD, 34 Discovery RP Amide C_{16}, 35 Hypersil BDS C_8, 36 HydroBond AQ, 37 Lichrospher Select B, 38 Allure Ultra IBD, 39 Platinum C_{18}, 40 Nova-Pak C_8, 41 Capcell Pak C_{18}, 42 Alltima C_8, 43 Discovery RP Amide C_{16}, 44 XTerra RP_{18}, 45 Symmetry300 C_{18}, 46 Spectrum, 47 Zorbax Bonus RP, 48 Supelcosil LC-ABZ Plus, 50 SymmetryShield RP_8, 51 Lichrosorb Select B, 52 PolyEncap A, 53 Prism, 54 Supelcosil LC-ABZ+, 55 Supelcosil LC-ABZ, 56 Luna C_8(2), 57 Inertsil C_8, 58 Kromasil C_8, 59**

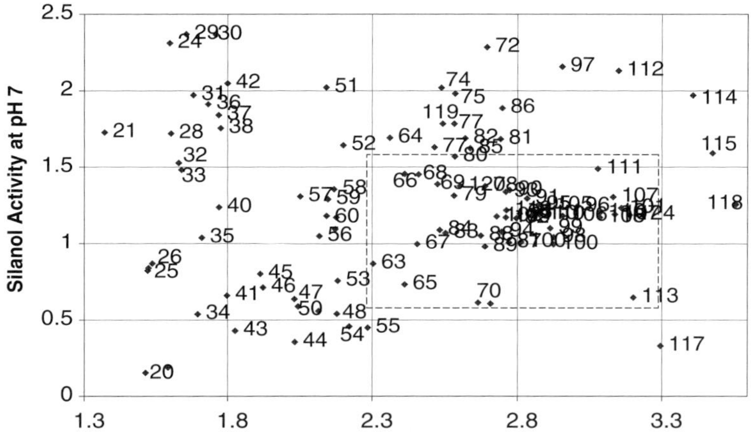

(c)

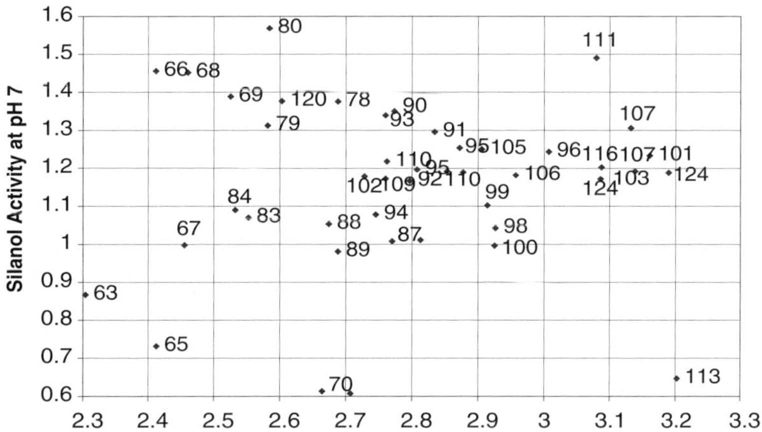

(d)

Zorbax Eclipse XDB C_8, 60 Symmetry C_8, 63 Hypersil HyPurity Elite C_{18}, 64 Hypersil ODS, 65 Polaris C_{18}-A, 66 Luna Phenyl-Hexyl, 67 Hypersil BDS C_{18}, 68 Supelcosil LC DB-C_{18}, 69 Aqua C_{18}, 70 SymmetryShield RP_{18}, 72 Nucleosil C_{18}, 73 Waters Spherisorb ODS-2, 74 Waters Spherisorb ODSB, 75 YMC J'Sphere M80, 77 Zorbax SB-C_{18}, 78 Synergi Max RP, 79 YMC Hydrosphere C_{18}, 80 Nova-Pak C_{18}, 81 PolyEncap C_{18}, 82 TSK-Gel 80Ts, 83 Ace C_{18}, 84 XTerra MS C_{18}, 85 Fluophase PFP, 86 Purospher RP_{18}, 87 Develosil C30 UG 5, 88 Develosil ODS UG 5, 89 Hypersil Elite C_{18}, 90 Zorbax Rx C_{18}, 91 Zorbax Eclipse XDB C_{18}, 92 L-Column ODS, 93 YMC ODS AQ, 94 Prodigy C_{18}, 95 Luna C_{18}(2), 96 Kromasil C_{18}, 97 Allure PFP Propyl, 98 Discovery HS C_{18}, 99 Inertsil ODS-2, 100 Symmetry C_{18}, 101 L-column ODS, 102 Puresil C_{18}, 103 Cadenza CD-C_{18}, 105 Luna C_{18}, 106 Zorbax Extend C_{18}, 107 Inertsil ODS-3, 109 Zorbax Eclipse XDB C_{18}, 110 YMC Pack Pro C_{18}, 111 Purospher RP_{18}e, 112 Alltima C_{18}, 113 ODPerfect, 114 YMC J'Sphere H80, 115 Develosil ODS SR 5, 116 Nucleodur Gravity C_{18}, 117 Inertsil ODS-EP, 118 YMC-Pack Pro C_{18} RS, 119 Discovery HS F5, 120 Atlantis dC_{18}, 121 Discovery HS PEG, 122 Discovery Cyano, 123 Luna CN, 124 Cadenza CD-C_{18}, 125 experimental carbamate CN packing, 126 experimental carbamate phenyl packing, 127 Imtakt Presto FT C_{18}, 128 Aquasil C_{18}.

right, one finds hydrophobic C_{18} packings. On the upper part of the chart, packings with a high silanol activity are located, and the lower part is populated with packings with a low silanol activity. It is interesting to note that all packings with a low silanol activity are packings with an embedded polar group or other unusual ligands. We have therefore drawn a horizontal line between the standard packings and the packings with an embedded polar group.

With this orientation, we can now explore Figure 17 in more detail. In total, this chart contains 117 commercially available packings, and two experimental packings. The selection consists of columns from all over the world, but the emphasis has been on packings available in the US. Modern packings based on high-purity silica have been the focus of this collection, but older ones are included as well. In a few instances, several columns of the same type have been tested to check the reproducibility of a packing. There are a few packings that we have tested, but have not included in this chart, because the silanol activity was too high. Examples of this were Resolve C_{18}, Zorbax ODS, Aquasil C_{18}, or the YMC C_{30} packing, which is used for carotenoid analysis.

Let us now look at individual packings in Figure 17b, which is drawn to the same scale as Figure 17a. An example of a rather polar packing is Nova-Pak CN HP, #1. It has an average silanol activity. A cyano packing with a very low silanol activity is the experimental packing #125 which contains an embedded carbamate group. Another packing with a low silanol activity and low retention is the pegylated packing #127. Packings with a higher hydrophobicity but a high silanol activity are Waters Spherisorb Phenyl, #4, or Waters Spherisorb ODS-2, #73. The most hydrophobic packings on this chart are Develosil ODS SR5, #115, YMC J'Sphere H80, #114 and YMC-Pack Pro C_{18} RS, #118. Packings with a low silanol activity are XTerra RP_8, #20, Discovery RPAmide C_{16}, #43, XTerra RP_{18}, #44, SymmetryShield RP_{18}, #70 and Inertsil ODS EP, #117, all packings with embedded polar groups. Packing #113 is OD Perfect, a non-silica C_{18} packing. Packing #126 is an experimental phenyl packing with incorporated carbamate group.

Figure 17c shows the section of Figure 17b that covers most C_8 and C_{18} packings, with Zorbax Rx C_8, #24, Zorbax SB C_8, #29 and µBondapak C_{18}, #30, in the upper left-hand corner. Figure 13d is a further expansion with Luna Phenylhexyl, #66, Supelcosil LC DB C_{18}, # 68 and Aqua C_{18}, #69, in the upper left-hand corner, Hypersil Hypurity Elite, #63, on the left, and Shodex ODPerfect, #113, on the lower right.

If we are interested in packings with about the same hydrophobicity, we should select packings that are located on a vertical line in this chart. Let us say that we would like to find packings that result in an overall hydrophobic retention similar to Spherisorb ODS-2, #73. We could select, in sequence of reduced silanol activity, the following packings: Nucleosil C_{18}, #72, Purospher RP_{18}, #86, Polyencap C_{18}, #81, Develosil ODS UG5,

#88, Hypersil Elite C_{18}, #89 or SymmetryShield RP_{18}, #70. However, if secondary interactions play a role, different selectivities will be found. This is especially true for packings with an embedded polar group.

Let us assume that our favorite packing is the Discovery HS C_{18}, #98. We would like to find a packing that promises to be most similar from the standpoint of hydrophobic retention and silanol activity. We would look at this chart and find that Inertsil ODS-2, #99, or Symmetry C_{18}, #100, are located close to this packing. Thus the probability is high that these three packings will deliver very similar separations. However, we must understand that this is not guaranteed, since other influences on selectivity, such as hydrogen-bond interaction or steric selectivity, have been ignored. But since we have covered the two dominant interactions on reversed-phase packings, chances are high that the separations are rather alike.

On the other hand, if we are working on a separation that is nearly complete, but requires a bit of fine-tuning that cannot be accomplished with mobile phase manipulations, we would select a packing that is somewhat dissimilar but not too different. For example, if our primary separation was done on a Purospher RP18e column, #111, we could select a Hypersil Hypurity Elite, #63, a Polaris C_{18}-A, # 65 or a SymmetryShield RP_{18} column, #70, as alternatives. These alternatives would change the selectivity of the separation somewhat, but not too drastically.

The hydrogen-bond donor and acceptor properties of a packing are another dimension of column selectivity. They are independent of the silanol ion-exchange activity described above, and affect the selectivity of a packing in a separate way. We have measured these properties by the relative retention between butylparaben, a hydrogen-bond donor, and dipropylphthalate. The same property, or at least a very similar property, has been called hydrogen bond basicity by Snyder.[40] In our case, it has been defined as follows:[39]

$$P = \ln\left(k_{\text{butylparaben}}\right) - 0.8962 \times \ln\left(k_{\text{dipropylphthalate}}\right) \tag{4}$$

In Figure 18 this property is plotted vs. the hydrophobicity of a packing as defined by the logarithm of the retention factor of acenaphthene, as in Figure 17. A horizontal line can be drawn which divides the columns into a group of classical C_8 and C_{18} packings below the line, and special packings above the line.[24] At the upper end of the chart, one finds the polyethylene-glycol-based packing Discovery HS PEG, #121, or Fluofix 120N, #5, an unendcapped fluorinated packing. Far above the line are packings with an embedded polar group, such as SymmetryShield RP_{18}, #70, with an embedded carbamate group, or Discovery RP Amide C_{16}, #34, with an amide group, or Spectrum, #46, and Prism, #53, with urea groups. Closer to the line are packings with special endcapping procedures such as Purospher RP_{18}, #86, with its amine endcapping or packings with special

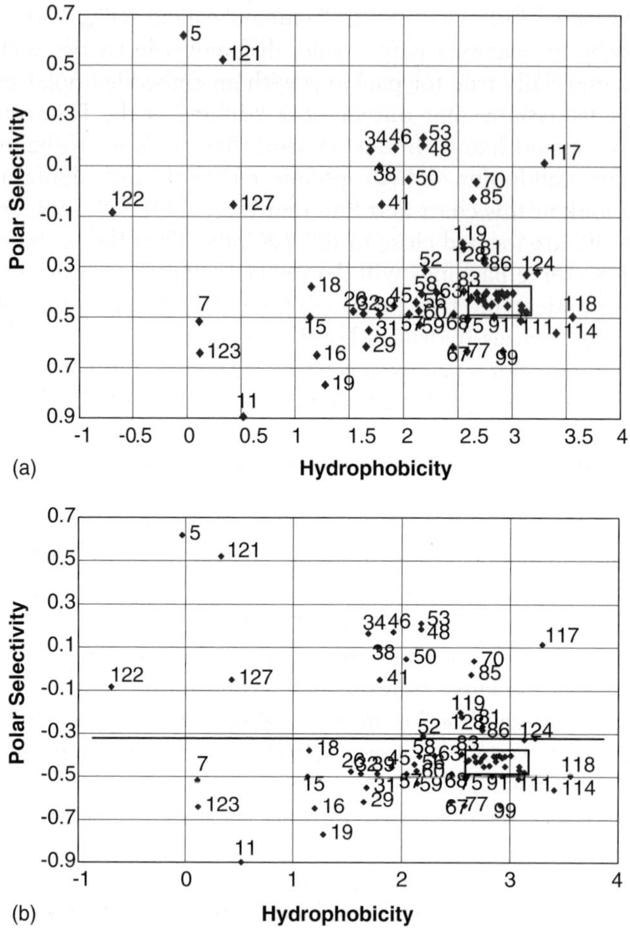

FIGURE 18 **Plot of polar selectivity *P* vs. the logarithm of the retention factor of acenaphthene. Column designations as in Figure 17. The cluster in the rectangle contains the packings 78, 82, 88, 89, 91, 94, 95, 96, 98, 97, 100, 102, 103, 105, 106, 107, 110, 116 and 120.**

preparation techniques such as Poly Encap C_{18}, #81 and Polyencap A, #52. The box contains a cluster of standard C_{18} packings. At the bottom of the chart one finds another fluorinated packing Fluophase RP, #11, and Synergi Polar-RP, #19, with its ether linkage.

This hydrogen bonding capacity expressed by the polar selectivity is not limited to an increase in retention for phenols. Packings with embedded polar groups that excel in this feature also exhibit increased retention for analytes with other functional groups. Specifically, carboxylic acids at acidic pH and with acetonitrile as the organic modifier and compounds with sulfonamide functions at acidic and neutral pH exhibit significant enhanced retention on these packings. Due to the suppression of silanol effects, the retention of compounds with amine functions is decreased on these packings as well. This phenomenon is more pronounced in methanol than in acetonitrile. In general, larger and more specific selectivity effects are found when comparing a classical column to a packing with an embedded polar group relative to a comparison between two classical packings. The specific knowledge outlined here is helpful, if one needs to develop a specific separation for compounds carrying the functional groups mentioned. However, more subtle effects are found for other analytes as well. Generally, the largest differences in selectivity can be found when comparing packings with an embedded polar group to classical packings.

How can we use this information to our advantage? Once again, the largest impact of the difference between a classical phase and a phase with an embedded polar group is on method development.[19] For example, one can do a simultaneous screening of the separation problem on both types of phases early on in the method development cycle, potentially with a simultaneous manipulation of the pH of the mobile phase. Then one can decide which of the combinations of mobile phase and stationary phase give the best chance for the development of an optimized separation. Or, one can attempt to develop a method on one stationary phase, and then select the alternative type of phase, if the method development stalls. In this case, the substitution of a classical phase with a phase with an embedded polar group (or vice versa) is highly likely to change the selectivity sufficiently to provide a new beginning.

IV. SUMMARY

In these few paragraphs, we have given an overview of the choices in column configuration and column chemistry. In the section on column configuration, we emphasized that column performance should be looked at from the standpoint of the ratio of column length to particle size. The overall performance characteristics of a column remain constant when this ratio is kept constant: the same maximum resolving power and the same shortest analysis time can be achieved. We also touched on the subjects of elevated temperature, column diameter, high-speed analysis, and monolithic columns. In the section on column chemistry, we described silica-based stationary phases in great detail, including new developments such as water-wettable stationary phases and packings with embedded polar groups. We gave an overview of commonly used surface chemistries of packings used in reversed-phase applications. Separate paragraphs were dedicated to hybrid packings and zirconia-based packings. Because the long-term reproducibility of an assay is of utmost importance in the pharmaceutical industry, we reviewed the reproducibility of stationary phases briefly. Since many modern chemical entities are rather polar compounds, we briefly discussed hydrophilic interaction chromatography as a suitable technique for these compounds. Finally, we presented a synopsis of what is currently understood about column selectivity in reversed-phase HPLC. Here, we emphasized the advantages of modern packings based on high-purity silicas and with embedded polar groups.

REFERENCES

1. Neue, U. D., *HPLC-Columns—Theory, Technology, and Practice,* Wiley-VCH, New York, 1997.
2. Snyder, L. R., Kirkland, J. J. and Glajch, J. L., *Practical HPLC Method Development,* Wiley, New York, 1997.
3. Meyer, V. R., *Practical High-Performance Liquid Chromatography,* Wiley, New York, 1999.
4. Poole, C. F. and Poole, S. K., *Chromatography Today,* Elsevier, Amsterdam, 1991.
5. Poole, C. F., *The Essence of Chromatography,* Elsevier, Amsterdam, 2003.
6. Doyle, C. A. and Dorsey, J. G., Reversed-Phase HPLC: Preparation and Characterization of Reversed-Phase Stationary Phases, In *Handbook of HPLC,* Katz, E., Eksteen, R., Schoenmakers, P. and Miller, N., Eds., Marcel Dekker, New York, pp. 293–323, 1998.
7. Neue, U. D., Silica Gel and its Derivatization for Liquid Chromatography, In *Encyclopedia of Analytical Chemistry,* Meyers, R. A., Ed., John Wiley & Sons, Ltd., Chichester, pp. 11450–11472, 2000.
8. Giddings, J. C., *Unified Separation Science,* John Wiley & Sons, Inc., New York, 1991.
9. Neue, U. D. and Mazzeo, J. R., A Theoretical study of the Optimization of Gradients at Elevated Temperature. *J. Sep. Sci.,* 24:921–929, 2001.
10. Neue, U. D., Carmody, J. L., Cheng, Y.-F., Lu, Z., Phoebe, C. H. and Wheat, T. E., Design of Rapid Gradient Methods for the Analysis of Combinatorial Chemistry Libraries and the Preparation of Pure Compounds, In *Advances in Chromatography,* Grushka, E. and Brown, P., Eds., Vol. 41, Marcel Dekker, New York, Basel, pp. 93–136, 2001.
11. Cheng, Y.-F., Lu, Z. and Neue, U. D., Ultrafast LV/UV and LC/MS Analysis, *Rapid Commun. Mass Spectrom.* 15:141–151, 2001.
12. Tiller, P. R., Romanyshyn, L. A. and Neue, U. D., Fast LC/MS in the Analysis of Small Molecules, *Anal. Bioanal. Chem.,* 377:788–802, 2003.
13. Dong, M. W. and Gant, J. R., Short Three-Micron Columns: Applications in High-Speed Liquid Chromatography, *LC-GC,* 2(4):294–303, 1984.
14. Dong, M., The Time is Now For Fast LC, *Today's Chemist Work,* 9(2):46–51, 2000.
15. Cabrera, K., Lubda, D., Eggenweiler, H.-M., Minakuchi, H. and Nakanishi, K., A New Monolithic-type HPLC Column for Fast Separations, *J. High Resolut. Chromatogr.,* 23:93–99, 2000.
16. Tanaka, N., Kobayashi, H., Nakanishi, K., Minakuchi, H. and Ishizuka, N., Monolithic LC Columns, *Anal. Chem.,* 73:420A–429A, 2001.
17. Kele, M. and Guiochon, G. Repeatability and Reproducibility of Retention Data and Band Profiles on Six Batches of Monolithic Columns, *J. Chromatogr.,* 960:19–49, 2002.
18. Kirkland, J. J., Glajch, J. L. and Farlee, R. D., Synthesis and Characterization of Highly Stable Bonded Phases for HPLC Column Packings, *Anal. Chem.,* 61:2–11, 1989.
19. Neue, U. D., Grumbach, E. S., Mazzeo, J. R., Tran, K. and Wagrowski-Diehl, D. M., Method Development in Reversed-phase Chromatography, In *Handbook of Analytical Separations,* Vol. 4, Wilson, I. D., Ed., Elsevier Science, Amsterdam, pp. 185–214, 2003.

20. O'Gara, J. E., Walsh, D. P., Alden, B. A., Casellini, P. and Walter, T. H., Systematic Study of the Chromatographic Behavior vs Alkyl Chain Length for HPLC Phases Containing and Embedded Carbamate Group, *Anal. Chem.,* 71:2992–2997, 1999.
21. Wirth, M. J., Fairbank, R. W. P. and Fatunmbi, H. O., Science, 275:44–47, 1997.
22. Wyndham, K. D., O'Gara, J. E., Walter, T. H., Glose, K. H., Lawrence, N. L., Alden, B. A., Izzo, G. S., Hudalla, C. J. and Iraneta, P. C. Characterization and Evaluation of C_{18} HPLC Stationary Phases Based on Ethyl-Bridged Hybrid Organic/Inorganic Particles, *Anal. Chem.* 75(24):6781–6788, 2003.
23. Méndez, A., Bosch, E., Rosés, M. and Neue, U. D., Comparison of the acidity of residual silanol groups in several liquid chromatography columns. *J. Chromatogr. A* 986:33–44, 2003.
24. Neue, U. D., Cheng, Y.-F., Lu, Z., Alden, B. A., Iraneta, P. C., Phoebe, C. H. and Tran, K., Properties of Reversed-Phase Packings with an Embedded Polar Group, *Chromatographia,* 54:169–177, 2001.
25. Dunlap, C. J., Carr, P. W., McNeff, C. V. and Stoll, D., Zirconia Stationary Phases for Extreme Separations. *Anal. Chem.,* 73:598A–607A, 2001.
26. Neue, U. D. and Murphy, B. P., unpublished results, 1979.
27. Meyer, E. and Engelhardt, H., *Fresenius Z. Anal. Chem.,* 333:734, 1989.
28. Kele, M. and Guiochon, G., Repeatability and Reproducibility of Retention Data and Band Profiles on Reversed-phase Liquid Chromatography Columns. I. Experimental Protocol. *J. Chromatogr.* A, 830:41–54, 1999.
29. Kele, M. and Guiochon, G., Repeatability and Reproducibility of Retention Data and Band Profiles on Reversed-phase Liquid Chromatography Columns. II. Results Obtained with Symmetry C_{18} Columns, *J. Chromatogr. A,* 830:55–79, 1999.
30. Kele, M. and Guiochon, G., Repeatability and Reproducibility of Retention Data and Band Profiles on Reversed-phase Liquid Chromatography Columns. III. Results Obtained with Kromasil C_{18} Columns, *J. Chromatogr. A,* 855:423–453, 1999.
31. Kele, M. and Guiochon, G., Repeatability and Reproducibility of Retention Data and Band Profiles on Reversed-phase Liquid Chromatography Columns. IV. Results Obtained with Luna C18 (2) Columns, *J. Chromatogr. A,* 869:181–209, 2000.
32. Kele, M. and Guiochon, G. Repeatability and Reproducibility of Retention Data and Band Profiles on Reversed-phase Liquid Chromatography Columns. V. Results Obtained with Vydac 218TP C18 Columns, *J. Chromatogr. A,* 913:89–112, 2001.
33. Felinger, A., Kele, M. and Guiochon, G., Identification of the factors that influence the reproducibility of chromatographic retention data. *J. Chromatogr. A,* 913:23–48, 2001.
34. Neue, U. D., Serowik, E., Iraneta, P., Alden, B. A. and Walter, T. H., A Universal Procedure for the Assessment of the Reproducibility and the Classification of Silica-Based Reversed-Phase Packings; 1. Assessment of the Reproducibility of Reversed-Phase Packings. *J. Chromatogr. A,* 849:87–100, 1999.
35. Engelhardt, H. and Jungheim, M., Comparison and Characterization of Reversed Phases, *Chromatographia,* 29:59–68, 1985.
36. Kazuhiro, K., Tanaka, N., Iwaguchi, K., Onishi, S., Jinno, K., Eksteen, R., Hosoya, K. and Araki, M., Chromatographic Characterization of Silica C_{18} Packing Materials. Correlation between a Preparation Method and Retention Behavior of Stationary Phase. *J. Chromatogr. Sci.,* 27:721–728, 1989.
37. Euerby, M. R. and Petersson, P., Chromatographic Classification and Comparison of Commercially Available Reversed-phase Liquid Chromatographic Columns Using Principal Component Analysis, *J. Chromatogr. A,* 994:13–36, 2003.
38. Neue, U. D., Alden, B. A. and Walter, T. H., A Universal Procedure for the Assessment of the Reproducibility and the Classification of Silica-Based Reversed-Phase Packings; 2. Classification of Reversed-Phase Packings. *J. Chromatogr. A,* 849:101–116, 1999.
39. Neue, U. D., Tran, K., Iraneta, P. C. and Alden, B. A., Characterization of HPLC Packings, *J. Sep. Sci.,* 26:174–186, 2003.

40. Wilson, N. S., Nelson, M. D., Dolan, J. W., Snyder, L. R., Wolcott, R. G. and Carr, P. W., Column Selectivity in Reversed-phase Liquid Chromatography: I. A General Quantitative Relationship, *J. Chromatogr. A,* 961:171–193, 2002.
41. Wilson, N. S., Nelson, M. D., Dolan, J. W., Snyder, L. R. and Carr, P. W., Column Selectivity in Reversed-phase Liquid Chromatography: II. Effect of a Change in Conditions, *J. Chromatogr. A,* 961:195–215, 2002.
42. Wilson, N. S., Nelson, M. D., Dolan, J. W., Snyder, L. R., Carr, P. W. and Sander, L. C., Column Selectivity in Reversed-phase Liquid Chromatography: III. The Physico-Chemical Basis of Selectivity, *J. Chromatogr. A,* 961:217–236, 2002.
43. Gilroy, J. J., Dolan, J. W. and Snyder, L. R., Column Selectivity in Reversed-phase Chromatography: IV. Type-B Alkyl-silica Columns, *J. Chromatogr. A,* 1000:757–778, 2003.
44. Gilroy, J. J., Dolan, J. W., Carr, P. W. and Snyder, L. R., Column Selectivity in Reversed-phase Liquid Chromatography: V. Higher Metal Content (type-A) Alkyl-silica Columns, *J. Chromatogr. A,* 1026:77–89, 2004.
45. Wilson, N. S., Gilroy, J. J., Dolan, J. W. and Snyder, L. R., Column Selectivity in Reversed-phase Liquid Chromatography: VI. Columns with Embedded or End-capped Polar Groups, *J. Chromatogr. A,* 1026:91–100, 2004.
46. Lopez, L. A. and Rutan, S. R., Comparison of Methods for Characterization of Reversed-phase Liquid Chromatograpic Selectivity, *J. Chromatogr. A,* 965:301–314, 2002.
47. Nawrocki, J., The Silanol Group and its Role in Liquid Chromatography, *J. Chromatogr. A,* 779:30–71, 1997.

5 SAMPLE PREPARATION FOR HPLC ANALYSIS OF DRUG PRODUCTS

CLARK K. CHOI AND MICHAEL W. DONG

Purdue Pharma, Ardsley, NY

ABSTRACT

This chapter provides the novice and the experienced analyst with an overview of sample preparation techniques focusing on solid dosage forms. It describes the "best practices" in the "dilute and shoot" approach, and the "tricks of the trade" in grinding, mixing, sonication, dilution and filtration of drug products. Selected case studies of sample preparations for assays and impurity testing are used to illustrate the strategies, trade-offs

Handbook of Pharmaceutical Analysis by HPLC,
S. Ahuja and M.W. Dong, editors.

and potential pitfalls encountered during method development. Future trends in sample preparation are also discussed.

I. INTRODUCTION

A. Scope

Sample preparation (SP) is generally not the focus of attention in discussing pharmaceutical analytical methods even though its proper execution is paramount in achieving accurate quantitation. Non-robust SP procedures, poor techniques or incomplete extraction are the major causes of out-of-trend and out-of-specification results. This chapter is written for the beginner as well as the experienced analyst. It reviews common SP techniques with a strong focus on tablets or capsules. Detailed descriptions of SP methods for assays and impurity testing are illustrated with selected case studies of single and multicomponent products.

B. Background

SP refers to a family of solid/liquid handling techniques to extract or to enrich analytes from sample matrices into an analyzable format, namely, the final analyte solution. While SP techniques are well documented,[1–5] few publications address the specific requirements for drug product preparations, most of which tend to employ the simple "dilute and shoot" approach. A more elaborate SP is often needed for complex sample matrices (e.g., lotions and creams). Many newer SP technologies such as solid-phase extraction (SPE),[6–8] supercritical fluid extraction (SFE),[9–10] pressurized fluid extraction or accelerated solvent extraction (ASE)[11–13] and robotics[14–15] are topics of numerous research papers, symposia and commercial promotion. However, for reasons discussed later, these newer developments have had little impact on the way pharmaceutical laboratories conduct their SP for drug products today.

C. Characteristics of Typical SP of Solid Dosage Forms

The term "dilute and shoot" is quite literal for the analysis of drug substances. However, for drug products, a more elaborate process of "grind → extract → dilute → filter", shown in Figure 1, is generally employed to extract the active pharmaceutical ingredient (API) from the solid matrices (Tablets) of inactive excipients.[4,12] As high accuracy and precision (<2–3% relative standard deviation or RSD) are mandatory in assays for drug products, the volumetric flask (i.e., 25–500 mL) is used as the main platform to attain very precise and sufficient volumes for solubilizing the API in the tablet(s). This approach is straightforward for

FIGURE 1 The common sequence of sample preparation steps for drug products: grinding → extraction → dilution → filtration.

immediate-release dosage forms but can be challenging for controlled-release products and formulations of low-solubility APIs. Further sample clean-up or fractionation is typically not needed and is to be avoided to maintain a high level of assay simplicity and accuracy.

D. Ease of SP of Various Dosage Forms

In general, the SP of drug substances, parenteral products and even some syrups[16] is easy, whereas tablets or capsules require some method development and optimization. Dosage forms such as suppositories, lotions or creams containing a preponderance of hydrophobic matrices might require more elaborate SP and sample clean-up such as SPE or liquid–liquid extraction.[12] Other difficult samples are the over-the-counter (OTC) products containing multiple APIs or natural products. Sample preparation for other non-dosage sample types varies from easy (for dissolution media and cleaning validation samples) to complex (for fermentation broths and biological fluids). The output of pharmaceutical SP is typically a high-pressure liquid chromatography (HPLC) vial containing the final analyte solution ready for HPLC analysis.

II. SAMPLE PREPARATION OVERVIEW

This section provides a brief but systematic overview of the common SP techniques, organized into three major categories: (A) Solid-handling techniques; (B) Extraction techniques; (C) Dilution, isolation, enrichment and derivatization techniques. For further reading, analysts are referred to a more detailed discussion of these techniques in various textbooks and key references.[1–5] Common techniques for drug products are covered in more depth through discussions of "best practices," extraction strategies, device selection and potential pitfalls. These techniques include grinding, extraction (i.e., mixing, sonication, vortexing, etc.), pipetting and filtration.

A. Solid-handling Techniques

Solid-handling techniques are typically physical methods, such as grinding and milling, that reduce solid dosage forms into small particles (i.e., fine powders) to facilitate extraction.[4] Grinding multiple tablets (composite) into a homogeneous pulverized form (typically 10–20 tablets for assay and >5 tablets for impurity testing) also provides a more representative sample for the batch.

1. Intact Formulations

Many immediate-release tablets or disintegrating dosages can be dissolved directly as intact entities; however, most formulations require some grinding or rough crushing to ensure complete extraction in a timely fashion.

2. Manual Grinding and Rough Crushing

Grinding or crushing is performed manually with a mortar and pestle made from ceramic, glass, agate, diamond or other inert materials. The composite grind can be transferred quantitatively into a large volumetric flask for dilution. Alternately, a representative portion equal to the average tablet weight (ATW) can be transferred into a small volumetric flask for extraction. If only rough crushing is needed, one or more tablets can be wrapped in a piece of weighing paper and crushed with a pestle or a screwdriver handle. This simple approach facilitates recovery of the powders and is useful for impurity testing and for the rapid screening of extraction solvents during method development.

3. Milling

Tablets can be ground rapidly by a variety of laboratory mills or blenders as shown in Figure 2. Care should be exercised to avoid sample degradation from heat generated during prolonged grinding. Soft, semisolid and very labile samples can be ground inside a freezer mill in liquid nitrogen temperatures, which renders the samples brittle. Freezer milling is particularly useful for the determination of residual solvents or volatile components to prevent analyte loss during sample processing.

B. Extraction Techniques

Since most APIs have some aqueous solubility, extraction is typically performed at room temperature with aqueous or a mixture of aqueous and organic media.[12,17] Extraction can be performed in one step with a single solvent or in multiple steps executed sequentially with different solvents. The extraction solvent must be able to solubilize the API and be compatible with the HPLC mobile phase for the final analysis. The pH, buffer concentration and organic solvent composition of the extraction

Laval Lab ball mill | BelArt micro mill | Spex 6750 freezer mill

FIGURE 2 **Common devices for laboratory milling.**

media are evaluated for extraction efficiency and for signs of sample degradation. Parameters such as extraction time and volume (occasionally extraction temperature) should be optimized to ensure quantitative recovery in 30 min or less if possible. Since many APIs are acids or bases, knowledge of their pK_a and their pH-solubility profiles is important for selecting the optimum extraction medium. Multi-step extractions are common with controlled-release products to ensure the initial dissolution of the polymer matrice. For example, controlled-release tablets/capsules can be sonicated with an organic solvent such as methanol to dissolve or soften the polymer matrice followed by subsequent dilution with an aqueous buffer for further extraction.

1. Dissolution, Shaking, Vortexing, and Sonication

Dissolution of the API in solid dosages is expedited by agitation of whole tablets or ground powder by mixing, inversion, shaking, vortexing, or in most cases by sonication using various devices as shown in Figure 3. Mixing is performed manually by inverting the volumetric flask in an up-and-down fashion. More vigorous mixing is accomplished by vortexing the flask using a flat-top vortex. Prolonged mixing is performed by shaker tables or wrist-action shakers equipped with a timer. An adequate amount of solvent (typically 50% of the container volume) is added into the volumetric flask to ensure complete mixing of the contents and extraction of the API in solution when using the shakers.

Sonication using ultrasonic cleaner baths remains a popular extraction approach particularly for controlled-release products. In sonication, an ultrasonic wave of 20–40 kHz generated by a piezoelectric transducer is used to produce the formation and collapse of thousands of microscopic bubbles (cavitations) in the water bath to facilitate the break up of the solid particles and the subsequent dissolution of the API. Note that parameters such as the wattage power of the sonicator, presence of the perforated tray, depth of the water level, bath temperature and the number of sample flasks sonicated might all affect the extraction rate.[4,18] For

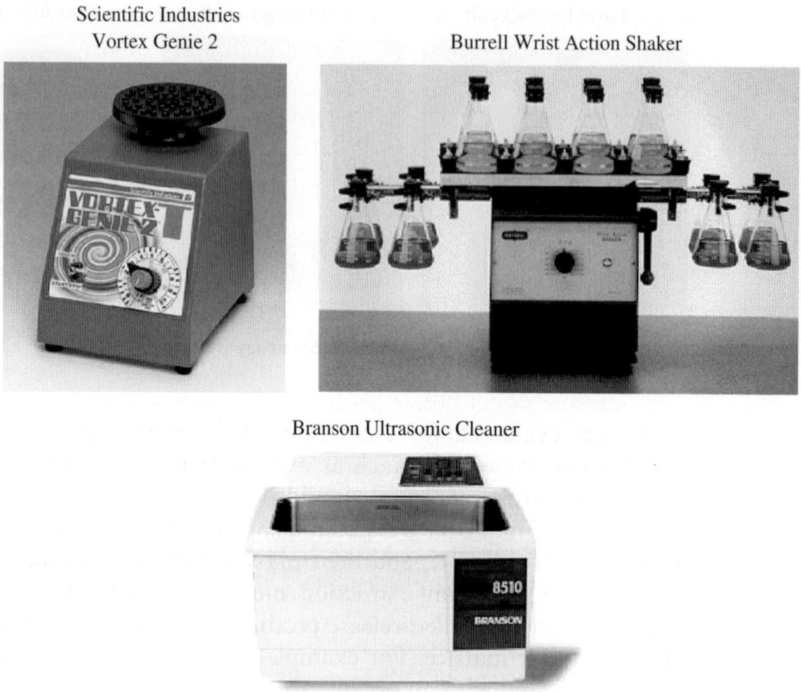

FIGURE 3 Common devices for mixing and extraction: Scientific Industries (Vortex Genie 2), Burrell wrist-action shaker and Branson 8510 sonicator.

laboratory extraction, the bath is typically filled with only 0.5–1 inch of water to maximize extraction efficiency. The bath is rarely filled to the "operating level" indicated by the manufacturer as intended for the "normal operation" of these baths such as for jewelry-cleaning purposes. The extraction efficiency can increase dramatically if the perforated tray is not placed inside the ultrasonic bath; under these conditions, the lifetime of the transducer might be reduced. Thus, the extraction time and flask volume are usually optimized empirically with the specific sonicator during method development. In some cases, sonication might generate sufficient heat to cause API oxidation and/or degradation and to produce artifact peaks in impurity testing.

2. Homogenizing and Liquid–liquid Extraction

Homogenizers, equipped with a set of motorized blades, combine wet grinding/shredding/shearing of the sample matrice together with extraction by swirling/agitation of the sample solution. They are popular in Karl Fischer analyzers, robotic systems and tissue homogenizers.

Liquid–liquid extraction[12] with two immiscible liquids is useful for sample clean-up of complex samples such as creams, suppositories, fermentation broths or natural product formulations. It takes advantage of the relative solubilities of the solutes in immiscible solvents as they concentrate in the solvent with the highest solubility, commonly governed by their partition or distribution coefficients. Liquid–liquid extraction is commonly performed in separatory funnels, though Erlenmeyers, capped test tubes or even HPLC vials may be substituted. For quantitative recovery of the analytes, multiple extractions (2–3 times) are often needed unless the partition coefficient is unusually large. For acidic or basic analytes, a sequence of "extraction → pH adjustment → back extraction" offers an effective generic approach for sample clean-up. For instance, a complex sample containing acidic analytes of interest can be extracted under basic pH with methylene chloride to eliminate extraneous hydrophobic materials. The acidic analytes (ionized) remain in the basic aqueous layer. After discarding the organic layer, the pH of the aqueous phase is adjusted to acidic followed by back extraction with methylene chloride. Since the acidic analytes are now protonated (unionized), they preferentially partition into the organic layer leaving any hydrophilic interferences in the aqueous layer. The methylene chloride layers are pooled and can be further concentrated or evaporated before analysis. Liquid–liquid extraction is labor-intensive and is commonly used for samples such as suppositories,[19] creams[20] or fermentation broths.[21]

3. Other Extraction Techniques (SPE, SFE, ASE and Refluxing)

SPE using packed sorbents in disposable syringe barrels was developed in the 1970s. SPE in 96-well plates, designed and popularized in the 1990s, is popular for the clean-up of bioanalytical fluids, combinatorial libraries or proteomics samples.[7] For quantitative bioanalytical analysis, internal standards such as deuterated analogs of the API or metabolite are often used to compensate for recovery loss. SPE is rarely used for tablets because filtration is typically adequate for the most common single-component drug products. Other SP techniques popular in environmental and food analysis such as refluxing,[11,17] Soxhlet extraction,[22,23] SFE[9,10,24] and ASE[11–13,25] are more amenable to non-polar and non-labile compounds. They are seldom used in drug product assays due to concerns with solubility, thermal degradation, recovery or process precision.

C. Dilution, Isolation, Enrichments, and Derivatization Techniques

1. Pipetting and Liquid handling

Pipetting is used for the accurate sampling of liquid samples and for sample dilution. Pipetting is performed manually using a variety of glass or disposable pipettes, auto-pipettes with disposable tips, electronic dispensing systems or a myriad of sophisticated automated liquid-handling systems (see Figure 4). Class-A volumetric pipettes have an accuracy greater than 0.1% (i.e., $25 \pm 0.03\,\text{mL}$) and do not require any calibration.

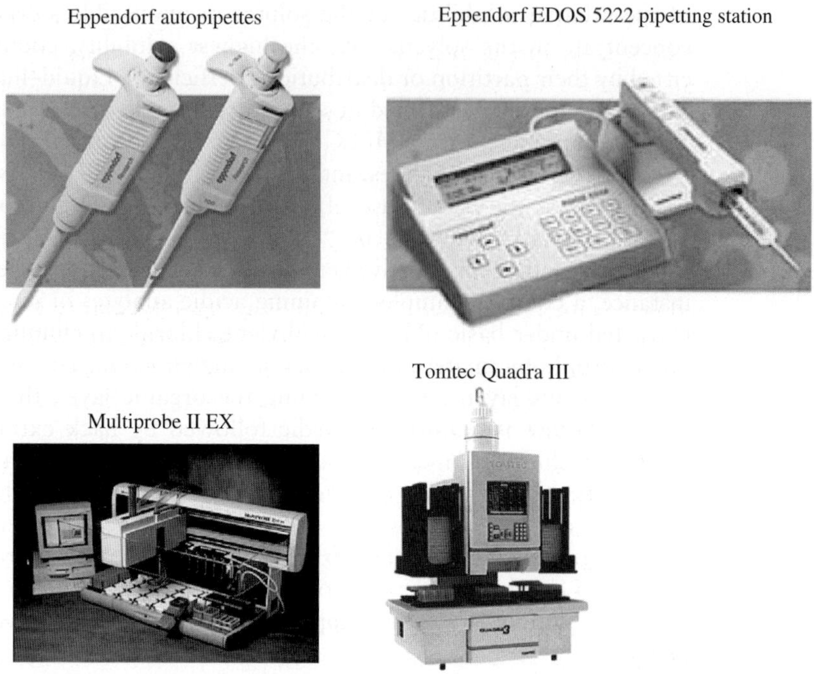

FIGURE 4 Pipetting devices: Eppendorf pipettes, Eppendorf EDOS 5222 electronic dispensing system, PerkinElmer Packard MultiProbe II EX and Tomtec Quadra 3 96-well plate liquid handling system.

Other auto-pipettes need periodic calibration for Good Manufacturing Practice (GMP) operation.

There are three kinds of pipettes: (A) TD (to deliver) pipettes, which deliver the final volume before the tip (no rings marked on the top of the pipette); (B) TD pipettes, which deliver the final volume to the tip and are designed to be blown out (marked with two double rings at the top of the pipette); (C) TC (to contain) pipettes, which deliver the final volume by touching the tip of the pipette to the side of the beaker or flask and leaving drop of solution in the pipette. TC pipettes are rare as opposed to TD pipettes. Volume is measured at the bottom of the meniscus.

Sample dilution is used to reduce the analyte concentration or the solvent strength of the extraction media to eliminate potential chromatographic anomalies (i.e., peak distortion, splitting and broadening) with the injection of a solvent stronger than the mobile phase. Sample dilution is performed by pipetting an aliquot of the concentrated solution from one volumetric flask into a second volumetric flask and filling the second flask to volume. Alternately, a two-step extraction–dilution scheme can be devised by partially filling the flask with one solvent, performing the extraction and finally filling the flask to volume with a second solvent.

Pipetting is also the first step of sampling liquid dosages in parenterals, syrups, emulsions or suspensions. In sampling viscous syrups, adequate time must be allowed to let the sample layer adhering to the pipette sidewall to drain out completely. This might take up to 2–3 min. Alternately, an auto-pipette using tips with a positive displacement inner piston such as Eppendorf Combi-Tips can be used. Sampling liquid dispersions and suspensions can be particularly challenging. Some shaking is required to ensure a homogenous sample but vigorous shaking might introduce too many air bubbles into the sample. A practical approach is to determine the density of the sample using a pycnometer that can vent the air bubbles from the top and to sample an accurate volume by measuring the weight of the sample in an analytical balance.

2. Filtration and Centrifugation

Filtration is the final step in most SP to separate the final analyte solution from the solid particles, which might clog the HPLC system. In most cases, the tablet extracts are filtered through syringe filters directly into the HPLC sample vial. Syringe membrane filters are made in various pore sizes (0.2 or 0.45 μm), media and dimensions (4, 13 or 25 mm). The most popular are the 25-mm nylon (NYL), polyvinylidene fluoride (PVDF) and polytetrafluoroethylene (PTFE) 0.45-μm filters available from Whatman, Millipore, Pall, Gelman and other suppliers. Other membrane media and their applications are listed in Table 1. The filter membrane media must be compatible with the extraction solvent. The casing of the outer cartridge is typically made from high-density polypropylene with low extractibles. Filter validation studies are conducted to ensure quantitative recovery of the APIs and related substances

TABLE 1 Common Filter Media Available Commercially from Suppliers

Types of filter media[1]	Sample preparation applications
Cellulose acetate (CA)	Aqueous and some organic samples
Glass microfiber filter (GMF)	Aqueous and/or organic samples; high loading capacity
Nylon (NYL)	Aqueous and/or organic samples esp. hydrophilic; resistant to chemicals and HPLC solvents
Polypropylene (PP)	Aqueous and organic samples; high flow throughput
Polysulfone (PSU)	Aqueous samples esp. biological sera and fluids; low protein binding
Polytetrafluoroethylene (PTFE)	Organic samples esp. hydrophobic
Polyvinylidene fluoride (PVDF)	Aqueous and/or organic samples; resistant to common HPLC solvents

[1]Adapted from http://www.whatman.com

from the recommended filters specified in the analytical method. The first several milliliters of the filtrate are usually discarded. Clogged filters are a common problem with tablet extracts due to a preponderance of insoluble matrice materials. This problem can be minimized by using membranes with prefilters such as Whatman GD/X filters (see Figure 5). Small sample volumes (<2 mL) might benefit from smaller diameter 13- or 4-mm filters. An innovative design is the Whatman UniPrep HPLC vial which incorporates a built-in plunger filter. Centrifugation is an alternate technique for difficult-to-filter samples and for filtration evaluation studies to ensure non-adsorption of the analytes by the filter.

3. Evaporation and Freeze-drying

Sample concentration is sometimes needed to increase the analyte concentration or to eliminate the extraction solvents, which might be incompatible with the HPLC mobile phase. Evaporation is carried out in open hoods, ovens, rotary evaporators, Kuderna-Danish evaporators,[26] freeze driers or lyophilizers.

4. Sample Clean-up

Sample clean-up or fractionation is required for complex samples to enrich the analytes and eliminate interferences. This is accomplished by, but not limited to, a number of techniques such as liquid–liquid extraction,

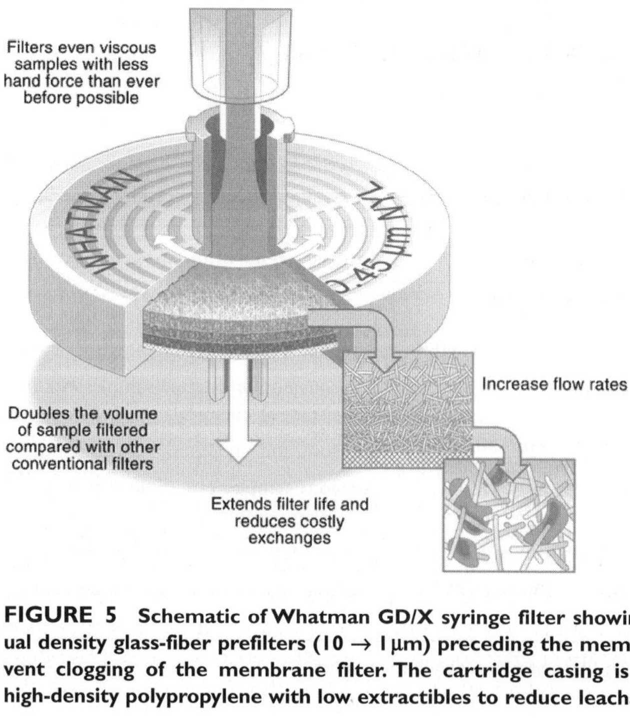

FIGURE 5 **Schematic of Whatman GD/X syringe filter showing the use of gradual density glass-fiber prefilters (10 → 1 µm) preceding the membrane filter to prevent clogging of the membrane filter. The cartridge casing is constructed from high-density polypropylene with low extractibles to reduce leachable components.**

SPE, chromatography (i.e., thin-layer chromatography, gel-permeation chromatography and HPLC) and on-line multi-dimensional chromatography. These techniques are rarely used in routine tablet assays due to their technical difficulties, labor intensity and problems with recovery.

5. Derivatization

Chemical derivatization is used to convert the non-UV absorbing analytes into forms that are easily chromatographed or detected with high sensitivity.[27] Due to the problematic nature of this process (incomplete reaction and formation of multiple derivatives) and the additional complexity of the methodology leading to low recovery and poor chromatographic reproducibility, derivatization is often the last resort. In pharmaceutical analysis, derivatization is commonly used if the API does not have a chromophore for UV detection or a fluorophore for fluorescence detection. Common reagents associated with derivatization involve alkyl bromides, acid chlorides, *N*-hydroxysuccinimidyl esters and chlorides, alkylchloroformates and trimethylsilyl esters, which react to the functional group(s) of the non-UV absorbing analyte.

Automation in derivatization has led to the development of pre-column and post-column HPLC techniques. In post-column derivatization, the separated components eluting from the column are derivatized on-line in a heated mixing coil prior to entering the detector. Pre-column derivatization can be performed in many autosamplers or automated liquid handlers.

III. PRIMARY SP CONCERNS FOR TYPES OF ANALYSIS

A. Potency Assay

The assay of potency involves the determination of the API to ensure conformance to label claim. For tablets or capsules, a composite assay requires 10–20 units to minimize tablet-to-tablet variation. In assays, it is critical for the quantitative extraction and recovery of the API meet the typical specification limit between 90.0% and 110.0% of label claim.

B. Content Uniformity

The content uniformity test is an assay of individual tablets/capsules to ensure processing consistency. Generally, 10 single-tablet assays are required to achieve statistical significance. Content uniformity is mandatory for all film-coated tablets, transdermal systems, suspensions in single-unit containers or in soft capsules, metered-dose inhalers, sterile and non-sterile solids that are packaged in unit-dose containers and suppositories. On the other hand, the test of weight variation can be substituted for content uniformity if the product tested contains >50 mg of an API and comprises >50% by weight of the dosage form unit. It is easier to carry out weight variation measurement than to assay individual dosage units. The criteria proposed in USP 26-NF 21 of <905> Uniformity of Dosage Units, for example, mandate that each of the dosage units must lie between 85.0% and 115.0% of the label claim and that the RSD of 10 dosage units be less than or equal to 6.0% for the product to pass specification. As in the assay, the quantitative extraction and recovery of the API is the most critical parameter in SP considerations.

C. Impurity Analysis

The measuring of process impurities is important to assure drug product purity and stability. The Food and Drug Administration (FDA) has published updated guidelines that outline the need for stability testing on human drugs and biologics.[28] In addition, the International Conference on Harmonisation (ICH) has recently revised its guidelines for the validation of analytical methods to measure and to report impurities and degradants in new chemical entities.[29,30] For SP, precision performance is less important for the determination of trace degradants than for assays. Rather, the near complete extraction of the parent drug and impurities, and the elimination of contaminants and artifact peaks in the SP procedure is more critical since any unknown peaks above the ICH guideline of ~0.1% might require reporting, identification or qualification. This is typically evaluated by running a procedural blank and minimized by using clean glassware, high-purity solvents and accessories with low extractibles (i.e., filters). Method sensitivity (limit of quantitation or LOQ <0.03%) and elimination of artifact or contaminant peaks are the most important concerns of the SP methodology for impurity testing.[31,32]

IV. SAMPLE PREPARATION TRENDS IN PHARMACEUTICAL ANALYSIS

As stated earlier, the SP trend in the pharmaceutical analysis of solid dosage forms is dominated by manual batch techniques such as grinding, sonication, dilution in volumetric flask and filtration. The unique SP requirements for solid dosages in grinding and handling large volumetrics are particularly difficult to automate.

The application of robotics to enhance laboratory productivity has met with some success with many early adapters and in laboratories with high-volume repetitive routine assays. However, the requirements for extensive installation, development, validation and continuing maintenance effort have slowed the growth of robotics in pharmaceutical laboratories. Many robots in pharmaceutical R&D labs have been abandoned or "mothballed" because they were not sufficiently flexible or reliable at automating the varying applications to warrant the continuing upkeep. The ideal dream of full automation where one simply inputs samples (tablets) and receives viable data with close-to absolute reliability without frequent troubleshooting has generally not been realized by laboratory robotics.[5] However, the use of a variety of semi-automated equipment (e.g., shakers and sampling/dilution dispensers) to minimize error and reduce the most labor-intensive work is quite popular. There is probably a strong market for more universal automation strategies and more reliable technologies for the oral tablet and other formulations.

Other newer techniques such as SPE, SFE and ASE, widely successful in bioanalytical, environment or food laboratories, are not readily amenable for tablet assays. The miniaturization trend of using smaller sample size and lower sample volumes is also not practical since the minimum unit is a single tablet, which requires at least 10–100 mL of solvent for total dissolution. The 96-well plate SPE format, which revolutionized SP in bioanalytical and high throughput applications, is not compatible for drug products due to this requirement for higher extraction volumes. The requirement for extraction with aqueous solvents for most oral tablets precludes the general application of SFE. The high precision extraction requirements (typically <2–3% RSD) also render the successful application of ASE more difficult. Future trends in SP of drug products actually might indicate its elimination by using in-process (at-line) non-destructive analytical technology such as near infrared spectroscopy (NIR) for assays or the minimization of SP for impurity testing using more sensitive and specific analytical techniques such as LC/MS/MS.

V. CASE STUDIES

A. Assay of a Single API Component Product

1. Optimization Considerations for SP

This case study illustrates the SP method development of an assay method for a controlled-release analgesic tablet with a single API. Certain considerations were taken into account. First, the analyte within the tablet matrice core had to be extracted quantitatively. Second, the analyte was diluted into a final solution that was compatible with the HPLC mobile phase. Third, short SP time was required (i.e., 30 min) to maximize productivity of the work scheme for processing a large number of samples.

2. Sample Preparation

The drug substance was a water-soluble base in salt form with a pK_a of about 9 that was also soluble in most organic solvents. The polymer in the tablet was, however, soluble only in an organic medium like methanol. Other organic solvents such as acetonitrile and dimethylformamide were not chosen because they were less compatible with the HPLC mobile phase in which the analyte is stable. Crushing or milling could speed up the solubilization process but would require additional SP steps. Table 2 summarizes the results of an extraction study by comparing the extraction efficiency of using methanol to dissolve intact and crushed tablets by sonication and shaking. While crushed tablets required less extraction time and a more gentle process such as shaking, a sonication time of 30 min for the intact tablets was selected because of its simplicity and quantitative extraction. Figure 6 shows the comparative HPLC chromatogram of the tablet extracts using whole and crushed tablets by sonication, and crushed tablets by shaking. The final two-step extraction method was developed as follows:

1. Transfer 10 tablets into a 500-mL volumetric flask and add approximately 200 mL of methanol.
2. Sonicate the tablet stock solution in ultrasonic bath (>320 W) for 30 min at ambient temperature or until all tablets are solubilized.
3. Cool the solution to room temperature and dilute to volume with methanol. Mix the stock solution by flask inversion.
4. Pipette exactly 10 mL of this sample stock and 10 mL of internal standard solution into a 200-mL volumetric flask and dilute to volume with mobile phase.
5. Mix thoroughly the working sample solution via flask inversion and filter aliquots of the solution through a Whatman 25-mm GD/X 0.45-μm PP membrane syringe filter. Discard the first few milliliters of the filtrate and then transfer the subsequent filtrate into a capped HPLC vial for analysis. Inject 20 μL of the filtrate into the HPLC.

TABLE 2 Composite Assay Data of Controlled-release Tablets Using Various Solid-handling and Extraction Techniques

	Solid-handling techniques		Extraction techniques			
No.	**Intact**	**Crushed**	**Sonication**	**Shaking**	**Extraction time (min)**	**% Label claim**
1	X		X		5	87.0
2	X		X		10	89.8
3	X		X		20	97.6
4	X		X		30	99.2
5		X		X	5	98.1
6		X		X	10	98.7
7		X		X	20	99.0
8		X		X	30	98.8
9		X	X		5	98.8

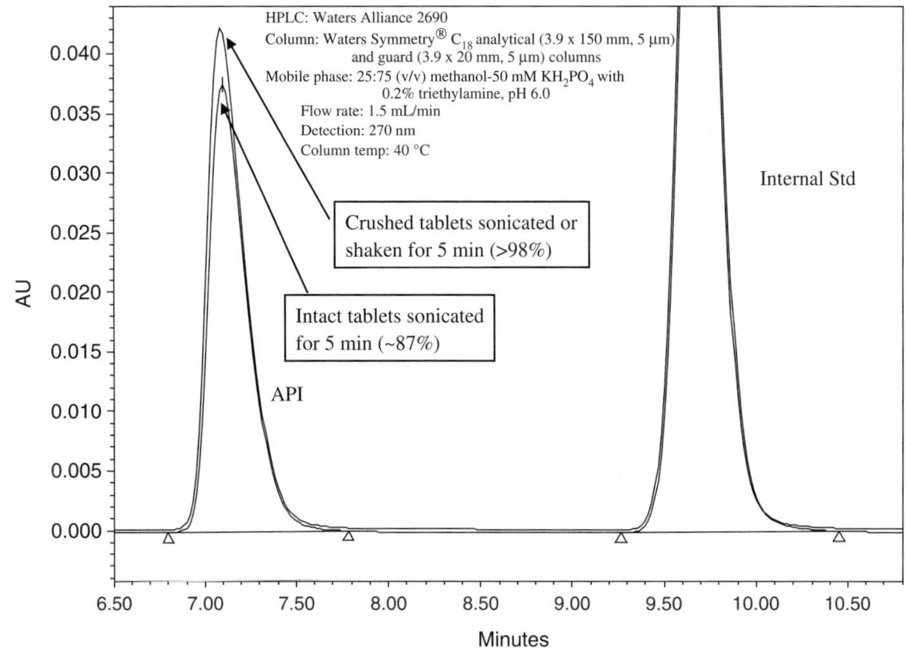

FIGURE 6 HPLC chromatogram of an extract of a controlled-release tablet with one API showing the extraction efficiency between whole tablets by sonication and crushed tablets by shaking and sonication.

TABLE 3 Composite Assay Data of Controlled-release Tablets by Various Filter Types from Different Suppliers

Filter suppliers and types	% Label claim
Whatman PP GD/X[1]	99.0
Whatman NYL GD/X	100.0
Whatman PVDF GD/X	98.8
Whatman PTFE GD/X	98.8
Whatman GMF GD/X	98.8
Whatman PP (no GD/X)	98.9
Millipore PVDF Millex-HV[1]	98.9
Pall Glass Acrodisc	98.8
Gelman GHP Acrodisc GF	98.9

[1]The filters used for the filtration procedure are performed as per validated method.

The method yielded >99% label claim of the API for the composite tablet assay. Table 3 lists the results of a filter validation study showing the quantitative recovery of the API from most membrane media.

B. Assay of Multiple Components in an OTC Product

3. Assay of Water-soluble Vitamins (WSVs) in Multivitamin Tablets

The SP procedure of water-soluble vitamins from multivitamin tablets is particularly challenging due to the diverse analytes of varied hydrophobicities and pK_a. Water-soluble vitamins (WSVs) include ascorbic acid (vitamin C), niacin, niacinamide, pyridoxine (vitamin B_6), thiamine (vitamin B_1), folic acid, riboflavin (vitamin B_2) and others. While most WSVs are highly water soluble, riboflavin is quite hydrophobic and insoluble in water. Folic acid is acidic while pyridoxine and thiamine are basic. In addition, ascorbic acid is light sensitive and easily oxidized. The extraction strategy employed was a two-step approach using mixed solvents of different polarity and acidity as follows:[33]

1. Grind one multivitamin tablet with a mortar and pestle. Then transfer the contents into a 125-mL Erlenmeyer flask.
2. Pour exactly 10 mL of 1% ammonia in dimethylsulfoxide (DMSO) into the ground powder and sonicate it in an ultrasonic bath for two minutes. Add exactly 90 mL of 2% acetic acid in water to the mixture. Stir it with a magnetic stir bar for 1 min.
3. Sonicate the mixture at about 40°C for 5 min.
4. Filter the extract while it is still warm, through a 0.45-μm NYL membrane filter into amber vials. Analyze it immediately by injecting 10 μL of the sample solution into the HPLC.

4. Assay of Fat-soluble Vitamins (FSVs) in Multivitamin Tablets

The SP for FSVs A, E and D presented a different set of challenges. The strategy was to use a two-phase liquid–liquid extraction system to isolate the FSV into hexane while retaining other analytes and excipients in the water layer. The platform was a capped centrifuge test tube rather than the separatory funnel to minimize extraction time and solvent volumes. The final hexane solution was injected directly into the HPLC system with a mobile phase consisting of 85% methanol/water. Chromatographic anomalies from injecting a stronger solvent were avoided by reducing the injection volume to 5 μL.[33] The steps were:

1. Grind one multivitamin tablet with mortar and pestle. Transfer the tablet powder into a 15-mL capped test tube.
2. Pour exactly 8 mL of extracting solution (1% citric acid in 80% ethanol/water) followed by 4 mL of hexane into the test tube. Cap the test tube and shake the mixture, then vortex the mixture for 1 min.
3. Sonicate the mixture in an ultrasonic bath (>200 W) at about 40°C for 5 min. Centrifuge the mixture for 1 min to separate layers. Transfer the hexane layer into a 10-mL volumetric flask.

4. With a spatula, loosen the tablet powder at the bottom of the tube. Repeat the hexane extraction at least once to obtain exactly 10 mL of hexane extract. Dilute to volume with hexane if necessary. Analyze immediately by injecting 5 μL of the extract into the HPLC.

Both WSV and FSV methods were rapid, with extraction and analysis completed in an hour. A single tablet was used to minimize sample volume and solvent use. This is more acceptable for OTC products. Further details and data were published elsewhere.[33] Table 4 shows assay data on precision and recovery and Figure 7 shows HPLC chromatograms of WSV in standard solution and in multivitamin extracts.

C. Assay of an OTC Product from a Natural Material

The case study of the SP development of an HPLC assay method for an OTC product derived from natural materials illustrates the importance of optimizing both the extraction and filtration conditions to fully recover all the labile active components in the formulation. The HPLC gradient assay method, which separates all active ingredients from other

TABLE 4 Analysis Data of WSVs and FSVs in a Multivitamin Tablet

	Tablet assay						
WSV	#1 (mg)	#2 (mg)	#3 (mg)	Average (mg)	Label (mg)	Precision (RSD %)	Recovery of Spikes (%)
Vitamin C	62.1	74.0	70.2	68.8	60	8.8	101
Niacin	0	0	0	0	0	0	96
Niacinamide	19.2	18.50	18.1	18.6	20	2.8	93
Pyridoxine	2.52	2.51	2.41	2.48	2.0	2.6	98
Thiamine	1.51	1.45	1.45	1.47	1.5	2.3	95
Folic acid	0.416	0.422	0.416	0.418	0.40	0.8	95
Riboflavin	1.67	1.62	1.70	1.66	1.7	2.5	92
FSV							
Vitamin A acetate	4.60	4.45	4.40	4.48	[1]	2.3	91
Vitamin D2	0.013	0.014	0.014	0.014	0.010	4.2	94
Vitamin E acetate	32.4	31.5	32.8	32.2	30.0	2.0	87

Note: In recovery experiments, the spike levels were 150 mg for vitamin C, 2–20 mg for the other WSVs, 5.0 mg of vitamin A acetate, 0.5 mg of vitamin D_2 and 25 mg of vitamin E acetate.

Table reprinted with permission from Reference 33.

[1]Tablet contains 500 IU of vitamin A as vitamin A acetate and β-carotene.

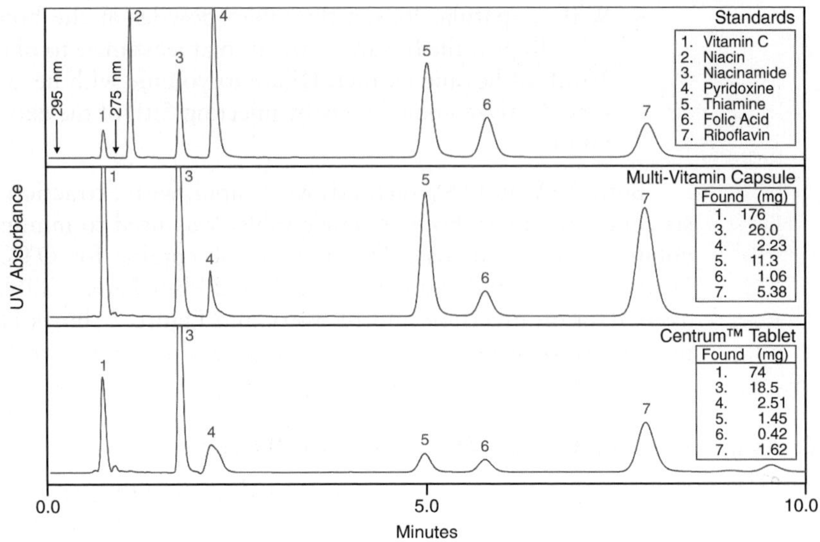

FIGURE 7 HPLC chromatogram of water-soluble vitamin in multivitamin tablets. See Reference 33 for details. Reprinted with permission from Reference 33.

additives and extraneous components, was developed to supplement a standard wet spectrophotometric method to provide more information for process enhancement and stability information.[34] Since the active ingredients are not very water-soluble, a two-step extraction procedure was developed:

1. Grind or mill 20 tablets into a fine powder and transfer the powder equivalent to the average tablet weight into a 100-mL volumetric flask.
2. Add 50 mL of an extraction fluid consisting of 50% acetonitrile/0.1N phosphoric acid into the flask and sonicate for 15 min.
3. Fill to volume with mobile phase A (20 mM ammonium acetate buffer at pH 3.75) and filter an aliquot of the extract through a Whatman GDX PTFE membrane filter into an HPLC vial after discarding the first 2 mL of the filtrate.
4. Inject 5 μL into the HPLC.

Figure 8 shows a typical HPLC chromatogram of the tablet extract illustrating the complexity of the sample, which contains numerous natural components, excipients and additives. During the SP method development, it was found that extraction with a neutral aqueous buffer was problematic due the loss of one of the active ingredients by hydrolysis. The use of the PTFE filter was also important because this hydrolysis product, which is highly hydrophobic, was absorbed by nylon filters under the filtration conditions.

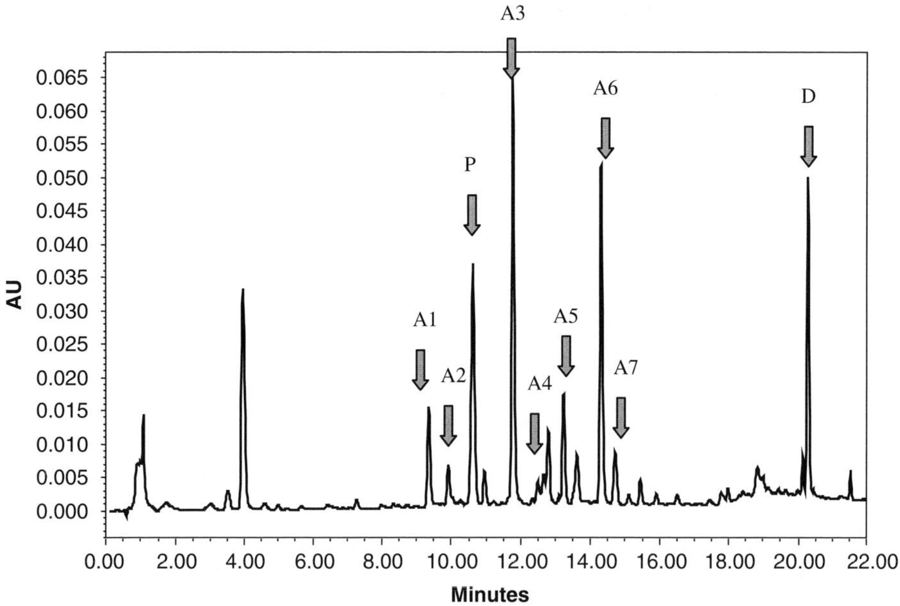

FIGURE 8 **HPLC chromatogram of a tablet extract of an OTC product from natural material showing many natural components including active ingredient (peaks labeled "A"), preservatives "P" and degradants "D". Note that Peak A2 was found to hydrolyzed into "D" under neutral extraction conditions and component "D" was found to be retained by Nylon filters.**

D. Impurity Testing of a Controlled-release Tablet with Two APIs

The case study on the SP development of an impurity method for a controlled-release tablet with two APIs illustrates the importance of finding an extraction condition that fully recovers both parent drugs and their respective impurities and degradants while avoiding any procedural artifacts or contaminants.[32] The two APIs are water-soluble bases with pK_a around 8. Extraction in acidic buffer with sonication of crushed tablets for 30 min yielded quantitative recovery of all analytes but produced an artifact peak, which was later identified to be the dihydroxy derivative of the second API. Sonication at neutral pH lowered the artifact peak and reducing sonication time yielded further reduction of the artifact. Substituting sonication by shaking or vortexing yielded no artifact peak but did yield a lower than acceptable recovery of 80–90% of the parent drugs. In our final extraction method, five tablets were wrapped in weighing paper, crushed, and the powder transferred into a 100-mL volumetric flask. Fifty milliliters of a pH 6.7 buffer were added and the flask was vortexed for 2 min followed by shaking for 30 min in a wrist-action shaker to extract all API from the polymer. The flask was then filled to volume, vortexed for another minute, and filtered with a Whatman 25-mm GD/X 0.45-μm PTFE or NYL

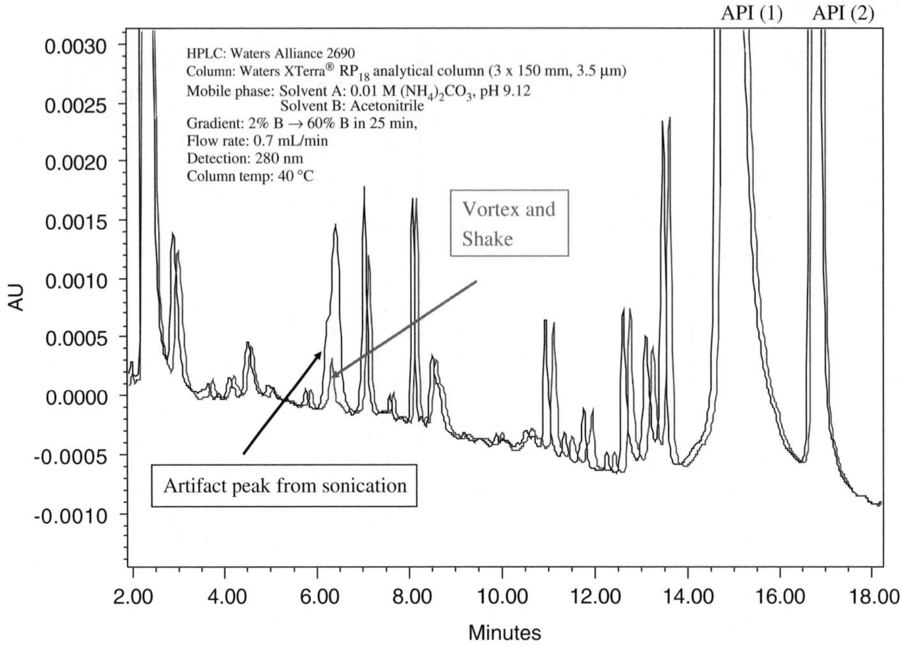

FIGURE 9 HPLC chromatogram of an extract of a controlled-release tablet with two APIs showing the presence of an artifact peak stemming from sonication.

membrane syringe filter into a HPLC vial after discarding the first milliliters of filtrate. The procedure produced no artifact and yielded >97% recovery of both parent drugs. Figure 9 shows the comparative HPLC chromatograms of the tablet extracts using sonication and vortexing.

VI. SUMMARY AND CONCLUSION

This chapter reviews the myriad SP techniques used in pharmaceutical analysis and focuses discussion on those commonly used for pharmaceutical products, such as grinding, mixing, sonication, dilution and filtration. The best practices and technical judgments used in developing SP procedures are illustrated with several case studies of assays and impurity testing.

ACKNOWLEDGMENTS

The authors acknowledge helpful ideas and suggestions from C. Davidson and L. Gehrlein, from Purdue Pharma L.P., A. Wong from Alza Corporation and R. Majors from Agilent Technologies.

REFERENCES

1. Pawliszyn, J., Ed., *Sampling and Sample Preparation for Field and Laboratory: Fundamentals and New Directions in Sample Preparation,* Vol. XXXVII, Elsevier Science, Amsterdam, Netherlands, 2002.
2. Mitra, S., Ed., *Sample Preparation Techniques in Analytical Chemistry,* Vol. 162, John Wiley and Sons, New York, 2001.
3. Lo, T. C., Baird, M. H. I. and Hanson, C., Eds., *Handbook of Solvent Extraction,* Krieger Publishing Company, Melbourne, FL, 1991.
4. Majors, R. E., *LC.GC,* 17:S8–S13, 1999.
5. Majors, R. E., *LC.GC,* 13:742–749, 1995.
6. Poole, C. F., Principles and Practice of Solid-phase Extraction, In *Sampling and Sample Preparation for Field and Laboratory: Fundamentals and New Directions in Sample Preparation,* Pawliszyn, J., Ed., Vol. XXXVII, Elsevier Science, Amsterdam, Netherlands, pp. 341–387, 2002.
7. Wells, D. A., *High Throughput Bioanalytical Sample Preparation: Methods and Automation Strategies,* Elsevier Science, Amsterdam, Netherlands, 2003.
8. Wells, D. A. and Lloyd, T. L., Automation of Sample Preparation for Pharmaceutical and Clinical Analysis, In *Sampling and Sample Preparation for Field and Laboratory: Fundamentals and New Directions in Sample Preparation,* Pawliszyn, J., Ed., Vol. XXXVII, Elsevier Science, Amsterdam, Netherlands, pp. 837–868, 2002.
9. Taylor, L. T., *Supercritical Fluid Extraction: Techniques in Analytical Chemistry,* John Wiley and Sons, New York, 1996.
10. Westwood, S. A., Ed., *Supercritical Fluid Extraction and its Use in Chromatographic Sample Preparation,* CRC Press, Boca Raton, FL, 1993.
11. Dean, J. R. and Cresswell, S. L., Extraction Techniques for Solid Samples, In *Sampling and Sample Preparation for Field and Laboratory: Fundamentals and New Directions in Sample Preparation,* Pawliszyn, J., Ed., Vol. XXXVII, Elsevier Science, Amsterdam, Netherlands, pp. 559–586, 2002.
12. Kataoka, H. and Lord, H. L., Sampling and Sample Preparation for Clinical and Pharmaceutical Analysis. In *Sampling and Sample Preparation for Field and Laboratory: Fundamentals and New Directions in Sample Preparation,* Pawliszyn, J., Ed., Vol. XXXVII, Elsevier Science, Amsterdam, Netherlands, pp. 779–836, 2002.
13. Björklund, E., Järemo, M., Mathiasson, L., Karlsson, L., Strode III, J. T., Eriksson, J. and Torstensson, A., *J. Liq. Chrom. Rel. Technol.,* 21:535–549, 1998.
14. Dong, M. W., *J. Liq. Chromatogr.,* 9:3063–3092, 1986.
15. Shamrock, W. F., Reilly, K. and Lloyd, D. K., *J. Pharm. Biomed. Anal.,* 21:1225–1232, 2000.
16. Signoretti, E. C., D'Arpino, C. and La Torre, F., *J. Chromatogr.,* 473:301–304, 1989.
17. Pawliszyn, J., Unified Theory of Extraction, In *Sampling and Sample Preparation for Field and Laboratory: Fundamentals and New Directions in Sample Preparation,* Pawliszyn, J., Ed., Vol. XXXVII, Elsevier Science, Amsterdam, Netherlands, pp. 253–278, 2002.
18. Berliner III, S., Ultrasonic Cleaning Page, http://home.att.net/~Berliner-Ultrasonics/usclean.html.
19. Priego-Capote, F. and Luque de Castro, M. D., *Anal. Chim. Acta,* 489:223–232, 2003.
20. Kaale, E., Van Schepdael, A., Roets, E. and Hoogmartens, J., *J. Pharm. Biomed. Anal.,* 30:1331–1337, 2002.
21. Jarvinen, M., Myllykoski, L., Keiski, R. and Sohlo, J., *Bioseparation,* 9:163–166, 2000.
22. Luque-Garcia, J. L. and de Castro, L., *J. Chromatogr., A* 998:21–29, 2003.
23. Majors, R. E., *LC.GC,* 19:1120–1130, 2001.
24. Chandrasekar, R., Calbraro, R. and Myer, L. J. D., *J. AOAC Inter.,* 84:466–471, 2001.
25. Accelerated Solvent Extraction, http://www.dionex.com.

26. Luke, M. A., Froberg, J. E., Doose, G. M. and Masumoto, H. T., *J. AOAC Inter.*, 64:1187–1195, 1981.
27. Rosenfeld, J. M., Recent Developments in the Chemistry and Application of Analytical Derivatizations, In *Sampling and Sample Preparation for Field and Laboratory: Fundamentals and New Directions in Sample Preparation,* Pawliszyn, J., Ed., Vol. XXXVII, Elsevier Science, Amsterdam, Netherlands, pp. 609–668, 2002.
28. FDA Guidelines for Center for Drug Evaluation and Research (CDER), http://www.fda.gov/cder.
29. ICH Guidelines for Method Validation: Q2B Validation of Analytical Procedures: Methodology, 1996, http://www.ich.org/pdfICH/Q2B.pdf.
30. ICH Guidelines for Impurity Testing: Q3B(R) Impurities in New Products, 2003, http://www.ich.org/pdfICH/Q3BRStep4.pdf.
31. Dong, M. W., Passalacqua, P. V. and Choudhury, D., *J. Liq. Chromatogr.*, 13:2135–2160, 1990.
32. Dong, M. W., Miller, G. D. and Paul, R. K., *J. Chromatogr., A* 987:283–290, 2003.
33. Dong, M. W. and Pace, J. L., *LC.GC,* 14:794–803, 1996.
34. Lister, A., *Method Equivalency,* presented at PhRMA Meeting at Washington DC, September 2003.

INTERNET RESOURCES

http://www.brinkman.com.
http://www.waters.com.
http://www.chem.agilent.com.
http://www.beckman.com.
http://www.varianinc.com.
http://www.dionex.com.
http://www.gilson.com.
http://www.zymark.com.
http://www.cetac.com.
http://www.whatman.com.
http://www.perkinelmer.com.
http://benchtop-ultrasonics.com.
http://www.sampleprep.com/index.html.
http://www.viscoalpha.com/extractionmenu.html.

6 HPLC METHOD DEVELOPMENT

HENRIK T. RASMUSSEN, WEIYONG LI, DIRK REDLICH, AND M. ILIAS JIMIDAR

Johnson & Johnson Pharmaceutical Research & Development, Raritan, NJ

ABSTRACT

High-pressure liquid chromatography method development of pharmaceuticals is an iterative process required to support successive phases of pharmaceutical development and clinical studies. This chapter details the approach currently in use in our laboratories, from receipt of a new chemical entity to post transfer support.

Handbook of Pharmaceutical Analysis by HPLC,
S. Ahuja and M.W. Dong, editors.

I. INTRODUCTION

High-pressure liquid chromatography (HPLC) is used in analytical development to quantitate the active pharmaceutical ingredient (API) and to evaluate impurity and degradation product profiles of drug substances (DS) and drug products (DP). Additional uses of HPLC include the determination of content uniformity of dosage forms, monitoring of dissolution profiles, determination of antioxidant and microbial preservative content, and support of cleaning validations. By and large, these latter uses are relatively simple applications in that the demands placed on the chromatographic separations are minimal. Separations of these types require only a monitoring of one, or a limited number of, predefined components. A significantly larger challenge is presented in the composite assay of DS and DP where the goal is to quantitate API and relevant impurities and degradation products in a single chromatographic run. This chapter will focus on approaches to composite assays by HPLC and provide guidance as to what method attributes are required during progressive stages of pharmaceutical development.

Key performance requirements for composite assay methods are provided in ICH guidelines Q2A, Q2B, Q3A and Q3B.[1–4] ICH Guidelines are discussed further in Chapter 10. To summarize briefly, and with some generalizations, HPLC methods applicable to the analysis of DS synthesized via the commercial route and final formulations of DP, should meet the following criteria:

For DS: Methods should separate the API, synthetic process impurities, and DS degradation products. Methods should be able to detect impurities and degradation products present at levels greater than 0.05% relative to the API. Impurities and degradation products present at levels greater than 0.1% should be identified and specifications should be placed on limits.

For DP: Methods should separate the API and DP degradation products from excipients. DP methods are not required to monitor synthetic process impurities, unless they are also DP degradation products. Methods should be able to detect degradation products present at levels greater than 0.1% relative to the API. Degradation products present at levels greater than 0.2% should be identified and specifications should be placed on limits.

To validate the suitability of methods, experiments to establish specificity, accuracy, precision, limits of detection, limits of quantitation, linearity, range, and robustness should be conducted. Furthermore, appropriate system suitability criteria should be developed to assure that the method is performing to pre-established criteria at the time of use.

It is unrealistic to envision that a single method can be developed for the determination of API and related substances in both DS and DP during the early phases of pharmaceutical development, and maintain applicability during the later phases of development. During development, there are changes in synthetic routes and DP formulations and as a result the set of analytes that the method is required to separate may change.[5]

II. PHASE-APPROPRIATE METHOD DEVELOPMENT

In the approach described here method development is broken down into discrete stages as shown in Figure 1. The levels described are correlated with clinical and regulatory phases in Table 1 to provide rough guidance as to when in the pharmaceutical development process each level is applicable.

Each level of development will be described in detail in this chapter. In brief, method development starts with preliminary screening experiments

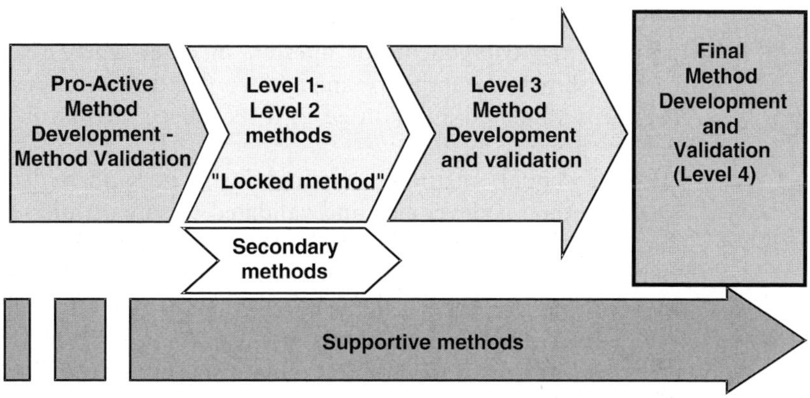

FIGURE 1 **Schematic of method types used to support successive phases of pharmaceutical development.**

TABLE 1 **Correlation Between Level of Method Development and Clinical and Regulatory Phases of Pharmaceutical Development**

Development level	Level 1	Level 2	Level 3/Level 4
Clinical phase	Phase 1	Phase 2	Phase 3
Regulatory phase	CTA/SAD/IND	IND/CTA	CTA/NDA/MAA/PAS

CTA: clinical trial application;
IND: investigational new drug;
MAA: marketing authorization application;
NDA: new drug application;
PAS: post-approval support;
SAD: safety assessment document.

to separate (and identify, as applicable) mixtures of the API and impurities and degradation products generated via preliminary synthetic schemes and forced decomposition, respectively. This step is herein referred to as proactive method development and should result in the generation of numerous orthogonal conditions; i.e., conditions that provide different separation selectivity. In practice this is typically achieved using different HPLC columns with different mobile phases (varying in pH or organic modifier). The methods should ideally be compatible with HPLC-MS and HPLC-NMR (see Chapter 18) and should include broad gradients to assure, to the degree possible, that all components are separated from the solvent front and are eluted from the column.[5]

Following proactive development, a single method is identified and is validated as a level 1–level 2 method to support early phase pharmaceutical development. Of great importance at this juncture, an orthogonal method, identified during proactive development, should also be available, and should be used to evaluate DS generated via new synthetic schemes and to evaluate new formulations. The orthogonal method serves to assure that the primary method continues to be viable for separating all the components of interest. In cases where the orthogonal method shows the primary method to be deficient, the orthogonal method is validated and is used as a secondary method in conjunction with the primary method.

If a secondary method is required during early phase development, a level 3 method is developed and validated to support phase 2b or early phase 3 clinical studies. This method should be capable of separating all components of interest identified up to this stage of pharmaceutical development. In cases where the initial primary method is still viable, the level 1–level 2 method may be maintained. As in early development, the use of an orthogonal method to evaluate DS generated via new synthetic schemes and to evaluate new formulations remains an important means of assuring that the primary method is sufficient for characterizing DS and DP.

The level 3 method is used until synthetic routes and formulations have been finalized and forced degradation and preliminary stability studies have been conducted; i.e., until the components that need to be separated in the final DS and in the final DP have been clearly determined. At this juncture, the focus shifts to the development of fast, robust and transferable final methods to be used for primary stability studies and post-approval analyses. Frequently, separate methods are developed for DS and DP since the goals of each method are different (see Section I). Orthogonal methods continue to be of importance to troubleshoot any questions that may arise during the subsequent life cycle of the drug.

Notably, new methods may be required if there are post-approval changes in either the DS synthesis or in the DP formulation.

III. PROACTIVE METHOD DEVELOPMENT

A. General Considerations for Early Phase Development

During early phase development there is limited knowledge about the chemistry of the new chemical entity (NCE) with respect to synthetic impurities and degradation pathways and kinetics. It is, therefore, desirable to develop an array of methods that show applicability to a broad range of potential impurities, degradation products, and excipients. The methods are intended to provide the information necessary to guide the improvement of a synthesis route or a new drug formulation.

As a starting point to method development, the physicochemical properties of the API such as structure(s), solubility/stability in different solvents, pK_a(s), spectra (UV-, IR-, MS-, and NMR-), and chirality should be determined. It is also extremely valuable to know the targeted potency (if known), the synthetic routes used in production (to postulate potential synthetic impurities), any information regarding intended formulations, and to review any literature documenting the analyses of similar compounds.[5,6] Using examples from a recent review by Bakshi and Singh,[6] it is known that functional groups such as amides, esters, lactams, and lactones undergo hydrolysis, that thiols and thioethers undergo oxidation, and that olefins, aryl halo derivatives, aryl acetic acids, and N-oxides undergo photodegradation. It is also pointed out by these authors that most NCEs are congeners of existing drugs. Therefore degradation behavior may be postulated based on the behavior of similar drugs.

One main purpose in initiating method development is to obtain or generate samples that can be used for method development. Ideally, the samples should collectively contain the API, all significant synthetic impurities and degradation products, and all excipients of interest. The synthetic impurities are typically obtained using early batches of drug substance, along with critical intermediates and starting materials. A general way to obtain samples of degradation products is to place the DS or an early formulation under a variety of stress conditions. Guidance for conducting forced decomposition studies is given in ICH guideline Q1A.[7] In general, the drug is subjected to acidic, neutral, and basic solutions, as well as to high temperatures, light and oxidative conditions (hydrogen peroxide solution). A recent review of approaches to conducting forced decomposition studies[8] shows a large variability in the experimental approach within the pharmaceutical industry. Usually the goal is to degrade the parent drug by 10% to 20%. Further degradation increases the risk of forming secondary degradation products, which may not occur under real conditions. In fact, one shortcoming of many forced decomposition studies is the formation of degradation products that are not observed under normal storage conditions. More definitive studies are typically conducted later in the development process, and are discussed in this chapter in Section IV. C. If candidate excipients are known, these should be included as well in samples to be evaluated for method development.

As the approach detailed above can result in the generation of numerous samples to be screened, it is frequently possible to combine some solutions into a set of "selectivity" solutions for some method development activities. However, there is also merit in analyzing the individual solutions separately to obtain information on degradation pathways and DS impurities that can facilitate the understanding of drug chemistry.

To analyze the samples generated for method development, reversed-phase HPLC is generally the method of choice. (About 75% of current HPLC analyses are performed using reversed-phase columns.)[9] A typical reversed-phase liquid chromatographic (RPLC) system consists of a relatively polar mobile phase and non-polar stationary phase. Water (or a buffer) is used as the weak solvent and acetonitrile, methanol, or, to a lesser extent, tetrahydrofuran, as the strong solvent. The retention and separation of solutes are based on their partition between the two phases. Variations in silica purity, bonding chemistry, and surface chemistry (end-capping) will affect selectivity of the stationary phase. Many types of RPLC columns are available, including classic C8, C18, phenyl, and cyano phases with different forms of silica (base-deactivated, high purity, etc.) as well as polar-embedded group columns.[10] RPLC systems offer a wide range of selectivity. Generally, the traditional C8/C18 phases are less effective in separation of positional isomers than the recently marketed polar-embedded columns. Zirconia-based stationary phases are also becoming popular because of their much wider usable pH range compared with silica.[11,12] Zirconia has an amphoteric nature at extreme pHs and can act as a cation or anion exchanger. This property can sometimes be problematic but at other times it is responsible for unique selectivity.

Additional modes of HPLC include normal phase, where the stationary phase is relatively polar and the mobile phase is relatively non-polar. Silica, diol, cyano, or amino bonded phases are typically used as the stationary phase and hexane (weak solvent) in combination with ethyl acetate, propanol, or butanol (strong solvent) as the mobile phase. The retention and separation of solutes are achieved through adsorption/desorption. Normal phase systems usually show better selectivity for positional isomers and can provide orthogonal selectivity compared with classical RPLC. Hydrophilic interaction chromatography (HILIC), first reported by Alpert in 1990, is potentially another viable approach for developing separations that are orthogonal to RPLC.[13] In the HILIC mode, an aqueous-organic mobile phase is used with a polar stationary phase to provide normal phase retention behavior. Typical stationary phases include silica, diol, or amino phases. Diluted acid or a buffer usually is needed in the mobile phase to control the pH and ensure the reproducibility of retention times. The use of HILIC is currently limited to the separation of very polar small molecules. Examples of applications include pharmaceutical peptides[14] and polar compounds such as pyrimidines, purines, and amides.[15]

In summary, the use of RPLC is ideal for pharmaceutical analyses because of the broad range of commercially available stationary phases; because the most common RPLC mobile phases (buffers with acetonitrile or methanol) have low UV cut-off wavelengths, which facilitate high sensitivity detection for quantitation of low-level impurities; and because selectivity can readily be controlled via mobile phase optimization. Additionally, the samples generated for selectivity screening (as detailed above) are typically aqueous based. In subsequent phases of pharmaceutical development, aqueous-based sample solvents are ideal for sample preparation and are, under limited constraints, compatible with MS detection required to identify impurities and degradation products.

B. Method Screening Experiments

Classical approaches[16] to HPLC method development include approaches based on sequential isocratic steps and gradient elution optimization. For sequential isocratic development, one strategy is to start method development by using 100% acetonitrile as mobile phase to ascertain that all components elute. Subsequently, the percentage of acetonitrile is reduced in 20% increments until the k' range of the solutes is approximately 1–20. If this criterion is met, but the separation of all components is not achieved, methanol/water or THF/water can be evaluated to improve selectivity. An equivalent solvent strength (iso-elutropic series) conversion table (Table 2) can be used to determine the right percentage of the organic modifiers without going through incremental adjustments

TABLE 2 Equal Elution Strength of Mobile Phases for RPLC

% MeOH in water	% ACN in water	% THF in water
0	0	0
10	6	4
20	14	10
30	22	17
40	32	23
50	40	30
60	50	37
70	60	45
80	73	53
90	86	63
100	100	72

MeOH = methanol, ACN = acetonitrile, THF = tetrahydrofuran.

of mobile phase. For example, if 60% acetonitrile will allow all components in a sample to elute within a suitable retention time window, using 70% methanol or 45% THF should give equivalent retention.

A more effective approach is to use the retention data from gradient runs to predict the proper separation conditions.[17] Compared with isocratic elution, a gradient method varies mobile phase composition with time. It starts with a weak solvent to elute the less retained solutes. Then the mobile phase strength is increased successively to elute more strongly retained solutes. One of the characteristics of gradient elution is that all components will have similar peak widths. This is a great advantage in detecting late eluting impurities. A gradient method is preferable for analyzing early-phase DS as it allows a relatively larger number of peaks to be separated in a single run.

Method development is greatly facilitated by the use of chromatographic software. DryLab[18–24] is the most widely used computer-assisted chromatography optimization software package. It uses retention data from scouting runs for subsequent retention and resolution prediction via simulation. However, additional computer-assisted chromatography methods have been developed, including ChromSword,[25,26] EluEx,[27–29] and LabExpert.[30] ChromSword uses structure fragments and dipole–dipole interactions to predict retention behavior; Eluex suggests initial experimental conditions based on chemical structures. The more recently introduced LabExpert can plan experiments, collect and evaluate results, and adjust chromatographic conditions in real time according to predefined decision schemes until a satisfactory separation is achieved.

To use DryLab, in the most conventional approach, a short scouting gradient run (e.g., 20 min) and a long gradient run (e.g., 60 min) are needed for each column/mobile phase combination. The collected retention times are then entered into DryLab for simulation. Based on the retention data, the software can generate a resolution map, identifying critical pairs of peaks, and allowing the optimization of the gradient profile on a computer screen. The method development chemist can identify the best separation condition without running additional experiments. Other functions of DryLab include the optimization of mobile phase pH and temperature, and the optimization of isocratic methods based on gradient scouting runs.[31]

With the help of DryLab, the method development chemist can quickly screen a large number of column/mobile phase combinations to identify the best chromatographic conditions for a gradient method. The screening experiments can be easily automated using instruments equipped with a column-switching device,[32,33] which can select from multiple columns. This approach is highly efficient and can significantly reduce the number of runs required for method development.

For a complicated separation problem, screening alone may not result in a satisfactory separation and further optimization may be needed. Two different approaches can be considered for further separation optimization. One is to utilize multiple organic modifiers, for example, a mixture of methanol and acetonitrile. The other is to modify additional variables, such as the pH of the mobile phase.[34,35] DryLab can be used in the same way as discussed above for these optimization steps.

To apply a screening approach to proactive method development, analyses of selectivity samples under a variety of mobile phase conditions are conducted on different HPLC columns. HPLC columns should be as orthogonal[12,36] as possible and variations in solvent composition should be designed to maximize the probability of selectivity differences. Alternate separation techniques, such as ion exchange chromatography (IC), supercritical fluid chromatography (SFC), or capillary electrophoresis (CE) may also be used to obtain orthogonality.[36–38]

Orthogonality between systems is based on differences in retention mechanisms and can be measured by plotting the solute *k*'s achieved in each system vs. each other. Generally, the lower the resulting correlation coefficient, the greater the orthogonality is between the systems.[36] While the orthogonality is readily measured, it is significantly more difficult to predict a priori which sets of conditions will be most orthogonal for a given set of compounds. Ideally, any set of orthogonal conditions established should have broad applicability to allow for the use of a single screening strategy for multiple samples.

As a starting point, several authors have devised classification schemes for various column types to allow for the selection of potentially orthogonal stationary phases.[39–41] However, columns also need to be chosen with regard to performance criteria, such as lot–to-lot reproducibility of retention time, selectivity and band-broadening.[42–46] Mobile phase (organic modifier, pH, etc.) and temperature variation, for a screening strategy, need to consider the relative expected selectivity changes associated with each variable.[47]

An example set of columns that may be used in a screening approach is provided in Table 3. For mobile phases, large differences in elution behavior are observed for compounds with acidic and basic groups, common to most pharmaceuticals, based on mobile phase pH (Figure 2). For the purposes of screening, broad gradients should be used to maximize the probability that all compounds elute from the column and that no components elute at the solvent front. However, limits are placed on the initial conditions to insure that compound solubility is maintained and that stationary phase collapse does not occur. Salt precipitation dictates the percent of organic solvent in the final conditions.[36] Additionally, mobile phases should be chosen so as to be MS-compatible, to allow for sample component identification and component tracking under different conditions.

On the basis of these considerations, a possible screening experiment uses each of the mobile phases described below with each of the columns

TABLE 3 Set of RPLC Columns Used for the Screening Approach Described

Column	Particle size (μm)	Length (mm)	ID (mm)	Pore size (Å)	Surface area (m^2/g)	Carbon load (%)	End capping	pH range
Luna phenyl-hexyl	3	100	4.6	100	400	17.5	Yes	1.5–10.0
Zorbax Bonus RP	3.5	100	4.6	80	180	9.5	Triple	2.0–9.0
Zorbax Extend C18	3.5	100	4.6	80	180	12.5	Double	2.0–11.5
YMC Pro C 18	3	100	4.6	120	340	17.0	Yes	2.0–8.0
Waters Symmtery-Shield RP 18	3.5	100	4.6	90	330	17.5	Polar shielded	2.0–8.0
Waters Xterra RP	3.5	100	4.6	125	180	15.0	Yes/polar	2.0–12.0
Waters Xterra MS	3.5	100	4.6	125	180	15.5	Yes	1.0–12.0
Hypersil BDS C 18	3	10	4.6	130	170	11.0	Yes	2.0–7.5

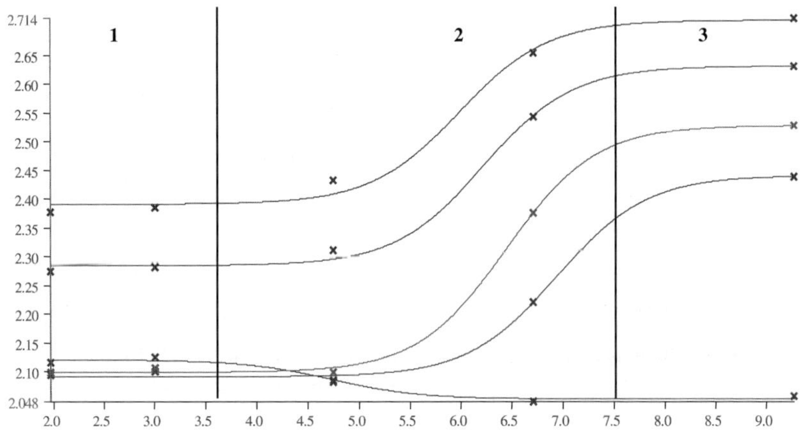

FIGURE 2 Plot showing the influence of pH on retention of five impurities.

noted in Table 3 to screen the selectivity solutions. Mobile phase salts are chosen based on MS compatibility and are used at a relatively low concentration (10 mM) to minimize risks of precipitation. Gradients are run as specified in Table 4 and detection is by absorbance using a photodiode array (PDA) detector and by MS. The screening module does not run the columns that are incompatible with basic mobile phases at pH 9.

Tested mobile phases:

Method A [pH = 2.5]

A: 10 mM NH_4OAc in water – CH_3CN (95/5, v/v) + 0.1% TFA (v/v)

B: 10 mM NH_4OAc in water – CH_3CN (15/85, v/v) + 0.1% TFA (v/v)

TABLE 4 Gradient Conditions for Screening Experiments

Time (min)	0	20	25
% A	100	0	0
% B	0	100	100

Flow: 0.8 mL/min; temperature: 35°C; injection volume: 10 μL; detection: record full spectrum with PDA; inspect chromatograms at wavelength of 220, 230, 240, 254, 260, 275, and 300 nm.

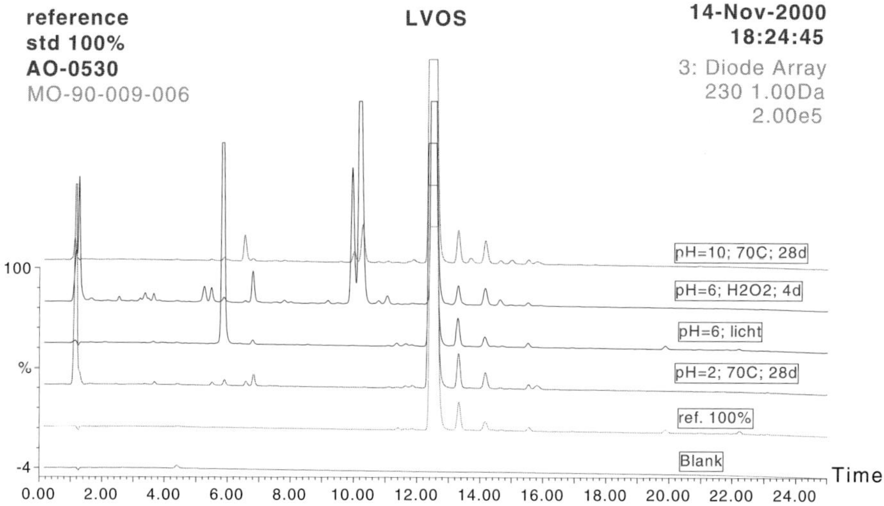

FIGURE 3 Overlays of chromatograms of DS obtained under different stress conditions.

Method B [pH = 4.8]

A: 10 mM NH_4OAc in water − CH_3CN (95/5, v/v) + 0.05% HOAc (v/v)

B: 10 mM NH_4OAc in water − CH_3CN (15/85, v/v) + 0.05% HOAc (v/v)

Method C [pH = 7]

A: 10 mM NH_4OAc in water − CH_3CN (95/5, v/v)

B: 10 mM NH_4OAc in water − CH_3CN (15/85, v/v)

Method D [pH = 9]

A: 10 mM $(NH_4)_2CO_3$ in water − CH_3CN (95/5, v/v)

B: 10 mM $(NH_4)_2CO_3$ in water − CH_3CN (15/85, v/v)

Typical results, under one set of conditions, are shown in Figure 3. Such results allow for the determination of which degradation products need to be monitored by the chromatographic method. Chromatograms of the same sample, generated under different chromatographic conditions may show significant changes in solute elution order, as shown in Figure 4.

C. Selection and Optimization of Candidate Methods

Based on the evaluation of the screening results, a "candidate" method is identified and will be optimized. This method is selected on the basis of its being able to monitor all components of interest. It is additionally ideal that no minor peaks elute in close proximity before and after the API.

A "supportive" method that is orthogonal to the candidate method is also selected on these terms. The elution order of the supportive method is by definition significantly different from the candidate method. Ideally the method screening experiments will provide two sets of conditions, as shown in Figure 4. The utilization of the supportive method maximizes the probability that new unknown related compounds, which possibly co-elute in the candidate method, will be detected and taken into account when evaluating results for subsequent batches of DS or new DP formulations.

Following selection of the candidate method, the method is optimized for resolution and run time. As discussed previously, this is readily performed using software tools such as DryLab. The approach may

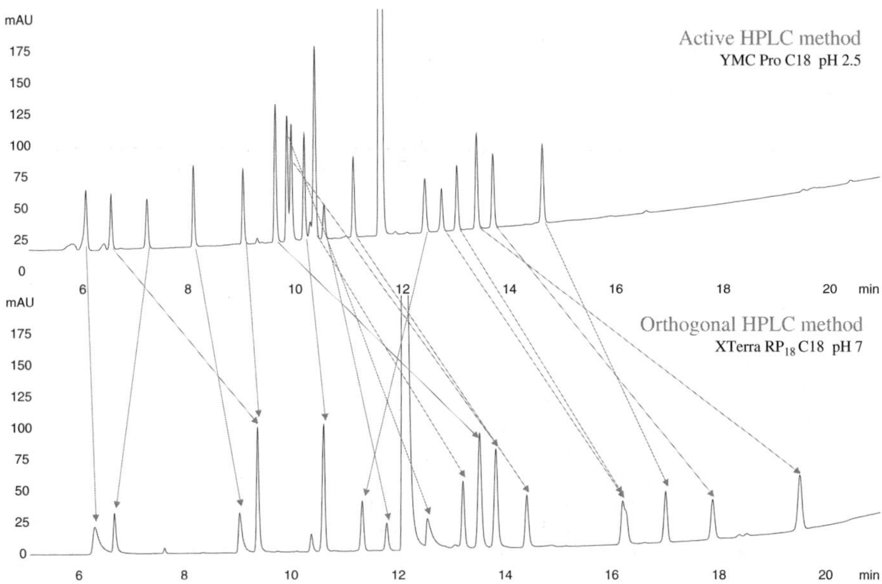

FIGURE 4 Comparison of two orthogonal HPLC methods for the same sample.

include two to as many as six or more runs depending on the number of variables (gradient time, buffer concentration, column temperature, pH, type of organic modifier, ternary eluent composition, etc.) to be investigated.

To illustrate the approach, consider the separation of eight components. As an initial optimization step, a one-dimensional optimization of gradient time was performed by the use of two calibration runs. As shown in Figure 5, the resulting resolution map indicates that optimum separation is achieved in 105 min with a resolution of ~1.6 between the critical pair. Since this is clearly unacceptable, an additional one-dimensional optimization was performed to optimize buffer concentration. Modeling, based on three calibration runs, again shows the chromatographic conditions to be unacceptable (Figure 6). Simultaneous optimization of the gradient time and the buffer concentration (Figure 7) results in an adequate separation for all compounds within an acceptable run time.

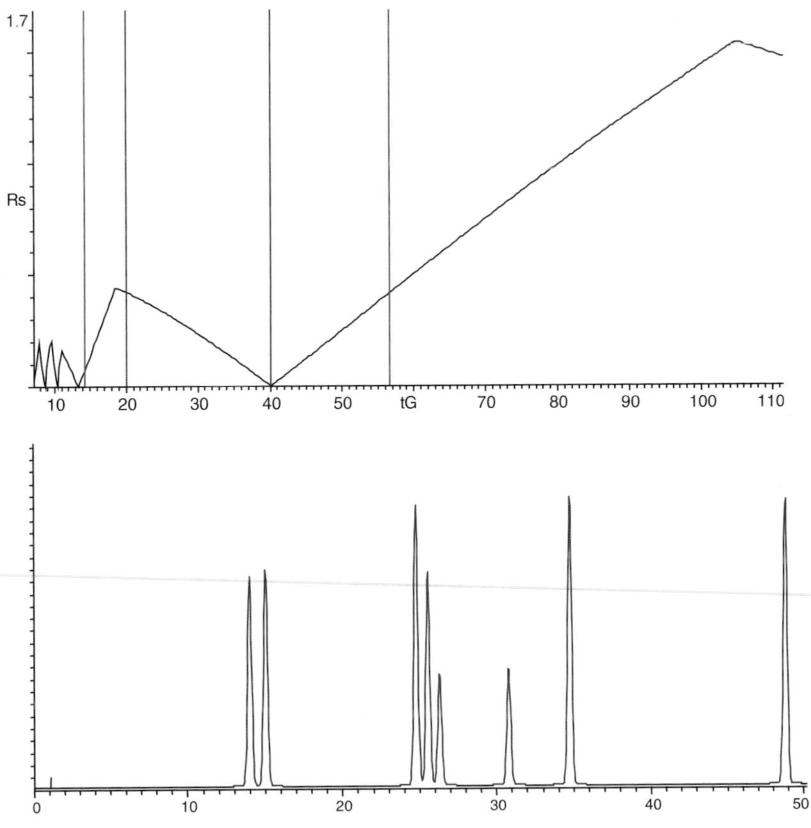

FIGURE 5 Optimization of gradient time (t_G): resolution map and chromatogram.

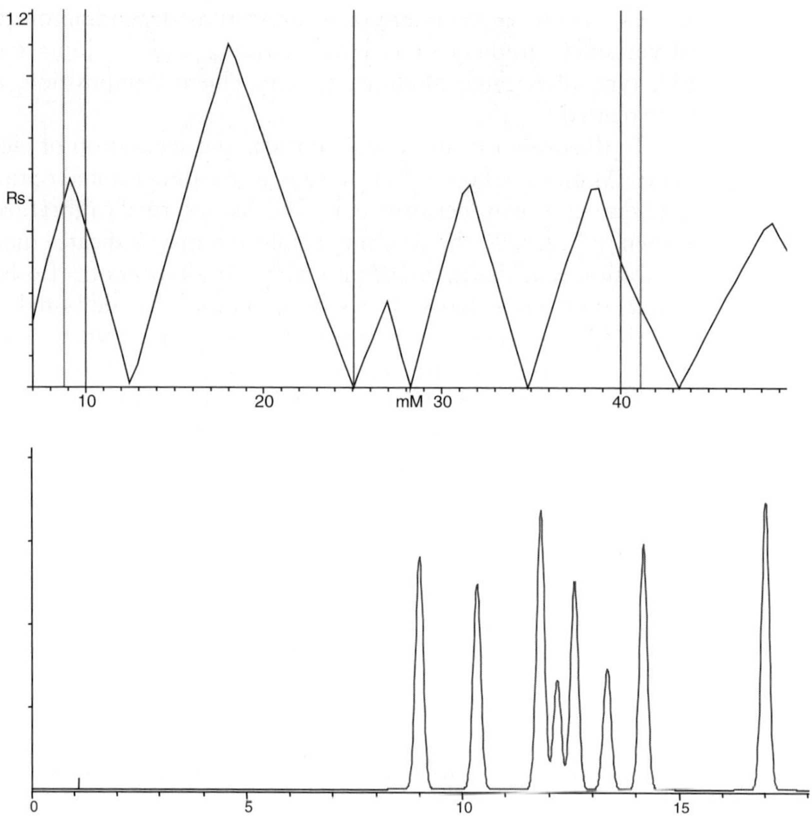

FIGURE 6 Optimization of buffer concentration: resolution map and chromatogram.

During proactive method development, special attention should be drawn to four key areas, as shown in Figure 8, prior to issuing the candidate method:[48]

(1) Co-elution of impurities.
(2) Highly retained compounds.
(3) Unretained compounds.
(4) Co-elution of components with the API.

The co-elution of minor components, such as impurities or degradation products, is generally investigated by examining peak purity using the PDA detector or by assuring that mass spectra are consistent across the peak.[48,49] The potential retention of non-eluting compounds such as dimers or oligomers can be investigated using non-aqueous RP-chromatography as demonstrated in Figure 9. The third critical region is the void volume. Since highly polar compounds elute in this area, LC-MS can be used to ensure the absence of non-retained substances.

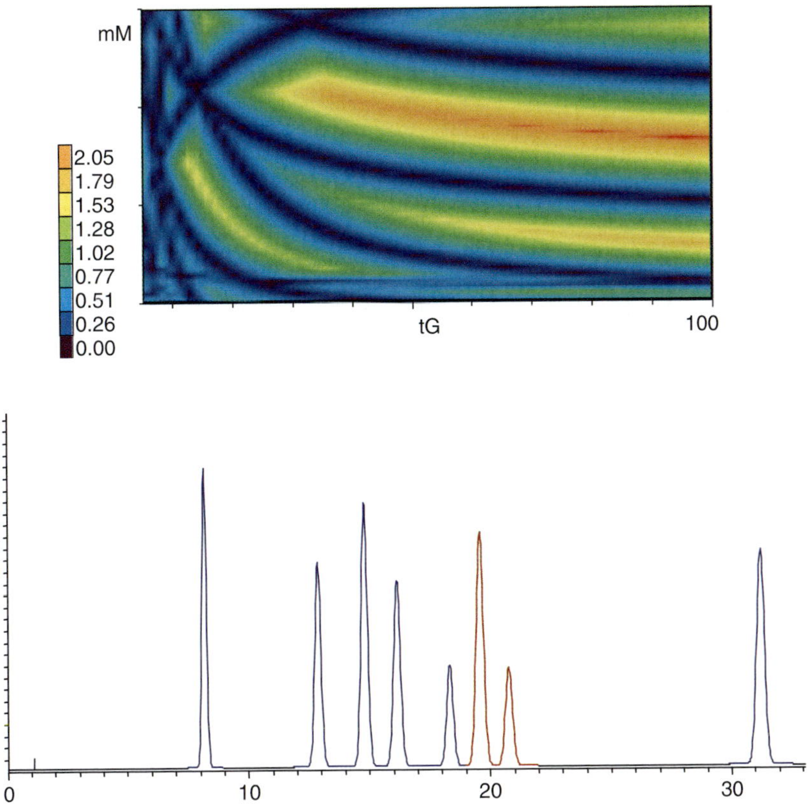

FIGURE 7 **Simultaneous optimization of gradient time (t_G) and buffer concentration: resolution map and chromatogram.**

The most challenging chromatographic problem is to assure that there is no co-elution of minor components with the API. The approach detailed above for examining peak purity of co-eluting impurities is frequently unsuccessful due to the large difference in concentration between the API and the impurity, and since co-eluting impurities are frequently structural isomers, spectra may be very similar. The most promising approach is peak fractionation, where the peak is collected and subsequently analyzed by an orthogonal method such as CE-MS.[48]

IV. DEVELOPMENT, VALIDATION, AND USE OF EARLY PHASE METHODS

A. Development of Additional Method Parameters

Following the development of a candidate separation method there are many method factors, such as detection wavelength, sample

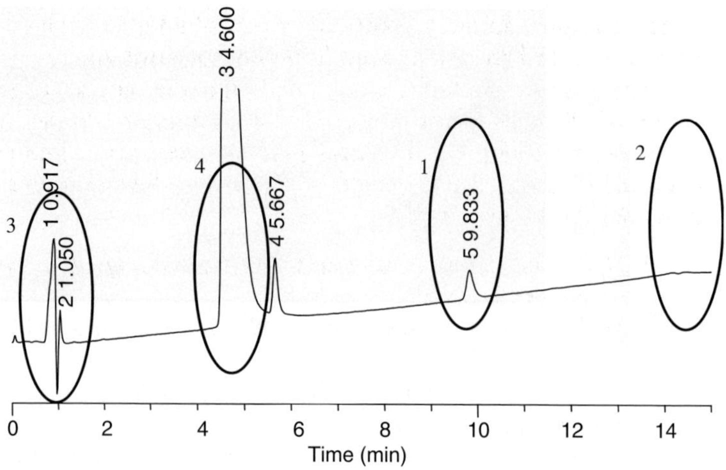

FIGURE 8 Critical regions in a chromatogram to evaluate: (1) co-elution of impurities, (2) highly retained compounds (apolarity issue), (3) non-retained compounds (polarity issue), and (4) co-elution with the API (isomers).

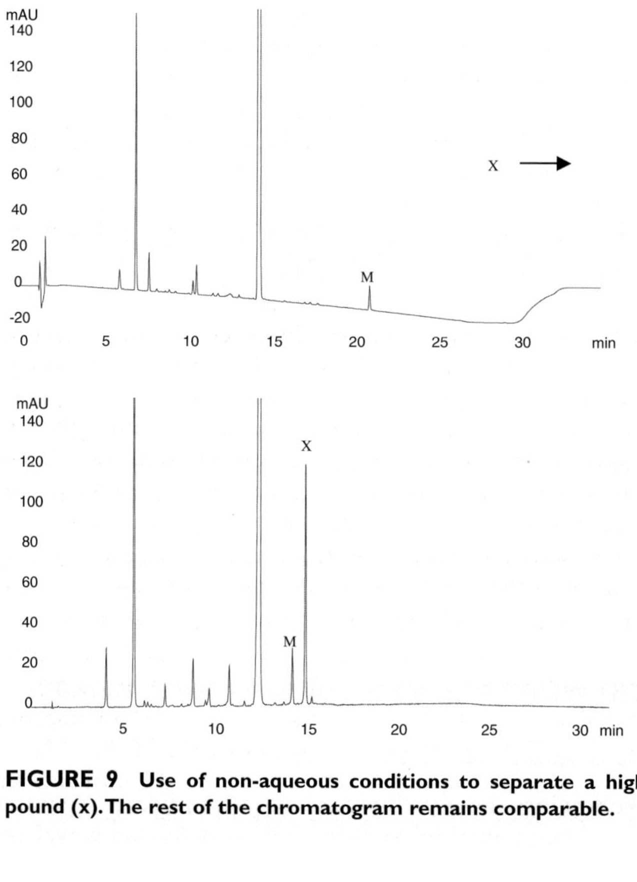

FIGURE 9 Use of non-aqueous conditions to separate a highly retained compound (x). The rest of the chromatogram remains comparable.

concentration, sample solvent, injection volume, etc., which also need to be optimized to ensure overall method performance.

The ultimate goal of an assay method is the separation and visualization of all components in a single chromatogram. Proper selection of detection wavelength is a critical part of method development. When choosing a detection wavelength, the following factors need to be taken into consideration:

- The major components must have suitable dynamic range at the selected wavelength.
- All impurities should be detected with suitable sensitivity at the selected wavelength.
- Mobile phase should not show strong background absorption at the selected wavelength.
- Detectability should be maximized for components of interest and minimized for those of excipients.

A diode array detector is well suited to achieve these goals. Full range spectra should be collected for the method development samples and evaluated with two-dimensional or three-dimensional visualization to determine the best detection wavelength.

With a modern variable-wavelength or PDA detector, it is safe to use a maximum absorbance of up to 1.0–1.5 absorbance units (AU). A typical UV detector today should have a baseline noise in the order of 10 micro absorbance units (10^{-5} AU). Therefore, modern UV detectors should have a dynamic range of 4–5 orders of magnitude. The sample concentration and injection volume can be adjusted to make it possible to quantify the major component and impurities at the 0.05% level in the same chromatographic run without changing detection wavelength, sample concentration, or injection volume.

Ideally, mobile phase, or a sample solvent of weaker elution strength, should be used as a sample solvent to allow for the largest injection volume without causing peak-shape deteriorations. However, very often this is not possible because of solubility problems of the APIs. Higher concentrations of organic solvent may have to be used to solubilize the APIs. This requires the method development chemist to properly balance sample solvent, sample concentration, and injection volume. Early-eluting peaks are more significantly affected by these factors. Improper factor selection will result in tailing or fronting peaks, which sometimes may affect the accuracy and precision of the method. An example is given in Figure 10, which shows the detection of a very polar impurity in an NCE. When the mobile phase (2.5% acetonitrile in water) was used as sample solvent, the peak shape was independent of injection volume (Figure 10a). Due to the poor solubility of the NCE, 10% acetonitrile had to be used as sample solvent. In this case the injection volume had to be ≤30 μL (Figure 10b) to avoid peak broadening.

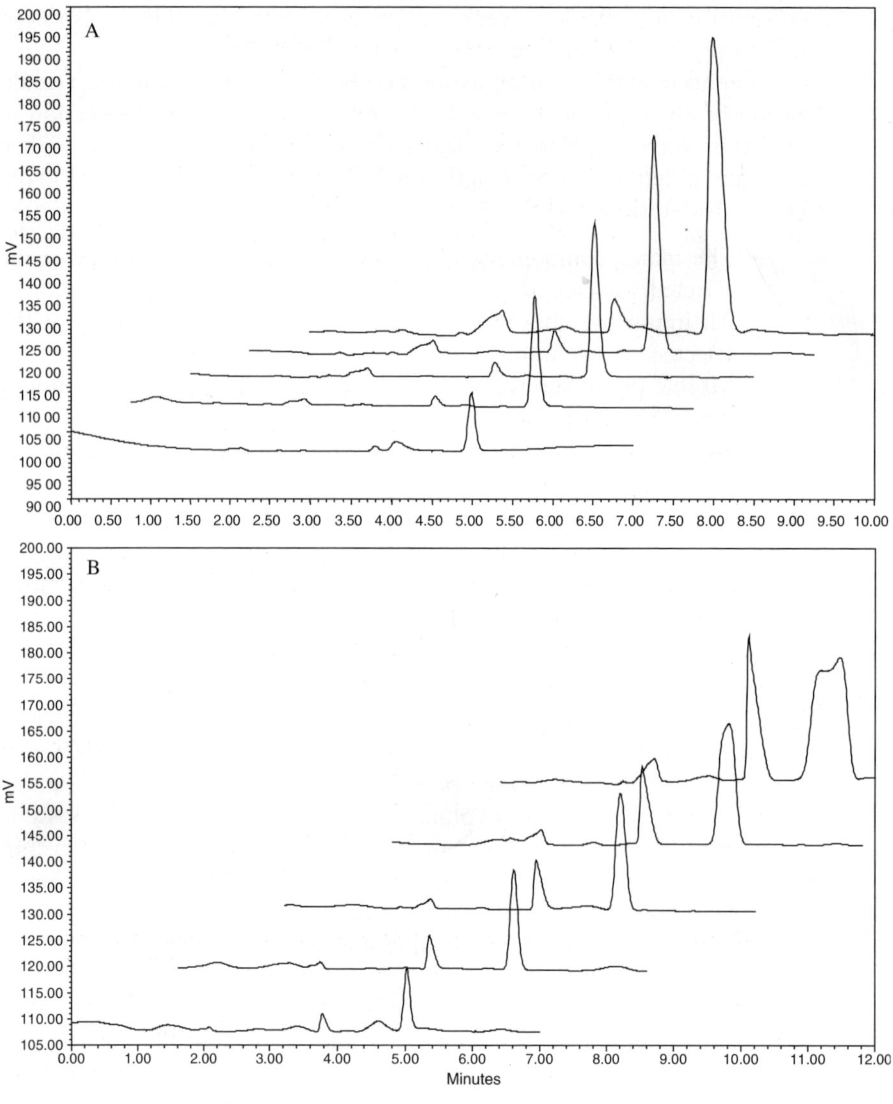

FIGURE 10 Chromatograms showing the effect of sample solvent and injection volume on peak shape. The peak at 5 min is an impurity in a NCE. Column: Supelco Discovery RP amide C16; mobile phase, 2.5% acetonitile in water; detection, UV 190 nm; injection: 10, 20, 30, 50 and 100 μL; sample solvent: (A) 2.5%; (B) 10% (acetonitrile in water).

B. Validation of Early Phase Methods

For early phase methods emphasis is placed on specificity and these methods generally require less extensive validation than those in final development. The following method parameters should be included in validation: specificity, accuracy, precision, linearity, range, sensitivity, and solution stability. Specificity of the method is demonstrated by separating all process impurities and degradation products from the API and from each other. This is inferred from the method development experiments conducted to generate the candidate method.

The accuracy of the method is assessed by preparing multiple samples (e.g., three) at multiple concentration levels (usually at 80%, 100%, and 120% of the nominal concentration). For DP methods, the sample solutions are also spiked with placebos. Similar experiments should also be conducted for impurities when authentic materials become available.

For early phase methods, the precision tests only include injection repeatability (also referred to as system repeatability) and method repeatability (also referred to as analysis repeatability). The former is demonstrated by repeating injections of a standard solution and the latter by preparing multiple samples over multiple concentration levels (usually at 80%, 100%, and 120% of the nominal concentration) from the same lot of a composite sample of the dosage form.

The linearity of the method is demonstrated using 3–5 concentration levels. For APIs, the correlation coefficient should be greater than 0.999. The range is derived from linearity, precision, and accuracy studies and should be established based on the intended application of the method.

Early phase methods usually adopt a simple way to establish method sensitivity by preparing a 0.05% solution of the API relative to the standard solution, which is injected multiple times during validation to obtain an RSD value ($\leq$15%, $N=6$). Later, this solution is injected every time the method is used to assure that adequate sensitivity is observed at the time of use.

An additional key validation criterion for early phase methods is an evaluation of solution stability[50] to establish that the API and related substances do not degrade in the solvent system used for sample preparation. This allows for limits to be set for solution lifetime.

C. Use of Early Phase Methods

Following validation, the candidate method is referred to as a level 1, 2, or 3 method depending on the level of pharmaceutical development that the method is intended to support (see Table 1).

The candidate method is used to support drug synthesis, excipient compatibility, and ultimately to evaluate candidate formulations. Such support typically involves analyses of stressed materials to identify degradation trends. These studies are conducted in the solid state by exposing the DS and DP to relative humidity, temperature, light, and oxidizing conditions that are closer to ICH stress conditions[7] than the harsh conditions used during forced decomposition studies. An intended goal is to minimize the formation of degradation products that are not obtained under ICH storage conditions.

Typical studies may be set up as detailed in Table 5.[5] Samples set up as indicated should be sampled and analyzed intermittently over the course of approximately 12 weeks, and the degradation products identified. To evaluate the influence of light, DS and DP should be exposed to a light source that produces an output similar to the D65/ID65 emission standard.[51] Samples should ideally be exposed at 50%, 100%, and 200% of the ICH specified 1.2 million lux hours at an integrated near-UV energy of not less than 200 Wh/m^2.[5]

The use of multiple time increments allows for the attribution of any observed peaks as degradation products if the peaks are observed to increase in size as a function of time. The use of multiple temperatures allows for the confirmation that degradation products observed at elevated temperatures are also observed at lower temperatures, albeit at lesser levels, and supports the theory that the degradation mechanism occurs at conditions approaching ambient. Further evaluation can be conducted using kinetic modeling and Arrhenius plots to estimate the formation of individual degradation products under ICH stress conditions.[5]

During this phase of development it is also critical to examine mass balance for the separation, i.e., to establish that the summation of API, impurities, and degradation products = 100% on an anhydrous basis (within experimental error). Several factors should be considered if mass balance is not achieved[6]:

(1) The API may degrade to a volatile component.
(2) Diffusive losses into or through containers may occur.
(3) The method may have elution or resolution problems.

TABLE 5 Stress Conditions for Solid-state Evaluation of DS and DP Stability

Stress parameter	Stress conditions
Temperature	80°C, 70°C, 60°C, 50°C
Hydrolysis	60°C/80% relative humidity 40°C/80% relative humidity
Oxidation (experiments performed in a Calorimetry bomb)	80°C/300 psi oxygen 70°C/300 psi oxygen 60°C/300 psi oxygen 50°C/300 psi oxygen

(4) Inappropriate response factors may have been used for quantitation of impurities/degradation products (i.e., molar extinction coefficients of degradation products are less than those of the API).

Research to elucidate factors (1) and (2) is highly dependent on the specific nature of the problem, and may have to be considered on a case-by-case basis. To address issues (3) and (4) it is recommended that the critical regions of the chromatogram as detailed in Section III. C be examined. It is particularly useful to conduct these experiments using additional detectors such as evaporative light scattering detectors (ELSD) and nitrogen chemiluminescence detectors (NCD)[52] to assure that all compounds are observed. ELSDs provide (near) equimolar responses to analytes and NCDs provide (near) equivalent responses to analytes containing nitrogen. Accordingly, the detectors may be used to detect components that were previously unaccounted for as a result of poor molar extinction coefficients at the UV wavelength(s) selected for analysis. Subsequent identification and synthesis (or isolation) of these compounds allows for the establishment of correct relative response factors and improved quantitative analyses.

The supportive method developed during proactive method development is used:

- When a change is made in the synthesis route.
- When a change is made in suppliers of raw materials and excipients.
- When there is a change in the formulation.
- When a new degradation product is observed.
- To assess stressed samples at pivotal time points.

In other words, it is used to verify that the specificity of the levels 1–3 method is maintained throughout development. In cases where the orthogonal method shows the primary method to be deficient, the orthogonal method is validated and is used as a secondary method in conjunction with the primary method. Additional new methods may also need to be developed to address specific issues encountered in stress studies, such as when mass balance is not observed.

If a secondary method is required during level 1–level 2 development, a level 3 method is developed and validated to support phase 2b or early phase 3 clinical studies. This method should be capable of separating all components of interest identified up to this stage of pharmaceutical development.

An output of the impurity profiling of DS and stress studies of DP is an understanding of drug chemistry, particularly with respect to which synthetic impurities are expected to continue to be present in optimized synthetic lots and what degradation products are formed in DS and DP. Significant impurities and degradation products, formed during these evaluative studies, are identified and synthesized as authentic substances. As the substances become available, the supporting methods are validated for the determination of the impurities or degradation products.

V. FINAL METHOD DEVELOPMENT AND VALIDATION

A. Overview of Requirements for Final Methods

Final methods are developed for transfer to operational quality control (QC) laboratories for the release testing of production batches. Additionally, the methods are intended to be applied during Registration Stability studies and for the release of the DP or DS validation batches during the pre-approval development stage. The analytical methods should last for the entire product lifetime; therefore, the aim of final method development is to generate fast, robust, reliable, and transferable HPLC methods (preferably isocratic and at low cost).

As has been noted by Renger,[53] many HPLC methods intended for QC are incapable of adequately controlling quality within the typical European expectations of 95–105% of label claim for DP and 98–100.5% for DS assays. This is due in part to basing method precision on injection repeatability (system suitability), and in setting the acceptance criteria for this evaluation too liberally.[54] An evaluation of method performance must include all sources of error. According to Renger,[53] HPLC assay results, derived from the injection of a single injection and/or sample solution may generate up to 10% out-of-specification results. Performing replicate analyses can reduce these errors. However, it is also important to assess and minimize error through method development.

Final methods are ideally developed in close collaboration with the customers (QC-labs) and it is highly recommended to involve the customer right from the beginning. To address performance concerns, a Six Sigma road map was followed to troubleshoot and improve the existing method development process in our laboratories. By this process: (i) the voice of the customer was captured, (ii) key process input variables were identified, (iii) critical to quality (CTQ) factors were determined, (iv) several method verification tests were designed, (v) proactive evaluation of method performance during development was established, (vi) continuous customer involvement and focus was institutionalized, and (vii) method capability assessment (suitability to be applied for release testing against specification limits) was introduced. An outcome was the recognition of the importance of customers in the method development process. The involvement of the customer is clearly demonstrated in Figure 11. The final development process is performed in different steps[55]: Planning, Method Development, Method Evaluation, Method Validation, Method Transfer and Method Performance Monitoring.

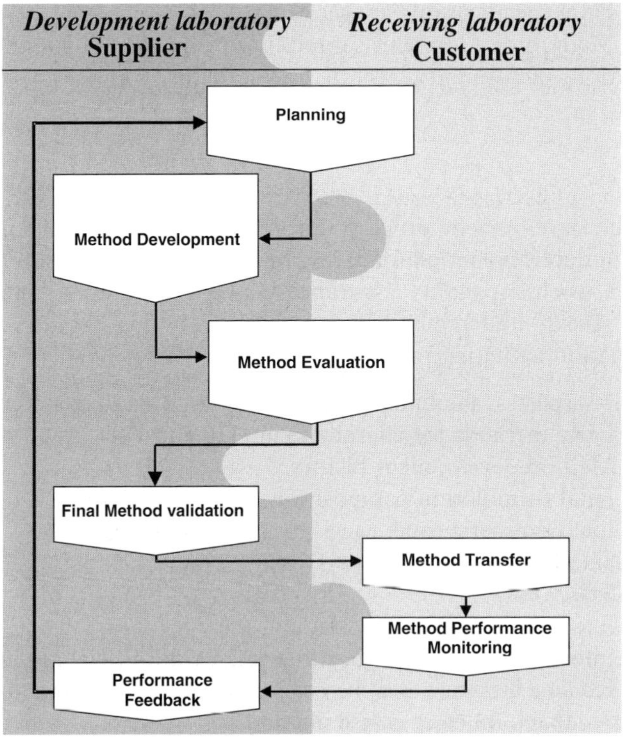

FIGURE 11 **Diagram illustrating the method development process for final methods.**

B. Prerequisites to Final Method Development

Final method development should not proceed until several prerequisites are fulfilled. Specifically:

- The synthesis route is locked (finalized) and all critical intermediates and starting materials are defined for a DS method.
- Formulation and dosage forms compositions are locked for a DP method.
- Relevant impurities and degradation products are known and available as certified standards. It is very demanding, but important to know prior to development which impurities and degradation products are specified to be limited by specifications. This knowledge will determine the complexity of the development task. The specified impurities and degradation products need to be separated fully from all other potential impurities and degradation products.
- Certified standard of the API and a selectivity batch or selectivity samples (i.e., samples containing all components of interest) are available.

- Final specifications are set in draft.
- Method development samples (appropriate DS batches, formulations (active and placebo), and stability samples) are available.

C. Planning

The development process starts with a thorough planning step in which a development plan is generated by the collaboration of all stakeholders, such as quality assurance (QA), QC sourcing sites, stability, development, and regulatory functions. Planning should also include information gathering to obtain:

- UV spectra, solubility, p*K*a, stability of API and related compounds.
- Early methods for characterizing DP and DS.
- Method development history.
- Final formulation composition.
- Final synthetic route.
- Stability data for DS and DP.
- Safety data.
- Excipient compatibility data.
- Information in regulatory files (e.g., clinical trial applications).
- Existing literature and current compliance guidelines and procedures.
- Feedback on early phase methods.

The outcome of planning should include the generation of a method definition requirement document (MDRD) in which all stake holders agree, prior to method development, on the critical attributes of the method. Considerations such as what impurities/degradation products should be monitored, requisite reporting thresholds, the need for an MS-compatible method, if identical methods for DP and DS are required, etc., should be clearly decided and agreed upon by all stakeholders.

It is also useful to assure that internal (company specific) practices are adhered to in method development by making these considerations part of the MDRD. Based on the voice of the customer surveys, several technical and practical method attributes have been identified within our organization.

It is generally required that all methods allow for the monitoring of API and impurities/degradation products in the same chromatographic run, that run times per sample should not be too long, and that for precise and robust quantitative analyses, the separation of the peaks of interest should have target resolutions of >2.0. To allow for easy transfer, the detector response for the nominal concentration of the API (100%, w/w) should be about 75% of the qualified linear dynamic range of the detector. Methods should be temperature controlled (e.g., at 35°C, since this is slightly higher than ambient temperature and can be achieved by ordinary HPLC column ovens).

When feasible, isocratic elution is preferred for final methods because of better reproducibility. When isocratic conditions cannot be used, linear gradients are preferred. Step gradients may be used for the reduction of analysis time if there is a late eluting compound with a high capacity factor. Additionally, the initial percentage of organic modifier in the mobile phase should preferably not be less than 5%, which will allow for better mixing properties of the mobile phase and will thereby avoid precipitation issues. Premixing of the mobile phase is preferred for robust performance. For example, when in a gradient elution method two solvents A and B are used as the mobile phase and the gradient program is such that it starts at 30% B and ends at 60% B, a (pre-)mixture of solvents A (70%) and B (30%) is prepared and hooked in line 1 of the HPLC pump. Another (pre-)mixture of solvents A (40%) and B (60%) is prepared and hooked in line 2 of the HPLC pump. The gradient will now be programmed to start at 100% solvent in line 1 and end with 100% solvent in line 2. In this way mixing incompatibilities are avoided during the HPLC run. The inclusion of a rinsing step at the end of the gradient should always be considered, especially for DP analyses, where placebo ingredients may retain in the stationary phase and potentially result in ghost peaks in subsequent runs.

The mobile phase composition at the starting conditions of the gradient should preferably also be used as sample solvent. Differences in solvent strengths are usually the cause of peak distortion (splitting) in the early part of the chromatogram, as discussed in Section IV. A.

Buffer systems should be prepared by accurately weighing or pipetting the components of the buffer. Adjusting the pH during buffer preparation should be avoided, if possible. The maximum concentration of the buffers should not be more than 40 mM because of the risk of precipitation during mixing with organic solvents in a gradient run. Tetrahydrofuran (THF) in the mobile phase should be avoided where possible.

A suitable, globally available HPLC column should be used in accordance with the manufacturer's specifications and its durability should be tested using the specified test conditions. An alternative column to be used in the method should be considered and appropriately validated. The initial pressure for a typical new column of 15-cm length and 5-µm particles, using typical HPLC conditions like a flow rate of 0.5–1.0 mL and ordinary buffer systems should be <200 bars.

Sample preparation is also a crucial step in the analysis. A typical sample preparation scheme consists of a sampling step, weighing into suitable glassware, a disintegration/milling step, an extraction step that is usually facilitated by shaking, stirring, etc., a dilution step, and finally a filtration or centrifugation step. For reasons of poor reproducibility small volume pipettes are avoided and volumetric flasks in the

25–500 mL volume range are used. The weighed reference standard amounts for DS and DP are preferably >100 mg (to meet precision requirements). For DP, at least 10 units are used in the preparation of the sample solutions. Additional sample preparation considerations are discussed in Chapter 5.

D. Method Development

Final method development is started with a review of the available methods from early development. The methods are evaluated against the method requirements set in the MDRD. Generally, the method is optimized or re-developed in order to fulfill the requirements, using the approaches detailed previously in this chapter.

A typical example of method development during this stage of pharmaceutical development is provided in Figure 12. Figure 12a shows a chromatogram of the separation of an API and its four impurities by a gradient elution method on a C-18 BDS column using ammonium acetate–acetonitrile in the mobile phase. An impurity elutes on the tail of the API, indicating that method optimization is required for a more robust separation. Using DryLab, the gradient profile and the column temperature were optimized simultaneously. The minimum resolution plot for this application as a function of the gradient time and the column temperature is shown in Figure 12b.

Acceptable separations are predicted at conditions with a gradient time >15 min. Temperature is very important since at lower gradient times, higher temperatures are required for better resolution. Optimal separation conditions were selected at a gradient time of 18 min and a column temperature of 40°C. The chromatogram at these conditions is presented in Figure 12c. The impurity is adequately separated from the API using the optimized conditions. Additionally, a new unknown impurity is detected that co-eluted with the main compound peak in the previous method. It is important to note that peak tracking is crucial when applying computer-assisted method development. The correct assignment of peak identity will determine the accuracy of the predicted separation. As can be seen in Table 6, the DryLab-predicted retention times corresponded very well with the experimental values. The separation conditions should be optimized further for robustness and transferability.

E. Method Evaluation

As shown in Figure 11, before extensive validation, the performance of the method is evaluated appropriately. Column durability tests, robustness testing for the chromatographic and sample preparation conditions, analytical method evaluation ring tests (AMERTs), method capability assessments, and pre-validation studies are applied to

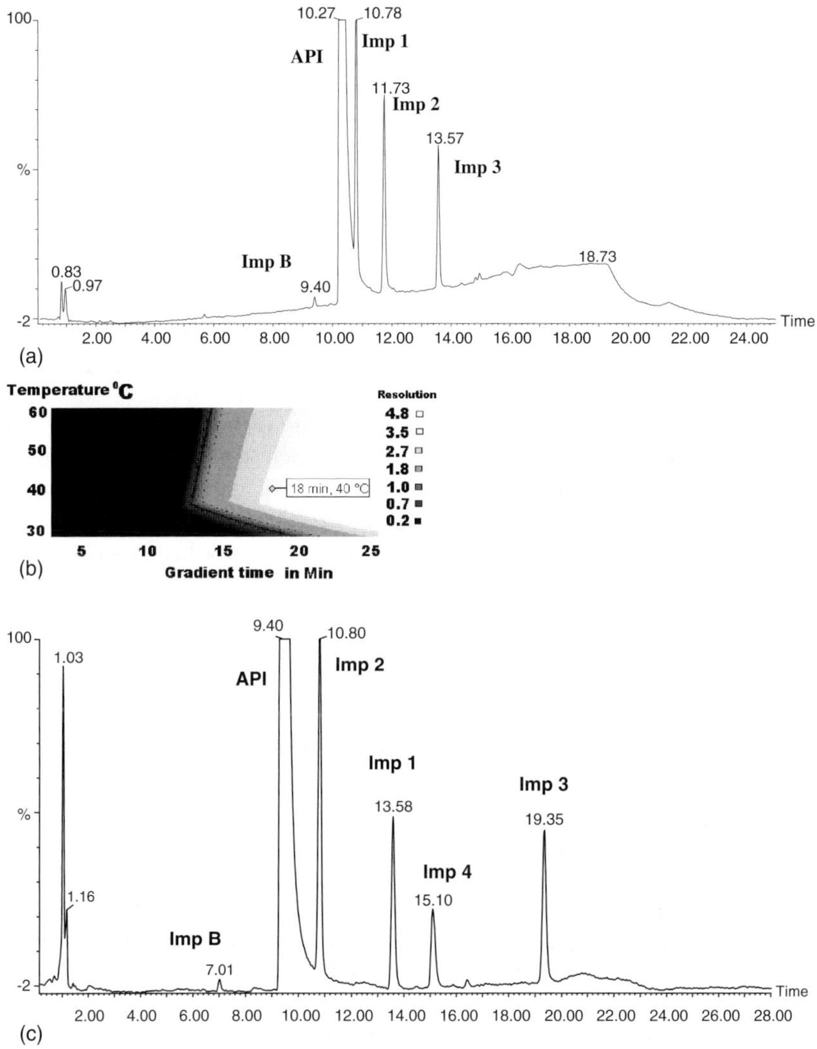

FIGURE 12 (a) Early development HPLC gradient elution method for an API and four related impurities. The separation between the API and Imp 1 required further optimization. Courtesy of Dr. Koen Vanhoutte, Johnson & Johnson Pharmaceutical Research and Development. (b) DryLab minimum resolution plot for the simultaneous optimization of temperature and gradient time. Optimum conditions are predicted at gradient times of more than 15 min and at higher temperatures. At lower temperatures longer gradient times are required. Optimal conditions were selected at 18 min and 40°C. (c) Experimentally obtained chromatogram at the predicted optimum conditions of 18 min. gradient time and 40°C.

challenge the method. During method evaluation studies the QC-labs play an important role, since they are the ultimate end-user of the method.

TABLE 6 Comparison of Retention Times and Relative Retention Times (RRT) Between a Simulated Chromatogram and an Experimentally Obtained Chromatogram

Peak	DryLab prediction (RRT)	Experiment (RRT)
Imp B	7.70 min (0.77)	7.00 min (0.74)
API	9.99 min (1.00)	9.40 min (1.00)
Imp 2	11.28 min (1.13)	10.80 min (1.15)
Imp 1	13.88 min (1.39)	13.58 min (1.44)
Imp 4	15.04 min (1.51)	15.10 min (1.61)
Imp 3	19.35 min (1.94)	19.35 min (2.06)

Once the chromatographic method is developed, the selectivity of the optimized separation is checked on at least three different lots of the selected stationary phase for the selected column to determine whether the stationary phase is suitable for use in a final method.

The results are qualitatively evaluated by making overlay plots. An example of an HPLC gradient separation for a typical assay and impurities analysis is shown in Figure 13a. The selectivity is evaluated on five different lots of the selected stationary phase. As can be observed, apart from two minor selectivity issues (indicated by the arrows in Figure 13a), the reproducibility on different lots of stationary phases is acceptable.

In addition to the study of different lots of the stationary phase, a durability test is performed on one column by making at least 100 consecutive injection runs of a reference solution and a sample (preferably an aged sample) solution. The result is qualitatively evaluated by making overlays of the sample solutions. Figure 13b shows an example in which more than 300 injections were performed continuously in a single sequence. In the chromatogram, the API and its related compounds are separated using a gradient elution method with a Waters XTerra^{TM} MS C_{18} 3.5-μm particle size column (4.6 × 100 mm); with a pH 9.0 mobile phase (ammonium acetate–diethylamine) as solvent A and a mixture of acetonitrile and 2-propanol as solvent B. The flow rate was set at 1 mL/min, the column temperature at 35°C, injection volume at 10 μL and UV detection at 275 nm. As can be observed in the overlay chromatograms there is little discernable peak drift, indicating stable run conditions for the chromatographic method.

A method is considered to be robust when the effect of typical fluctuations in the method parameters on the outcome of the procedure is demonstrated to be negligible. Robustness is demonstrated by a test in which the effect of deliberate changes in the method parameters is studied on a number of method responses. The extent of the changes should

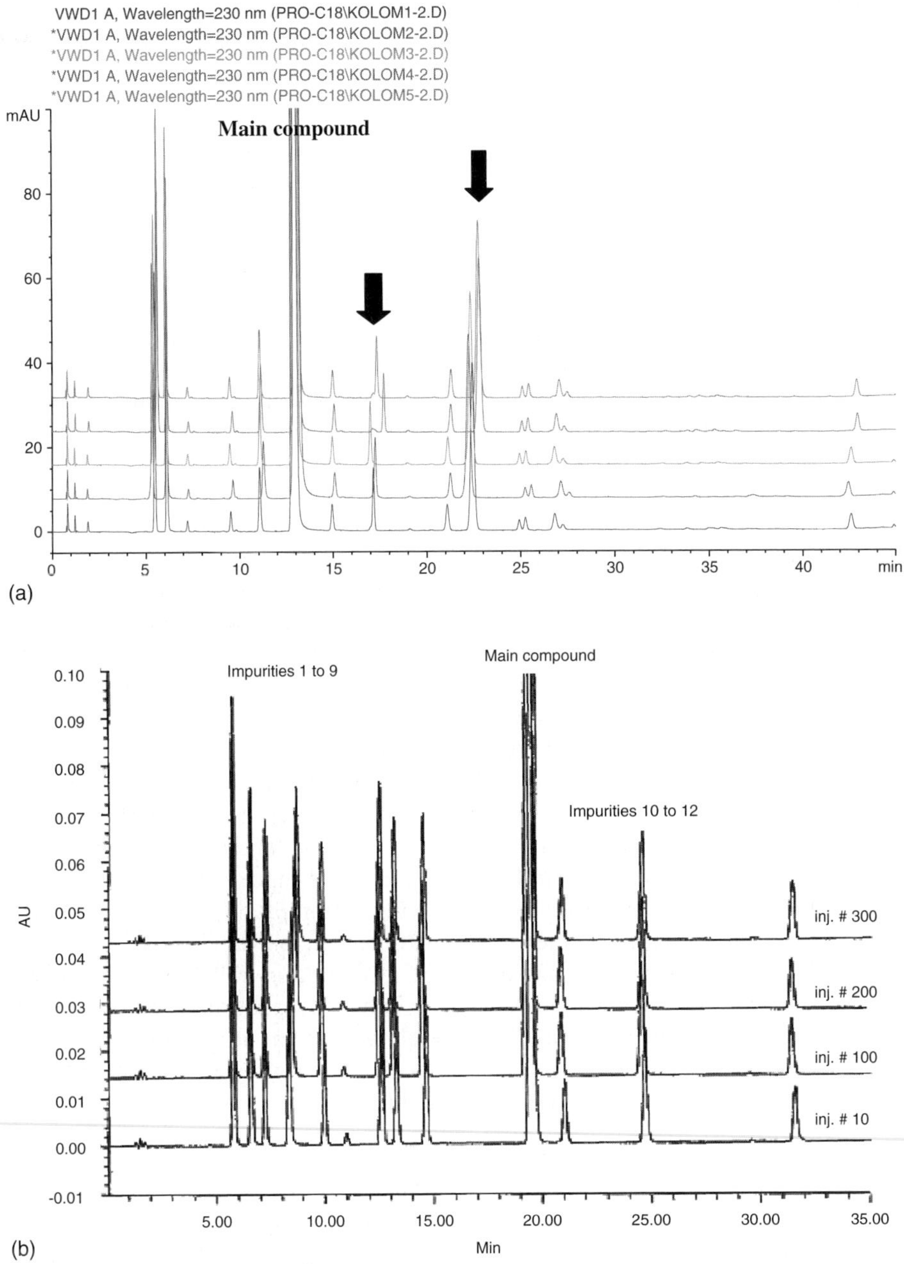

FIGURE 13 (a) Overlay chromatograms of an API and its related impurities on five HPLC columns with different lots of the stationary phase. There are only slight shifts (arrows) in retention times detected. (b) Overlays of the 10th, 100th, 200th, and 300th chromatograms on a single HPLC column.

reflect the normal variation in the method parameters that is observed during their application in different labs. In general, the purpose of robustness testing is to indicate factors that can significantly influence the outcome of the studied responses.[56–58] This gives an idea of the potential problems that might occur when the method is repeated in different laboratories (transfer). In this way potential problems can be anticipated by controlling the critical factors adequately, for example by including a "precautionary statement" (ICH)[1,2] in the method description stating the accurate limits for the critical factors. However, the proposed limits should be feasible or re-development of the method should be considered. Robustness testing may also be used to determine system suitability limits for the system suitability parameters.

Robustness testing of the chromatographic and the sample preparation factors is conducted either by an experimental design or a step-by-step approach (variables are evaluated sequentially). An experimental design approach is preferred since as much relevant information as possible is obtained in the shortest possible time, from a limited number of experiments. The choice of the experimental design requires a decision as to which factors and responses will be investigated and determination of whether only the main effects of the factors or also the interactions will be examined. Typically, a Plackett–Burman design (Table 7) will be sufficient.[57–60] Notably, in addition to the variables to be examined, this design includes the use of dummy (dum) factors to measure noise and thereby provide an evaluation of the statistical significance of any noted effects.

A tentative list of factors that may be investigated in the robustness test is presented in Table 8. This list is not complete and additional factors may be added. The limits for the factor levels are proposals and should be evaluated case by case. The significance of the effects of the factors on the responses such as the resolution of all peak pairs, the tailing factor, retention times, analysis time, etc., is evaluated.

A tentative list of factors to be evaluated in robustness testing for sample preparation is presented in Table 9. Additional factors can be added. The limits for the factor levels are typical and should be considered as proposals. Responses to be considered are usually related to the quantitative performance of the method.

Criteria for Assessing Performance:

The recovery of the main compound (100%) and all relevant impurities (spiked at, for example, the 0.5% wt/wt level) in a reference solution are calculated against a freshly prepared reference solution as described in the method description. This is done to demonstrate effects of the studied parameters on the recovery.

The recovery of the main compound (100%) and all relevant impurities spiked with, for example, 0.5% wt/wt in a recent and worst–case (old) sample batch, is calculated against a freshly prepared reference solution as

TABLE 7 A Plackett-Burman Design in Which Eight Factors are Studied Simultaneously in 12 Experiments

Experiment	Dum1	Column	pH	Temp	Grad_begin	Dum2	Grad_end	Flow	Wavelength	buff_conc	Dum3
Start	0	0	0	0	0	0	0	0	0	0	0
1.	1	1	1	−1	1	1	−1	1	−1	−1	−1
2.	1	1	−1	1	−1	−1	−1	1	1	1	−1
3.	1	−1	1	1	−1	1	−1	−1	−1	1	1
4.	1	−1	−1	−1	1	1	1	−1	1	1	−1
5.	1	−1	1	−1	−1	−1	1	1	1	−1	1
6.	−1	1	1	1	−1	1	1	−1	1	−1	−1
7.	−1	1	−1	−1	−1	1	1	1	−1	1	1
8.	−1	−1	−1	1	1	1	−1	1	1	−1	1
9.	−1	−1	1	1	1	−1	1	1	−1	1	−1
10.	−1	1	1	−1	1	−1	−1	−1	1	1	1
11.	1	1	−1	1	1	−1	1	−1	−1	−1	1
12.	−1	−1	−1	−1	−1	−1	−1	−1	−1	−1	−1
End	0	0	0	0	0	0	0	0	0	0	0

TABLE 8 Chromatographic Factors Potentially Investigated in a Robustness Study

Factor	Units	Limits of factor levels
1. Column	Different lots	Preferentially two different lots of which at least one is new
2. Column temperature	°C	± 5 absolute
3. pH of the buffer	Units	± 0.2 absolute
4. Percentage organic modifier in the mobile phase at the start of the gradient or isocratic conditions.	%	± 1 % absolute
5. Slope of the gradient	%/min	± 5% relative (expressed in gradient time)
6. Mobile phase flow rate	mL/min	± 0.1 absolute
7. Concentration of the buffer	mM	± 10% relative
8. Injection volume	μL	± 25% relative
9. Concentration of additives	mM or %	± 10% relative
10. Detection wavelength	nm	± 3 nm absolute
11. Source of the organic solvent	Vendors	Two different vendors
12. Equipment	Vendors	Two different vendors of the same quality equipment
13. Others, ...		

described in the method description. This is done to demonstrate effects of the studied parameters on the recovery in the presence of sample matrix.

The recovery of the main compound (100%) and all relevant impurities spiked with, for example, 0.5% wt/wt of a recent and worst-case sample batch, is calculated against a reference solution prepared according to the conditions as described per experimental run of the design. This is done to demonstrate effects of the studied parameters on the extraction of the main compound and the relevant impurities.

In addition, depending on the situation, other responses such as reporting threshold recovery, etc. may also be considered. The injection of the selectivity batch solution is used to test the suitability of the system.

After data analysis, the statistically significant effects induced by certain factors are identified. This is typically achieved using standardized Pareto plots (Figure 14). The plot shows the standard values of the effects in descending order of magnitude. The length of each bar is proportional to the standardized effect, which is equal to the effect divided by the standard error. This is identical to the calculation of a *t*-value for

TABLE 9 Sample Preparation Factors Potentially Investigated in a Robustness Study

Factor	Units	Limits of factor levels
1. Percentage of organic modifier in the extraction solvent	%	± 10% relative
2. Sample matrix concentration	Mg or units	± 10% relative or ±10 mg absolute or adding yes/no of 1 placebo solid unit
3. pH of the extraction buffer	Units	± 0.1 absolute
4. Concentration of the extraction buffer	mM or mg	± 10% relative or ±10 mg absolute
5. Mixing and equilibration times	min	± 10% relative
6. Filter type	Vendors	Two different vendors
7. Extraction temperature	°C	± 5 absolute
8. Dilution ratio	Different volume ratio's with similar final concentration	A minimal volume ratio and a maximal volume ratio (e.g., 1 mL stock/10 mL final volume and 10 mL stock/ 100 mL final volume
9. Type of mixing	Mixing procedure	Shaking/ sonication
10. Shaking speed	RPM	± 10% relative
11. Blending time	min	± 10% relative
12. Concentration of active	in mg	± 10% relative or ±5 mg absolute
13. Filtration pressure	Filtration purge	Gradual/fast
14. Discard volume at filtration	mL	1 to 6 mL, absolute
15. Equipment used in sample preparation (mechanic shakers, blenders, etc.)	Different vendors	Two world wide available vendors
16. Others, ………		

each effect in a Student's *t*-test. The vertical line in the plot is used to decide which effects are significant. This line corresponds with the boundary of a 99% reliability level. Each effect that transgresses this vertical line is considered to be significant.

A statistical significant effect is not always relevant from the practical point of view. Therefore, a worst-case level combination experiment with regard to the studied response (e.g., resolution) is determined and performed with replicates. In this experiment, only the method parameters with major effects (both statistically significant and almost significant) are considered. As can be seen in Figure 15, the major effect, temperature and pH, are easily detected. The worst-case combinations

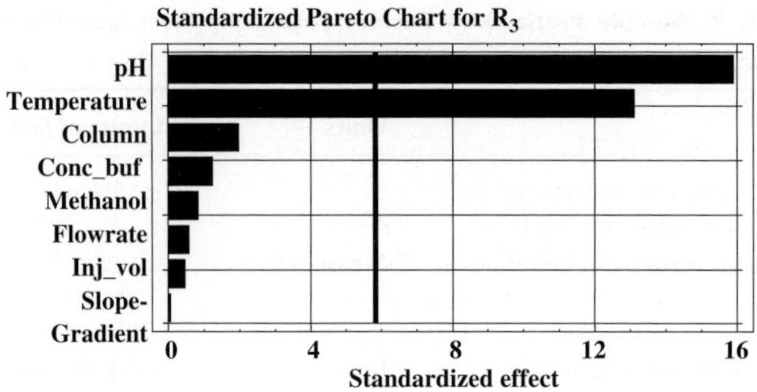

FIGURE 14 **Standardized Pareto plot. The standardized effects are plotted as a function of their extends. The dark line is the border of decision according to a t-test. Critical effects exceed this line.**

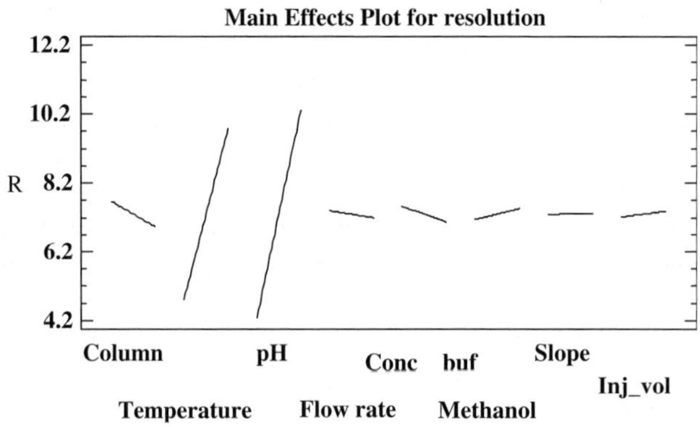

FIGURE 15 **An example of a main effect plot. The average responses at low level and high level of the factors are plotted.**

for resolution are conditions of low temperature and pH (decrease in resolution). The response for the worst-case experiment is then compared with the system suitability limit (if available) by a *t*-test to investigate whether the system suitability limit will be violated. When this occurs, the suspected factors are investigated quantitatively in order to describe their influence in more detail. In such a case limits of tolerable factor levels are defined clearly for the critical factors so that unexpected problems can be avoided. In the case where the system suitability limits are not yet set, it is possible to use the response of the worst-case experiment to determine appropriate limits that are more practical.

After method optimization the most important aspects of the validation are checked in a pre-validation study to examine whether the method is ready for evaluation at the customer site. If necessary, method optimization may need to be repeated. Suggested parameters to be checked include: specificity/selectivity against placebo; stressed placebo; stressed samples; relevant impurities and degradation products; verification of the reporting threshold; linearity evaluation for the API; filtration studies; stability of solution studies; method comparison by evaluation of batch results obtained with the early and final development methods; and a small precision study performed by analyzing a batch in triplicate. It is recommended to check peak purity of the API by using LC-PDA, LC-MS, or orthogonal methods.

The concept of performing the AMERT was introduced by Crowther et al.[55] and has been applied extensively within our laboratories. The AMERT is a critical component of the analytical method development process as it allows one to check whether a method performs adequately for the intended purposes and complies with the specific country requirements with respect to analytical methods. An AMERT enables the development laboratory to take into account valuable feedback from the receiving laboratories on the completeness of the method description and on the performance of the test. It is therefore scheduled prior to the time-consuming validation exercise. At this stage the acceptability of the test method description by the customer should be challenged. Method development is continued only after approval of the method description by the receiving laboratory. An additional goal is to facilitate the transfer of the methods immediately following the validation.

Measurement systems analysis (MSA)[61] aims to assess the appropriateness of the analysis method in its application environment. It is explicitly chosen to refer to a system, because the outcome of an analysis is determined not only by the quality of the method description, but also by the equipment, reagents, the lab environment, well-trained analysts, a good understanding of procedures, definitions, nomenclature, etc. MSA is performed by applying a Gage R&R study that investigates the precision and the discriminative ability of the measurement method and relates this with process variation and the product specifications. The Gage R&R study will tell whether the analytical method can discriminate differences in the measured properties (e.g., assay value) of the batches (parts) and therefore can be used for process improvement/control and for acceptance testing. The repeatability and reproducibility of the measurement system is typically estimated from a designed trial across different batches and labs. Since products are tested according to analytical method descriptions, the complete analytical method, i.e., from sample receipt to result, has to be investigated. In the Gage R&R study a randomized block design is applied, in which the two major factors (analyst and batch) are crossed and randomized. As shown in Figure 16, the analytical variance that is observed in analytical results obtained after the analysis of different production batches is the total variance due to the method and to the production process. In order to determine the

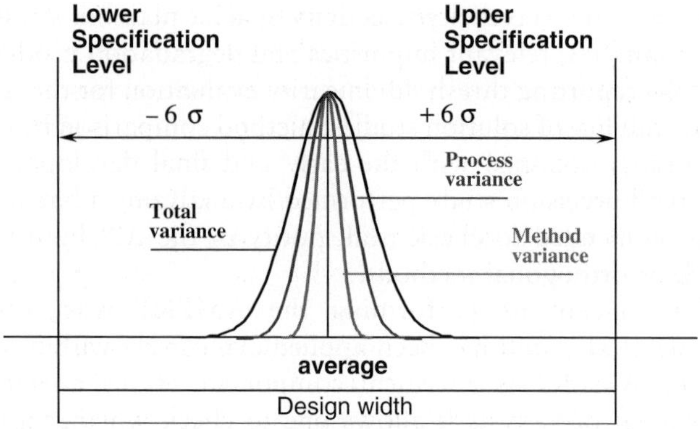

FIGURE 16 **Diagram indicating that the total variance in the analysis results equals the sum of the method variance and the process variance. A capable measurement system has a method variance that is less than 30% of the design width (difference between upper specification level and lower specification level). The production process is considered to be under full control when the average assay value is centered at Six Sigma values away from the lower and upper specification levels.**

contribution of the method variance from that of the production process variance an ANOVA method is applied.

By applying a Gage R&R study it is possible to assess the reproducibility and the method capability for process improvement/control and for acceptance testing against specifications. The following metrics are calculated:

- The % *R&R*-value (repeatability and reproducibility), addresses the portion of the observed total process variation that is taken up by the measurement error. The % R&R-value is an important metric for the method capability towards application in process improvement studies.

$$\%R\&R = (s_{\mathrm{MS}}/s_{\mathrm{Total}}) \times 100$$

where s_{MS} is the overall standard deviation of analytical method and s_{Total} is the observed total process variation (method component included).

- % *P/T*-value (precision to tolerance ratio), addresses what portion of the specification is taken up by the measurement error. The % *P/T*-value is an important metric for the method capability towards application in release testing against specifications.

$$\%P/T = 5.15 \times s_{\mathrm{MS}} \times 100/(\mathrm{USL} - \mathrm{LSL})$$

where s_{MS} is the overall standard deviation of analytical method, USL is the upper specification limit, and LSL is the lower specification limit.

The samples (minimum four batches) included in the study should cover the expected normal variation of the process (target ±3 sigma). If the batches used do not represent the full and normal process variation, the calculations are based on a historical value for process variation. The same batches are analyzed by multiple analysts (minimum 2) in different laboratories (minimum 3) using their own instruments, reagents, and solvents. Each analyst performs the entire method as described. Every sample should be analyzed at least twice (with independent sample preparation) in the same run. The replicates should also be blinded and randomly tested.

As an example, the results obtained for method precision and capability are presented in Table 10 for a Gage R&R study performed for an oral film-coated tablet in four different labs, using six sample batches, and analyzed in replicate by two analysts in each lab. The %*P/T* metric was calculated according to the Ph. EUR[62] specifications: 95–105% and according to the USP[63] specifications: 90–110%. The method standard deviation (apart from the process standard deviation) is also presented, together with average assay results and confidence intervals thereof per lab and for all labs together.

The %*R&R* of 52.9% implies that approximately half of the observed total process variation is taken up by the measurement error. This percentage is mostly caused by the repeatability (the difference between the replicates by the same analyst). The %*P/T* of 50.4% implies that when testing against the specification limit of 95.0–105.0%, half of the specification width is taken up by the method imprecision. This means that there will be a high risk of unnecessarily rejecting batches when analyzed with this method against the considered specifications. The method is more capable of being applied for testing according to USP specifications (90.0–110.0%), since a %*P/T* of 25.2% is obtained. For lab 1, which was the development lab, a better method capability is obtained compared to the application labs 2, 3, and 4. Lab 1 has developed the method and therefore knows the method in full detail. Usually

TABLE 10 Results of a Gage R&R Study for a Typical Oral Film-coated Tablet Assay Procedure

Laboratory	%R&R	%P/T (Ph. Eur) for 95–105%	%P/T (USP) for 90–110%	STDEV method	Mean assay	95% CI
Lab 1	32.6	28.3	14.1	0.55	97.2	±1.10
Lab 2	49.6	55.0	27.5	1.07	96.6	±2.14
Lab 3	55.1	48.8	24.4	0.95	96.3	±1.90
Lab 4	56.1	46.0	23.0	0.89	96.1	±1.78
Total	52.9	50.4	25.2	0.98	96.5	±1.96

STDEV = standard deviation; CI = confidence interval.

the development lab works under the most optimal conditions for the method, resulting in better performance. The difference between the two types of labs may be an opportunity for the application labs. With proper training, guidance, and support both types of labs should be able to achieve comparable performance.

The overall method standard deviation is quite good (0.98%) and is mostly caused by within analyst variation. Such a level of precision is typical for HPLC DP assay methods and will be difficult to improve further. The required effort and costs of reducing the method variance further will be high and technically very demanding to the QC labs (increased effort for little extra quality). This means that such an HPLC assay is less capable of being applied for QC release testing against the Ph. EUR specifications. One way of reducing the analytical variation further is by increasing the number of sample preparations. For example, to detect differences of 2% between batches using a standard deviation of the analytical method of 1.0% and a product variation of 0.7% (resulting in a total standard deviation of 1.2), a sample size of 6 should be used. For differences of 3%, four samples are sufficient and for 4%, three samples should be used. Moreover, when the average assay value of the produced batches is at 100% the impact may be minor, but when there is an offset in the content of API in the produced batches there is a higher probability that batches will be rejected.

An example of the Gage Run Chart is shown in Figure 17, in which all measurement raw data are depicted in one graph. This is an example

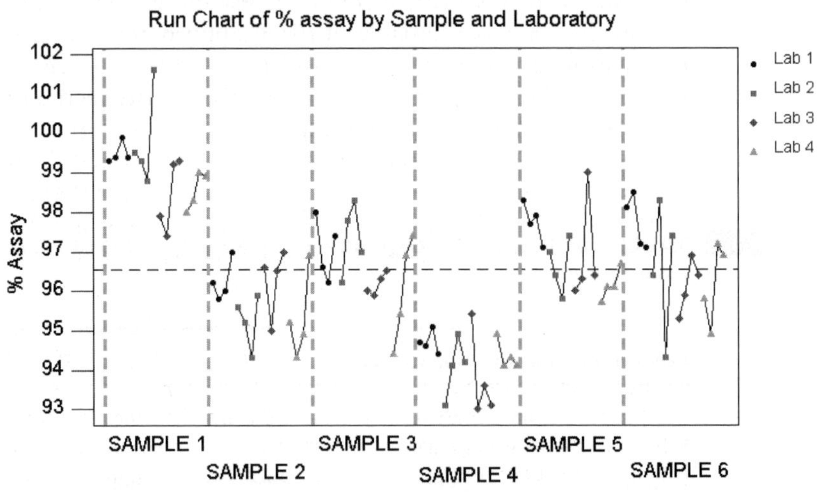

FIGURE 17 Gage Run Chart of an HPLC method for the determination of API in an oral tablet formulation. Six samples are analyzed with: two replicates by two analysts per lab at four different sites. For each lab four results are plotted, the first two are from analyst 1 and last two results are from analyst 2.

for a Gage R&R study of an HPLC method for the determination of active ingredient in an oral film-coated tablet formulation. The method was intended to be transferred to four different QC-labs. The study is performed on six samples with two replicates by eight analysts (two different analysts in four different laboratories). From this visualization the reproducibility and repeatability of the method is immediately observed. For example it can be observed that all lab average results are well correlated. The repeatability of the method appears to be less satisfactory (e.g., results of analyst 2 in lab 2).

F. Final Method Validation

Validation is the process of collecting documented evidence that the method performs according to the intended purpose.[1,2,56,64,65] The validation characteristics and the acceptance criteria to be applied in validation of HPLC methods for MAA/NDA filings and marketed products should comply with the international guidelines on method validation.[56,63,64] Table 11 details validation activities to be conducted for type 1, type 2, and type 3 methods:

- Type 1: assay and identification of the main component(s) (active, preservatives and key excipients).
- Type 2: assay and identification of the impurities.
- Type 3: limit tests.

Further discussion of method validation can be found in Chapter 7. However, it should be noted from Table 11 that it is frequently desirable to perform validation experiments beyond ICH[1–4] requirements. While ICH addresses specificity, accuracy, precision, detection limit, quantitation limit, linearity, and range, we have found it useful to additionally examine stability of solutions, reporting threshold, robustness (as detailed above), filtration, relative response factors (RRF), system suitability tests, and where applicable method comparison tests.

The stability of solutions is tested by determining the period of time during which the analytes of interest remain stable under the described conditions. The acceptance criteria are based on relative differences between the initial value and the value at the specified solution stability time. For the API, this relative difference should, for example, be ≤2.0%. For the impurities the relative differences may range from ≤50.0% at the reporting threshold level to ≤10.0% at 2× the specification limit. Additionally, there should be no new degradation compounds formed during storage at levels higher than the reporting threshold.

Filtration studies are conducted to investigate the filtration process (e.g., the binding of the analytes of interest to the applied filtration device, filtration discard volume, and the extraction of contaminants).

TABLE 11 Validation Parameters for Different Types of Analytical Test Methods

Test type	Type 1	Type 2	Type 3
Validation characteristics mandatory			
Specificity[1,2]	+	+	+
Accuracy	+	+	–
Precision			
System repeatability	+	+[3]	–
Analysis repeatability	+	+	–
Intermediate precision[4]	+	+	–
Reproducibility	+	+	–
Detection limit	–	–[5]	+
Quantitation limit	–	+	–
Linearity	+	+	–
Range	+	+	–
Additional validation characteristics			
Stability of solutions	+	+	+
Reporting threshold	–	+	–[6]
Robustness[7]	+	+	–
Filtration study	+	+	+
Relative Response Factor	–	+	–
System Suitability Tests	+	+	+
Method Comparison	+	+	+

–, Performance characteristic not evaluated; +, performance characteristic is evaluated.
[1]A combination of two or more analytical procedures can compensate for the lack of specificity for an individual identity test.
[2]Lack of specificity for an assay used for release may be compensated by testing for impurities.
[3]Only required when assay of main compound and purity determinations are not combined in one method.
[4]Intermediate precision may not be necessary if reproducibility has been evaluated.
[5]May be needed in some cases.
[6]May be needed in some cases.
[7]Method robustness may be evaluated during method development.

Filtered and unfiltered samples are compared for the API and the impurities. Additionally, there should be no new impurities higher than the reporting threshold detected.

The RRF is calculated and reported. For impurities for which no authentic reference material is available and for unknown peaks, the RRF is assumed to be 1.00.

The reporting threshold is a limit above which (≥) an impurity needs to be reported. The reporting threshold is determined as described by the

ICH guidelines,[3,4] and is generally 0.05% for DS and 0.10% for DP methods. Any impurity above this level should be reported. Especially for stability indicating methods, it may be valuable to apply a lower level as the reporting threshold to allow for a better trending process for degradation products at levels below 0.05% and 0.10% for DS and DP methods, respectively.

System suitability test characteristics and limits are recommended as a component of any analytical method. This ensures that both methodology and instrumentation are performing within expectations prior to the analysis of test samples. The test characteristics are inferred from robustness studies and evaluated during the validation experiments.

A method comparison study is performed to demonstrate that the new final method is adequate for the purpose by comparing its analyses results with the previously existing methods. It is expected that the newly developed final method will be better than the previous methods. Differences in analysis performance should be evaluated according to sound scientific judgment.

Changes in the analytical method or manufacturing processes may necessitate re-validation to ensure that the analytical method maintains its performance characteristics. The degree of re-validation depends on the nature of the change and should be assessed on a case-by-case basis.

In the process we have outlined, method validation is not considered to be the most time-consuming activity in the method development process (only ±15% of the development time).

G. Method Transfer

While method transfer remains a formal process to demonstrate equal method performance between the development and the application laboratories, the continuous involvement of the customers during method development greatly facilitates the process. The final method is not "new" for the application laboratories at the time of transfer and the transfer process is not the primary challenge of the method.

For each test it is advised to analyze the samples with replicates (e.g., n=6). The acceptance criteria are based on the method purpose and on the validation characteristics. A recommended approach would be to propose acceptance limits based on the results of the Gage R&R study. If this study is not performed the Horwitz equation[59] can be used to relate method repeatability with method reproducibility. Typically the difference in average assay values for DS methods should be within 2.0% and the precision should be less than 1.0% RSD in each laboratory. For DP, these limits are 3.0% and 2.0% for average assay difference and for the precision in each lab, respectively. The impurities are usually considered at a 0.5% level and the typically allowed difference between labs should be less than 0.2% (absolute) with a within lab precision of less than 15% RSD in each lab.

H. Method Performance Monitoring and Feedback

Monitoring the performance of methods during real time analysis will generate a historical source of objective data and will be the basis for future discussions on the method between the application and development labs. For the development lab the objective data serve as a valuable source to detect method shortcomings and to find out which aspects of the method should be improved. Moreover, the method feedback will be used to steer the method development process in general. The feedback on method performance is discussed regularly, but it is recommended that this be done at least once a year. Methods applied during phase 3 and post-approval should be monitored in order to track and record their performance in real applications. During the monitoring a number of key performance indicators are recorded and filled out in feedback sheets by the application labs (Stability and Operational labs) each time the method is applied. The method feedback sheet is sent together with the method description to the application labs at transfer. A completed example of such a feedback sheet is shown in Figure 18.

Several method performance indicators are tracked, monitored and recorded. Items that are recorded include: the date of analysis, identification of the HPLC system, identification of the analyst, number and type of samples analyzed, the system precision, the critical resolution or

Laboratory: Stability Evaluation and Release
Method: - Drug product
Date: Start July 2001

Key Process Output Variable Feedback Sheet

General					SST								OOS	Deviations SST failure			
Date	System type	Initials Analyst	number of samples	Sample type	RSD	R	T	RT-check	Ref 2	Control references	Selectivity (separation pass/fail?)	Blank peaks (Y/N)	OOS reference number	NC reference number	Carry over Y/N	Other causes	Comments
23-Jul-01	AO-0442	KMAE	4	F32_F34	0.20	6.6	1.6	98	99.8	99.8	PASS	Y			N		KB101113
06-Aug-01	AO-0435	DDEK	6	F32_F34	0.10	4.2	1.6	89	99.4	99.5	PASS	N			N		KB101118
07-Aug-01	AO-0435	DDEK	18	F32_F34	0.10	4.3	1.6	95	99.6	99.9	PASS	N			N		KB101119
21-Aug-01	AO-0445	DDEK	6	F32_F34	0.10	6.2	1.9	80	100.0	99.8	PASS	N			N		KB101126
04-Sep-01	AO-0444	ADES	12	F32_F34	0.10	7.0	1.7	114	100.9	99.8	PASS	Y			N		KB101138
12-Sep-01	AO-0445	DDEK	12	F32_F34	0.10	6.4	1.5	101	99.8	99.8	PASS	Y			N		KB101140
18-Sep-01	AO-0445	DDEK	4	F32_F34	0.30	6.3	1.6	98	99.8	100.0	PASS	Y			N		KB101144
18-Sep-01	AO-0442	DDEK	6	F32 F34	0.20	6.7	1.6	93	99.7	100.2	PASS	Y			N		KB101145
19-Sep-01	AO-0445	DDEK	6	F32	0.10	6.2	1.6	104	99.8	100.1	PASS	Y			N		KB101147
20-Sep-01	AO-0445	DDEK	18	F32	0.20	6.1	1.6	98	99.6	99.8	PASS	Y			N		KB101149
20-Sep-01	AO-0442	DDEK	6	F34	0.10	6.6	1.5	93	99.8	100.0	PASS	Y			N		KB101150
21-Sep-01	AO-0442	DDEK	18	F34	0.20	6.7	1.5	99	99.8	100.2	PASS	Y			N		KB101152
27-Sep-01	AO-0445	DDEK	16	F32-F34	0.20	6.2	1.6	104	100.0	99.9	PASS	Y	LO111059		N		KB101156
02-Oct-01	AO-0445	DDEK	4	F32-F34	0.20	6.5	1.6	102	99.7	99.8	PASS	Y			N		KB101159
03-Oct-01	AO-0445	DDEK	12	F32-F34	0.20	6.4	1.6	106	99.9	99.8	PASS	Y			N		KB101160
12-Oct-01	AO-0445	DDEK	8	F32-F34	0.20	6.4	1.6	98	99.9	100.2	PASS	Y			N		KB101167
15-Oct-01	AO-0445	DDEK	6	F32-F34	0.50	6.5	1.6	103	99.7	100.0	PASS	Y			N		KB101168
16-Oct-01	AO-0445	DDEK	18	F32-F34	0.20	6.4	1.5	100	99.5	100.2	PASS	Y			N		KB101169
26-Oct-01	AO-0445	DDEK	2	F32	0.10	6.3	1.5	101	99.8	99.9	PASS	Y		NC112440	Y		KB101178
14-Nov-01	AO-0445	DDEK	10	F32-F34	0.10	6.3	1.6	103	99.6	99.7	PASS	Y			N		KB101206
28-Nov-01	AO-0445	DDEK	18	F32-F34	0.40	6.4	1.6	100	99.8	100.4	PASS	Y			N		KB101224
03-Dec-01	AO-0445	DDEK	18	F32-F34	0.30	6.4	1.6	98	100.0	99.8	PASS	Y			N		KB101229
05-Dec-01	AO-0444	DDEK	8	F32-F34	0.10	6.6	1.7	105	99.6	99.8	PASS	N			N		KB101231
		Upper Limit			2.00		2.6	150.0	102.0	102.0							
		Lower limit				3.0		50.0	98.0	98.0							

FIGURE 18 Example of a completed feedback sheet.

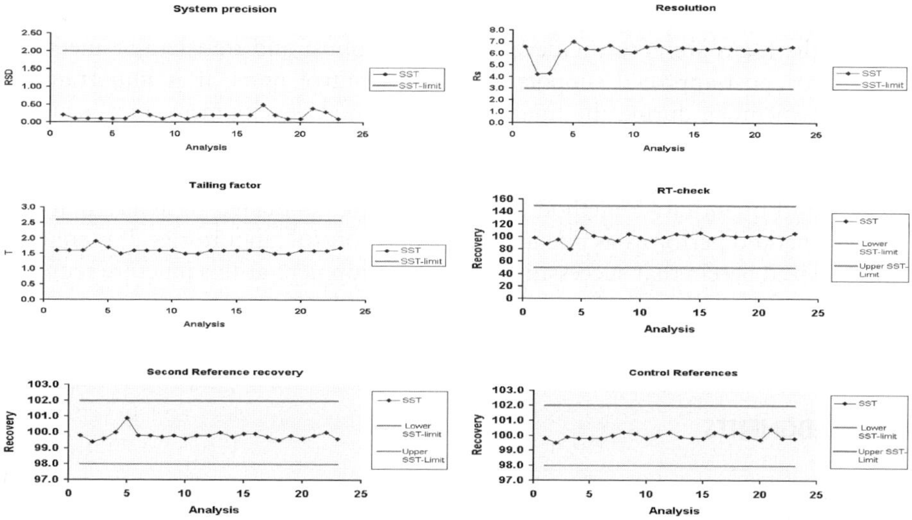

FIGURE 19 **Control charts of six quantitative method performance indicators.**

tailing factor, the recovery at the reporting threshold level, the recovery of a check standard, the recovery for the control references (repeated reference injections for evaluation of system drift), the separation quality, blank issues, out-of-specification issues, carry-over issues, and other non-conformances. The quantitative indicators are additionally visualized by plotting on control charts (Figure 19). As can be observed in Figure 19, the system precision for a half-year time period (23 analyses sequences) is always below 0.5% RSD. Additionally, the second reference, the control reference and the reporting threshold level have quite stable recovery values.

VI. CONCLUSIONS

HPLC method development of pharmaceuticals begins with the generation of an orthogonal array of methods suited for the separation of early DS lots, synthetic intermediates and starting materials, excipients (where known) and products from forced decomposition studies. From the array, two orthogonal methods are selected to support subsequent stages of pharmaceutical development. One of these methods is optimized to separate all known related substances and excipients using chromatographic optimization software such as DryLab. The other method is used to analyze pivotal lots of DS and DP formulations to assure that the primary method continues to be viable for the separation of all relevant components.

Following finalization of DS synthetic routes and DP formulation, the focus shifts to the development of robust and transferable methods for post-approval support at quality control units. It is important to remember, during the final stage of method development, that achievement of separation conditions is only one parameter necessary for successful method implementation. Extensive studies to measure robustness and quantitative method performance are conducted to assure that the method performs as intended in quality control laboratories. It is emphasized herein that successful method development at this juncture requires extensive cooperation between the development laboratory and the receiving quality control laboratories.

ACKNOWLEDGMENTS

The authors acknowledge the leadership of Dr. Maurits De Smet in establishing the internal processes described herein. We thank Drs. Greg Worosila, Xiande Wang and Kelly Swinney for critical review of the manuscript.

REFERENCES

1. ICH Guideline Q2A: Text on Validation of Analytical Procedures, 1994. http://www.ich.org/pdfICH/Q2A.pdf
2. ICH Guideline Q2B: Validation of Analytical Procedures: Methodology, 1996. http://www.ich.org/pdfICH/Q2B.pdf
3. ICH Guideline Q3A(R): Impurities in New Drug Substances (revised guideline), 2002. http://www.ich.org/pdfICH/Q3ARStep4.pdf
4. ICH Guideline Q3B(R): Impurities in New Drug Products (revised guideline), 2003. http://www.ich.org/pdfICH/Q3Bstep4.pdf.
5. Rasmussen, H. T., Method Development, Chapter 10. In *Handbook of Modern Pharmaceutical Analysis,* Ahuja, S. and Scypinski, S., Eds., Academic Press, New York, 2001.
6. Bakshi, M. and Singh, S., *J. Pharm. Biomed. Anal.*, 28:1011, 2002.
7. Stability Testing of New Drug Substances and Products, ICH Guideline Q1A.
8. Singh, S. and Bakshi, M., *Pharm. Tech.*, On-Line, April, 1, 2000 www.pharmportal.com.
9. Majors, R. E., *LC-GC,* 16:336, 1998.
10. Neue, U. D., Alden, B. A. and Walter, T. H., *J. Chromatogr., A*, 849:101, 1999.
11. Rigney, M. P., Weber, T. P. and Carr, P. W., *J. Chromatogr.*, 484:273, 1989.
12. Zhao, J. and Carr, P. W., *Anal. Chem.,* 72:302, 2000.
13. Alpert, A. J., *J. Chromatogr.*, 499:177, 1990.
14. Oyler, A. R., Armstrong, B. L., Cha, J. Y., Zhou, M. X., Yang, Q., Robinson, R. I., Dunphy, R. and Burinsky, D. J., *J. Chromatogr.*, 724:378, 1996.
15. Olsen, B. A., *J. Chromatogr.,* 913:113, 2001.
16. Snyder, L. R., Kirkland, J. J. and Glajch, J. L., *Practical HPLC Method Development,* 2nd ed., Wiley, New York, 1997.
17. Dolan, J. W., *LC-GC* 18:478, 2000.
18. Snyder, L. R., Dolan, J. W. and Rigney, M. P., *LC-GC* 4:921, 1986.

19. Snyder, L. R. and Dolan, J. W., *J. Chromatogr.*, 721:3, 1996.
20. Lewis, J. A., Snyder, L. R. and Dolan, J. W., *J. Chromatogr.*, 721:15, 1996.
21. Dolan, J. W., Snyder, L. R., Djordevic, N. M., Hill, D. W. and Waeghe, T. J., *J. Chromatogr.*, 857:1, 1999.
22. Dolan, J. W., Snyder, L. R., Djordevic, N. M., Hill, D. W. and Waeghe, T. J., *J. Chromatogr.*, 857:21, 1999.
23. Dolan, J. W., Snyder, L. R., Wolcott, R. G., Haber, P., Baczek, T., Kaliszan, R., Sander, L. C., *J. Chromatogr.*, 857:41, 1999.
24. Snyder, L. C. and Dolan, J. W., *J. Chromatogr.*, 892:107, 2000.
25. Galushko, S. V. and Spez, G. I. T., *Chromatographia*, 16:88, 1996.
26. Galushko, S. V., Deguchi, K. and Ito, M., *Chromatography (Japan)*, 18:298, 1997.
27. Csokan, P., Valko, K., Darvas, F. and Csizmadia, F., *LC-GC*, 6:361, 1993.
28. Fekete, J., Morovjan, G., Csizmadia, F. and Darvas, F., *J. Chromatogr.*, 660:33, 1994.
29. Csokan, P., Darvas, F., Csizmadia, F. and Valko, K., *LC-GC*, 12:1, 1994.
30. Smith, T. P. I, R., Guhan, S., Taksen, K., Vavra, M., Myers, D. and Hearn, M. T. W., *J. Chromatogr.*, 972:27, 2002.
31. Stadalius, M. A., Ghrist, B. F. D. and Snyder, L. R., *J. Chromatogr.*, 387:21, 1987.
32. Catalono, T., Madsen, G. and Demarest, C., *Am. Pharm. Rev.*, 3:62, 2000.
33. Letter, W. S., *LC-GC*, 15:508, 1997.
34. Glajch, J. L., Kirkland, J. J., Squire, K. M. and Minor Jr., J.M., *J. Chromatogr.*, 199: 57, 1980.
35. Kiel, J. S., Morgan, S. L. and Abramson, R. K., *J. Chromatogr.*, 485:585, 1989.
36. Van Gyseghem, E., Van Hemelryck, S., Daszykowski, M., Questier, F., Massert, D. L. and Vander Heyden, Y., *J. Chromatogr.*, 988:77, 2003.
37. Meyer, V., *Anal. Chem.*, 72:392A, 2000.
38. Swartz, M. E., *J. Liq. Chromatogr.*, 14:923, 1991.
39. Neue, U. D., Alden, B. A. and Walter, T. H., *J. Chromatogr.*, 849:101, 1999.
40. Visky, D., Vander Heyden, Y., Ivanyi, T., Baten, P., De Beer, J., Kovacs, Z., Noszal, B., Roets, E., Massart, D. L. and Hoogmartens, J., *J. Chromatogr.*, 977:39, 2002.
41. Euerby, M. and Petersson, P., *J. Chromatogr.*, 994:13, 2003.
42. Kele, M. and Guiochon, G., *J. Chromatogr.*, 830:41, 1999.
43. Kele, M. and Guiochon, G., *J. Chromatogr.*, 830:55, 1999.
44. Kele, M. and Guiochon, G., *J. Chromatogr.*, 855:423, 1999.
45. Kele, M. and Guiochon, G., *J. Chromatogr.*, 869:181, 2000.
46. Felinger, A., Kele, M. and Guiochon, G. *J. Chromatogr.*, 913:23, 2001.
47. Snyder, L. R., *J. Chromatogr.*, B 689:105, 1997.
48. Jimidar, M., De Smet, M., Sneyers, R., Van Ael, W., Janssens, W., Redlich, D. and Cockaerts, P., *J. Cap. Elec. and Microchip Tech.*, 8:45, 2003.
49. Antonovich, R. S. and Keller, P. R., *J. Chromatogr.*, 971:159, 2002.
50. Shabir, G. A., *J. Chromatogr.*, 987:57, 2003.
51. ICH Guideline Q1B: Photostability Testing of New Drug Substances and Products, 1996. http://www.ich.org/pdfICH/Q1B.pdf.
52. Allgeier, M. C., Nussbaum, M. A. and Risley, D. S., LC-GC (*North America*) 21:376, 2003.
53. Renger, B., *J. Chromatogr. B*, 745:167, 2000.
54. Kristensen, E. and Varney, M., *Pharm. Forum*, 23:4952, 1997.
55. Crowther, J. B., Salomons, P. and Callaghan, C., Analytical Method Development for Assay and Impurity Determination in Drug Substances and Drug Products, Chapter 12. In *Analytical Chemistry in a GMP Environment*, Miller, J. M. and Crowther, J. B., Eds., John Wiley & Sons, Inc., New York, 2000.
56. Crowther, J. B., Jimidar, M. I., Niemeier, N. and Salomons, P., Qualification of Laboratory Instrumentation, Validation and Transfer of Analytical Methods. Chapter 15. In *Analytical Chemistry in a GMP Environment*, Miller, J. M. and Crowther, J. B., Eds., John Wiley & Sons, Inc., New York, 2000.

57. Jimidar, M., Niemeier, N., Peeters, R. and Hoogmartens, J., *J. Pharm. Biomed. Anal.*, 18:479, 1998.
58. Vander Heyden, Y., Jimidar, M., Hund, E., Niemeijer, N., Peeters, R., Smeyers-Verbeke, J., Massart, D. L. and Hoogmartens, J., *J. Chromatogr.*, 845:145, 1999.
59. Massart, D. L., Vandeginste, B. G. M., Deming, S. N., Michotte, Y. and Kaufman, L., *Chemometrics: A Textbook, Data Handling in Science and Technology,* Vol. 2, Elsevier Science Publishers, Amsterdam, Netherlands, 1998.
60. Morgan, E., *Chemometrics: Experimental Design, Analytical Chemistry By Open Learning*, John Wiley and Sons, Chichester, 1991.
61. Wheelerand, D. J. and Lyday, R. W., *Evaluating the Measurement Process*, SPC Press Inc., Knoxville, TN, 1989.
62. European Pharmacopoeia, 4th ed., Directorate for the Quality of Medicines of the Council of Europe, Strasbourg, France, 2002.
63. The United States Pharmacopeia, National Formulary 26, United States Pharmacopeial Convention, Rockville, USA, 2002.
64. FDA Draft Guidance for Industry on Analytical Procedures and Methods Validation: Chemistry, Manufacturing, and Controls Documentation, 2000, http://www.fda.gov/cder
65. Green, C., *J. Val. Tech*., 6:622, 2000.

7 VALIDATION OF HPLC METHODS IN PHARMACEUTICAL ANALYSIS

ASHLEY S. LISTER

Analytical Sciences, Worldwide Research & Development, Purdue Pharma, Ardsley, NY

Handbook of Pharmaceutical Analysis by HPLC,
S. Ahuja and M.W. Dong, editors.

ABSTRACT

This chapter focuses on approaches to the validation of high-performance liquid chromatography methods based on regulatory guidance documents and accepted industry practices. The information in this chapter gives a brief review of the reasons for performing method validation and the regulations that describe this activity. Individual validation parameters are discussed in relation to the type of method to be validated. Examples of typical validation conditions are presented with references to additional information on individual topics. This chapter was written to help analysts responsible for method validation.

I. INTRODUCTION

Method validation is the process of ensuring that a test procedure performs within acceptable standards of reliability, accuracy, and precision for its intended purpose. In short, validation is the act of confirming that a method does what it is intended to do. Although validation is required by law and is needed for all regulatory methods, the actual execution of validation activities is open to interpretation and practices differ greatly between organizations. This chapter will focus on validation as it relates to regulatory methods using high-performance liquid chromatography (HPLC) and is intended as a general guide for the analyst performing method validation.

A. Background

Why do we have to worry about method validation in the first place? The most compelling reason is that the law requires it. Title 21, Part 211 of the Code of Federal Regulations on Current Good Manufacturing Practice for Finished Pharmaceuticals states:

> The accuracy, sensitivity, specificity, and reproducibility of test methods employed by the firm shall be established and documented.[1]

The Food and Drug Administration (FDA) is the federal agency authorized by Congress to enforce the standards of strength, quality, purity, packaging and labeling as defined by 21CFR Part 211 and the Federal Food, Drug and Cosmetic Act.[1,2] Within the FDA, the Center for Drug Evaluation and Research (CDER) is charged with assuring that safe and effective drugs are available to the public. CDER is responsible for reviewing and approving (or denying) New Drug Applications (NDA) submitted to the FDA. In addition, CDER has published a number of guidance documents that represent the FDA's current position on a particular subject.

B. Regulatory Guidance Documents

Among the guidance documents published by CDER are several that deal with method validation.[3–9] These documents provide guidelines on the validation of regulated products and establish policies intended to achieve consistency in the FDA's regulatory approach and establish inspection and enforcement procedures. Because guidance documents are not regulations or laws, they are not enforceable either through administrative actions or through the courts. In addition to the FDA, the International Conference on Harmonisation (ICH) of Technical Requirements for Registration of Pharmaceuticals for Human Use has published guidance documents for method validation.[3,4] The ICH guidances were developed to harmonize the registration requirements of method validation between the United States, Japan, and the European Union, and they do not necessarily cover all requirements that may be required in other parts of the world. These are general documents that seek to provide clear definitions of terms between the different compendia and regulatory agencies within the European Union, the United States, and Japan. The United States Pharmacopoeia (USP) is published by a non-government organization whose publications of standards for the analysis of pharmaceuticals (*US Pharmacopoeia/National Formulary*) are officially recognized by the Federal Food, Drug and Cosmetic Act. General Chapter <1225> within the USP covers the requirements for validation of compendial methods and is generally included in any discussion of guidance documents relating to method validation.[9] Guidances are not considered, nor should they be used as directions on how to perform validation. Table 1 provides a current list of the published guidance documents pertaining to method validation.

TABLE 1 Regulatory Guidance Documents on the Validation of Analytical Methods

Agency	Title	Published
ICH	ICH-Q2A "Text on Validation of Analytical Procedure"	1994
	ICH-Q2B "Validation of Analytical Procedures: Methodology"	1995
CDER	Reviewer Guidance: Validation of Chromatographic Methods	1994
	Submitting Samples and Analytical Data for Method Validations	1987
	Analytical Procedures and Method Validation	2000
	Bioanalytical Method Validation for Human Studies	1999
USP	<1225> Validation of Compendial Methods	2003

1. Online Resources

All of the guidance documents described above are available online through the Internet and can be downloaded free of charge. The ICH tripartite guidance documents can be found at *http://www.ich.org/*. The FDA website at http://www.fda.gov/cder/ contains a full listing of all guidance documents, including those that relate to methods validation as described in Table 1. The entire text of the USP/NF is available through online subscription at http://www.uspnf.com/. Information about the subscription price and terms can be found at http://www.usp.org/.

2. References for Method Validation

In addition to online resources, other texts and references have discussed the process of validation for methods used in the pharmaceutical industry in relation to the regulatory guidance documents. These guides include discussions on method development in relation to method validation, the validation of non-chromatographic methods and stability indicating methods.[10–13]

II. VALIDATION REQUIREMENTS FOR METHOD TYPE

The amount and type of work required to validate an HPLC method varies depending on the purpose of the method. Validation activities should be organized so that the experiments planned complement the method it is intended to support. A method for monitoring degradation products in a finished product will require extensive validation of the limits of detection (LOD) and limits of quantitation (LOQ) due to the fact that these components are present in trace amounts. However, for the assay of the active pharmaceutical ingredient (API) there is no need to determine the LOD and LOQ since the method was never intended to operate at such low detection levels. The following sections describe the appropriate validation work required for each common method type. In the event that a method does not fit into one of the following categories, it is up to the analysts to use their discretion and sound scientific judgment to determine the appropriate amount of validation to support the method. Table 2 lists the validation requirements for each general method type as discussed by the USP. The ICH tripartite guidance documents only address methods in Categories I, II, and IV. Discussion of validation for Category III methods is not addressed in the ICH texts and may be included in subsequent documents.

A. Category I

Methods for the quantitation of major components of bulk drug substances or APIs, including preservatives, in finished drug products are classified in Category I. Assay and content uniformity methods fall into this category. These methods demonstrate that the label claim of the drug

TABLE 2 Validation Requirements for Each Type of Analytical Method

Validation parameter	Assay Category I	Assay Category II Quantitiative	Assay Category II Limit test	Assay Category III	Identification
Linearity	Yes	Yes	No	Yes	No
Accuracy	Yes	Yes	1	Yes	No
Precision	Yes	Yes	No	Yes	No
Range	Yes	Yes	1	Yes	No
Specificity	Yes	Yes	Yes	Yes	Yes
LOD	No	Yes	Yes	1	No
LOQ	No	Yes	No	1	No
Robustness	Yes	Yes	No	1	No

[1]May be required, depending on the nature of the method.

product meets current specifications. Assay methods are generally performed using a composite sample (e.g., 20 tablets) while content uniformity is performed on a single dosage unit. Trace analysis is usually not required in an assay or content uniformity method and determining LOD and LOQ is not necessary. However, LOD and LOQ may be important if the assay method is used to quantitate multiple active components present in different amounts (e.g., a product with a second active component at 0.5% relative to the main component). According to ICH, when HPLC is used in the measurement of dissolution, the method validation required is the same as in Category I.

B. Category II

Category II methods are intended to determine impurities in bulk drug substance, degradation products in finished drug product, or for verification of cleaning processes. These methods are further subdivided into quantitative tests and limit tests.

1. Quantitiative

The determination of process impurities and degradation products involves trace analysis where components present at less than 1% of the analytical concentration of the bulk drug substance or the label claim of the finished drug product. All validation parameters required for Category I methods and the LOQ should be determined for methods of this type. Cleaning methods are a special case and arguments can be made about where to place them within this category. For some purposes a limit test is appropriate for cleaning methods, but many times a quantitative result is required. Here, cleaning methods will be treated using the more rigorous criteria for Category II quantitative methods. Both the analytical method and the sampling method should be challenged to ensure whether contaminants can be recovered from the cleaning surface and to what level. Recovery of the contaminant from the rinse solution, the sampling device, and the cleaning surface must be evaluated depending on the cleaning procedure used. It has been suggested that the LOQ for cleaning methods should be at least two times more sensitive than the cleaning limit required for the cleaning process.[14]

2. Limit Tests

In the case of limit test, no quantitation is involved. A sample is run against a standard prepared at the specification level. The response of the sample is determined to be either above or below the standard amount and the results either pass or fail the specification. The LOQ is not required for this method type, but the LOD is needed. Specificity is the only other parameter required for validation, although accuracy and range may be determined depending on the nature of the method.

C. Category III

Methods used to determine the performance characteristics of finished products fall into Category III. Dissolution tests (excluding measurement) and drug release tests are examples of these types of methods. Precision is the only parameter required for these methods according to the regulatory guidances, although all validation parameters may be determined based on the intent of the method.

D. Category IV

Identification (ID) tests in Category IV require only specificity for their validation. Identification by HPLC usually involves comparison of the retention time (t_R) or relative retention time (RRT) of a sample and standard injection. The increasing use of photodiode array (PDA) detectors in HPLC methods also allows identification by comparison of UV spectra for standards and samples, in addition to retention characteristics. The information required for either ID test by HPLC can be gathered while performing any other HPLC method for a given sample. Identification tests are often incorporated into the assay method and the satisfactory completion of specificity for the assay will meet the requirements for ID as well.

III. PROCEDURES AND PROTOCOLS

Validation should always proceed according to a validation plan. A well-defined validation strategy saves time and effort on the part of the analyst by providing a structured work plan and prevents problems from regulatory agencies concerning validation activities. Inconsistent and poorly documented validation efforts may appear to be suspect even though the validation of the method is not in question. Consistent validation approaches signal to regulatory agencies that validation is a well-controlled and planned process.

Although there is no specific discussion of validation protocols in any of the guidance documents, ICH Q2B does mention that the applicant is responsible for generating an adequate validation protocol suitable for the drug substance or product. Two approaches have been used for planning validation activities. The first is to develop a single, broad-scope procedure such as a standard operating procedure (SOP) that covers all validation activities. This has the advantage of being a static document, with the result that all validation activities are similar in scope and execution. Validation follows a pre-set formula and the resulting validation reports are consistent within the respective company. The disadvantage of this approach is that the document must be extremely well written taking into account every conceivable possibility that may be encountered during the drug development process. A document that is too general may not provide enough information to guide the analysts in their validation efforts. One that is overly detailed may tie the hands of the analyst by forcing him/her to adopt validation requirements that are overly strict and unrealistic for the method in question.

A second approach is to prepare validation protocols for each method prior to the commencement of validation activities. This has the advantage of performing method-appropriate experiments while eliminating those portions of validation work that do not apply. It also allows the analysts to set more specific acceptance criteria based on the performance of the individual method instead. A drawback is that the protocol adds to the workload of the analyst since it must be prepared for each validation. This approach also allows for the possibility of greater variance in the approach to validation within the company.

IV. VALIDATION PARAMETERS

A. Specificity

Specificity is the ability of a method to discriminate between the analyte(s) of interest and other components that are present in the sample. Studies are designed to evaluate the degree of interference, if any, which can be attributed to other analytes, impurities, degradation products, reagent "blanks" and excipients. This provides the analyst with a degree of certainty that the response observed is due to the single analyte of interest. The degree of specificity testing varies depending on the method type and the stage of validation. Specificity should be evaluated continually through the drug development process. Specificity is sometimes used interchangeably with the term "selectivity". The argument over which term is more correct is one of semantics. Although there is some dissention, the term "specificity" has been adopted by the regulatory guidance documents and should be used to prevent further confusion.

1. Non-Interference of Placebo

This portion of specificity evaluation applies to the finished drug product only. Excipients present in the formulation should be evaluated and must not interfere with the detection of the analyte. Individual solutions of each excipient prepared at several times the normal concentration of the component in the drug product ensure that any detector response from the excipient will be readily visible. Injecting individual solutions of each excipient into the HPLC system in comparison with a standard solution of the analyte is one means of performing this experiment. The absence of a peak eluting at the retention time of the active ingredient is sufficient to demonstrate specificity for excipients. Figure 1 shows a sample chromatogram demonstrating the absence of a peak for the main component for an injection of a mixture of the excipients.

2. Challenge Study

Injecting solutions of known process impurities, degradation products, intermediates, homologues, dimers, etc. further challenges the specificity of a method. Identification of these compounds may require an extensive search in order to identify all possible species that may be

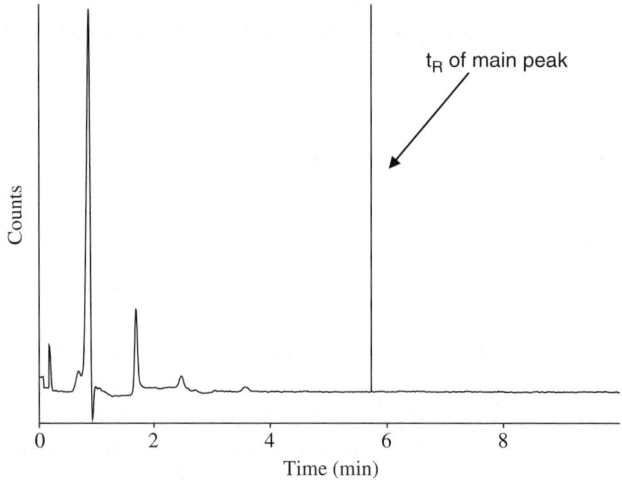

FIGURE 1 Example chromatogram demonstrating the absence of interference from placebo at the retention time (t_R) of the active component.

present in the sample. For new chemical entities (NCE), this information may not be readily available. Probable suspects should be identified by careful review of the synthetic route and manufacturing process to identify any likely species that may be present in the sample.

If reference materials of the suspected impurities are available, the drug substance or finished drug product should be spiked at an appropriate level to demonstrate that the result is unaffected by the addition of the impurity. Figure 2 shows examples of individual chromatograms of the API and three known process impurities. As shown here, none of the three process impurities interfere with the API peak, although peaks for impurities A and C appear to overlap and could co-elute if both were present in the sample. This specificity may be acceptable if the method was designated as an assay method for the quantitation of the API. For a method intended to quantitate process impurities, the overlap of these two components would in most cases be unacceptable due to the inability of the method to accurately measure the two individual components.

If reference materials are not available, the challenge study lives up to its name. Specificity may still be demonstrated by a comparison of test results containing the impurities of interest to a second, well-characterized procedure (e.g., USP method). If a secondary method is unavailable, peak purity evaluation may be used to further demonstrate specificity of the method.

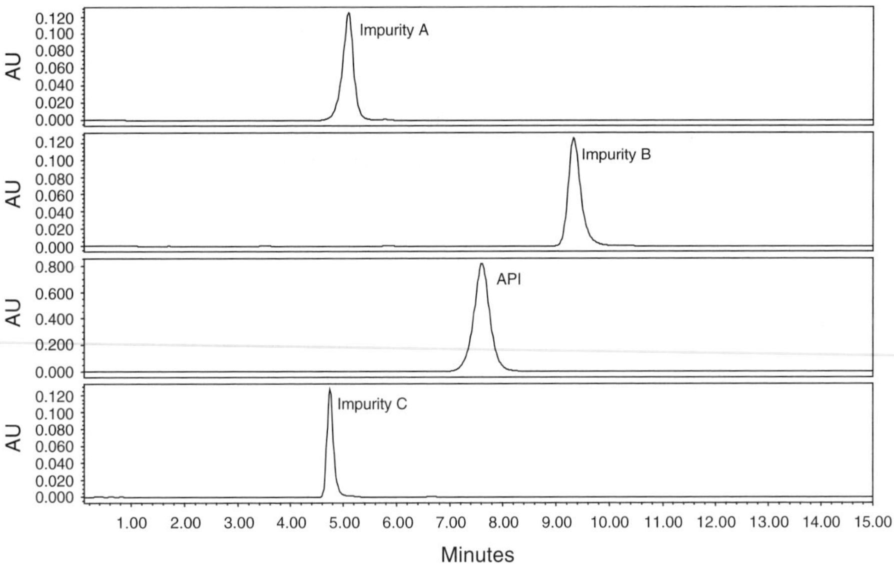

FIGURE 2 Examples of chromatograms showing selectivity between the peak for an API and its impurities.

3. Degradation Studies

Degradation studies involve exposing the sample to a variety of stressed conditions to further evaluate the specificity of degradation products. In this study, the drug substance, drug product, and the combined excipients (or placebos) are each exposed to the stressed conditions. These may include, but are not limited to, heat, light, acidic media, alkaline media, and oxidative environments. Other conditions may be used depending on the nature and chemistry of the test subject. Forced degradation is usually evaluated with not more than 20% degradation of the drug substance, although more may be acceptable depending on the particular properties of the drug. A reasonable effort should be made to degrade samples in order to identify possible degradation products. If the planned experiments do not show any appreciable degradation, the strength and/or exposure time of the stress condition may be increased, but degradation is not required for every condition studied. There is a point beyond which the stress condition becomes extreme and unrealistic. Sound scientific judgment should be used to determine the extent and degree of degradation studies. Table 3 shows suggested conditions for forced degradation studies.

4. Peak Purity Evaluation

Peak purity tests are used to demonstrate that an observed chromatographic peak is attributable to a single component. Mass spectrometry is the most sensitive and accurate technique to use for peak purity evaluation because of the specific information derived from the analysis. However, a good number of HPLC methods use mobile phase conditions that are incompatible with mass spectrometry detection. In this case, PDA spectrophotometers using peak purity algorithms may be used to support the specificity of the method. Almost all commercially available diode array detectors are equipped with proprietary software that will perform these calculations. Although this technique is more universal in application to HPLC methods, the data provided is neither particularly

TABLE 3 Suggested Conditions for Performing Forced Degradation Studies

Forced degradation	Suggested conditions[1]
Acidic	24 h in 1N HCl solution
Basic	24 h in 1N NaOH solution
Oxidative	24 h in 10% H_2O_2 solution
Heat	24 h in 105°C (oven)
Light	1 h exposure to UV lamp (200–300 nm)

[1]Exposure duration and strength of the degradant condition may be increased or decreased depending on the performance of the product.

plausible nor conclusive and can only suggest the gross purity of the analyte peak in question since spectrally similar species may not be resolved using this technique.

B. Linearity

The linearity of a method is its ability to obtain test results that are directly proportional to the sample concentration over a given range. For HPLC methods, the relationship between sample concentration and detector response (peak area or height) is used to make this determination.

1. Concentration Ranges

The concentration range used for linearity should be large enough to encompass the desired range of the method. A minimum of five concentration ranges should be investigated and a plot of the detector response vs. the sample concentration should be generated. It is important that the concentration ranges selected for the linearity study are relatively equally spaced throughout the range of the method (e.g., 50%, 75%, 100%, 125% and 150%), and not clustered, as this will provide a skewed estimation of linearity. When using a single-point calibration, regression of the linearity data should not be forced through zero, as this is not a true measure of the linearity of the method. Figure 3 shows a comparison of an appropriate linearity study and one that is not appropriate.

2. Acceptance Criteria

Acceptance criteria should be evaluated to ensure that they are meaningful when compared with the performance of the method. Table 4 gives a list of suggested acceptance criteria for use in evaluating method linearity. The ranges in Table 4 are suggestions only and should be adjusted to ensure that all specification limits are within the validated linear range for any given method.

3. Statistical Analysis

Linearity data should be evaluated using appropriate statistical methods. A simple regression line of the detector response *vs.* the sample concentration is the most common means of evaluation. Regulatory agencies require the submission of the correlation coefficient, *y*-intercept, slope of the regression line, and the residual sum of squares for linearity evaluation. A graphical representation of the linearity data should also be generated. Additional analysis of the deviation of the actual values from the regression line is suggested, especially when the method uses a single-point calibration standard. The percent *y*-intercept is calculated by dividing the *y*-intercept by the detector response at the nominal concentration expressed as a percentage. For single-point calibration, this value should be less than 1–2% to ensure accurate results.

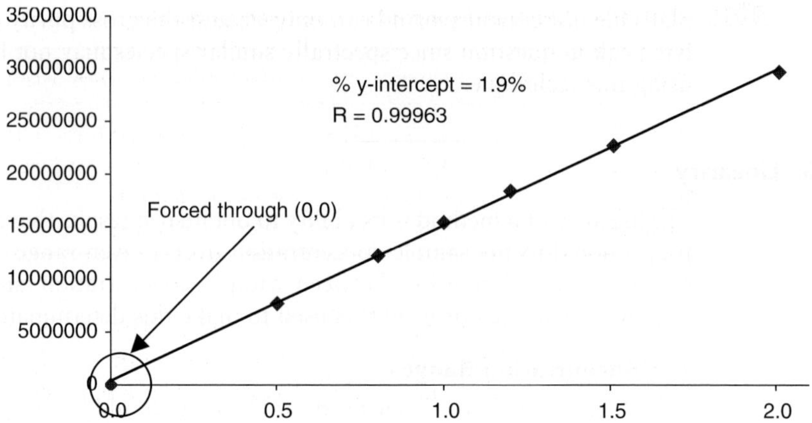

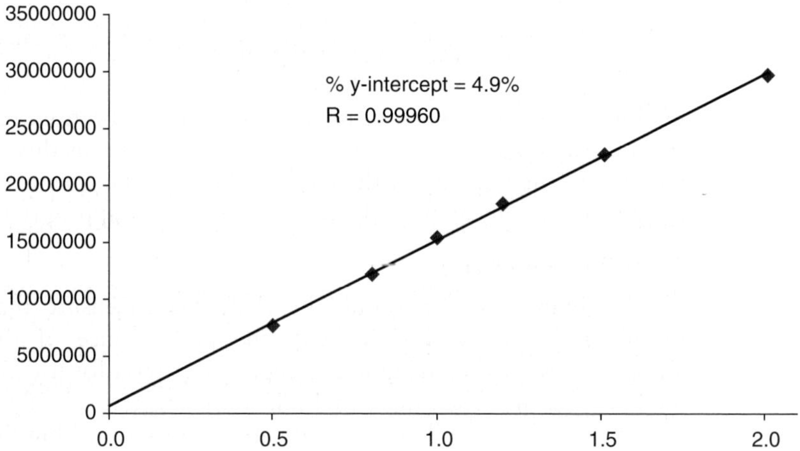

FIGURE 3 Examples of incorrect (upper) and correct (lower) regression plots for determining linearity of a method.

C. Accuracy

Accuracy is the closeness in agreement between the accepted true value or a reference value and the actual result obtained. Accuracy studies are usually evaluated by determining the recovery of a spiked sample of the analyte into the matrix of the sample to be analyzed.

1. Drug Substance

For a drug substance, the accuracy experiment should demonstrate that the method is free from the interference of process impurities and

TABLE 4 Levels and Ranges for Determining Linearity

Test	Levels	Ranges	Acceptance criteria
Assay	5	50% to 150%	$R \geq 0.999$, %y-intercept ≤ 2.0%
Dissolution	5–8	10% to 150%	$R \geq 0.99$, %y-intercept ≤ 5.0%
Impurity	5	LOQ to 2%	$R \geq 0.98$
Cleaning	5	LOQ to 20 times LOQ	$R \geq 0.98$

that the response of the sample is free from bias and/or interference. There are several acceptable ways of performing this study. Accuracy of a method can be demonstrated by comparison to a reference standard of known purity. A second approach is to compare the results for the proposed method with a second well-defined procedure (e.g., USP method) of known accuracy. In the absence of either of these options, accuracy can be demonstrated by internal normalization (area percent) combined with a thorough analysis using supporting techniques such as moisture, heavy metals, differential scanning calorimetry (DSC), thermal gravimetric analysis (TGA), etc.

2. Drug Product

Accuracy is measured in drug products by spiking known amounts of the analyte into the excipients and calculating the percent recovered. If excipients are not available, accuracy can be measured by spiking the analyte into the drug product by means of standard additions. A third approach compares the results of the method with a second, well-characterized method of a known or stated accuracy. Table 5 shows a list of ranges and suggested acceptance criteria for use in evaluating accuracy during method validation.

D. Limit of Detection

The LOD is the smallest amount of analyte that can be detected, but not necessarily quantitated using a given method. This parameter is important in the use of limit tests as it sets the level below which the method cannot function.

1. Calculations

There are several means of calculating LOD for HPLC methods. The most common approach is to determine the sample amount that provides a signal-to-noise ratio of either 2:1 or 3:1. In most cases, a signal-to-noise ratio of 3:1 is preferable as the LOD. Many chromatographic data-acquisition systems provide integrated functions to determine the signal-to-noise ratio that are easily employed by the trained analyst. An

TABLE 5 Levels and Ranges for Determining Accuracy

Test	Accuracy		
	Level	Range	Acceptance criteria
Assay/CU	3 levels in triplicate	70%, 100%, 130%	±2% (98.0% to 102.0%) average for each level
Dissolution	3 levels in triplicate	20–35%, 50–80%, and 100–130%	±5% (95.0% to 105.0%) average for each level
Related substances	1 level in triplicate	LOQ to 1% of analyte concentration	±20.0% (80.0% to 120.0%) average for each level
Cleaning surface validation	3 levels in triplicate	LOQ to 20 times LOQ	±50.0% (50.0% to 150.0%) average for each level

alternative to the signal-to-noise approach is to estimate LOD based on the standard deviation of response. For this calculation, LOD = 3.3(SD/*S*), where SD is the standard deviation of the response based on the standard deviation of the blank, the residual standard deviation of the regression line, and the standard deviation of the *y*-intercepts of the regression line, and S is the slope of the calibration curve. And, as many former students remember, signal-to-noise levels can be calculated the old-fashioned way—measuring the peaks by hand. Although this is a legitimate means of determining LOD, it is highly subjective and can be influenced by the skill and experience of the person performing the measurement. If this approach is used, the chromatograms should be expanded as much as possible in order to obtain an accurate measurement of the peak-to-peak noise. The LOD should be reported as a concentration. It is not necessary to report the actual LOD for a method and a sufficiently low value (e.g., 0.05%) may be designated as the LOD even though the actual value is much lower.

E. Limit of Quantification

The LOQ is the lowest level that an analyte can be quantitated with any degree of certainty. The determination of LOQ is similar to that for LOD and there are several accepted approaches, although one should be consistent in the manner in which these figures of merit are calculated within a given method. Because the calculation of LOD and LOQ are similar and one can be inferred from the other, it is common to see LOD and LOQ reported together even though the method type may not require both values.

1. Calculations

The LOQ can be determined by a signal-to-noise ratio of 10:1, or approximated by multiplying the LOD by 3.3. As with LOD, this function is easily obtained from current data-acquisition software. Similarly, LOQ can be estimated by the equation LOQ = 10(SD/S) and by hand calculation as well. LOQ should be reported as a concentration and the precision and accuracy of this value should also be reported. As for LOD, measurement of the actual LOQ value may not be necessary if the method is shown to perform at a level that is sufficiently low (e.g., 0.1%). Figure 4 shows an example of an estimate of LOQ for an HPLC sample chromatogram.

F. Precision

Precision is a measure of the ability of the method to generate reproducible results. The precision of a method is evaluated using three separate determinations for repeatability, intermediate precision, and reproducibility.

1. Repeatability

Repeatability is a measure of the ability of the method to generate similar results for multiple preparations of the same sample. This determination is carried out by a single analyst and varies only the number of sample preparations. Repeatability may be measured by making six sample determinations at 100% concentration, or by preparing three samples at 80%, 100%, and 120% concentration levels. This portion of precision

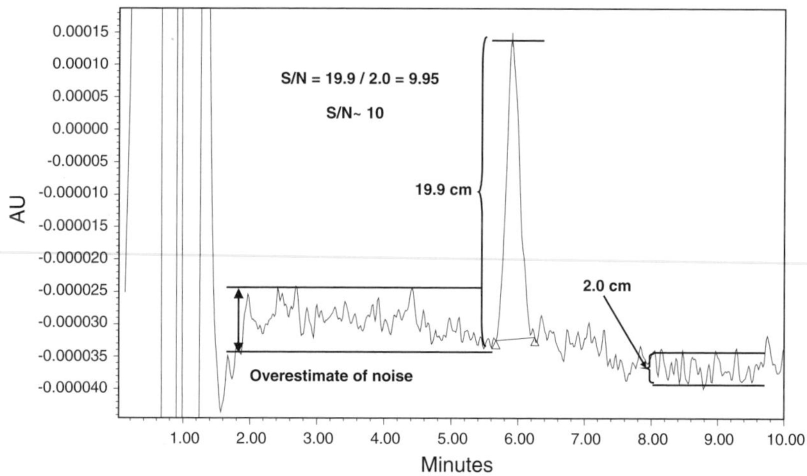

FIGURE 4 **Example chromatogram showing the determination of the LOQ for an HPLC method.**

evaluates the performance of the method conditions and estimates the amount of variability that can be expected for a single analyst and HPLC system for a given sample.

2. Intermediate Precision

Intermediate precision is another measure of the performance of the method where samples are tested and compared using different analysts, different equipment, different days, etc. This study is a measure of inter-lab variability and is a measure of the precision that can be expected within a laboratory. Intermediate precision is not required if a reproducibility study has been performed. Table 6 lists the ranges and suggested acceptance criteria for evaluation of precision during method development.

3. Reproducibility

Reproducibility is the third and final portion of precision testing. Here, samples are prepared and compared between testing sites. This usually occurs at the time of technology transfer. Samples are prepared in a similar manner between the two sites and are compared to a previously agreed-upon set of acceptance criteria. As per the ICH, reproducibility studies are not a part of submission filings, but should be performed as a confirmation of the ability of each testing site to perform the method reliably. An executed reproducibility study can be used in place of intermediate precision for the validation of a method, although there is no problem if both validation parameters are evaluated.

G. Range

The range of an analytical method is the interval between the upper and lower analytical concentration of a sample where the method has been shown to demonstrate acceptable accuracy, precision, and linearity.

TABLE 6 Levels and Ranges for Determining Precision

	Repeatability and intermediate precision	
Test	**Levels and ranges or units**	**Acceptance criteria**
Assay/content uniformity	3 levels in triplicate 70%, 100%, 130% or 6 determinations at 100%	Relative standard deviation NMT 2%
Dissolution	12 sample units	Relative standard deviation NMT 20%
Related substances	6 replicates at LOQ	Relative standard deviation NMT 20%
Cleaning validation	6 replicates at 10 × LOQ	Relative standard deviation NMT 20%

Range is not a parameter that can be measured independently, but must be inferred from a review of the data collected for accuracy, precision, and linearity. The validation of an assay method may demonstrate linearity from 50% to 150% of the sample concentration. Accuracy and precision data may have only been collected from 70% to 130% of the sample concentration. The range of this particular method would be from 70% to 130% where all three parameters have been evaluated.

For assay methods, a range of 80% to 120% of the test concentration is suggested. Content uniformity methods should cover a wider range of 70% to 130% of the test concentration, unless a wider range can be justified due to the nature of the dosage form (e.g., metered dose). Dissolution testing should demonstrate a range of ±20% over the specified range of the method, meaning the validated range should be 20% less than the lowest specification value and 20% higher than the highest specification value. Impurities should demonstrate a range from the reporting level up to 120% of the specification level.

H. Robustness Vs. Ruggedness

Any discussion of robustness will almost certainly lead to some measure of confusion between two similar but distinctly different words. Although it is a common practice for many people to use the terms "robustness" and "ruggedness" interchangeably, it is incorrect.[15] The similarity between the definitions of the two words and the parameters that they describe leads to this confusion. *Robustness* is a measure of the performance of a method when small, deliberate changes are made to the method conditions. The intent of this validation parameter is to identify which, if any, of the method conditions are the most critical to the successful performance of the method. *Ruggedness* is an older term that has been replaced by *intermediate precision*, which is a measure of how well the method performs under normal conditions from laboratory to laboratory, instrument to instrument, and analyst to analyst (see Section IV. F.2). The confusion between these two terms has been carried over into other reference texts that cover method validation. The USP Chapter <1225> on validation of compendial methods continues to use ruggedness to describe the validation parameter's intermediate precision as previously discussed in this chapter. To further complicate matters, the discussions on ruggedness and robustness are located next to each other in the text of USP <1225>. Other regulatory guidance documents seem to be moving away from the term ruggedness to prevent further confusion.

A well-designed robustness experiment can be the difference between a successful method and a poor method. Thorough robustness data can provide the flexibility needed to perform method adjustments if required.

For example, if an HPLC system does not pass system suitability because of a capacity factor (k') that is too low, the results of robustness testing could tell the analyst that the amount of organic can be changed by ±10% without affecting the result. If evaluated at the time of validation, the analyst can make the adjustment with no additional validation. If this particular study has not been performed, the analyst has no other choice but to verify the validity of the method at the new condition.

For a method that was developed according to a well-planned experimental approach, the majority, if not all, of the robustness data should already be available to the analyst. It makes sense to perform robustness testing as early as possible so that critical parameters can be identified and method optimization can take place while minimizing the amount of revalidation required. Too often robustness testing is one of the last validation parameters investigated when changes to method conditions are more difficult to make. If during robustness testing a significant issue arises that requires a change to the method conditions when most of the validation work has been completed, it is possible that the previous validation work was in vain and must be repeated. It is unfortunate that most texts discussing robustness (including this one) leave robustness as one of the last validation parameters discussed.

1. Typical Variations in Method Parameters

For HPLC, the typical parameters that are investigated during robustness testing are listed in Table 7.

The parameters to be evaluated should be selected based on the intent of the method and its known performance characteristics. The ranges tested should also take into account the tolerance of the system. Changing the detector wavelength setting by ±2 nm is insufficient if the tolerance of the detector calibration is ±3 nm. The change should at least be the same, if not more than the allowance for the calibration of the detector to determine the effect for robustness testing. However, changes that are too large can prove to be as useless as changes that are

TABLE 7 Typical Parameters and Ranges Evaluated During Method Robustness

Parameter	Nominal	Low value	High value
Percent organic in mobile phase	25	20	30
Flow rate (mL/min)	1.0	0.8	1.2
Buffer pH	7.2	6.8	7.6
Buffer concentration (mM)	20	15	25
Column temperature (°C)	30	25	35
Detector wavelength (nm)	220	210	230
HPLC column	Different batches for same column type		

too small. The analyst should choose ranges that are practical and that make sense when designing a robustness study.

2. Experimental Design

The use of experimental design as described by Plackett and Burman is a valuable tool in the estimation of method robustness.[16,17] Validation parameters are evaluated using a two-level approach for each variable to be tested. For HPLC methods, chromatographic figures of merit such as capacity factor, resolution, peak tailing, theoretical plates, etc. are the responses used to determine the effect of the changes in the parameters. Table 8 shows a sample of a Plackett–Burman experimental design to evaluate the effect of changes to seven parameters using eight experiments.

3. One Factor at a Time

Although experimental design is the preferred and accepted means for determining method robustness, the use of one factor at a time (OFAT) testing can also accomplish this task. In this case, changes to the validation parameters are made one at a time and the number of experiments required is the product of the number of parameters and the number of levels to be tested. For the example in Table 9, six parameters at two levels would require 12 individual experiments. Although this approach is acceptable for robustness testing, its disadvantage is that it is not able to evaluate the cumulative effect of changes to multiple validation parameters.

I. System Suitability

System suitability is probably one of the least understood validation parameters. The well-trained analyst should know that setting proper and

TABLE 8 Example of a Plackett–Burman Experimental Design to Evaluate the Effect of varying Seven Conditions on Method Robustness

Condition/factor	A	B	C	D	E	F	G
1	−	+	+	+	−	−	+
2	+	−	+	+	+	−	−
3	−	+	−	+	+	+	−
4	−	−	+	−	+	+	+
5	+	−	−	+	−	+	+
6	+	+	−	−	+	−	+
7	+	+	+	−	−	+	−
8	−	−	−	−	−	−	−

TABLE 9 Example of Experiments Required to Evaluate the Effect of varying Seven Conditions on Method Robustness using OFAT

Experiment	Condition
1	Column temperature +5°C
2	Column temperature −5°C
3	Mobile phase +5% organic
4	Mobile phase −5% organic
5	Flow rate +0.2 mL/min
6	Flow rate −0.2 mL/min
7	pH +0.2 units
8	pH −0.2 units
9	Buffer concentration +5 mM
10	Buffer concentration −5 mM
11	Detector wavelength +5 nm
12	Detector wavelength −5 nm

appropriate system suitability requirements will, in the long run, benefit the person using the method. The purpose of system suitability is to define a set of parameters that are measured prior to each experiment that will tell the analyst if the system is performing adequately. Table 10 gives a list of typical system suitability parameters that are evaluated for HPLC methods.

Far too often, system suitability parameters are set too high or too broad to be useful. This is not done with the intention of generating data that is poor, but more out of a concern for setting the specifications too tight so that other analysts cannot perform the method routinely. System suitability requirements should be set by a review of a significant amount of data collected using the method.[18] This is more easily done when the bulk of the validation work has been completed. Robustness data should also be considered when setting the criteria for system suitability. Acceptance criteria should be set so that they reflect the actual performance of the method and not an arbitrary number. For example, if a method typically performs with a peak area RSD of 0.2%, an acceptance criterion of NMT 2% for system suitability is not appropriate. These criteria should be set based on statistical analysis of a significant historical data set collected using the method (3σ) rather than historical precedence or a desire to make all experiments pass the criteria.

J. Solution Stability

The stability of analytical solutions is an important factor to consider during method validation. This information can be invaluable to

TABLE 10 Parameters and Acceptance Criteria for System Suitability

Parameter	Acceptance criteria
Peak area reproducibility	% RSD of replicate injections
Capacity factor (k')	k' large enough to provide adequate retention
Tailing factor	T_f=1 demonstrates no tailing, acceptable range shown
Resolution (R_s)	Minimum resolution required for critical pair
Theoretical plates (N)	Minimum theoretical plates allowed for separation

the QC laboratory where multiple analyses are performed day in and day out. The difference in preparing a standard solution fresh daily vs. using the same solution for a period of days or weeks can mean savings of thousands of dollars in laboratory costs. Any solution used to quantitate a result in the method should be evaluated for solution stability. A general acceptance criterion for this study is that the aged solution must not change by more than 2% from the initial analysis. The value of 2% is an estimate of the day-to-day instrumental variance, and any change greater than this can be attributed to the sample. It is preferable that solutions are compared to fresh standards when possible. A study where the test solutions are injected repeatedly over a period of time and the detector response is compared may be acceptable, but care must be taken to ensure that the performance of the detector is reproducible over the course of the study. A dedicated instrument should be employed if this approach is used or the analyst risks assigning a shorter expiration date to a solution based on variations in the detector response rather than a true degradation or loss of the standard solution.

Once the solution stability has been determined, a storage period should be determined based on the use of the method under real-world conditions. Data may show a standard solution to be very stable for 6 months at room temperature, but a quality control (QC) lab may not want to use this solution for such a long time.

V. METHOD VALIDATION BY PHASE OF DEVELOPMENT

Validation is a continuous activity that begins during method development and continues throughout the drug development process and well past the regulatory approval of the NDA. As a method develops and matures within the drug development cycle, the validation also progresses by incorporating more of the analytical performance parameters. Table 11 shows the validation requirements for methods at each stage within the product-development cycle.

TABLE 11 Validation Required at Each Stage of the Product Development Cycle

Product development stage	Validation requirements
Pre-clinical (Prior to human studies)	Specificity, linearity
Initiation of phase 1 clinical studies	Specificity, linearity, range, accuracy, precision, LOD, LOQ
Initiation of registration batches, NDA submission	Specificity, linearity, range, accuracy, solution stability, precision, LOD, LOQ
Completion of tech transfer, prior to product launch	Specificity, linearity, range, accuracy, solution stability, precision, LOD, LOQ, robustness

VI. REFERENCE STANDARDS

HPLC methods rely on reference standards in order to provide accurate data. The quality and purity of reference standards are critical and these materials should be highly purified and well characterized. Reference standards from the USP/NF are "pure" materials by legal definition unless a purity value other than 100% has been assigned by the USP, and no further characterization is required prior to their use. If USP reference standards are not available, a non-compendial primary reference standard may be certified for use. These should be of the highest purity available and well characterized to assure the purity, strength, identity, and quality of the material.[19] Methods using non-compendial reference standards must incorporate any purity correction factor into the calculations. It is important that any and all reference materials used for method validation should be well documented in the validation report. Working reference standards are usually materials that have been characterized and have had their purity established against a compendial reference standard. These may be used in cases where it is more cost effective to certify an in-house lot than to purchase USP reference materials for routine analysis.

VII. METHOD TRANSFER

Once a method has been developed and validated, it should be transferred to each site that intends to use the method. A typical method transfer would take place between the research group that developed and validated the method and the QC group that will use the method to release the finished commercial product, although method transfer may occur at any point where knowledge moves from one group to another. As in the case of method validation, method transfer should be performed under the control of a protocol that details the steps required for the study.

The Society for Life Science Professionals (ISPE) has recently published a manual that presents a standardized process for transferring technology between two parties with recommendations for a minimum base of documentation in support of a transfer request. It is divided into three segments: APIs, Dosage Forms, and Analytical Methods. The manual provides guidance and insight into the activities and documentation required to move a product, process, or method from one unit to another. ISPE has published the Guide in collaboration with the FDA and the American Association of Pharmaceutical Scientists (AAPS), with input from the European regulatory authorities and submission to the Japanese Ministry of Health and Welfare (MHW).[20]

VIII. DOCUMENTATION

It is generally understood that "if you didn't write it down, it didn't happen". In this age of GMPs, it is assumed that anyone performing validation data is operating under GMP conditions that include appropriate documentation practices. As mentioned previously, validation should proceed according to a protocol. Once the experimental work is complete, a final report should be written that includes the test data and a summary of the results. When validation is complete, a summary validation report should be written that includes all validation data collected for the method in question. One approach to simplifying the documentation of method validation is to incorporate data tables into the validation protocol so that data can be entered as the validation progresses. Once completed, a summary of the data can be added and raw data can be attached. These types of protocols can serve as templates for future validation work. A second approach uses method validation software as a means of collecting and presenting validation data (see Section X). Some companies attach copies of the method and/or the method development report to the validation report, while others keep these as separate documents. Regardless of the practice, it is important that each of these documents reference the other and that they are readily available.

IX. METHOD REVALIDATION

Method validation is not a one-time process. Companies that rely on regulatory methods should review and update method validations on a regular basis to ensure that validation data are up to date and meet current regulations. A review of the method and validation every few years can ensure that the validity of the method remains in good shape. Failure to do so may result in costly revalidation work, or worse, regulatory findings in the form of 483s and/or warning letter citations. In addition to a periodic review for method revisions, there are some instances when revalidation efforts should at least be considered after the initial method validation has been completed.[21] The following sections describe these scenarios.

A. New API Source

Upon identification of a new source for API, or in the event that the current API supplier makes any significant change to the approved synthetic process, revalidation should be considered. At a minimum, specificity of the method should be re-evaluated to ensure that any new process impurities and/or synthetic intermediates, precursors, etc. do not interfere with the analyte of interest. Revalidation is complete if the specificity study demonstrates that the change to the API has no adverse affect on the performance of the method. If the method is affected and changes are required, revalidation should proceed according to an original plan.

B. Formulation Change

The extent of revalidation required for formulation changes should be determined on a case-by-case basis. Slight adjustments to the formulation may not require further validation work. This would include an adjustment of the excipient ratios, a change in tablet shape, etc. Specificity and accuracy should be re-evaluated for the inclusion of a new excipient into the formulation (e.g., antioxidants, dyes, preservatives). A change in the formulation such as going from a tablet to a capsule or from a liquid to a solid would mean a significant change to the formulation and complete validation should be performed.

C. New Dosage Strength

If the new dosage strength is proportional to the existing formulation, limited revalidation is required. Linearity should be re-evaluated if the concentration of the sample falls outside of the original validated linear range. Accuracy tests should be repeated to ensure the compete extraction of the analyte from the new dosage form. If the new dosage strength differs significantly from the existing formulations, complete revalidation should be performed.

D. Instrumentation Changes

In many cases the purchase of new equipment for use with validated methods is not considered as a means for revalidation. Instruments that are reportedly "equivalent" to models used during validation have repeatedly caused problems due to unknown or minor changes. For detectors especially, linearity, LOD and LOQ may be affected by an equipment change due to differences in the linear range of the instrument. It is a good practice to evaluate any new instrumentation for use with validated methods to ensure that the performance is acceptable and appropriate for a given validated method.

E. Outdated Technology

Eventually, even well-validated methods become dated and obsolete due to improvements in technology. When this occurs, development of a new method using the new technology should occur and start a new round of method-validation activities. Any change to the method should take into account the effect on the long-term data for the product. Comparison of the validation for the old and new procedures should verify that the change will provide some benefits before it is accepted.

X. METHOD-VALIDATION SOFTWARE

A number of computer software packages are available to the analyst to assist in the planning and execution of both method development and validation experiments.[22] The attraction of these systems is that they can automate the validation process from planning the experiment to test execution to the presentation of the data in a final report form.

Several chromatography vendors market method-validation software. Waters' (www.waters.com) chromatographic data system Empower has a method-validation option called Elsa.[32] Agilent's (www.chem.agilent.com) Method Validation Pack is an add-on software module for the Agilent ChemStation Plus data system. In the US, Hitachi (www.hitachi-hta.com) markets Validation Manager which works with the D-7000 HPLC System Manager Software. Although each system differs in options and functionality, they all operate in a similar manner. Once validation data has been collected according to a method-validation protocol, the system automatically calculates parameters such as linearity, range, accuracy, repeatability, intermediate precision, detection limits, quantitation limits, sensitivity, and robustness. Additional features include electronic import of data from other vendors' equipment or from other analytical techniques, and integrated statistical evaluation functionality. Each system provides audit trails for revision management and documents results with a direct transfer of chromatography data into a summary report in a form appropriate to regulatory agencies. These software packages offer the advantage of collecting, storing, and reporting the validation data in an electronic format that is in compliance with 21 CFR Part 11. As with any computerized system used in a GMP environment, it is up to the analyst and the firm responsible for performing method validation to ensure that any software used in method validation meets their own standards for functionality and compliance.

Other chromatographic systems such as DryLab® 2000 Plus from LC Resources (http://www.lcresources.com) incorporate aspects of method development and validation through simulation software combined with actual experiments to determine optimal operating conditions. This system is designed to generate efficient method development and method optimization experiments. Results from these experiments can be used for the evaluation of robustness as part of the validation for methods developed using this type of software.

XI. SUMMARY

This chapter has attempted to discuss the concept of method validation as it relates to regulatory methods using HPLC analysis. Validation is a process required by law and the concept is described by regulatory agencies in guidance documents. The analyst performing method validation is responsible for interpreting those guidances into acceptable practices. Establishing a clear, well-planned validation program that incorporates the concepts discussed here is critical to success. Method validation is not a one-time process that can be ignored once completed, but a constant, evolving process. A well-executed and documented validation serves as evidence to regulatory agencies that the method in question is compliant and performs as intended. Periodic reviews should occur to evaluate validation data for compliance with the current regulatory requirements. To be successful, applicants should remain up to date with the current thinking of regulatory agencies and anticipate the changes that will occur in the regulations. Validation programs that are able to incorporate these changes quickly and efficiently will save both time and money for the applicant and will streamline the process to an approvable application.

ACKNOWLEDGMENTS

The author would like to thank Lane Gehrlein, Phil Palermo, Michael Dong, and Mike Brockner of Purdue Pharma, LP for their help in the preparation of this chapter.

REFERENCES

1. Federal Register 21 CFR 211, 165(e).
2. Title 21, USC Chapter 9, The Food, Drug and Cosmetic Act.

3. International Conference on Harmonisation (ICH) Q2A, Text on validation of analytical procedures; March 1995.
4. International Conference on Harmonisation (ICH) Q2B, Validation of analytical procedures: Methodology; November 1996.
5. Reviewer Guidance; Validation of Chromatographic Methods, November 1994. Center for Drug Evaluation and Research, FDA, Department of Health and Human Services, 1994.
6. Guideline for Submitting Samples and Analytical Data for Methods Validation, February 1987. Center for Drugs and Biologics, FDA, Department of Health and Human Services, 1987.
7. Guidance for Industry, Analytical Procedures and Method Validation: Chemistry, Manufacturing and Controls Documentation, Draft, August 2000, Center for Drug Evaluation and Research, Center for Biologics Evaluation and Research, FDA, Department of Health and Human Services, 2000.
8. Guidance for Industry, Bioanalytical Method Validation, May 2001, Center for Drug Evaluation and Research, Center for Veterinary Medicine, FDA, Department of Health and Human Services, 2001.
9. USP 26/NF 21; General Chapter <1225>, Validation of Compendial Methods. Rockville, MD, United States Pharmacopoeial Convention, 2003.
10. Hong, D. D. and Shah, M., *Development and Validation of HPLC Stability Indicating Assays, Drug Stability Principles and Practices*, 3rd ed., Marcel Decker, Inc., New York, 2000.
11. Crowther, J. B., Validation of Pharmaceutical Test Methods, In *Handbook of Modern Pharmaceutical Analysis*, Vol. 3, (Ahuja, S. and Scypinski, S., Eds.), Separation Science and Technology, Academic Press, New York, pp. 415–443, 2001.
12. Swartz, M. E. and Krull, I. S., *Analytical Method Development and Validation*, Marcel Dekker, Inc., NY, USA, 1997.
13. Green, M. J., A Practical Guide to Analytical Method Validation, *Anal. Chem. News Features*, 68:305–309A, 1996.
14. Kirsch, R. B., *Validation of Analytical Methods Used In Pharmaceutical Cleaning Assessment and Validation; Analytical Validation in the Pharmaceutical Industry*, Supplement to Pharmaceutical Technology; pp. 40–46, 1998.
15. Krull, I. and Swartz, M., Validation Viewpoint, Introduction: National and International Guidelines, *LC-GC*, 15(6):534–539, 1997.
16. Plackett, R. L. and Burman, J. P., The Design of Optimum Multifactorial Experiments, *Biometrika*, 33:305–325, 1943–1946.
17. ASTM E1169-89 Standard Guide for Conducting Ruggedness Tests (Plackett-Burman Design), American Society for Testing and Materials (ASTM), 100 Barr Harbor Drive, West Conshohocken, PA 19428-2959.
18. Dong, M., Paul, R. and Gershanov, L., Getting the Peaks Perfect: System Suitability for HPLC, *Today's Chemist at Work*, 10(09):38–40, 42, 2001.
19. Guidance for Industry, Submitting Supporting Documentation in Drug Applications for the Manufacture of Drug Substances, February 1987, Center for Drug Evaluation and Research, FDA, Department of Health and Human Services, 1987.
20. ISPE Good Practice Guide: Technology Transfer, March 2003, The Society for Life Science Professionals (ISPE), 2003.
21. Hokanson, G. C., A Life Cycle Approach to the Validation of Analytical Methods During Pharmaceutical Product Development, Part II: Changes and the Need for Additional Validation, *Pharm. Tech.*, 18(16):92–100, 1994.
22. Rozenman, Y. and Di Bussolo, J. M., Developing Robust HPLC Separations Using an Automated Method Development System, *Pharm. Tech.*, 10–19, 1998.

8
ION CHROMATOGRAPHY

CHRISTOPHER POHL

Dionex Corporation, Sunnyvale, CA

ABSTRACT

This chapter reviews the underlying principles of ion chromatography and its applications in pharmaceutical analysis. An overview of eluent systems, applications of gradients, electrolytic eluent generation, suppressors and stationary phases is provided. Application of ion chromatography to the confirmation of counterions, active ingredient analysis, competitive analysis and development work are discussed.

I. INTRODUCTION

Ion chromatography has become an essential tool of the pharmaceutical analytical chemist. The high sensitivity of the technique, coupled with the wide dynamic operating range made possible with modern high-capacity stationary phases makes it ideal for the analysis of ions in pharmaceutical applications. The combination of gradients and suppressed conductivity detection provides a powerful screening

Handbook of Pharmaceutical Analysis by HPLC,
S. Ahuja and M.W. Dong, editors.

tool for the analysis of ions in drug substances and in pharmaceutical formulations, providing a basis for the analysis of counterions, additives and manufacturing by-products.

Ion chromatography as we know it today began with the pioneering work of Small, Stevens and Bauman.[1] The topic has been well covered in a number of books[2–5] and reviews.[6,7] It has been expanded on a number of fronts, now covering an extremely wide variety of analytes, and has seen significant improvements in stationary phase design, chromatographic performance, detection sensitivity, suppressor design and electrolytic generation of eluents. But with all of these improvements, ion chromatography continues to be much more than simply ion exchange chromatography by another name.

Ion exchange chromatography was widely used in HPLC well before the advent of ion chromatography. And even now there are those who take the perspective that ion chromatography is simply a marketing ploy to carve out a niche for an HPLC instrument with an ion analysis focus. However, the truth is that while ion analysis can certainly be accomplished on a conventional HPLC, optimal performance and reliability dictate special design considerations which ultimately result in fundamental instrumentation differences. For example, it is well known that stainless-steel is not readily corroded by alkaline solutions so it is tempting to suggest that anion analysis using carbonate-bicarbonate buffered eluent systems should be readily amenable to use with standard stainless-steel HPLC instrumentation. But, in fact, the protective oxide film on the surface of passivated stainless-steel components is attacked by repeated cycles between acidic and alkaline solutions, ultimately removing the passivation layer and rendering the stainless-steel susceptible to pitting corrosion. Consequently, dedicated instrumentation suitable for use with both acidic and basic eluent systems perform much more reliably when all wetted surfaces are composed of inert materials such as PEEK, KelF, etc. For this reason, ion chromatography has come to be a subdiscipline of HPLC with specialized instrumentation, consumables and detectors. Furthermore, ion chromatography has come to mean HPLC separation of inorganic anions, small organic acids, inorganic cations and simple amines. Accordingly, the following will focus on the special instrumentation and separation aspects of this class of compounds as it relates to what has come to be known as ion chromatography.

II. ION CHROMATOGRAPHY INSTRUMENTATION

As mentioned above, instrumentation designed for ion chromatography is constructed so that all wetted surfaces are composed of inert materials such as PEEK, KelF and Teflon. For short-term operation, it is possible to make use of standard HPLC equipment. But in this case it is generally advisable to frequently passivate all stainless-steel components in order to minimize the contamination of stationary phase with corrosion products and to minimize corrosion damage to the equipment. As a general rule, it is better to utilize equipment designed specifically for ion chromatography when a given application is expected to be performed for an extended period of time.

UV detection has limited application in ion chromatography. As discussed below, it can be used for detection of ions with a UV chromophore and it can also be used in the indirect detection mode. However, indirect detection generally suffers from high sensitivity to environmental factors, and provides limited linear working ranges. Generally, the most useful detector in ion chromatography is the conductivity detector. There are a wide variety of conductivity detectors available commercially, ranging from crudely designed "bargain basement" detectors to detectors with digital auto ranging, built-in suppressor power supplies and electrochemical detector capabilities. When considering conductivity detectors, an important point is suppressor capability. Some vendors such as Dionex provide detectors with a suppressor power built in, while others require the purchase of a separate module in order to provide power to the suppressor and/or to provide a chemical regenerant. The cost of the secondary module can in some cases, exceed the cost of the conductivity detector itself, making the conductivity detector with a built-in power supply an attractive option.

III. CHROMATOGRAPHIC COMPONENTS

In ion chromatography, the approach to method selection is largely dominated by the combination of analytes of interest and the desired level of method sensitivity. Ions such as sulfate, phosphate or trimethyl amine, which may be present in pharmaceutical samples, contain no UV chromophore and as such cannot be detected with conventional HPLC equipment. The need to analyze these species, therefore, dictates the use of ion chromatography equipment. On the other hand, a number of analytes of pharmaceutical interest can be analyzed using either conventional HPLC equipment or ion chromatography equipment depending upon the sensitivity required for these analytes. For example, citrate absorbs weakly in the UV at low wavelengths and can be analyzed using HPLC with UV detection when high sensitivity is not required. However, since the sensitivity for citrate is limited using UV detection, ion chromatography methods are necessary in cases where sensitive detection is required.

In HPLC, where methods are nearly exclusively based on "generic" reversed phase columns, methods can generally be transferred from one column to another with minimal modification. In ion chromatography, such is rarely the case. Instead, methods for specific analyte ions are frequently specific to the column. Fortunately, most major manufacturers provide substantial specific application information and eluent system guidance as part of the column manual. It is highly recommended that the reader take the time to investigate the wealth of information available in column manuals and column selection materials from leading ion chromatography column vendors before pursuing the development of new applications. This will help prevent unnecessary wasted effort as commercial products are generally available for most analyte ions of interest.

A. Stationary Phases

One distinguishing area of ion chromatography is the strong interplay between stationary phase design and detection. In reversed phase HPLC, relatively little attention is paid to the detection process when designing new stationary phases. However, in ion chromatography there is a strong motivation to consider the details of the detection process when designing the stationary phase. When working with suppressor-based ion chromatography, the stationary phase must be selected to be compatible with eluents suitable for achieving optimal detection sensitivity. When working without a suppressor, ion exchange capacity must be kept low in order to minimize the total background conductivity of the eluent. In addition, when working with suppressor-based ion chromatography, eluents are typically acidic in cation-exchange chromatography and basic in anion-exchange chromatography. Accordingly, a stationary phase compatible with extremes in pH is a general requirement for ion chromatography phases.

1. Silane Modified Porous Silica Substrates

There are seven basic stationary phase architectures used in ion chromatography: silane-based modification of porous silica substrates, electrostatic agglomerated films on nonporous substrates, electrostatic agglomerated films on ultra-wide-pore substrates, polymer grafted films on porous substrates, chemically derivatized polymeric substrates, polymer-encapsulated substrates and adsorbed polymer-coated substrates. Of these seven approaches, silane-based modification of porous silica substrates (SMPSS), although one of the first approaches to be employed, is now rarely employed in the separation of small ions. Their relatively poor hydrolytic stability coupled with the tendency of the ion exchange materials to bring the stationary phase pH to unsafe values, has limited the use of such materials in applications where column life is of significant importance.

2. Electrostatic Agglomerated Nonporous Substrates

The second stationary phase architecture (electrostatic agglomerated films on nonporous substrates, EANPS) was one of the first types of

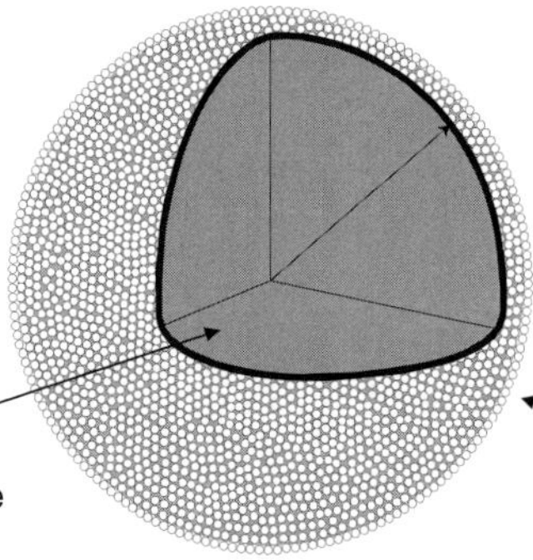

FIGURE 1 Schematic diagram illustrating the construction of an anion exchange latex coated, surface-sulfonated, nonporous cross-linked polyethylvinylbenzene–divinylbenzene resin bead.

materials to be employed in separation of small ions (see Figure 1). Developed in the early 1970s and first described by Hamish Small of Dow Chemical, these materials have been the mainstay of suppressor based ion chromatography for many years. Originally developed as a convenient means of producing low capacity hydrolytically stable materials when first generation suppressors had limited capacity, recently developed products utilizing this type of stationary phase have been limited in their application to guard columns and concentrator columns where capacity is not a major factor in column design. Nonetheless, materials of this design are still widely used in ion chromatography (see Table 1). Early materials used a low cross-link nonporous substrate but modern materials of this type make use of high cross-link substrate to render the stationary phase compatible with all common HPLC solvents. The substrate is then derivatized so as to introduce charged groups into the surface of the substrate. Following this, the substrate is brought into contact with a suspension of oppositely charged colloidal particles to produce the final product. Although this construction might sound like something that would be inherently unstable, in fact such materials are nearly indestructible when built using a styrenic substrate with a colloidal film of vinyl-aromatic ion exchange material.

3. Electrostatic Agglomerated Ultra-Wide-pore Substrates

For the most part, materials of the architecture described above have been replaced by a higher capacity version of the material: electrostatic agglomerated films on ultra-wide-pore substrates (EAWPS, see Figure 2). Using architecture similar to that described above but making use of substrates with pore sizes in the 1000–3000 Å range, it is possible to construct substantially higher capacity materials. For example, in a typical application with an ultra-wide-pore substrate, pore sizes are large enough to accommodate a coating of ion exchange colloid on the interior and

TABLE I Anion-Exchange Ion Chromatography Columns

Column	Source	pH range	Maximm solvent (%)	Capacity	Ionic type	Structure	Stationary phase	Preferred eluent
IonPac AS4A	Dionex	0–14	5	Low	SAX	EANPS	Styrenic	Carbonate
IonPac AS4A-SC	Dionex	0–14	100	Low	SAX	EANPS	Styrenic	Carbonate
IonPac AS5	Dionex	0–14	5	Low	SAX	EANPS	Styrenic	Hydroxide
IonPac AS5A	Dionex	0–14	5	Low	SAX	EANPS	Styrenic	Hydroxide
IonPac AS7	Dionex	0–14	5	Moderate	SAX	EANPS	Styrenic	Nitric Acid
IonPac AS9-SC	Dionex	2–11	100	Low	SAX	EANPS	Methacrylic	Carbonate
IonPac AS9-HC	Dionex	0–12	100	High	SAX	EAWPS	Methacrylic	Carbonate
IonPac AS10	Dionex	0–14	100	High	SAX	EAWPS	Styrenic	Hydroxide
IonPac AS11	Dionex	0–14	100	Moderate	SAX	EANPS	Styrenic	Hydroxide
IonPac AS11-HC	Dionex	0–14	100	High	SAX	EAWPS	Styrenic	Hydroxide
IonPac AS12A	Dionex	0–14	100	Moderate	SAX	EAWPS	Styrenic	Carbonate
IonPac AS14	Dionex	0–12	100	Moderate	SAX	PGPS	Methacrylic	Carbonate
IonPac AS14A	Dionex	0–14	100	Moderate	SAX	PGPS	Styrenic	Carbonate
IonPac AS15	Dionex	0–14	100	Moderate	SAX	PGPS	Styrenic	Hydroxide
IonPac AS16	Dionex	0–14	100	High	SAX	EAWPS	Styrenic	Hydroxide
IonPac AS17	Dionex	0–14	100	Low	SAX	EANPS	Styrenic	Hydroxide
IonPac AS18	Dionex	0–14	100	High	SAX	EAWPS	Styrenic	Hydroxide
IonPac Cryptand A1	Dionex	0–14	100	Moderate	Cryptand	PGPS	Styrenic	Hydroxide
Allsep Anion	Alltech	2–10	100	?	SAX	CMS?	Methacrylic	Carbonate
Allsep A-1 Anion	Alltech	2–11	100	?	SAX	CMS?	Methacrylic	Carbonate
Novosep A-1 Anion	Alltech	2–11	100	?	SAX	CMS?	Methacrylic	Carbonate
Novosep A-2 Anion	Alltech	3–12	10	?	SAX	CMS?	Vinyl alcohol	Carbonate
Anion/S	Alltech	2–5.5	100	High	SAX	SMPSS	Silane	Phthalate
Anion/R	Alltech	2–12	100	Moderate	SAX	CMS?	Styrenic	Carbonate

Super Sep IC	Metrohm	1–13	20	Moderate	SAX	CMS?	Methacrylic	Phthalate
Metrosep Anion Dual 1 IC	Metrohm	2–12	10	Low	SAX	CMS?	Methacrylic	Phthalate / carbonate
Metrosep Anion Dual 2 IC	Metrohm	1–12	20	Moderate	SAX	CMS?	Methacrylic	Phthalate / carbonate
Metrosep A Supp 1 IC	Metrohm	1–13	100	Low	SAX	CMS?	Styrenic	Carbonate
Metrosep A Supp 2 IC	Metrohm	1–13	0	Low	SAX	APCS?	Styrenic	Carbonate
Metrosep A Supp 3 IC	Metrohm	1–13	100	Low	SAX	CMS?	Styrenic	Carbonate
Metrosep A Supp 4 IC	Metrohm	3–12	100	Moderate	SAX	CMS?	Vinyl alcohol	Carbonate
Metrosep A Supp 5 IC	Metrohm	3–12	100	Moderate	SAX	CMS?	Vinyl alcohol	Carbonate
IC I-524A	Showdex	?	?	?	SAX	CMS?	?	Phthalate
IC NI-424	Showdex	?	?	?	SAX	CMS?	?	Multiple
IC SI-90 4E	Showdex	?	?	?	SAX	CMS?	Vinyl alcohol	Carbonate
IC SI-50 4E	Showdex	?	?	?	SAX	CMS?	Vinyl alcohol	Carbonate
IC SI-52 4E	Showdex	?	?	?	SAX	CMS?	Vinyl alcohol	Carbonate
PRP-X100	Hamilton	1–13	100	Moderate	SAX	CMS	Styrenic	*p*-Hydroxy-benzoate
PRP-X110	Hamilton	1–13	100	Moderate	SAX	CMS	Styrenic	*p*-Hydroxy-benzoate
PRP-X110S	Hamilton	1–13	100	Moderate	SAX	CMS	Styrenic	Carbonate / SCN^-
Star Ion A300	Phenomenex	1–12	0	Moderate	SAX	APCS	Styrenic	Carbonate
AN1	Cetac	1–12	0	Moderate	SAX	APCS	Styrenic	Carbonate
AN300	Cetac	1–12	0	Moderate	SAX	APCS	Styrenic	Carbonate
ANSC	Cetac	1–12	100	Moderate	SAX	CMS	Styrenic	Carbonate

Capacity: low = < 20 μeq g^{-1}; moderate = 21–200 μeq g^{-1}; high = > 200 μeq g^{-1}.
Structure: EANPS = electrostatic agglomerated nonporous substrate, EAWPS = electrostatic agglomerated wide-pore substrate, PGPS = polymer grafted porous substrate, SMPSS = silane modified porous silica substrate, CMS = chemically modified substrate, APCS = adsorbed polymer coated substrate.

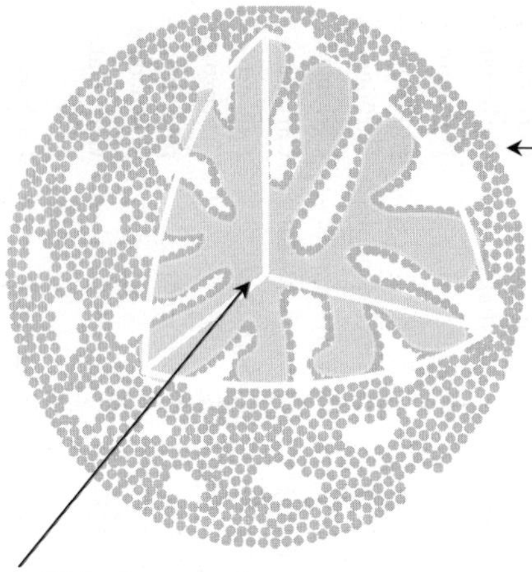

FIGURE 2 **Schematic diagram illustrating the construction of an anion exchange latex coated, surface-sulfonated, wide-pore cross-linked polyethylvinylbenzene–divinylbenzene resin bead.**

exterior surfaces. The resulting material will exhibit 6–8 times the capacity achievable on an identical particle size nonporous analog (i.e., 30–150 μeq mL^{-1} for the ultra-wide-pore format vs. 5–30 μeq mL^{-1} for the nonporous format). Given the increasing importance of high capacity chromatographic materials and the increasing use of high capacity suppressor devices, stationary phases of this type have seen wide use (see Table 1).

4. Polymer Grafted Porous Substrates

The fourth stationary phase architecture involves polymer grafted films on porous substrates (PGPS). This type of material is widely used to prepare high capacity materials where cross-link is not required for selectivity control. Materials of this sort are prepared through attachment of polymer strands to the surface of a substrate. To prepare such materials, the substrate is either prepared with polymerizable groups on the surface or the surface is modified to introduce polymerizable groups onto the surface. A slurry of resin, monomer(s) and initiator is then allowed to react to produce the grafted composite. Incorporation of a cross-linking monomer in the above reaction mixture will produce a gel with substrate particles suspended in the gel, much like fruit cocktail suspended in a gelatin fruit salad. As particles suspended in such a gel cannot be used as column packing material, this type of stationary phase synthesis precludes the use of cross-linking agents for selectivity control (unless the cross-linker is added after the graft step). In theory, such materials could be prepared from either polymer or silica substrates but in practice only polymeric substrates are in commercial use.

5. Chemically Modified Substrates

The fifth stationary phase architecture involves chemically derivatized polymeric substrates (CMS). This type of material tends to involve proprietary chemistry, so the actual chemistry used for the derivatization reaction is usually unknown. In general, materials of this sort are of rather substantial capacity, so they have come into vogue in recent years with the general shift toward materials of increasing capacity. The critical difficulty with such materials is the requirement that the derivitization must be constrained to the surface. Reactions that take place beneath the surface in the dense polymer matrix of the substrate will exhibit sluggish mass transport and relatively poor chromatographic performance. Early examples of this stationary phase architecture exhibited relatively poor performance but newer materials such as Showa Denko's IC SI-52 4E illustrate that high-performance materials can indeed be constructed in this manner.

6. Polymer-Encapsulated Substrates

The sixth stationary phase architecture involves polymer-encapsulated substrates (PES). Professor Schomberg pioneered this type of material as a means of preparing reversed phase materials using alumina as the base material. The synthesis of polymer-encapsulated materials is accomplished by combining the substrate, a preformed polymer with residual double bonds and a suitable free radical initiator dissolved in solvent, stripping off the solvent to leave a polymer film on the substrate and then curing the film at an elevated temperature to yield a cross-linked film permanently encapsulating the substrate. The technique was later adapted by Schomberg's group as a means of preparing a weak cation exchange phase using a preformed butadiene–maleic acid copolymer as the encapsulating polymer. Introduction of this material fundamentally changed the focus of stationary phase design for inorganic cations, shifting the emphasis from strong cation exchange materials to weak cation-exchange materials for most applications. This design is particularly popular for nonsuppressed applications (see Table 2).

7. Adsorbed Polymer-Coated Substrates

The seventh stationary phase architecture entails adsorbed polymer-coated substrates (APCS). This class of stationary phase materials is prepared by absorbing an ion exchange material on the surface of a support particle. Although in theory such a material could be prepared on either polymeric or silica base substrates, the only known commercial examples of such materials are all prepared from polymeric substrates. The exact composition of the coating material is unknown in the case of commercial products but the stability of the coating suggests that the absorbed coating is polymeric. A serious disadvantage of this stationary phase architecture is the catastrophic effect of solvents on stationary phase capacity. Even small amounts of solvent can quantitatively strip the

TABLE 2 Cation-Exchange Ion Chromatography Columns

Column	Source	pH range	Maximum solvent (%)	Capacity	Ionic type	Structure	Stationary phase	Preferred eluent
IonPac CS3	Dionex	0–14	5	Moderate	SCX	EANPS	Styrenic	HCl/DAP
IonPac CS5	Dionex	0–14	5	Low/low	SCX/SAX	EANPS	Styrenic	Oxalate/ PDCA
IonPac CS5A	Dionex	0–14	100	Low/low	SCX/SAX	EANPS	Styrenic	Oxalate/ PDCA
IonPac CS10	Dionex	0–14	100	Moderate	SCX	EANPS	Styrenic	HCl/DAP
IonPac CS11	Dionex	0–14	100	Moderate	SCX	EANPS	Styrenic	HCl/DAP
IonPac CS12	Dionex	0–14	100	High	WCX	PGPS	Acrylic	MSA
IonPac CS12A	Dionex	0–14	100	High	WCX	PGPS	Acrylic	MSA
IonPac CS14	Dionex	0–14	100	High	WCX	PGPS	Acrylic	MSA
IonPac CS15	Dionex	0–14	100	High	WCX	PGPS	Acrylic	MSA
IonPac CS16	Dionex	0–14	100	High	WCX	PGPS	Acrylic	MSA
IonPac CS17	Dionex	0–14	100	High	WCX	PGPS	Maleic	MSA
IonPac SCS 1	Dionex	2–7	100	High	WCX	PES	Maleic	MSA
Cation/R	Alltech	2–12	40	Moderate	SCX	CMS?	Styrenic	Nitric acid/ EDA
Universal Cation	Alltech	2–7	100	Moderate?	WCX	PES	Maleic	MSA
Universal Cation HR	Alltech	2–7	100	Moderate?	WCX	PES	Maleic	MSA
HyperREZ Monovalent	Metrohm	2–7	?	Moderate	SCX	CMS	Ethylenic	Nitric acid/ EDA

Metrosep Cation C 1	Metrohm	2–7	70	Moderate	WCX	PES	Maleic	Tartrate/ PDCA
Metrosep Cation 1-2	Metrohm	2–7	100	Moderate	WCX	PES?	Silane?	Tartrate/ PDCA
Metrosep Cation C 2	Metrohm	2–7	100	Moderate	WCX	PES?	Silane?	Tartrate/ PDCA
IC Y-521	Showdex	?	?	?	SCX	CMS?	Styrenic?	Nitric acid/ EDA
IC YK-421	Showdex	?	?	?	WCX	CMS?	Maleic?	Tartrate/ PDCA
IC T-521	Showdex	?	?	?	SCX	CMS?	Styrenic?	Multiple
IC R-621	Showdex	?	?	?	SCX	CMS?	Styrenic?	Multiple
PRP-X200	Hamilton	1–13	100	Moderate	SCX	CMS	Styrenic	Nitric acid/ EDA
PRP-X800	Hamilton	1–14	100	High	SCX	CMS	Itaconate	Cupric
Nucleosil 5SA	Macherey-Nagel	2–7	100	High	SCX	SMPSS	Silane	Multiple

Capacity: low = < 20 μeq g^{-1}; moderate = 21–200 μeq g^{-1}; high = > 200 μeq g^{-1}

Structure: EANPS = electrostatic agglomerated nonporous substrate, EAWPS = electrostatic agglomerated wide-pore substrate, PGPS = polymer grafted porous substrate, SMPSS = silane modified porous silica substrate, CMS = chemically modified substrate, APCS = adsorbed polymer-coated substrate, PES = polymer-encapsulated substrate.

absorbed coating, ruining the column. For that reason, there are relatively few commercial examples of this stationary phase architecture. Such materials can be readily identified by the prohibition of all solvent use.

8. Other Stationary Phase Architectures

In addition to the above ion exchange stationary phase architectures, there are a number of additional stationary phase architectures which are employed under special circumstances. Probably the most commonly used retention mechanism in this category is ion exclusion. Stationary phases for this application are invariably constructed from strong acid cation-exchange resin in the hydronium form. Such phases are used for separation of organic acids and weak inorganic acids. In addition, ion chromatography is sometimes accomplished through the use of reversed phase materials in the ion pair mode. Such separations can, of course, be accomplished on conventional silica-based reversed phase media but when utilizing such separations in conjunction with suppressors, polymeric reverse phase media are more commonly employed given the extreme pH of the mobile phase in such a system.

B. Column and Eluent Selection

In ion chromatography, column choice is intimately linked to the choice of eluent and the choice of detector. Although most ion chromatography is performed with a conductivity detector, there are many reports of use of other common HPLC detectors including UV detection,[8,9] refractive index detection,[10] evaporative light scattering detection[11] and mass spectrometry detection.[12,13] In the case of conductivity detection, one of the main demarcating factors in eluent choice revolves around whether or not the analyst will be utilizing a suppressor as part of the analytical method.

1. Column and Eluent Selection in Suppressed Ion Exchange Systems

When a suppressor is in use, eluents are typically chosen from a fairly limited set of useful eluent species compatible with optimal performance under suppressed conditions. For example, in suppressed anion ion chromatography, the most widely used eluent systems are carbonate/bicarbonate (see Figure 3), hydroxide (see Figure 4) and borate. There are a few other less widely used eluent systems which include *p*-cyanophenate[14] and Good's buffers.[15] Similarly in cation ion chromatography, the most widely used eluents systems are methanesulfonic acid (see Figure 5) and sulfuric acid, although older sulfonic acid based cation-exchange phases often make use of amino acid eluent additives such as diaminopropionic acid in order to facilitate the elution of divalent cations.

The lowest background, the lowest noise and the greatest sensitivity are achieved by utilizing hydroxide anion in the case of the anion-exchange

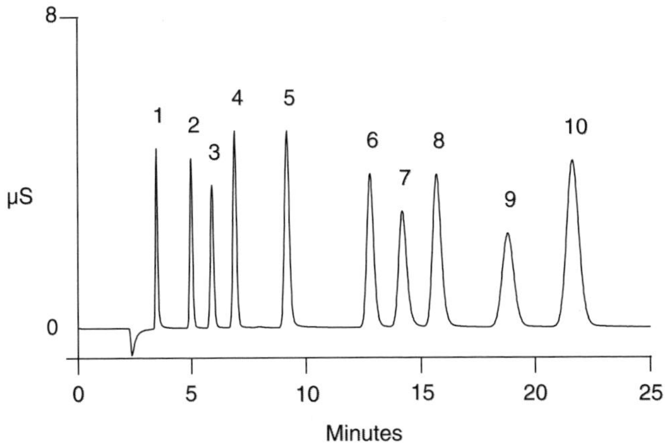

FIGURE 3 Separation of the common anions and disinfectant by-product anions on an IonPac® AS9-HC and an AG9-HC column. Columns: 4 × 250 mm IonPac AS9-HC and 4 × 50 mm AG9-HC. Eluent: 9 mM Sodium carbonate. Flow rate: 1 mL min^{-1}. Injection volume: 25 μL. Detection: suppressed conductivity utilizing the Anion Self Regenerating Suppressor (4 mm), recycle mode. Ions: 1—fluoride (3 mg L^{-1}); 2—chlorite (10 mg L^{-1}); 3—bromate (20 mg L^{-1}); 4—chloride (6 mg L^{-1}); 5—nitrite (15 mg L^{-1}); 6—bromide (25 mg L^{-1}); 7—chlorate (25 mg L^{-1}); 8—nitrate (25 mg L^{-1}); 9—phosphate (40 mg L^{-1}); 10—sulfate (30 mg L^{-1}).

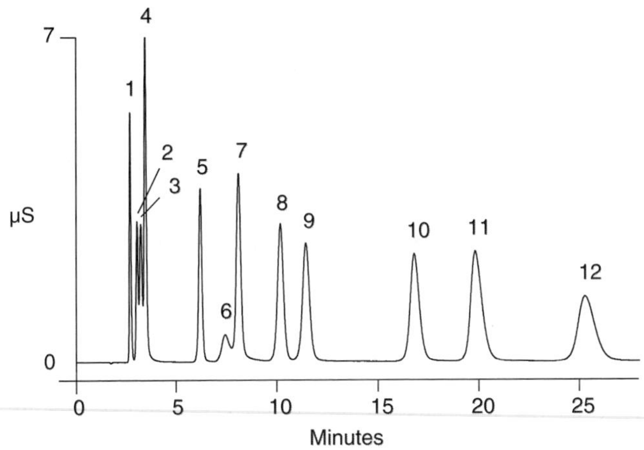

FIGURE 4 Isocratic separation of the common inorganic anions and organic acids on an IonPac® AS15 and an AG15 column. Columns: 4 × 250 mm IonPac AS15 and 4 × 50 mm AG15. Eluent: 38 mM KOH. Flow rate: 1.2 mL min^{-1}. Temperature: 30°C. Injection volume: 10 μL. Detection: suppressed conductivity utilizing the Anion Self Regenerating Suppressor (4 mm), recycle mode. Ions: 1—fluoride (2 mg L^{-1}); 2—glycolate (10 mg L^{-1}); 3—acetate (10 mg L^{-1}); 4—formate (10 mg L^{-1}); 5—chloride (5 mg L^{-1}); 6—carbonate (50 mg L^{-1}); 7—nitrite (10 mg L^{-1}); 8—sulfate (10 mg L^{-1}); 9—oxalate (10 mg L^{-1}); 10—bromide (20 mg L^{-1}); 11—nitrate (20 mg L^{-1}); 12—phosphate (30 mg L^{-1}).

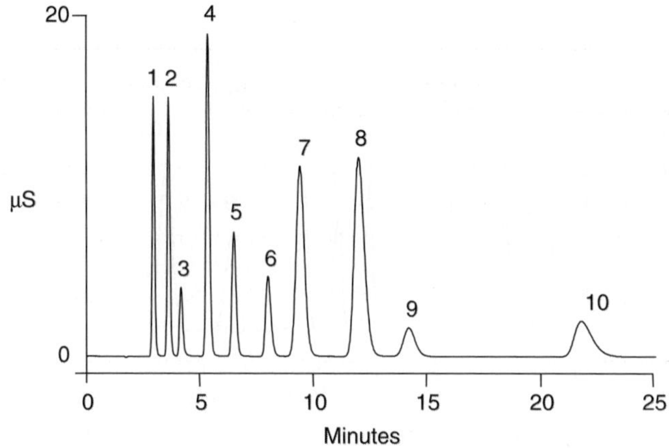

FIGURE 5 **Isocratic separation of the alkali metal cations, the alkaline earth cations and ammonium ion on an IonPac® CS12A column. Column: 4×250 mm IonPac CS12A. Eluent: 18 mM methanesulfonic acid. Flow rate: 1 mL min^{-1}. Injection volume: 25 μL. Detection: suppressed conductivity utilizing the Cation Self Regenerating Suppressor (4 mm), recycle mode. 1—lithium (1 mg L^{-1}); 2—sodium (4 mg L^{-1}); 3—ammonium (5 mg L^{-1}); 4—potassium (10 mg L^{-1}); 5—rubidium (10 mg L^{-1}); 6—cesium (10 mg L^{-1}); 7—magnesium (5 mg L^{-1}); 8—calcium (10 mg L^{-1}); 9—strontium (10 mg L^{-1}); 10—barium (10 mg L^{-1}).**

separations and hydronium cation in the case of the cation-exchange separations. Accordingly, these two eluent species comprise the preferred eluent ions with regard to detection. To make effective use of such eluent systems, it is necessary to choose a stationary phase that has been designed specifically for such eluent systems (see Tables 1 and 2 for a selection of columns suitable for use with hydroxide and hydronium eluent systems). Table 3 contains a listing of suppressor devices compatible with hydroxide and hydronium eluent systems. In general, such eluent systems require a suppressor with a high suppression capacity.

The specific column recommendation for work with hydroxide eluents depends somewhat upon the application. If working under isocratic conditions is of interest, a good column choice is the IonPac® AS18 column from Dionex. Using 23 mM potassium hydroxide at 1 mL per minute (with a 4 mm ID column) and 30°C, one can elute and separate common inorganic anions such as chloride, sulfate, nitrate and bromide in under 14 min. In addition, under the same conditions, one can elute and separate both formate and acetate. Under these conditions, trivalent anions such as phosphate have excessively long retention times, so if these ions are of interest, a better eluent system utilizes 39 mM potassium hydroxide (other conditions being the same as above). In this case, the common anions including phosphate can be eluted in under 10 min although the resolution of acetate and formate is somewhat degraded under these conditions. A good alternative column in cases where

TABLE 3 Suppressors

Suppressor	Source	Continuous	Solvent compatible	Capacity	Regeneration mode	Ion type	Regenerant source
AMMS	Dionex	Yes	Yes	High	Chemical	Anion	Reservoir
CMMS	Dionex	Yes	Yes	High	Chemical	Cation	Reservoir
ASRS	Dionex	Yes	Limited	High	Electrolytic/ chemical	Anion	Detector/ reservoir
CSRS	Dionex	Yes	Limited	High	Electrolytic/ chemical	Cation	Detector/ reservoir
Anion Atlas	Dionex	Yes	Limited	Moderate	Electrolytic	Antion	Detector
Cation Atlas	Dionex	Yes	Limited	Moderate	Electrolytic	Cation	Detector
MSM	Metrohm	No	Yes	Low	Chemical	Anion	Pump module
335 Suppressor	Alltech	No	Yes	Moderate	Disposable Columns	Anion	N/A
DS-Plus	Alltech	Yes	Yes	Moderate	Electrolytic	Anion	Suppressor module

polarizable anions such as iodide, thiocyanate, thiosulfate, chromate or perchlorate are of interest is the IonPac AS16 column. Using 35 mM potassium hydroxide at 1 mL per minute (with a 4 mm ID column) and 30°C, one can elute and separate these polarizable anions in under 20 min. Both the IonPac AS 16 and the IonPac AS 18 columns exemplify stationary phases prepared via electrostatic agglomeration of wide-pore substrates.

Other eluent systems in suppressed ion chromatography are typically chosen based on specific separation requirements. For routine analysis of monovalent and divalent anions, carbonate-based eluents represent a reasonable alternative to hydroxide-based eluent systems. Carbonate eluents are simple to prepare and can be useful in cases where anion analysis is only occasionally performed. It must be kept in mind, however, that carbonate lowers the detection sensitivity for anionic species and introduces significant nonlinearity into the analysis.[16]

While a wide variety of columns is available for carbonate eluent systems, there are several excellent choices depending on the specific requirements. For demanding applications where a high loading capacity and high resolving power are important, both the IonPac AS9-HC column from Dionex (see Figure 3) and the IC SI-50 4E from Showdex (also available as a private-label product from Metrohm under the name Metrosep A Supp 5 IC) are good choices. Both columns provide excellent chromatographic efficiency for most anions of interest although analysis time is somewhat long (approximately 25 min on either column). The IonPac AS9-HC is constructed via electrostatic agglomeration of wide-pore substrate while the IC SI-50 4E is most likely prepared via chemical modification of a polymeric substrate. When faster analysis is of interest, a better choice for carbonate eluent work is the IonPac AS4A-SC column from Dionex. An analysis in less than 8 min can be achieved working at 2 mL per minute (with a 4 mm ID column) with an eluent consisting of 1.8 mM sodium carbonate and 1.7 mM sodium bicarbonate. The IonPac AS4A-SC is constructed utilizing electrostatic agglomeration of nonporous substrate.

Other eluents systems are typically chosen for special sample types. For example, borate-based eluent systems are commonly employed within analyzing borated water samples found in nuclear power applications.[17] For samples of this type, the high concentration of borate present can compromise the quantitation of early eluting peaks when carbonate or hydroxide eluent systems are used. Another specialty eluent, *p*-cyanophenol is often applied to mask strong interaction between polarizable anions (see Figure 6) and pi electrons in the aromatic backbone of the anion-exchange material.

In the case of cation-exchange separations performed in conjunction with a suppressor, the most widely used eluents are based on sulfuric acid or methanesulfonic acid. While other mineral acids such as hydrochloric acid and nitric acid were frequently used in the early days of ion chromatography (see Figure 7), such eluent systems have largely been abandoned with the advent of modern electrolytic suppressors. Both hydrochloric acid and nitric acid cause significant damage to electrolytic suppressors due to the formation of strong oxidizing agents in the anode chamber of the suppressor. Consequently, electrochemically inert acids such as sulfuric acid or methanesulfonic acid have replaced hydrochloric acid and nitric acid in most cation-exchange ion chromatography applications.

More recently, strong acid cation-exchange columns have been replaced by weak acid cation-exchange columns for most cation ion chromatography applications.[18–22] Such phases have the advantage of eliminating the need for a divalent eluent species. Weak acid cation-exchange phases are operated at a pH low enough such that retention of divalent cations is comparable to that of monovalent cations. As a consequence, weak acid cation-exchange phases can be utilized without the need for incorporation of divalent eluent components such as diaminopropionic acid (DAP) or ethylenediamine (EDA) (see Figure 5).

A good general-purpose weak acid cation-exchange column is the IonPac CS12A column from Dionex (see Figure 5). This column utilizes 18 mM methanesulfonic acid at 1 mL per minute (with a 4 mm ID column) to perform the separation of lithium, sodium, potassium, ammonium, calcium and magnesium in under 14 min. The high capacity of this column makes it an ideal general-purpose column for the analysis of inorganic cations. However, when the analysis of amines is of interest a better choice is the IonPac CS17 column from Dionex (see Figure 12 and the discussion below under gradients for more details). Both IonPac CS12A

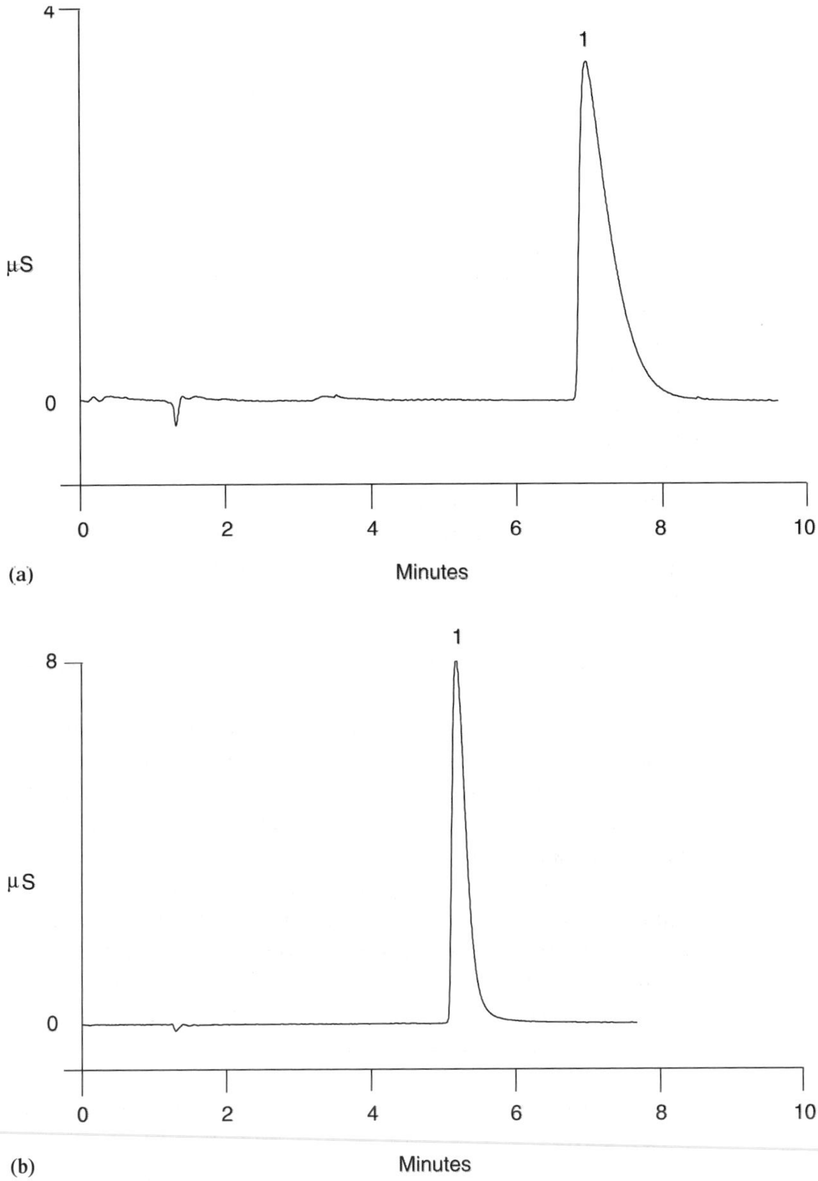

FIGURE 6 Effect of *p*-cyanophenol on the separation of perchlorate. Column: 4×250 mm IonPac® AS11. Flow rate: 1.0 mL min^{-1}. Injection volume: 25 µL. Detection: suppressed conductivity utilizing the Anion Self Regenerating Suppressor (4 mm), recycle mode. Ion: 1—perchlorate (20 mg L^{-1}). (a) Eluent: 100 mM NaOH. (b) Eluent: 50 mM NaOH and 5 mM *p*-cyanophenol.

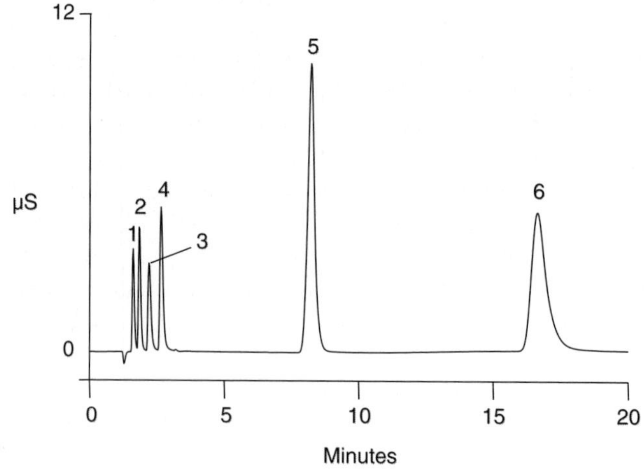

FIGURE 7 **Isocratic separation of the common cations on an IonPac® CS10 column. Column: 4×250 mm IonPac CS10. Eluent: 40 mM HCl and 4 mM diaminopropionic acid monohydrochloride. Flow rate: 1.0 mL min^{-1}. Detection: suppressed conductivity utilizing the Cation MicroMembrane Suppressor (4 mm). Regenerant: 100 mM tetrabutylammonium hydroxide. Regenerant flow rate: 7 mL min^{-1}. Injection volume: 25 μL. Ions: 1—lithium (0.5 mg L^{-1}); 2—sodium (2 mg L^{-1}); 3—ammonium (2.5 mg L^{-1}); 4—potassium (5 mg L^{-1}); 5—magnesium (15 mg L^{-1}); 6—calcium (25 mg L^{-1}).**

column and the IonPac CS17 column contain resin based on polymeric grafted porous substrate.

One fact to keep in mind with such phases is that weak acid cation-exchange materials based on carboxylic acid functional groups are subject to esterification in the presence of alcohol containing eluents. Even though typical eluent conditions (i.e., weakly acidic aqueous eluents containing alcohol) do not favor ester formation, such stationary phases typically exhibit slowly declining capacity when operated in the presence of alcohol-containing eluents. Consequently, such columns are normally operated with acetonitrile, tetrahydrofuran or acetone rather than with methanol, in order to avoid this problem.

2. Column and Eluent Selection in Non-suppressed Ion Exchange Systems

In the case of non-suppressed anion-exchange ion chromatography, eluent anions are chosen to minimize background conductivity since system noise is proportional to conductivity background and when operating in non-suppressed mode eluent background conductivity can be substantial. The strategy employed in order to minimize eluent background conductivity is to use a relatively low capacity anion-exchange column along with a fairly high potency eluent anion with a low equivalent conductance. The most frequently employed eluents used for anion separations in the non-suppressed mode include phthalate,[23] *p*-hydroxybenzoic acid and the borate–gluconate eluent system.[24] Eluents commonly used provide fairly limited dynamic capacity and detection limits that are generally 10 to 50-fold higher than comparable suppressed systems, so non-suppressed columns should only be used in applications where sensitivity and high capacity are not required.

A good general-purpose anion-exchange column for non-suppressed applications is the Hamilton PRP-X100 column. As the selection of eluents used with this column is quite varied, the reader is referred to Hamilton column literature for specific eluent recommendations. The Hamilton PRP-X100 column is most likely prepared via chemical modification of a porous polymeric substrate.

With strong acid cation-exchange phases, single column eluent systems typically utilize simple mineral acid eluent systems such as nitric acid.[25] While a number of eluent systems have been reported for elution of divalent cation from strong acid cation-exchange phases in a non-suppressed mode, most commonly employed eluent systems make use of EDA acidified with a suitable acid such as nitric or oxalic acid.[14] In case of carboxylic acid-based stationary phases, acidic eluent systems containing complexing agents are generally employed. These include tartaric acid, pyridine-2,6-dicarboxylic acid[26] (PDCA), oxalic acid and citric acid.

A good general-purpose weak acid cation-exchange column for non-suppressed applications is the Metrosep Cation C 2 column from Metrohm. As the selection of eluents used with this column is quite varied, the reader is referred to Metrohm column literature for specific eluent recommendations. The Metrosep Cation C 2 column is prepared via encapsulation of a porous silica substrate maliec acid copolymer.

C. Specialty Eluent Systems

1. Solvents in Ion Exchange Eluent Systems

Solvents are often utilized in ion chromatography. Although solvents do not directly affect retention in ion exchange, solvents can influence hydration of ions in the mobile phase and solvation of the ion exchange site in the stationary phase. In general, the effect of solvent is to reduce retention, predominantly affecting ions which are poorly hydrated or hydrophobic. In some cases, the addition of solvents can significantly alter selectivity, especially in the case of the coeluting analytes where one of the analytes contains a hydroxyl functional group (see Figure 8). However, solvents are generally only employed in cases where totally aqueous eluents systems are ineffective. Not only are solvent-based eluent systems more expensive, solvents can result in suppressor damage when the suppressor is used in the electrolytic mode: They can also generally result in decreased conductivity detection sensitivity and, in some cases, can cause significant damage to commercially available stationary phases (see Table 1 for details).

2. Eluent Systems for Alternate Retention Modes

Another commonly employed separation mode of ion chromatography is ion exclusion.[14] Although in theory both anions and cations can be separated by this mechanism, in practice the technique is used nearly exclusively for anion separations. Ion exclusion is frequently employed in cases where the analysis of organic acids is of interest and the concentration of organic acids present is low with respect to inorganic anions present in the sample. The advantage of this separation mechanism is that only weak acids exhibit significant retention. Since most inorganic anions are strong acids, the majority of the commonly occurring inorganic anions elute unretained, well resolved from most of the organic acids present in the sample. Figure 9 shows an example of ion exclusion separation. Eluent systems for ion exclusion separation of anions employ acidic eluents. The choice of eluent anion is dictated by the detection mode. When using conductivity detection, it is desirable to make use of acids with low equivalent conductance such as perfluorobutyric acid or octane sulfonic acid. The latter acid is also useful when utilizing UV detection but sulfuric acid is more commonly employed in this case.

Eluents employed for the separation of transition metals are generally chosen from compounds capable of forming complexes with transition metals including oxalate, citrate, tartrate, α-hydroxyisobutyric acid, diglycolic acid and PDCA. The choice of chelating agent employed is dictated by the specific metals of interest and the separation mode. Complexing agents capable of forming stable anionic complexes (such as PDCA) are best utilized in conjunction with an anion-exchange stationary phase whereas complexing agents which form weak anionic

FIGURE 8 Effect of methanol on the gradient separation of inorganic anions and organic acids on an IonPac® AS11-HC and an AG11-HC column. Columns: 4×250 mm IonPac AS11-HC and 4×50 mm AG11-HC. Flow rate: 1.5 mL min^{-1}. Temperature: 30°C. Injection volume: 10 µL. Detection: suppressed conductivity utilizing the Anion Self Regenerating Suppressor (4 mm), external water mode. Ions: 1—quinate (10 mg L^{-1}); 2—fluoride (3 mg L^{-1}); 3—lactate (10 mg L^{-1}); 4—acetate (10 mg L^{-1}); 5—glycolate (10 mg L^{-1}); 6—propionate (10 mg L^{-1}); 7—formate (10 mg L^{-1}); 8—butyrate (10 mg L^{-1}); 9—pyruvate (10 mg L^{-1}), 10—valerate (10 mg L^{-1}), 11—galacturonate (10 mg L^{-1}); 12—monochloroacetate (10 mg L^{-1}); 13—bromate (10 mg L^{-1}); 14—chloride (5 mg L^{-1}); 15—nitrite (10 mg L^{-1}); 16—sorbate (10 mg L^{-1}); 17—trifluoroacetate (10 mg L^{-1}); 18—bromide (10 mg L^{-1}); 19—nitrate (10 mg L^{-1}); 20—glutarate (10 mg L^{-1}); 21—succinate (15 mg L^{-1}); 22—malate (15 mg L^{-1}); 23—carbonate (20 mg L^{-1}); 24—malonate (15 mg L^{-1}); 25—tartrate (15 mg L^{-1}); 26—maleate (15 mg L^{-1}); 27—fumarate (15 mg L^{-1}); 28—sulfate (15 mg L^{-1}); 29—oxalate (15 mg L^{-1}); 30—ketomalonate (20 mg L^{-1}); 31—tungstate (20 mg L^{-1}); 32—phosphate (20 mg L^{-1}); 33—citrate (20 mg L^{-1}); 34—isocitrate (20 mg L^{-1}); 35—cis-aconitate and 36—*trans*-aconitate(20 mg L^{-1} combined). Eluent (a and b): sodium hydroxide: 1 mM from 0 to 8 min, 1 to 30 mM from 8 to 28 min, 30 to 60 mM from 28 to 38 min. (b) Methanol: 10% from 0 to 8 min, 10% to 20% from 8 to 18 min, 20% from 18 to 28 min, 20% to 10% from 28 to 38 min.

complexes (such as tartrate) are best utilized in conjunction with a cation-exchange stationary phase. A stationary phase providing both anion- and cation-exchange has been developed specifically to enable both modes of chromatography (see Table 2 and Figure 10). Detection of transition metals is generally accomplished by making use of post-column addition of a reagent suitable for the formation of colored complexes. While a number of colorimetric reagents are described in the

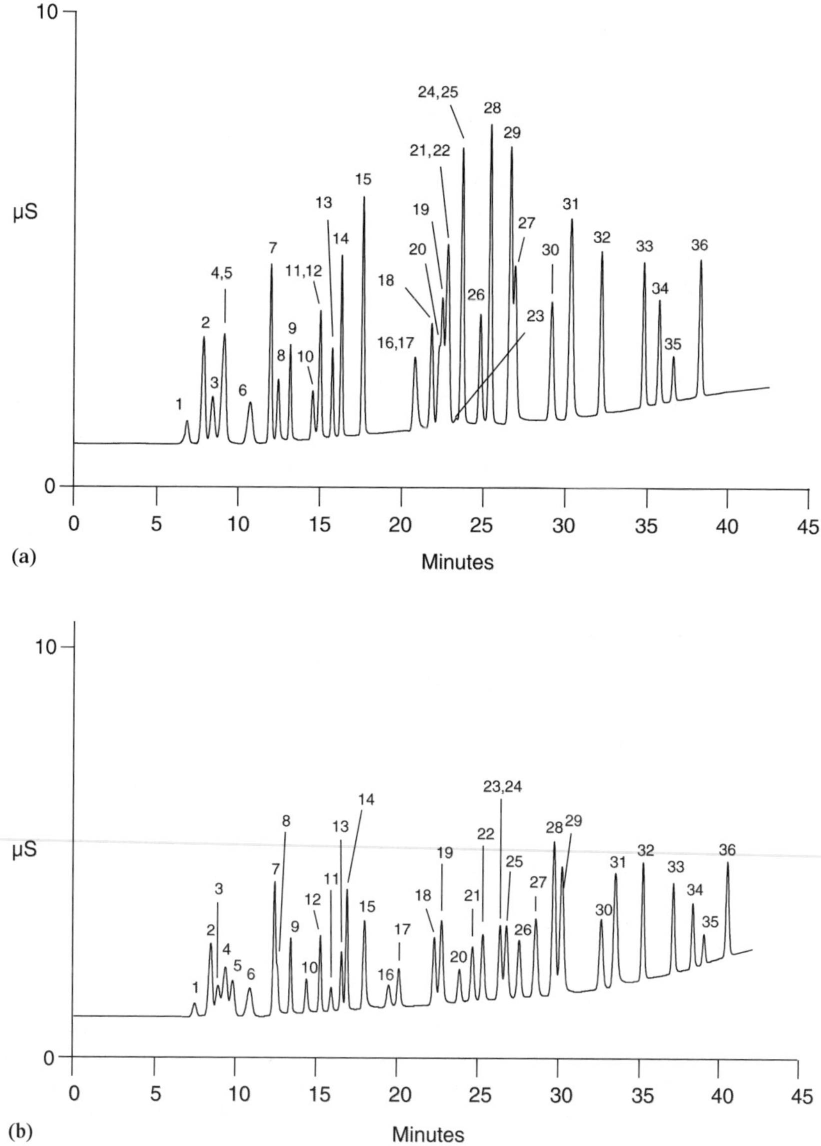

literature for this purpose, the most commonly employed reagent is 4-(2-pyidylazo)-resorcinol (PAR).

3. Eluents in Direct UV Detection

Eluents commonly used in the case of direct UV detection are generally selected from species that are transparent in the ultraviolet (see Table 4). Neither carbonate- nor hydroxide-based eluent systems are readily directly applied in the case of UV detection because of the strong absorbance of hydroxide anions. However, such eluent systems are frequently employed in conjunction with suppressor devices which effectively remove the hydroxide anion, thus producing a low background suitable for sensitive detection of UV-absorbing anions. In the case of cationic species, alkaline metals, alkaline earth metals, aliphatic amines and quaternary ammonium ions are all UV transparent. Several references in the literature suggest otherwise in the case of suppressed ion chromatography, but the actual UV-absorbing species in this case is the hydroxide counteranion introduced through the suppression reaction.

4. Eluents in Indirect UV Detection

For ions that are UV transparent, detection is possible through the use of indirect detection. A wide variety of different eluent systems have been described in the literature.[3] Eluents commonly used for indirect UV detection are similar to those used in non-suppressed conductivity detection: phthalate and *p*-hydroxybenzoic acid along with other

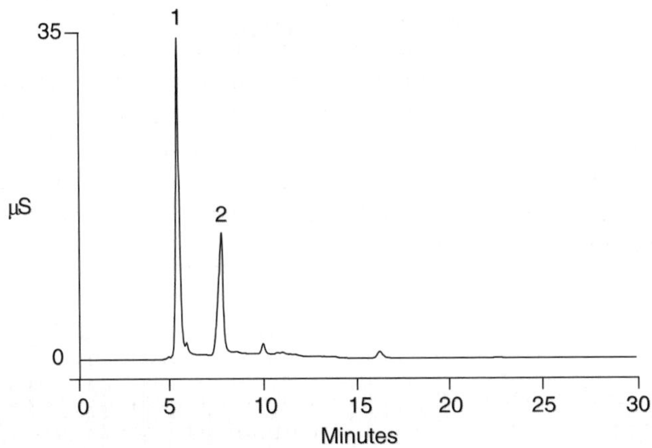

FIGURE 9 **Analysis of an analgesic by ion-exclusion chromatography. Column: 9×250 mm IonPac ICE-AS6. Eluent: 0.4 mM perfluorobutyric acid. Flow rate: 1.0 mL min^{-1}. Detection: suppressed conductivity utilizing the AMMS™-ICE MicroMembrane Suppressor. Regenerant: 5 mM tetrabutylammonium hydroxide. Regenerant flow rate: 7 mL min^{-1}. Injection volume: 25 µL. Sample: 0.6 g in 5 mL deionized water. Ions: 1—strong acid anions; 2—citrate.**

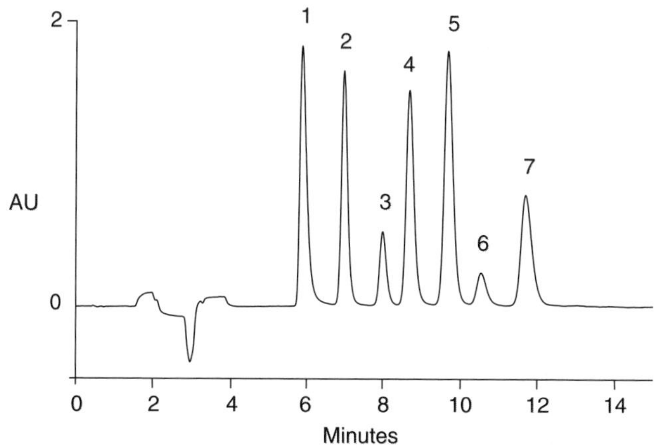

FIGURE 10 **Trace analysis of transition metals on a mixed mode column. Columns: 2×250 mm IonPac® CS5A and 2×50 mm CG5A. Eluent: 7 mM PDCA, 66 mM KOH, 5.6 mM K_2SO_4 and 74 mM formic acid (pH 4.2). Flow rate: 0.3 mL min^{-1}. Concentrator column: TCC-2. Concentrator flow rate: 2 mL min^{-1}. Concentration volume: 30 mL. Post column reagent: 0.3 mM PAR, 1 M 2-dimethylaminoethanol, 0.5 M ammonium hydroxide, 0.3 M $NaHCO_3$ (pH 10.4). Post column reagent flow rate: 0.15 mL/min.$^{-1}$ Detection: Vis., 530 nm. Ions: 1—iron (1 μg/L); 2—copper (1 μg/L); 3—nickel (1 μg/L); 4—zinc (1 μg/L); 5—cobalt (1 μg/L); 6—cadmium (1 μg/L); 7—manganese (1 μg/L).**

aromatic polycarboxylic acids such as trimesate[27] and pyromellitate[28] in the case of anion-exchange separations. For indirect UV detection of cations, commonly employed eluent cations include: benzyltrimethyl quaternary ions[29] and substituted pyridinium cations.[30] This mode of detection has considerable appeal for those interested in the detection of UV transparent ions. However, operating in this mode is of rather limited utility since the low sensitivity and limited dynamic capacity of indirect UV detection result in a fairly limited analytical operating range. As a consequence, indirect UV detection systems are of rather limited analytical utility.

D. Gradients

As with other modes of liquid chromatography, gradients are frequently employed in ion chromatography in order to provide adequate resolution of early eluting ions while still enabling the elution of strongly retained species. For UV absorbing species, gradient elution can be accomplished with the use of UV transparent eluent species (see Table 4 for examples of suitable eluent anions). However, in the case of conductivity detection-based ion chromatography special considerations limit options.

TABLE 4[1]

Anion	UV active	Anion	UV active	Anion	UV active
Fluoride	No	Tetrafluoroborate	No	Sulfite	Yes
Sulfide	Yes	Phosphite	No	Azide	Yes
Iodate	Yes	Iodide	Yes	Nitrate	Yes
Chlorite	Yes	Phosphate	No	Chlorate	Yes
Hypophosphite	No	Thiocyanate	Yes	Selenocyanate	Yes
Sulfamate	No	Arsenate	Yes	Perchlorate	No
Bromate	Yes	Thiosulfate	Yes	Sulfate	No
Chloride	Yes	Bromide	Yes	Selenate	Yes
Nitrite	Yes	Arsenite	Yes	Hydroxide	Yes[2]

[1]Adapted from: Williams, R. J., *Anal. Chem.*, 55:851,1983.
[2]This anion was not included in the original reference but is included here for completeness.

I. Gradients in Suppressed Systems

Because of the early dominance of carbonate based eluent systems in ion chromatography it is tempting to propose the use of carbonate for gradient applications. However, elevated background conductivity and relatively poor gradient performance make carbonate a poor choice for gradient applications. The problem of the residual background conductivity associated with the carbonic acid remaining in the suppressed eluent can be dealt with by passing the eluent through a carbon dioxide permeable membrane device. The first such devices were described in the early days of ion chromatography,[31,32] but the early devices provided unacceptable dispersion and were not widely adopted. One currently available suppressor (the DS Plus, see Table 3) combines the process of eluent suppression and carbon dioxide removal by incorporating a length of Teflon AF between the suppressor outlet and the conductivity cell inlet. However, in the hands of the author, this suppressor failed to provide reproducible carbon dioxide removal due to day-to-day variations in device split ratios. Considering the fact that variable levels of carbon dioxide in the suppressor effluent result in variable calibration response, such an approach is unlikely to result in reliable analytical measurements.

By far the most commonly employed eluent species in anion-exchange gradient separations is hydroxide (see Figure 11). Since the suppression by-product of this eluent is water, the suppressed eluent conductivity is independent of the eluent concentration. This makes hydroxide the ideal choice for gradient applications with very little baseline drift associated with such gradients. Likewise, the most commonly employed eluent species in gradient cation-exchange chromatography

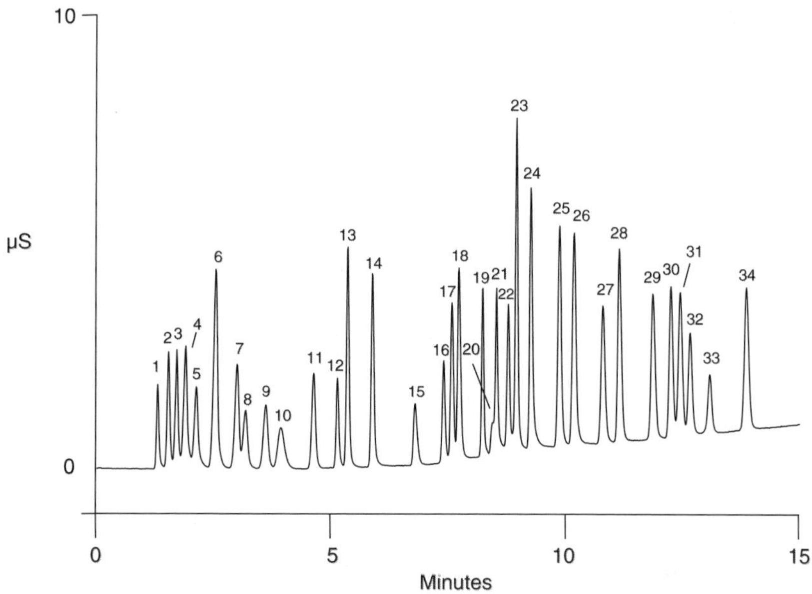

FIGURE 11 Fast gradient separation of inorganic anions and organic acids on an IonPac® AS11 column. Column: 4×250 mm IonPac AS11. Eluent: 0.5 mM NaOH for 2 min, followed by a linear gradient from 0.5 to 5 mM NaOH in 3 min and a second linear gradient from 5 to 38.25 mM NaOH for the last 10 min. Flow rate: 2.0 mL min^{-1}. Injection volume: 10 μL. Detection: suppressed conductivity utilizing the Anion Self Regenerating Suppressor (4 mm), recycle mode. Ions: 1—isopropylmethylphosphonate (5 mg L^{-1}); 2—quinate (5 mg L^{-1}); 3—fluoride (1 mg L^{-1}); 4—acetate (5 mg L^{-1}); 5—propionate (5 mg L^{-1}); 6—formate (5 mg L^{-1}); 7—methylsulfonate (5 mg L^{-1}); 8—pyruvate (5 mg L^{-1}); 9—chlorite (5 mg L^{-1}); 10—valerate (5 mg L^{-1}); 11—monochloroacetate (5 mg L^{-1}); 12—bromate (5 mg L^{-1}); 13—chloride (2 mg L^{-1}); 14—nitrite (5 mg L^{-1}); 15—trifluoroacetate (5 mg L^{-1}); 16—bromide (3 mg L^{-1}); 17—nitrate (3 mg L^{-1}); 18—chlorate (3 mg L^{-1}); 19—selenite (5 mg L^{-1}); 20—carbonate (5 mg L^{-1}); 21—malonate (5 mg L^{-1}); 22—maleate (5 mg L^{-1}); 23—sulfate (5 mg L^{-1}); 24—oxalate (5 mg L^{-1}); 25—ketomalonate (10 mg L^{-1}); 26—tungstate (10 mg L^{-1}); 27—phthalate (10 mg L^{-1}); 28—phosphate (10 mg L^{-1}); 29—chromate (10 mg L^{-1}); 30—citrate (10 mg L^{-1}); 31—tricarballylate (10 mg L^{-1}); 32—isocitrate (10 mg L^{-1}); 33—cis-aconitate and 34—*trans*-aconitate (10 mg L^{-1} combined).

separations is hydronium (see Figure 12). While other species are sometimes employed, the benefit of low background conductivity, high sensitivity and limited baseline drift make these two eluent ions the obvious choice for gradient applications in suppressor based ion chromatography.

The IonPac AS11 and the IonPac CS17 columns from Dionex are excellent choices for gradient work as illustrated in Figures 11 and 12 respectively. These columns provide for the rapid analysis of a wide variety of ions making them ideally suited to screening applications.

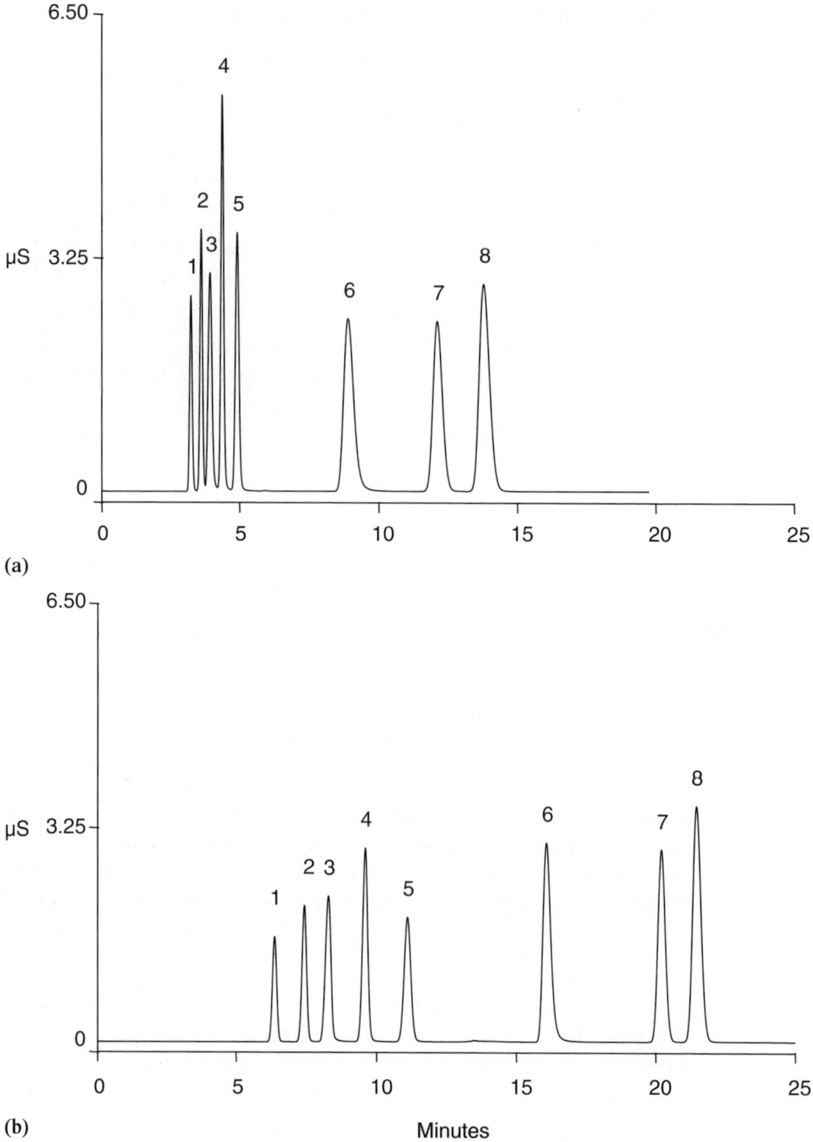

FIGURE 12 Comparison of isocratic and gradient separations of inorganic cations and amines on an IonPac® CS17. Column: 2×250 mm IonPac CS17. Flow rate: 0.25 mL min^{-1}. Temperature: 30°C. Injection volume: 25 µL. Detection: suppressed conductivity utilizing the Cation Atlas® Electrolytic suppressor in AutoSuppression® recycle mode (6 mA current setting). Ions: 1—lithium (0.1 mg L^{-1}); 2—sodium (0.4 mg L^{-1}); 3—ammonium (0.5 mg L^{-1}); 4—potassium (1 mg L^{-1}); 5—dimethylamine (1 mg L^{-1}); 6—triethylamine (2 mg L^{-1}); 7—magnesium (0.5 mg L^{-1}); 8—calcium (1 mg L^{-1}). Eluent: (a) 6 mM methanesulfonic acid (MSA), (b) 2 mM MSA with an eluent step change to 7 mM MSA at 11.1 min.

2. Gradients in Non-suppressed Systems

The method for performing gradients in non-suppressed ion chromatography makes use of isoconductive eluents.[33] In theory, by making use of a weak initial eluent and a second identical conductivity strong eluent, isoconductive eluents should enable gradients to be applied in non-suppressed applications. However, producing two separate eluents with precisely equal conductivity is nontrivial. This can only be achieved by preparing each solution separately and then going through the tedious process of adjusting the concentration of one eluent so that it matches the conductivity of the other eluent. Considering the fact that conductivity values must be matched to within <0.1% in order to produce useful gradients, one can easily see that preparation of such a matched set of eluents is not straightforward. Furthermore, unless the eluent species in both the weak eluent and the strong eluent have the same valency, stationary phase stoichiometry transitions will result in baseline disturbances in the midst of the gradient even if the two eluents are perfectly matched in terms of conductivity. Considering the problems given above, it is probably not surprising that gradients are almost never employed in the non-suppressed mode using conductivity detection.

E. Electrolytic Eluent Generation

Conventionally, ion chromatography eluents are directly prepared from reagent-grade chemicals. However, the preparation process for chromatographic eluents often introduces contaminants. For example, dilute NaOH solutions are easily contaminated by carbonate ions. Preparing carbonate-free NaOH eluents is difficult because carbonate can be introduced as an impurity from the reagents or by adsorption of carbon dioxide from air. The presence of carbonate in NaOH eluents often compromises the performance of an ion chromatographic method; it can cause undesirable chromatographic baseline drift during the hydroxide gradient or even irreproducible retention times of target analytes. In order to minimize baseline shifts during the hydroxide gradients, precautionary measures are utilized, such as thorough degassing of the DI water used in eluent preparation and the use of anion trap columns to remove carbonate.[34,35] Therefore, there is a need for a more convenient and reliable method to generate high-purity, carbonate-free hydroxide eluents for ion chromatography. Several approaches that utilize the electrolysis of water and charge-selective electromigration of ions through ion-exchange media have been investigated by researchers to purify or generate high-purity hydroxide solutions for use as ion chromatographic eluents.[36–39] In 1998, an automated electrolytic eluent generator was introduced commercially[40] (see Figure 13). This device produces high-purity acid or base eluents on-line using deionized water as the carrier stream for either isocratic or gradient ion

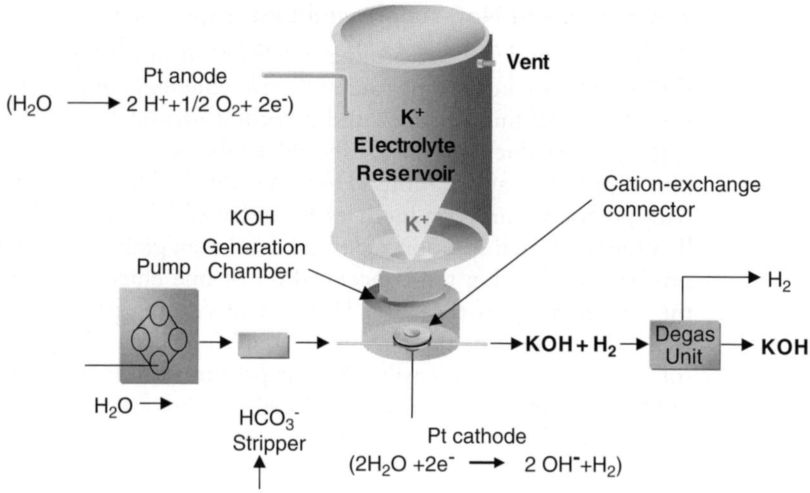

FIGURE 13 **Potassium hydroxide eluent generator schematic.**

chromatographic separations. Over the past several years, the use of electrolytic eluent generators has gained increasing popularity and is now widely used in suppressed ion chromatography.

F. Suppressors

The suppressor is a key component of the original invention of ion chromatography (the other key component being low capacity ion exchange phases required for practical suppressor implementation). The suppressor in ion chromatography serves a role without parallel in any other liquid chromatography methodology. The suppressor largely eliminates (or at least greatly minimizes) eluent conductivity through a neutralization reaction while simultaneously increasing detector response by replacing the analyte counterion with highly conductive ions. Together these two factors combine to provide a substantial reduction in baseline noise along with significant enhancement of analyte response. The original suppressors were columns packed with ion-exchange resins in appropriate ionic forms. Those packed-bed suppressors had a relatively large dead volume and required periodic off-line chemical regeneration. To overcome this problem, suppressors based on ion-exchange fibers and membranes were developed.[17,41] These suppressors can be continuously regenerated using either acid or base regenerant solutions. One disadvantage associated with such suppressors is that an external source of regenerant solution is required to maintain the suppressor in a fully regenerated state. Over the past 15 years, several designs of electrolytically-regenerated membrane suppressors have been developed to overcome the limitations associated with chemically-regenerated suppressors.[42–47] Figure 14 illustrates schematically the construction of an electrolytic suppressor. Electrolytic suppressors offer several advantages in ion chromatography. They provide continuous and simultaneous suppression of eluents, regeneration of the suppression bed, and sufficient suppression capacity for all common IC applications. They are easy to operate because the suppressed eluent or water is used to create regenerant ions electrolytically and there is no need to prepare regenerant solutions off-line. They are compatible with gradient separations. They have very low suppression zone volume, which makes it possible to achieve separations with very high chromatographic efficiency. The development of new and improved electrolytic suppressors remains an important area of research in ion chromatography.

G. Trap Columns

The removal of ionic contaminants from ion chromatographic eluents is often necessary and has traditionally been accomplished using trap columns that contain ion-exchange resins of appropriate functionality.

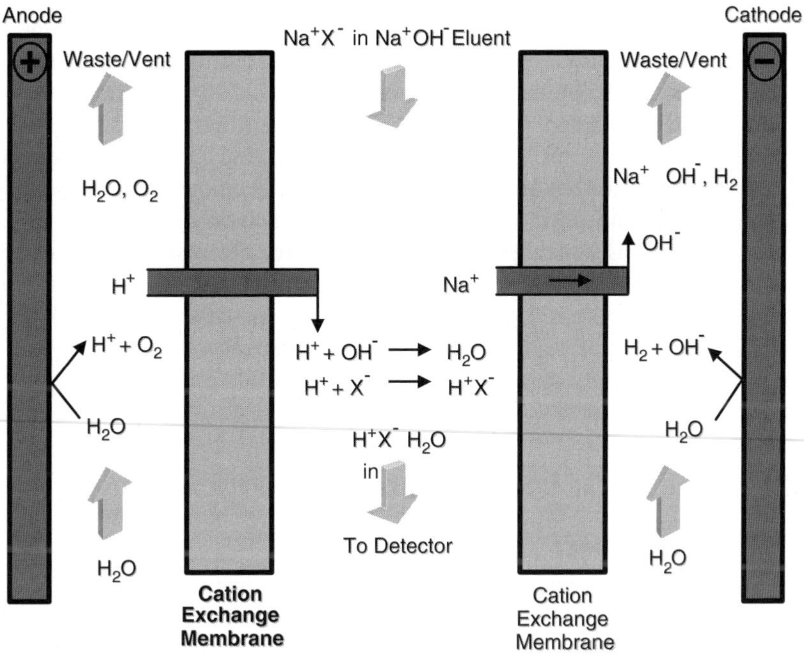

FIGURE 14 Electrolytic suppressor schematic.

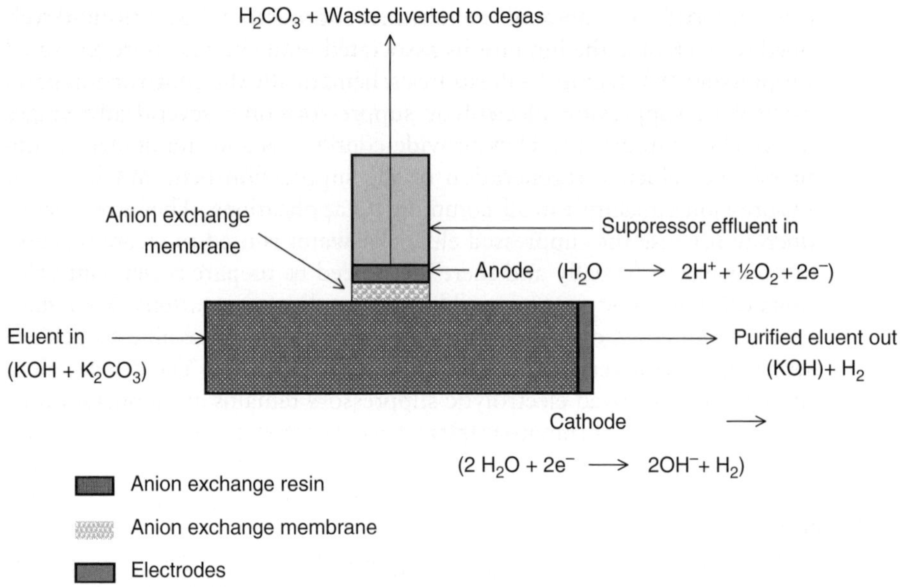

FIGURE 15 **Continuously regenerated trap column schematic.**

Trap columns must be washed frequently (depending on operating conditions) with concentrated acid or base solutions to regenerate their capacity for trapping the ionic contaminants. The regeneration of trap columns is performed off-line and is usually rather tedious. To overcome these problems, novel trap columns that can be regenerated continuously through the electrolytic process have been developed[48] (see Figure 15).

The developments of new electrolytic devices that utilize the electrolysis of water and charge-selective electromigration of ions through ion-exchange media have significantly changed the routine operation of ion chromatographic methods. The use of these electrolytic devices has made it a reality to perform various ion chromatographic separations using only deionized water as the mobile phase.

IV. PHARMACEUTICAL ION CHROMATOGRAPHY APPLICATIONS

Ion chromatography is widely employed in a number of areas of pharmaceutical analysis. The high sensitivity of suppressor-based ion chromatography makes it an ideal choice for the analysis of deionized water used in the preparation of solution pharmaceuticals and as a process fluid for the preparation of pharmaceuticals. Depending upon the sensitivity required, the analysis of deionized water can be accomplished by direct injection (see Figure 16) or by making use of a concentrator column. While concentrator columns provide an effective means of quantitating anions and cations at the low part-per-trillion level, direct injection analysis is far more convenient, especially when the analysis is performed on an instrument frequently used for other applications. While direct injection analysis limits the ultimate sensitivity of the technique, detection limits in the high part-per-trillion range for most common anions and cations are readily achieved using 0.5 mL injections with a 2-mm column system. Analysis of ions at this level requires special care in order to minimize environmental contamination problems. For example, the sample container used to transport high purity water should be selected to minimize contamination problems. Generally, glass containers are inadvisable for such applications as they contribute significant levels of chloride, sulfate and sodium to the sample even when exhaustively rinsed. Polystyrene culture flasks which have been well rinsed with high-purity water work best for this application. Likewise, when utilizing an autosampler for such an application, it is important to make sure that autosampler vials do not contaminate the sample. Again, polystyrene sample vials work best for trace analysis applications.

Another major ion chromatography application is the analysis of active ingredient counterions. Frequently, drug substances are formulated as salts in order to achieve specific pharmaceutical properties (such as improved solubility or control of dissolution rate). The analysis of

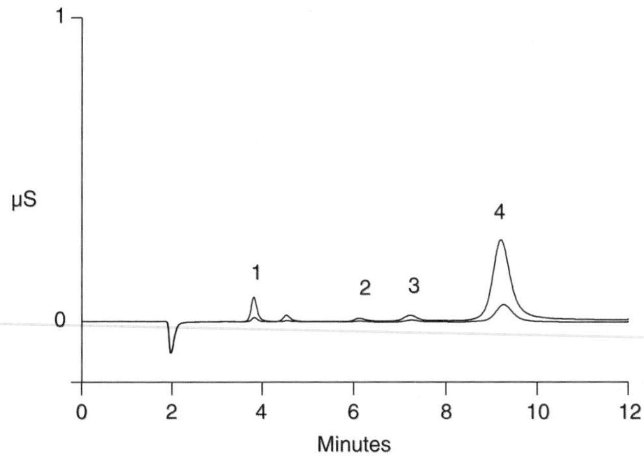

FIGURE 16 Water for injectables by cation-exchange chromatography. Columns: 4×250 mm IonPac® CS12, and 4×50 mm CG12 and CTC-1. Flow rate: 1 mL min^{-1}. Eluent: 20 mM MSA. Injection volume: 25 µL. Detection: suppressed conductivity, CSRS (4 mm), recycle mode. Ions: 1—sodium; 2—potassium; 3—magnesium; 4—calcium. Overlay of two water samples.

counterions is of interest because the drug substance may well be contaminated with other ionic species (either anionic or cationic depending upon the drug) as a by-product of the manufacturing process. Since a very wide variety of counterions are utilized in pharmaceutical materials, ion chromatography with suppressed conductivity detection provides a particular advantage for this application in that both UV absorbing and UV transparent counterions can be analyzed in a single analysis. In such an application, gradients are of particular advantage for quickly screening both intended and contaminant counterions since simple isocratic methods can rarely achieve elution of trivalent organic acids such as citrate, while still achieving retention of monovalent organic acids such as acetate. In such screening applications, it is generally advisable to utilize suppressed conductivity detection, because non-suppressed ion chromatography is generally impractical for gradient applications and even when the analyte ions are all suitable for detection via UV (such as in the case of organic acid counterions), there is always the chance that the drug substance is contaminated with UV transparent counterions, which would be missed using UV detection.

Ion chromatography is also used in the analysis of materials used in pharmaceutical preparations in order to assay for undesirable trace impurities which may be present as a by-product of the manufacturing process. An example of this is the analysis of trifluoroacetic acid in peptide and protein pharmaceuticals (see Figure 17). Trifluoroacetic acid is frequently present following chromatographic purification steps as it is a common eluent additive in reverse phase separations of peptides and proteins. Although trifluoroacetic acid does absorb weakly in the UV, the sensitivity is not adequate to provide the necessary detection limits. Because trifluoroacetic acid is strongly acidic, detection limits for the suppressed conductivity detection are 20–100 times lower than with UV detection. Another example is the use of ion chromatography for the analysis of contaminants introduced as a by-product of equipment cleaning. Figure 18 illustrates one such example where traces of surfactant residue were detected on a tablet press following a cleaning operation. While the analyte of interest is UV absorbing (dimethylbenzene sulfonate), the high sensitivity of ion chromatography allows for sensitive detection of even trace levels of this compound while allowing for simultaneous analysis of non-UV absorbing contaminants.

Another ion chromatography pharmaceutical application is the analysis of amines and amphoteric compounds such as choline (see Figure 19). Such compounds may be present either as counterions, as mentioned above, or as synthesis by-products. In either case, ion chromatography can be advantageously utilized for this class of compounds due to the limited utility of gas chromatography and the lack of a UV chromophore for such compounds. Again, screening for such compounds in pharmaceutical preparations can be best accomplished

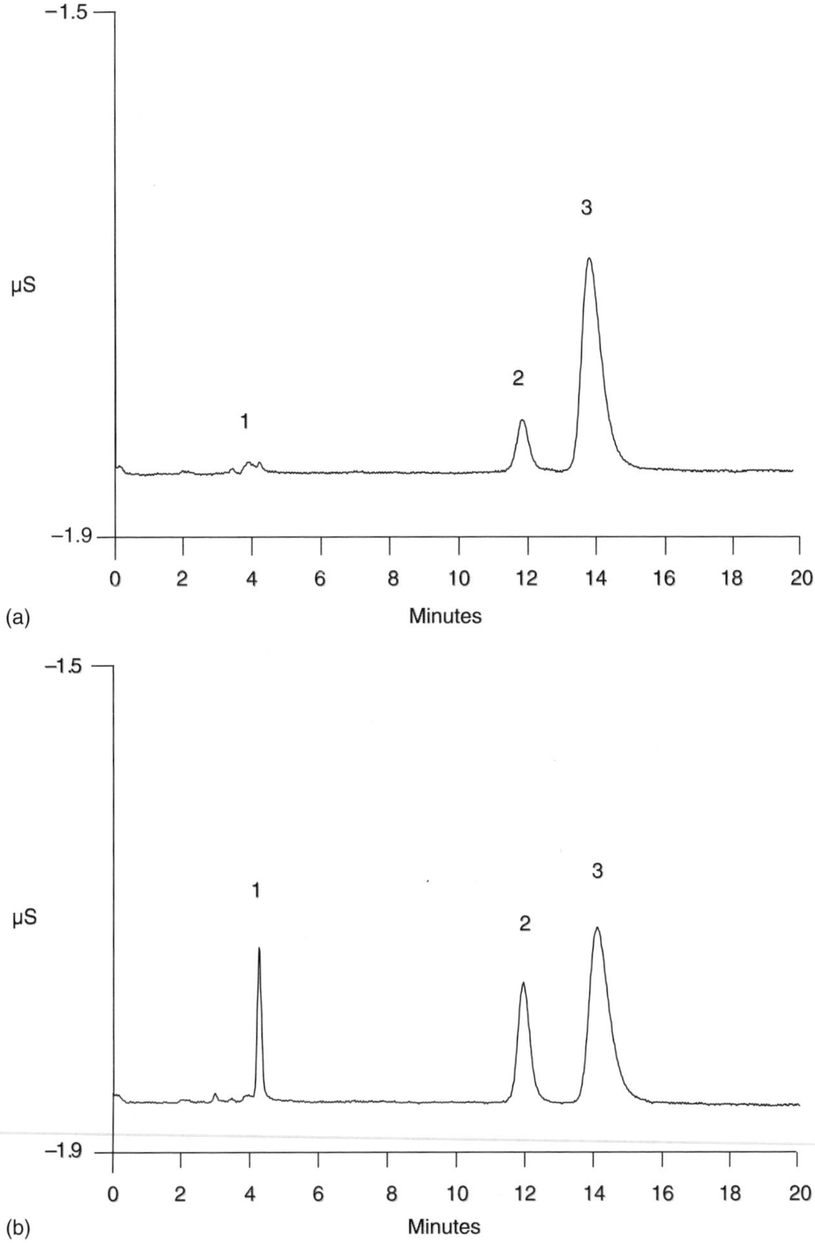

FIGURE 17 **Separation of inorganic anions and trifluoroacetic acid (TFA) on an IonPac® AS14 and an AG14 column. Columns: 4×250 mm IonPac AS14 and 4×50 mm AG14. Eluent: 3.5 mM Na_2CO_3 and 0.8 mM $NaHCO_3$. Flow rate: 1.2 mL min^{-1}. Injection volume: 10 µL. Detection: suppressed conductivity utilizing the ASRS (4 mm), recycle mode. Ions: 1—Chloride; 2—Sulfate; 3—TFA. Samples: 40 µg mL^{-1} (a) Crude peptide and (b) GPC-purified in-process peptide.**

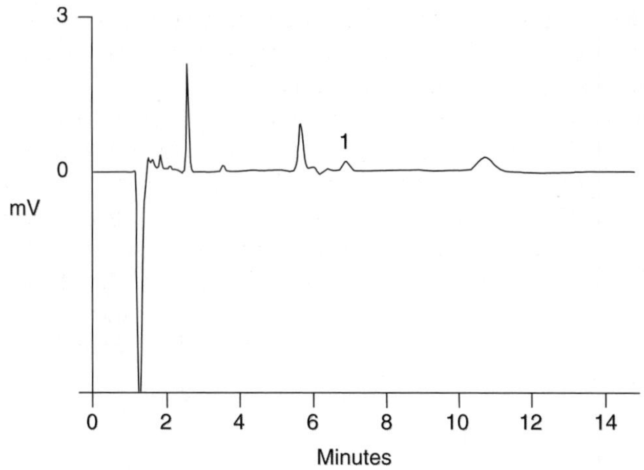

FIGURE 18 Chromatographic identification of tablet press contaminant. Column: 4×250 mm OmniPac PAX-100. Eluent: 1 mM disodium phosphate in 40% methanol. Flow rate: 1 mL min^{-1}. Injection volume: 10 μL. Detection: UV at 220 nm. Ion: 1—dimethylbenzene sulfonate.

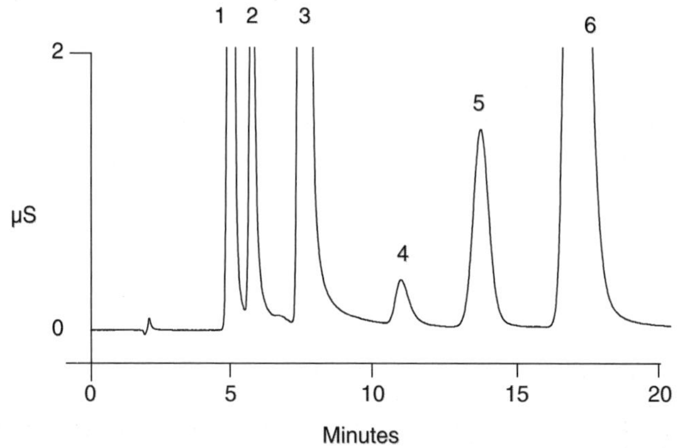

FIGURE 19 Choline in infant formula. Columns: 4×250 mm IonPac® CS12A, and 4×50 mm CG12A. Flow rate: 1 mL min^{-1}. Eluent: 18 mM MSA. Injection volume: 10 μL. Detection: suppressed conductivity, CSRS (4 mm), recycle mode. Ions: 1—sodium; 2—ammonium; 3—choline; 4—potassium; 5—magnesium; 6—calcium. Sample preparation: add 30 mL of 1 M HCl to 5 g sample, mix well, place in 70°C water bath for 3 h, cool, filter and dilute to 100 mL.

through the use of gradients in conjunction with suppressed conductivity detection.

To a limited extent, ion chromatography is also employed for the analysis of active ingredients. While most common drug substances are

UV active, many provide sufficient conductivity sensitivity to allow for simultaneous analysis of the active ingredient along with other ionic species present in a drug formulation.

Finally, ion chromatography is sometimes used for process applications, allowing for the tracking of the manufacturing process in order to optimize process variables and to allow for better control of process parameters. One example of this is the application of ion chromatography to the analysis of fermentation broths. Here ion chromatography is used both to measure the level of ionic nutrients in the fermentation broth in order to control the fermentation process and also to measure the level of fermentation by-product ions which may be indicative of problems with the fermentation process.

V. SUMMARY AND CONCLUSIONS

In summary, this chapter reviews ion chromatography and its applications to pharmaceutical analysis.

ACKNOWLEDGMENTS

The author acknowledges the helpful contributions of Yan Liu and Kannan Srinivasan. They supplied the original text for the sections on eluent generation, suppressors and trap columns, which was adapted for this chapter.

REFERENCES

1. Small, H., Stevens, T. S. and Bauman, W. C., *Anal. Chem.*, 47:1801, 1975.
2. Small, H., *Ion Chromatography*, Plenum, New York, 1989.
3. Haddad, P. R. and Jackson, P. E., Ion Chromatography Principles and Applications, *J. Chromatogr. Library*, 46:1990.
4. Fritz, J. S. and Gjerde, D. T., *Ion Chromatography,* Wiley-VCH, Weinheim, 2000.
5. Weiss, J., *Ionenchromatographie,* 3rd ed., Wiley-VCH, Weinheim, 2001.
6. Pohl, C. A., Stillian, J. R. and Jackson, P. E., *J. Chromatogr. A* 789:29–41, 1997.
7. Jackson, P. E and Pohl, C. A., *Trends Anal. Chem.,* 16(7):393–400, 1997.
8. Williams, R. J., *Anal. Chem.,* 55:851, 1983.
9. Small, H. and Miller, T. E., *Anal. Chem.,* 54:462, 1983.
10. Haddad, P. R. and Heckenberg, A. L., *J. Chromatogr.*, 252:177, 1982.
11. Mercier, J-P., Chaimbault, P., Elfakir, C. and Dreux, M., *Am. Lab. News,* October, 22, 1999.
12. Roehl, R., Slingsby, R., Avdalovic, N. and Jackson, P. E., *J. Chromatogr. A* 956:245–254, 2002.
13. Slingsby, R. and Schnute, B., *Recent Applications in LC-MS, LC.GC (Europe),* November, 15:50, 2002.

14. Weiss, J., Ion *Chromatography, 2nd ed.*, VCH, Weinheim, 1995.
15. Irgum, K. *Anal. Chem.*, 59:358, 1987.
16. Doury-Berthod, M., Giampaoli, P., Pitsch, H., Sella, C. and Poitrenaud, C., *Anal. Chem.*, 57:2257, 1985.
17. Stillian, J., *LC,* 3:802, 1985.
18. Kolla, P., Köhler, J. and Schomburg, G., *Chromatographia,* 23:465, 1987.
19. Jensen, D., Weiss, J., Rey, M. and Pohl, C., *J. Chromatogr. A,* 640:65, 1993.
20. Rey, M. and Pohl, C., *J. Chromatogr. A,* 739:87–97, 1996.
21. Pohl, C., Rey, M. Jensen, D. and Kerth, J., *J. Chromatogr. A,* 850:239–245, 1999.
22. Rey, M. A., *J. Chromatogr. A,* 920:61–68, 2001.
23. Haddad, P. R., Jackson, P. E. and Heckenberg, A. L., *J. Chromatogr.*, 346:139, 1985.
24. Haddad, P. R. and Heckenberg, A. L., *J. Chromatogr.*, 300:357, 1984.
25. Gjerde, D. T., *J. Chromatogr. A,* 439:49, 1988.
26. Rey, M., Pohl, C., Bordunov, A. and DeBorba, B., *Cation Exchange Chromatography with Conductivity Detection: Suppressed or Not Suppressed?* Poster B-6, International Ion Chromatography Symposium, 2003.
27. Laurent, A. and Bourdon, R., *Annal. Pharm. Franc.*, 36:453, 1978.
28. Jardy, A., Caude, M., Diop, A., Curvale, C. and Rosset, R., *J. Chromatogr.* 439:137, 1988.
29. Walker, T. A., *J. Chromatogr.* 466:438, 1989.
30. Haddad, P. R. and Foley, R. C., *Anal. Chem.* 61:1435, 1989.
31. Sunden, T., Cedergren, A. and Siemer, D. D., *Anal. Chem.* 56:1085, 1984.
32. Siemer, D. D. and Johnson, V. J., *Anal. Chem.* 56:1033, 1984.
33. Jones, W. R., Jandik, P. and Heckenberg, A. L., *Anal. Chem.,* 60:1977, 1988.
34. Technical Note No. 8, *The Use of Concentrator Columns in Ion Chromatography,* Dionex Corporation, Sunnyvale, CA, 1994.
35. Application Note No. 113, *Determination of Trace Anions in High Purity Waters by High Volume/Direct Injection Ion Chromatography,* Dionex Corporation, Sunnyvale, CA, 1996.
36. Strong, D. L., Dasgupta, P. K., Friedman, K. and Stillian, J. R., *Anal. Chem.*, 63:480, 1991.
37. Strong, D. L., Joung, C. U. and. Dasgupta, P. K., *J. Chromatogr.,* 546:159, 1988.
38. Strong D. L. and Dasgupta, P. K., *J. Membrane Sci.*, 57:321, 1991.
39. Small, H., Liu, Y. and N. Avdalovic, *Anal. Chem.*, 70:3629, 1998.
40. Liu, Y., Avdalovic, N., Pohl, C., Matt, R., Dhillon, H. and R. Kiser, *Am. Lab.*, November Issue: 48C-58C, 1998.
41. Stevens, T. S., Davis, J. C. and Small, H., *Anal. Chem.*, 53:1488, 1981.
42. Tian, Z. W., Hu, R. Z., Lin, H. S. and Wu, J. T., *J. Chromatogr.*, 439:159, 1988.
43. Strong, D. L. and Dasgupta, P. K., *Anal. Chem.*, 61:939, 1989.
44. Pohl, C., Slingsby, R., Stillian, J. and Gajek, R., *Modified Membrane Suppressor and Method for Use,* US Patent 4,999,098, 1991.
45. Stillian, J., Barreto, V., Friedman, K., Rabin, S. and Toofan, M. *Ion Chromatography System Using Electrochemical Suppression and Detection Effluent Recycle*, US Patent 5,352,360, 1994.
46. Hensall, A., Rabin, S., Statler, J. and Stillian, J., *Am. Lab.*, November Issue: 20R, 1992.
47. Rabin, S., Stillian, J., Barreto, V., Friedman, K. and Toofan, M., *J. Chromatogr.*, 640:97, 1993.
48. Srinivasan, K., Lin, R., Saini, S., Pohl, C. and Avdalovic, N., *A New Continuously Regenerate Trap Column for Ion Chromatography,* Paper 2020-2, Pittcon, 2003.

9 HOW TO BE MORE SUCCESSFUL WITH HPLC ANALYSIS: PRACTICAL ASPECTS IN HPLC OPERATION

MICHAEL W. DONG

Purdue Pharma, Ardsley, NY

ABSTRACT

This chapter describes how to be more successful in HPLC operation by summarizing a series of standard operating procedures representing the "best practices" of experienced HPLC analysts. It offers some brief guidelines on maintenance/troubleshooting and on means of enhancing HPLC assay precision.

Handbook of Pharmaceutical Analysis by HPLC,
S. Ahuja and M.W. Dong, editors.

I. INTRODUCTION

It takes many years of hands-on experience to be truly proficient in HPLC operation. Most analysts agree that the best way to learn HPLC is to be "apprenticed" to an expert chromatographer. Barring that, other effective ways include reference books,[1–4] user's manuals, training videos,[5] hands-on training courses, and computer-based training software.[6] One useful resource is John Dolan's recurring column in *LC.GC* magazine on various practical aspects of HPLC.

A. Scope

This chapter focuses on the practical aspects of HPLC including the "best practices" or operating guides for mobile phase preparation and HPLC system operation. Guidelines on maintenance and troubleshooting are highlighted as well as ways for improving HPLC assay precision. Our goal is to offer the analyst a concise overview and a guide on how to be more successful in HPLC analysis.

II. MOBILE PHASE PREPARATION

In premixing mobile phase, measure the volume of each solvent separately and combine them in the solvent reservoir. This is important because of the negative ΔV of mixing for many solvents. For instance, prepare 1 L of methanol/water (50:50) by measuring 500 mL of methanol and 500 mL of water separately in a measuring cylinder and combine them together. Do not pour 500 mL of methanol into a 1-L volumetric flask and fill it to volume with water as more than 500 mL of water is needed due to the shrinkage of solvents upon mixing. Figure 1 shows the results of actual retention times of an assay using various mobile phase preparations and on-line pump blending. Note that if a volumetric flask is used for preparing the mobile phase (such as in procedure B or C), the strength of the mobile phase and the resulting retention times can actually be too high or too low (when compared with procedure A) depending on the order of filling the flask. Note that the retention time of mobile phase preparation A is almost identical to that from procedure D (from pump blending).

A. Buffers

Buffers are commonly used to control the pH of the mobile phase, which greatly affects the retention of acidic or basic analytes. Table 1 summarizes the common buffers for HPLC and their respective pK_a and UV

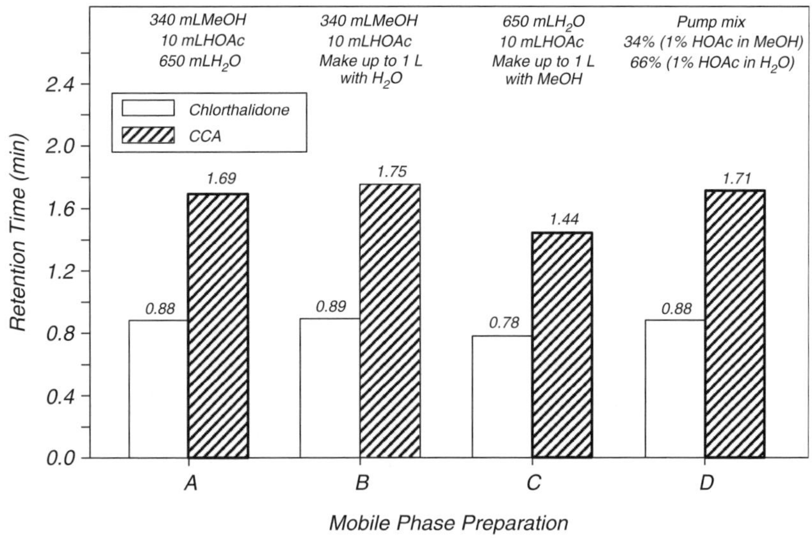

FIGURE 1 **Retention times of two analytes using various mobile preparations vs. pump blend. Note that procedures A and D are almost equivalent while column B or C (using volumetric flasks) yields retention times which are either too high or too low depending on whether methanol or water is filled first. This phenomenon is caused by the negative volume of mixing. Procedure A is recommended for all mobile phase preparation.**

TABLE 1 **Common HPLC Buffers and their Respective pK_a and UV Cutoff**

Buffer	pK_a	UV cutoff (nm)
Trifluoroacetic acid[1]	0.3	210
Phosphate	2.1, 7.2, 12.3	190
Citrate	3.1, 4.7, 5.4	225
Formate[1]	3.8	200
Acetate[1]	4.8	205
Carbonate[1]	6.4, 10.3	200
Tris(hydroxymethyl) aminomethane	8.3	210
Ammonia[1]	9.2	200
Borate	9.2	190
Diethylamine	10.5	235

[1]Volatile buffer systems, which are MS-compatible.

cutoffs. Volatile acids and their ammonium salts are used for mass spectrometer (MS) compatible methods. Since a buffer is only effective within ±1 pH unit of its pK_a, judicious selection of the proper buffer is paramount. For instance, phosphate has three pK_a values and is most effective for pH ranges of 1–3, 6–8, and 11–13. It is not effective at pH 4 because it does not have any buffering capacity at that pH. Note that citrate buffers are effective at pH 2–6.5 though their high UV cutoff of 225 nm and non-volatility can be a problem. Any pH adjustments should be made before mixing the aqueous buffer with any organic solvents. While many older HPLC methods use a "standard" buffer strength of ~50 mM, newer studies have suggested that a concentration of 10–20 mM is sufficient for most applications.[7] Precaution should be taken to prevent any possibility of buffer precipitation. While methanol is fairly miscible with most buffers, acetonitrile might precipitate some buffers at high concentrations (e.g., above 60–80%). Note that lower buffer strengths have fewer problems for buffer precipitation as well as substantially less background in LC/MS applications. An acidic pH of 2.5–3 is a good starting point for most pharmaceutical applications because the low pH suppresses the ionization of most acidic analytes (and provides a higher retention) and minimizes the interaction of basic analytes with surface silanols on the silica packing (because silanols are not ionized at that pH).

Follow these guidelines when using buffers:

- Do not use aqueous buffered mobile phase older than 10 days because of the increased risk of bacterial growth, which can damage the HPLC column.[8]
- Never let buffered mobile phases sit in the HPLC system due to the danger of precipitation. Run at low flow rate of 0.1 mL/min when idling.
- Always turn on piston seal wash feature to prolong seal life.
- Flush buffers from system with water (or 10% acetonitrile in water) before switching to organic mobile phase.

B. Filtration and Degassing

Aqueous mobile phase containing buffers or ion-pairing reagents must be filtered. However, when a high-purity buffer (>99.99%+) is used in performing trace analysis such as impurity testing, the filtration step might be skipped to reduce potential solvent contaminations that lead to gradient peaks.[9] Do not filter HPLC-grade solvents, which are typically pre-filtered by the manufacturers. Filter all organic or aqueous solvents through a 0.45-µm membrane filter (typically 47-mm diameter). Use cellulose acetate filters for aqueous solvents, PTFE filters for organic solvents, and nylon filters for either organic or aqueous solvents.

Mobile phase degassing is important to prevent pump problems and is critical for accurate pump blending or gradient operation. Low-pressure mixing pumps are particularly vulnerable to out-gassing problems. The most effective and convenient way for solvent degassing is by using an in-line vacuum degasser (see Chapter 3 on degassers). Note that degassing by vacuum filtration or sonication is much less effective and the solvent might "re-gas" within a few hours.

III. HPLC PUMP-OPERATING GUIDES

The following standard operating procedures are recommended for HPLC pump operation:

1. Place solvent line sinkers (10-µm filters) into intermediary solvent reservoirs when switching solvents to prevent cross-contamination or buffer precipitation. Cover the flask with parafilm to minimize atmospheric contamination and evaporation.
2. Turn on the on-line vacuum degasser and the seal wash pump system (if available).
3. Set the upper pressure limit (typically at 3500 to 4000 psi).
4. Perform "dry" prime by opening the prime/purge valve and draw out ~10 mL from each solvent line if the system has been idle for a long time or the pump has run dry.
5. Perform a wet prime for all lines when changing the solvent reservoirs. Note that many older vacuum degassers have an internal volume of 10 mL or more and must be purged out with the new solvents.
6. Turn on the column oven if required. Rinse the column with strong solvent (e.g., 100% MeOH) first then equilibrate column for 5–10 min with the new mobile phase until the baseline and pressure are stable.
7. Program the pump to purge out the column and shut down the pump automatically after the sample sequence.

IV. HPLC COLUMN-OPERATING GUIDES

The following operating guides are recommended for maintaining reversed-phase columns. Consult the vendor's column instructions for precautions and column regeneration guides.

1. Store the columns in acetonitrile or methanol or in a mixture of water and organic solvents. Cap them with "closed" fittings to prevent the columns from drying out.
2. Do not exceed the pH range of the column (typically pH of 2.5 to 8 for silica-based columns). Silica dissolves at pH greater than 8 and bonded groups can hydrolyze at pH less than 2. Note that many modern silica-based columns may have more durable bonding chemistries for extended pH usable range of 2–10 or even 1–12.[10,11]

3. Do not exceed the temperature limits of most columns, usually 60–80°C, without consulting the instructions from the column vendor.
4. Normal column lifetime is 3–24 months or 1000–3000 injections. Column performance decreases with time as signified by increased backpressure and peak widths. The appropriate regeneration procedure or back-flushing might restore some column performance.[12] Note that some columns cannot be back-flushed because the inlet frits might have larger pores to minimize plugging problems. Consult the vendor's column instructions before back-flushing.
5. Always flush the column with a strong solvent (methanol) before use to eliminate any highly retained analyte.
6. Never let buffers sit immobile inside pumps or columns. They can precipitate and cause severe damage to the piston and the piston seal.

The following operating guides are recommended for connecting HPLC columns:

1. Use 1/16" compression nuts and ferrules from vendors such as Parker, Swagelok, or Rheodyne. These have similar ferrule shapes, which are compatible with each other. Note that Waters and Valco fittings have different ferrule shape and seating depths and might not be interchangeable with others.
2. Use 1/16" o.d. and 0.010" i.d. Stainless or PEEK tubing for general applications and 0.005–0.007" for narrowbore (2-mm i.d.) and Fast LC columns (<5 cm long). Tubing should be square cut with a cutting wheel and have its cut end "deburred" or polished. Make sure that the tubing "bottoms out" inside the fitting when attaching a new ferrule. See Figure 2 for ferrule shapes and correct seating depths of the various fittings.
3. Alternately, plastic finger-tight fittings (e.g., UpChurch) are convenient and easy to use, but they do have a tendency to slip at a higher pressure.
4. Note that PEEK tubing has a lower pressure rating and might not be compatible with solvents such as dimethyl sulfoxide, tetrahydrofuran, inorganic acids, and methylene chloride.

V. DETECTOR-OPERATING GUIDES

The following guidelines are recommended for UV/Vis absorbance or photodiode array (PDA) detectors:

1. Turn the lamp on for at least 15 min to warm up before analysis.
2. Set to the appropriate wavelength and detector response time (i.e., 1–2 s)

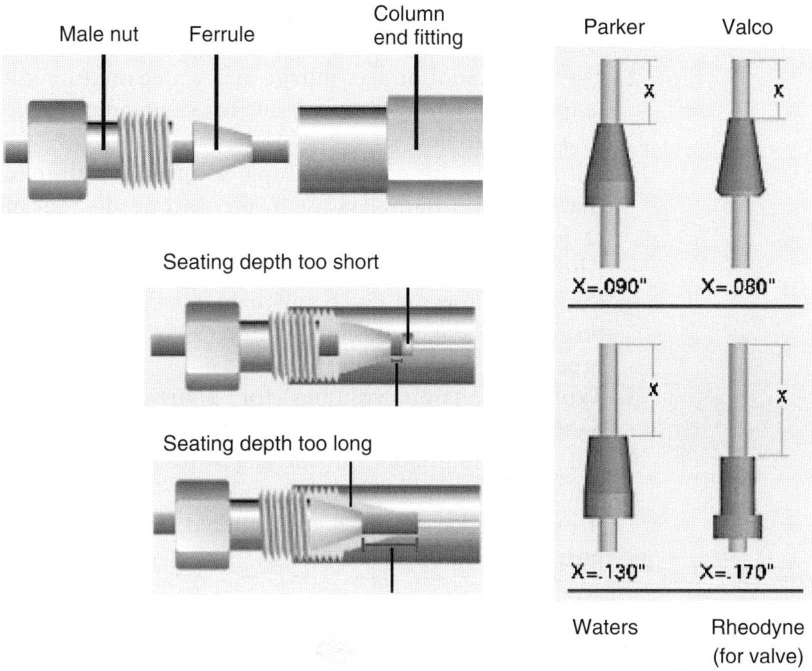

FIGURE 2 Schematic diagrams of column connection hardware and the importance of having the correct seating depths of the ferrule. If the seating depth is too low, additional void volume is created. If the seating depth is too high, leaks might occur. Note that various manufacturers offer fittings of different seating depths and ferrule shapes, which might not be interchangeable. Diagram reprinted from UpChurch.com.

3. Set scanning range of PDA (e.g., 200–400 nm) for method development. Use resolution of 1 nm high-resolution spectra and 5 nm for better chromatographic sensitivity.
4. Lamps typically last 12 months or over 1000 h. They should be replaced during annual preventive maintenance. Note that many long-life lamps with a guaranteed life of >2000 h are available.
5. Shut off lamps when not in use to increase lifetime.
6. If a nonstandard flow cell such as a semi-micro flow cell with a shorter path length is used, a label on the outside of the detector signifying the flow cell type and path length is recommended.

VI. AUTOSAMPLER-OPERATING GUIDES

The following guidelines are recommended for autosamplers:

1. Use the appropriate sample trays and vials. Avoid using airtight sample vials with silicone septa and overfilling the samples to

avoid the possibility of creating a partial vacuum during sample withdrawals.

2. Use a methanol or acetonitrile and water mixture without buffer (e.g., 50%) as the flush solvent for most autosamplers. Degas it to avoid bubbles.
3. Fill the vial with enough sample solution for all injections. Note that some autosamplers use a side-port needle that might require at least a 0.5 mL sample volume in a 2-mL vial (e.g., Waters Alliance).
4. Purge the injector daily and before sample analysis to remove bubbles in the sampling syringe. This is critical for sampling precision.
5. Typical injection volumes for pharmaceutical assays range 5–50 μL. Injection sizes of <5 μL might lead to poorer precision.
6. Put a warning label outside the autosampler if a non-standard sampling syringe or loop is installed.
7. If excessive carryover is encountered, explore the use of pre- or post-injection flush.

VII. QUANTITATION-OPERATING GUIDES

The following guidelines are recommended for HPLC quantitation:

1. Use Peak Area for quantitation which is less susceptible to flow variations.
2. Use "Normalized Area Percent" analysis for impurities and related substances testing. Index the area % to the parent drug (where parent = 100%). Apply a correction for relative response factors (RRF) if those are known.
3. Use "External Standardization" for assays and dissolution. Bracketed standards are preferred for improved accuracy.
4. Use Internal Standards for bioanalytical samples to compensate for sample loss during sample preparation. The internal standards should have similar structures to the analytes and in most cases are isotopically labeled analytes.
5. Use the same processing method and integration events to process the entire sample sequence. Always plot the integration baseline with each chromatogram in the report to highlight any integration problems. Particular attention must be paid to sloping baseline during gradient analysis or complex samples. The use of time events such as "valley-to-valley," "forced baseline by time," or "% touchdown" might be appropriate in many cases. Advanced integration algorithm such as "Apex Track" can be more effective for complex chromatograms with sloping baselines.

VIII. SYSTEM SHUTDOWN-OPERATING GUIDES

The following guidelines are recommended for system shutdown:

1. Turn off the column oven. Flush the buffer out of the system and stop the flow. Keep the idle column in methanol or methanol/water.
2. Turn off the lamp.
3. Log out of the data system.

IX. SUMMARY OF HPLC OPERATION

The following is a summary of important procedures or sequence of events for HPLC operation:

- Filter and degas mobile phase.
- Prime the pump, rinse the column with strong solvent, and equilibrate column.
- Purge injector and make sure there are no bubbles in the sampling syringe.
- Perform a system suitability test.
- Analyze the samples.
- Process and report data.
- Rinse the column and shutdown thc pump and lamp.

X. KEYS FOR SUCCESS IN HPLC

This section contains a list of steps the analyst can take to ensure successful HPLC operation and includes highlights of key maintenance and troubleshooting strategies.

- Maintain and calibrate the HPLC system periodically. Preventive maintenance performed by a service engineer annually and system calibrations performed every 6 months are recommended.[13]
- Check system performance using a reference sample with documented parameters.
- Always verify that pressure and baseline are normal before sample analysis. Check for leaks if the pressure reading is lower than expected (see Section X. B).
- Always use HPLC-grade solvents. Always filter and degas aqueous mobile phase.
- Clean-up and filter samples to prevent system and column damage. Dissolve or dilute the final sample solutions in mobile phase, if possible. If a final sample solvent is stronger than the mobile phase, inject smaller volumes to minimize peak distortion.

- Prime the pump and rinse the column with a strong solvent before sample analysis.
- Rinse the column/system before shutting down.

A. Common Maintenance Procedures

The following is a listing of commonly recommended maintenance procedures. Some are included in the manufacturer's preventive maintenance (PM) program, which should be conducted annually.

- LC pumps—replace or clean: piston seals, pistons, in-line filter, solvent filter sinkers; and check valves.
- On-line vacuum degasser and seal wash system—inspect and check functions.
- UV detector—replace deuterium source and clean/rebuild detector flow cell if contaminated.
- Autosampler or injector—replace rotor seal in injector valve or high-pressure needle seal pack, sampling needle, and sampling syringe.
- Column—regenerate or replace column.
- System test—turn off all modules and power-up to perform start-up diagnostics. Perform static leak test of the pump and compression test of the autosampler as suggested by the manufacturer.

B. Checking HPLC Performance by Using a Reference Procedure

A reference testing procedure should be performed periodically as a diagnostic test to ensure that the overall HPLC system is functioning properly. This test can be a system suitability test run routinely on the HPLC system or the autosampler precision test used in HPLC calibration. Figure 3 shows an example of a reference chromatogram of ethylparaben with typical parameters (retention time, pressure, plate count). This chromatogram documents the expected parameters under a set of standardized conditions when the system is functioning properly. It provides a powerful troubleshooting tool for diagnosing problem areas.

C. Some Troubleshooting Strategies

The following is a set of guidelines and strategies for troubleshooting HPLC system problems. Further details on HPLC troubleshooting can be found elsewhere.[14]

1. Verify that a problem exists—repeat the experiment.
2. Go back to a documented reference point by performing a system check with a test mix or system suitability sample that can lead to a quick diagnosis.

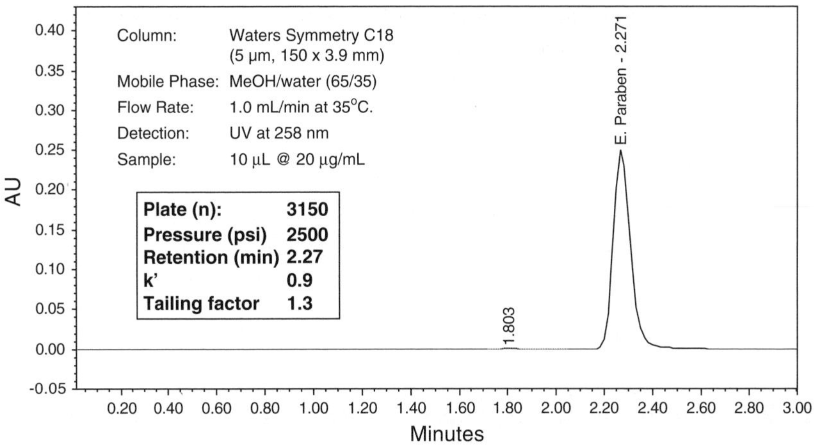

FIGURE 3 **An example of a reference chromatogram useful for troubleshooting.**

3. Isolate the problem area—is it the pump, the autosampler or the detector?
4. Inspect the equipment for leaks, loose connections, cables, etc.
5. Use the equipment's built-in diagnostics (e.g., power-up diagnostic, compression test).
6. Do the obvious or easy things first and change one component at a time.
7. Consult an expert or a manufacturer's support line. Do simple repairs yourself or call for service.

XI. ENHANCING HPLC PRECISION

HPLC precision is critical in pharmaceutical analysis.[15,16] For most pharmaceutical assays under a good manufacturing practice (GMP) environment, retention time and peak area precision of <2.0% RSD must be demonstrated before any samples can be analyzed.[15] This section reviews the fundamental principles of HPLC precision and offers practical guidelines for its enhancement. The reader is referred to Reference 18 for a more detailed treatment of this topic.

A. Precision in Retention Time

Retention time is the primary means for chromatographic peak identification. Retention time shifts are indicative of leaks, pump malfunctions, and changes in column temperature or mobile phases.

A theoretical equation on retention time precision has been proposed by Eli Grushka and Zamir:[17]

$$\frac{\delta t_r}{t_r} = \left[\left(\frac{\delta F}{F} \right)^2 + \left(\frac{\Delta\mu\delta T}{RT^2} \right)^2 \right]^{1/2}$$

where t_r is the retention time, F the flow rate, T the temperature, $\Delta\mu$ the change in chemical potential, and R is the gas constant.

According to the equation, column temperature is expected to be the dominant factor in controlling retention time precision because changes of retention are proportional to the inverse of the square of the absolute temperature.

B. Practical Guidelines for Improving Retention Time Precision

- Use a precise pump (flow precision <0.3%) and a column oven (±0.1°C).
- Use premixed mobile phases for isocratic analysis to reduce mobile phase variations.
- For pump blending or gradient applications, thorough mobile phase degassing is critical.
- The column can cause fluctuation of retention time if it is failing, not fully equilibrated or contaminated.
- For gradient applications at low flow rates (<0.2 mL/min), piston size and mixing volumes of the LC pumps can be the limiting factors. Use specialized pumps for microbore and capillary columns (see the pump section of Chapter 3).

C. Precision in Peak Area

Peak area precision is controlled by the sampling volume precision of the autosampler. In some specific instances, precision can be limited by the signal-to-noise ratio or the sampling rate as described by the equation below.[17]

$$\frac{\delta A}{A} = \frac{4}{\sqrt{2\pi}} \left(\frac{S}{N} \right)^{-1} \frac{W^{3/2}}{n}$$

where A is the peak area, S/N the signal-to-noise ratio, w the peak width, and n is the number of sampling points in the peak.

Figure 4 shows peak area precision vs. injection volume for a typical autosampler. Note that excellent peak area precision of 0.2% RSD was readily achievable for an injection volume >5 μL. Precision levels are much poorer (0.5–1% RSD) for sampling volumes <5 μL, attributable to the finite resolution of the sampling syringe and associated digital stepper motor.[18] To obtain optimum peak area precision, the analyst must avoid potential "problem" situations such as an "overly fast" sample

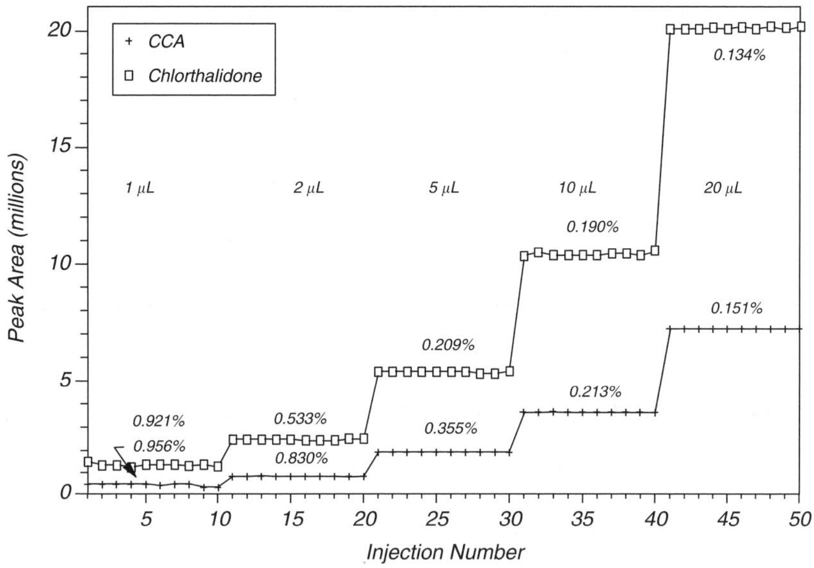

FIGURE 4 **Peak area precision study to evaluate the effect of injection volumes. The resolution of the sampling syringe was about 0.01 μL as determined by the digital resolution of the stepper motor and the size of the sampling syringe.**

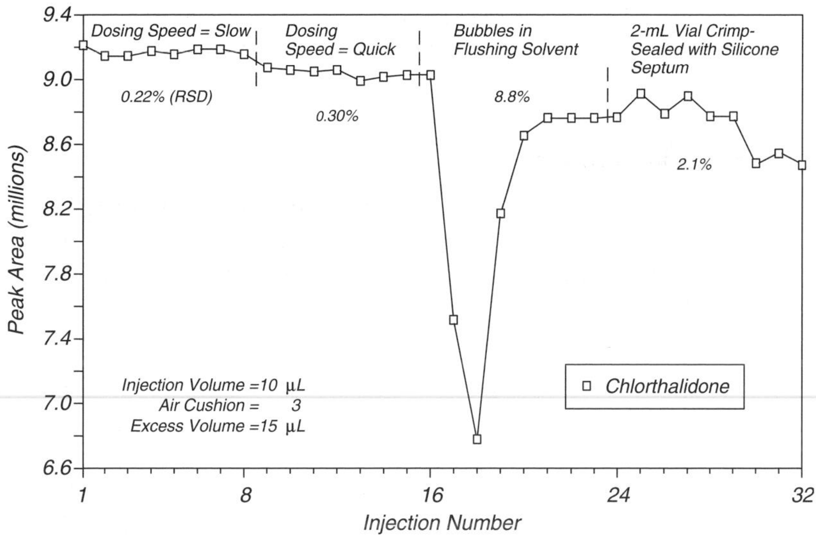

FIGURE 5 **Chart plotting the results of a peak area precision study showing the effect of sample dosing speed, bubbles in flush solvent and air-tight vial closure. Reprint from Reference 18 with permission from American Chemical Society.**

dosing speed, air bubbles in the sampling syringe, and airtight sample vials (which create a partial vacuum during sample withdrawals). Other situations include filling >50% of the sample loop, sample siphoning effects in the injection valve,[19] the sampling needle touching the vial bottoms, and sample adsorption into the injector rotor seals.[20] Figure 5 shows the results of a peak area precision study that illustrates how a perfectly functioning autosampler can be "marred" by some of these operating problems.

Theoretical considerations shown in the above equation also indicate that peak area precision is inversely proportional to the peak signal/noise ratio, and to the number of sampling points across the peak width. For very noisy peaks, the peak area precision is limited by random noise fluctuations (Figure 6). Figure 7 shows that the precision of the peak area degrades rapidly when the signal-to-noise ratio is less than 100. Statistical considerations also stipulate a minimum data sampling

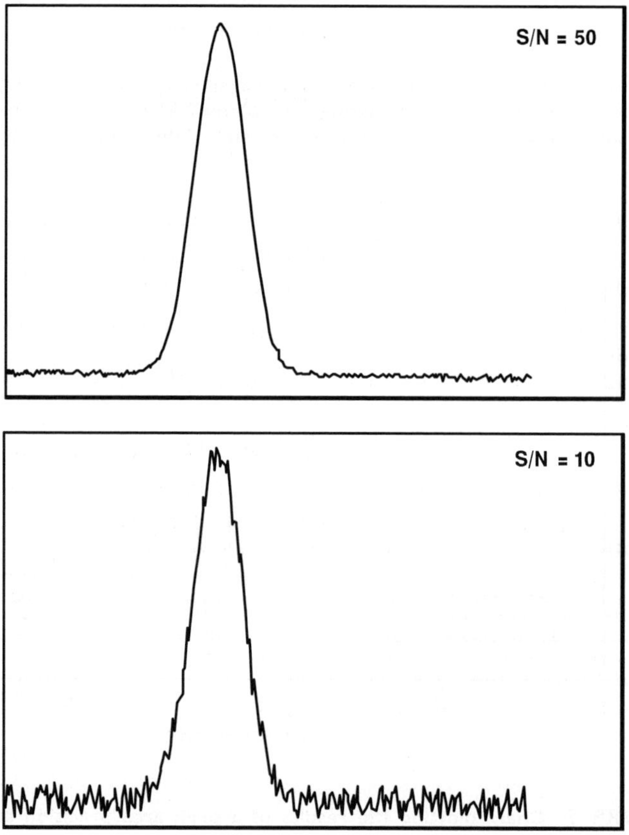

FIGURE 6 **Peak area precision study to delineate the effect of peak signal-to-noise ratio.**

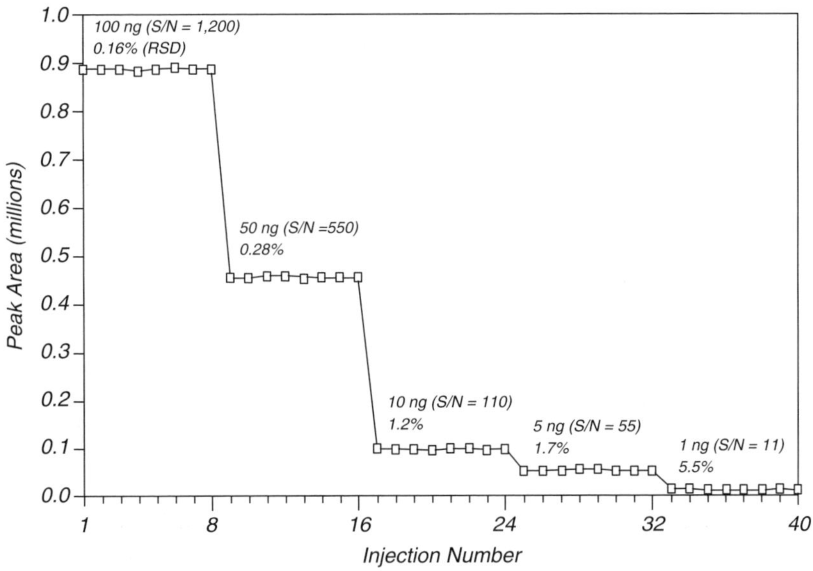

FIGURE 7 Comparison of two peaks with signal-to-noise ratio (*S/N*) of 50 and 10. The random distribution of noise in the noisy peak controls the variation of the peak area measurement.

rate of 8 points per peak. A low data sampling rate can severely limit the peak area precision of fast eluting peaks.

D. Guidelines for Improving Peak Area Precision

Recommendations for improving HPLC peak area precision are:

- Use a precise autosampler (<0.5% RSD) and injection volumes >5 μL. Degas flush solvents used for the autosampler.
- Eliminate potential pitfalls such as air bubbles in the sampling syringe, tightly sealed vial closures, an "overly fast" sample dosing speed for viscous samples, sample adsorption, siphoning, sample evaporation, etc.
- Improve *S/N* ratio of peaks to >50 if possible and use a data sampling rate of >8 points/peak.
- Plot baseline with chromatograms in reports to allow detection of integration problems.

XII. CONCLUSION AND SUMMARY

In summary, analysts can be much more effective and successful in performing HPLC assays by following guidelines and "best practice" procedures. Also, theoretical considerations and experimental data indicate that retention time precision is achieved by thermostating the column and effective degassing of the mobile phase while peak area precision is improved by the judicious operation of a precise autosampler while avoiding potential sampling pitfalls and noisy peaks.

ACKNOWLEDGMENTS

The author acknowledges helpful ideas and suggestions from R. Ornaf, D. Barker, G. Miller, C. Choi, J. Bonilla, and A. Lister of Purdue Pharma. This chapter has been presented in parts in short courses on "Advanced HPLC in Pharmaceutical Analysis" at Eastern Analytical Symposium 2001–2002, HPLC 2002, and Pittcon 2003. Numerous comments and suggestions from the attendees of these short courses were incorporated during the development of this chapter.

REFERENCES

1. Snyder, L. R. and Kirkland, J. J., *Introduction to Modern Liquid Chromatography*, John Wiley & Sons, New York, 1979.
2. Meyer, V. R., *Practical HPLC*, Wiley, New York, 1988.
3. Adamovics, J. A., *Chromatographic Analysis of Pharmaceuticals*, 2nd ed., Marcel Dekker, New York, 1997.
4. Katz, E., Eksteen, R., Schoenmakers, P. and Miller, N., *Handbook of HPLC*, Marcel Dekker, New York, 1998.
5. HPLC Video and Slides/Tapes Training Programs, Savant, Fullerton, CA.
6. HPLC equipment, CLC-30, Troubleshooting HPLC, CLC-40, Identification and Quantitation Techniques in HPLC, CLC-70, (Computer-based Instruction), Savant, Fullerton, CA.
7. Method Development by the Numbers Low, Mid and High-pH, presented by Agilent Technologies, 2001.
8. Dong, M. W., Gant, J. R. and Perrone, P. A., *LC.GC*, 3:786–795, 1985.
9. Dong, M. W., Miller, G. and Paul, R., *J. Chromatogr.* 987:283–290, 2003.
10. Neue, U. D., *HPLC Columns: Theory, Technology, and Practice*, Wiley-VCH, New York, 1997.
11. Kirkland, J. J., Van Straten, M. A. and Classens, H. A., *J. Chromatogr. A*, 797:111, 1998.
12. Dolan, J., *LC.GC*, 21(3):262, 2003.
13. Dong, M. W., Paul, R. and Roos, D., *Today's Chemist Work*, 10(2):42–48, 2001, http://pubs.acs.org/ subscribe/journals/tcaw/10/i02/html/02dong.html
14. Sadek, P., *Troubleshooting HPLC Systems: A Bench Manual*, Wiley-Interscience, New York, 1999.
15. Dong, M. W., Paul, R. and Gershanov, L., *Today's Chemist Work*, 10(9):38–42, 2001, http://pubs.acs.org/subscribe/journals/tcaw/10/i09/html/09dong.html.
16. Dong, M. W., *Today's Chemist Work*, 9(8):28–32, 2000, http://pubs.acs.org/hotartcl/ tcaw/00/aug/dong.html.

17. Grushka, E. and Zamir, I., Precision in HPLC, In *Chemical Analysis*, Brown, P. and Hartwick, R. Eds., Vol. 98, Wiley Interscience, New York, pp. 529–561, 1989.
18. Duggan, J., *Amer. Lab. News*, September Issue, 50F–50I, 1996.
19. Spruce, B. and Bakalyar, S. R., Troubleshooting Guides for HPLC Injector Problems, Rheodyne.com/Docs/Tech/Docs/TS1.html
20. Dolan, J., *LC.GC*, 9:2224, 1991.

10

REGULATORY CONSIDERATIONS IN HPLC ANALYSIS

CHRISTINE F. RICHARDSON[a] AND FRITZ ERNI[b]

[a]*Novartis Ringaskiddy, Co. Cork, Ireland*
[b]*Novartis Pharma AG, Basel, Switzerland*

ABSTRACT

This chapter discusses regulatory considerations in analysis of pharmaceuticals by HPLC.

I. INTRODUCTION

To guarantee the quality, clinical performance, efficacy, and safety of a pharmaceutical product, specifications are fixed and approved by the competent regulatory authorities of each country where the drug is marketed. Analytical testing ensures that these specifications are met by confirming the identity, purity, and potency of drug substances and drug products prior to their release for commercial use. For drug products, performance testing to ensure constant bioavailability is needed. This performance testing comes in the form of either dissolution or disintegration tests.

Handbook of Pharmaceutical Analysis by HPLC,
S. Ahuja and M.W. Dong, editors.

HPLC has become the most important separation technique for drug substance and drug product testing. The main reason for the growing importance of HPLC methods is that medicinal drug substances and products are becoming more complex in structure. The detection and quantification of chiral impurities has also become more significant. In today's laboratory the majority of testing is conducted by HPLC; most wet chemistry methods have become obsolete. For drug substances HPLC testing is focused on assay and impurity testing. Drug products also use HPLC techniques for dissolution testing (an indicator for bioavailability) and content uniformity (used to ensure that the content of the drug substance is uniform throughout a batch). For both drug substances and drug products there may be some products which still use older analytical techniques such as UV, titrations, and thin-layer chromatography in addition to HPLC or as independent methods. In most cases these can be replaced by HPLC methods. The technology surrounding HPLC instruments themselves and the software used to run them have advanced to such a degree that in many laborateries analyses will run overnight and in some cases 24 h a day, unattended. This chapter will review the regulatory guidelines and requirements for the successful use of HPLC in drug substance and drug product testing in development and routine quality control.

II. CHROMATOGRAPHY—SOME GENERALITIES

Separation by HPLC typically falls into one of six categories: chiral, ion exchange, ion pair, normal phase, reversed phase, or size exclusion. Reversed phase is the most common type employed, and although various detection principles can be used in HPLC, UV is the most typical.

In HPLC, a sample is separated into its components based on the interaction and partitioning of the different components of the sample between the liquid mobile phase and the stationary phase. In reversed phase HPLC, water is the primary solvent and a variety of organic solvents and modifiers are employed to change the selectivity of the separation. For ionizable components pH can play an important role in the separation. In addition, column temperature can effect the separation of some compounds. Quantitation of the interested components is achieved via comparison with an internal or external reference standard. Other standardization methods (normalization or 100% standardization) are of less importance in pharmaceutical quality control. External standards are analyzed on separate chromatograms from that of the sample while internal standards are added to the sample and thus appear on the same chromatogram.

For each item of the specification of a particular drug substance or drug product, a validated testing method describes the tests and the acceptance criteria. For each test, the method will be outlined in detail so that any analyst will be able to follow the testing instructions and carry out the analysis. Prior to running an analysis the HPLC system must be set up with the correct column, mobile phase, and detection method at the correct wavelength. Testing instructions will include directions on how the mobile phase is to be prepared in order to provide the required separation. The mobile phase will effect the separation, but the components may not be separated and/or may not appear at the correct retention times. Retention time is commonly used for identification of the drug substance and its related components. In these cases the testing instructions will have retention time ranges while others may rely on a relative retention time. Depending on how the mobile phase is prepared the retention times may drift. In general, slight mobile phase adjustments are allowed to bring retention times into the correct ranges.

Once the system is set up, the required standard solutions need to be prepared. Standard solutions are prepared from reference standards which have a known purity and content of the components likely to be present during the analysis. Their purity is important as their peak areas are used to quantify assay and impurity levels. The FDA and other regulatory authorities recognize two categories for reference standards: Pharmacopoeial standards (i.e., USP/NF or EP reference standards), which do not need further characterization, and non-compendial reference standards. Non-compendial reference standards are used for new products where compendial standards do not yet exist. Depending on the method, a reference standard may require drying to remove residual solvents or unbound/bound moisture prior to being used. This is particularly important for hygroscopic compounds and must be described in the method.

A. System Suitability Testing (SST)

Prior to running an analysis by HPLC suitability must be demonstrated indicating that the HPLC system is working properly. The criteria required for passing the system suitability test will vary depending on the product being analyzed and the chromatography. A good system suitability test (SST) is based on criteria defined from the knowledge gained during the development and optimization of the method. Those parameters that are critical to separation and quantitation should be carefully checked and adequate specifications should be fixed prior to an analysis being run. In general, three to five replicate injections of an SST solution or the standard solution will be run. The results of the SST should demonstrate that the performance of the HPLC system is adequate for the test to be run. For quantitative tests, the RSD of the peak areas between them must be less then a specified value. The reproducibility of the retention times may also be specified. The separation performance of the column may be specified in some methods by the number of theoretical plates obtained. For other methods, parameters such as tailing factor or minimal separations of critical components may also be included. For purity determinations, a test for detection limit or quantification limit should be included. A failing system suitability could be indicative of a number of problems including incorrect mobile phases or standard preparation, leaks in the system, and column degradation. All of these would indicate to the analyst that there is a problem which needs resolution in order to obtain quality data during the analysis. Passing system suitabilities are typically good for a specified time period (24 h) after which system suitability must be demonstrated again. The time period for system suitability should be specified in the method or SOP. In addition, during the analysis itself standards will be run between the samples as an additional check. Any problems which have occurred during the run will be clearly evident by the RSD and area counts between these injections. This is particularly important in content uniformity or assay testing where multiple samples for a batch or multiple batches are run sequentially. An HPLC run will typically follow a sequence where two standards will be injected followed by up to 10 samples and then another two standards and so on. This sequence of samples can be from different batches or a continuation of samples from the same batch. Having samples grouped by standards isolates any problems to a specific group and results from the entire run will not be affected. In this way if a problem occurred during the second set of 10 samples only that set would need to be investigated.

After the completion of an HPLC run, it is important to wash the system, particularly when reversed phase methods are employed. Mobile phases used in reversed phase chromatography usually contain salts which can accumulate on the column and in the system. Washing the system and equilibrating it at the most adequate pH using a sequence of water and the mobile phase without salt prolongs the column lifetime and minimizes the problems that can come up during analyses. A sample HPLC chromatogram is shown in Figure 1.

III. VALIDATION OF CHROMATOGRAPHIC METHODS

Analytical data generated in a testing laboratory are generally used for development, release, stability, or pharmacokinetic studies. Regardless of what the data are required for, the analytical method must be able to provide reliable data. Method validation (Chapter 7) is the demonstration that an analytical procedure is suitable for its intended use. During the validation, data are collected to show that the method meets requirements for accuracy, precision, specificity, detection limit, quantitation limit, linearity, range, and robustness. These characteristics are those recommended by the ICH and will be discussed first.

Accuracy is the measure of how close the experimental value is to the true value. In this case, the experimental value is the one obtained during

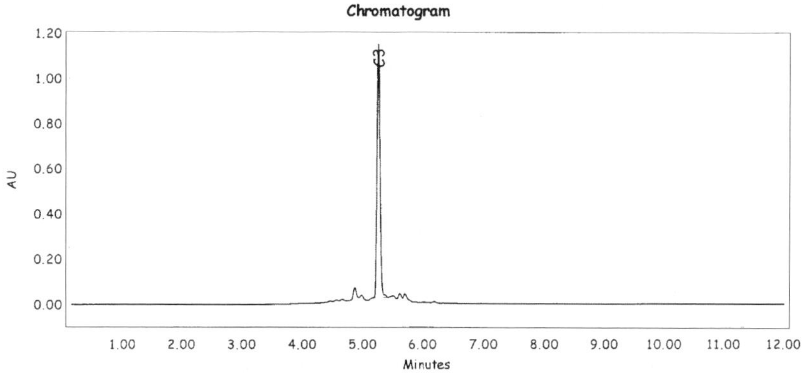

FIGURE 1 Sample HPLC chromatogram. HPLC of a C_3 *tris* amino fullerene after purification by flash chromatography using C_{18} silica gel. The corresponding TFA salt was subjected to flash chromatography using mixtures of water and acetonitrile with 1% trifluoroacetic acid. Samples were prepared with a concentration of 1 mg/mL dissolved in water and filtered through a 0.45 μm nylon filter. HPLC was carried out on an HP 1100 automated system using a gradient method. Mobile phase A consisted of 90% acetonitrile/10% water with 1% trifluoroacetic acid. Mobile phase B consisted of 90% water with 1% trifluoroacetic acid/10% acetonitrile. The LC was equipped with a PDA detector set at 300 nm.

the analysis and the true value for drug products is taken from the label claim (i.e., 50 mg tablets, 100 mg tablets, etc.). For drug substances the true value is 100% unless otherwise specified. In order to demonstrate accuracy it is recommended that studies are conducted at levels of 80%, 100%, and 120% of the label claim. The reason for this is that drug products do not contain the exact amount listed on the label. Most ranges for content uniformity will be within the 80–120% range. Therefore by carrying out accuracy measurements at the wider range of 80–120% it is ensured that results obtained in later analyses will be accurate.

Precision measures how close data points are to each other for a number of measurements under the same experimental conditions. According to the ICH, precision is made up of three components: repeatability, intermediate precision, and reproducibility. The term ruggedness, which has been used in the USP, incorporates intermediate precision, reproducibility, and robustness.

Repeatability can be divided into two areas: injection repeatability and analysis repeatability. Injection repeatability is measured by analyzing multiple injections of the same solution preparation. It is an indicator of the performance of the HPLC system under the specified conditions and at the time of the analyses. This information is included as part of the validation package and is also used during routine analysis in the form of a system suitability. During the validation the specification for % RSD will be set, which will determine the variation limit for the analysis. If the value is low the results can be expected to be more precise and sensitive to variation. During method validation a minimum of 10 replicate injections with an RSD of <1.0% is recommended. However for system suitability studies an RSD of <1.0% for three to five injections is desirable. Many methods have RSD requirements of <2.0%. Analysis repeatability is where multiple measurements of the same sample are carried out by the same analyst and is demonstrated in combination with accuracy studies. Intermediate precision is used to determine the reliability of a method in an environment different than that in which the method was developed. Methods are usually developed by an analytical group and then transferred to the testing groups where the routine analysis is carried out. It is important to ensure that the same results are obtainable in different labs, instruments and analysts. Demonstrating the accuracy of a method during two separate occasions is sufficient to ensure that intermediate precision is met. Finally, reproducibility is defined by the ICH as the precision between different labs. Depending on the size of the organization this may or may not be possible and is not expected if intermediate precision for the method has been proven.

Specificity is used to show that for the analyte in question the analytical method is able to resolve it from other components. One way of showing this is to add known impurities to a sample and show that they are resolved in the chromatogram. Alternatively, chromatograms from stress testing (acid/base hydrolysis, temperature, etc.) which demonstrate that any additional compounds which may form over time are baseline resolved from the main peak are also acceptable. In order to demonstrate that the peaks observed result from a single compound (as opposed to two compounds which co-elute) a chromatogram can be generated using a photodiode array detector (see Figures 2 and 3). Low levels of impurities

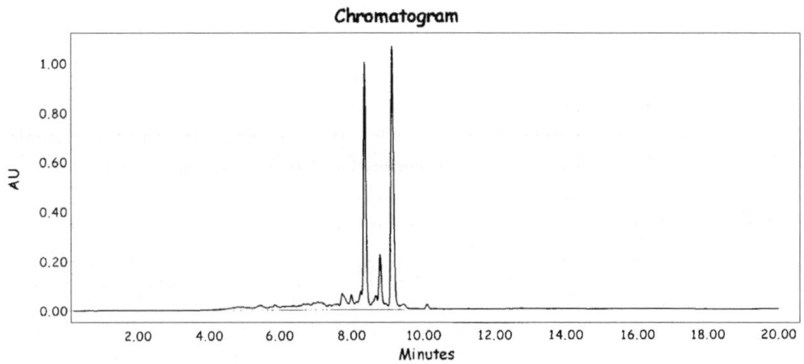

FIGURE 2 **HPLC chromatogram of the e,e,e half amine. The two large peaks represent two isomers of the compound. In order to prove that they were two isomers of the same compound, the UV spectrum of all peaks were compared to the known UV spectrum of the compound.**

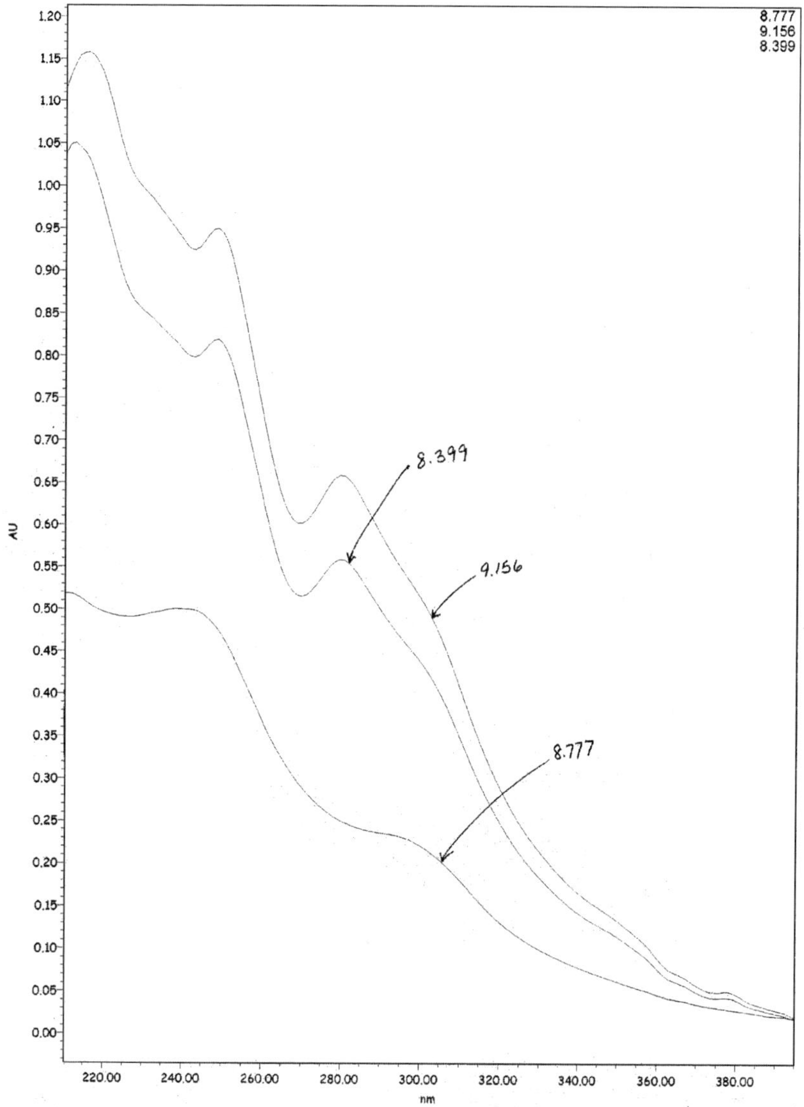

FIGURE 3 Corresponding UV spectrum for Figure 2. The top two spectra are for the two major peaks in the chromatogram while the bottom spectrum is for the smaller in between them. The top two spectra are indicative of fullerene derivatives with three substituents.

may not always affect the UV spectrum, however, and this should be kept in mind when using the data.

Limits for detection and quantitation are usually applied to related substances in drug substances or drug products. The detection limit is the lowest concentration of analyte in a sample that can be detected but not necessarily quantitated under the experimental conditions.

Quantitation limit is the lowest concentration of analyte in a sample that can be determined with acceptable precision and accuracy under the stated experimental conditions.

Linearity is the measurement of the linear range of detectability that obeys Beer's Law and is dependent on the compound analyzed and the detector used. In short there is a linear relationship between absorbance and concentration. To be within the linear range of the method you should be working within the absorbances and concentrations that form the linear part of the curve.

Range describes the area between the high and low levels of the analyte studied.

Robustness is defined as a measure of how well the method will remain unaffected by small variations in the parameters. Robustness can be assured by setting appropriate system suitability. However, it is important that these parameters be set properly. Some parameters which could be used to demonstrate robustness are the varying age of columns, column brands, temperature, pH of mobile phase, and the different amounts of mobile phase modifiers.

In addition to these validation characteristics any additional information that will aid in the analysis should be provided. For example, data to demonstrate the stability of the samples is recommended for the duration of the testing. This becomes particularly important for labile samples and also when the method becomes part of routine analysis. Some laboratories will have a general SOP governing the duration for which samples may be kept. Specific information for a particular method would be beneficial during times when analytical investigations are required. In addition, storage conditions for samples should be specified. Other additional information such as examples of chromatograms, representative calculations, information from stress testing, and impurity information are also recommended. Examples of chromatograms are useful and should be included in the testing document that will govern routine analysis, as they provide a nice way for the analysts to check their data. Impurity information in the form of names and locations on representative chromatograms is also recommended. The exact names of impurities and/or structures are beneficial to the analyst and can be valuable during investigations.

A. Case Studies Requiring Revalidation and New Method Development

During the life of a drug substance or drug product, changes in raw materials or processing may result in the current method becoming inadequate for its original purpose. This may require a revalidation of the current method or the development of a completely new method which will need to be validated. In other cases, while the method provides appropriate information, changes in the manufacturing process may cause results that are unexpected. We will approach this topic using some useful examples.

The first example comes from a case in which a drug substance was being produced by an alternative manufacturer. The drug substance was used in a variety of drug product formulations including different strengths of tablets and an ampoule dosage form. The structure is shown in Figure 4.

When an alternative process is being evaluated, the usual sequence of events from a technical perspective is to evaluate three separate batches using the current analytical method. If the material meets the given specifications it can be assumed that a drug product of acceptable quality will be produced. Several suppliers of this particular drug substance were evaluated and during these initial evaluations the new source chosen met all of the specifications and appeared to be of a high quality. There was one impurity that did not appear on the chromatogram using our method. In order for this material to be used our method needed to be altered to include that particular impurity on the chromatograms resulting in method revalidation. It is important to note that in a case like this it may not be necessary to revalidate for each criterion. Only those parameters likely to be affected would need to be revalidated.

In addition to method validation work, another problem existed with this drug substance. As mentioned previously, based on the analytical testing the drug substance appeared to be of a very high quality. The tablet dosage forms were all validated with no issues, producing a drug product of acceptable quality. During the validation of the ampoule form, problems were encountered and all three validation batches failed specifications for two specific impurities after sterilization (Table 1).

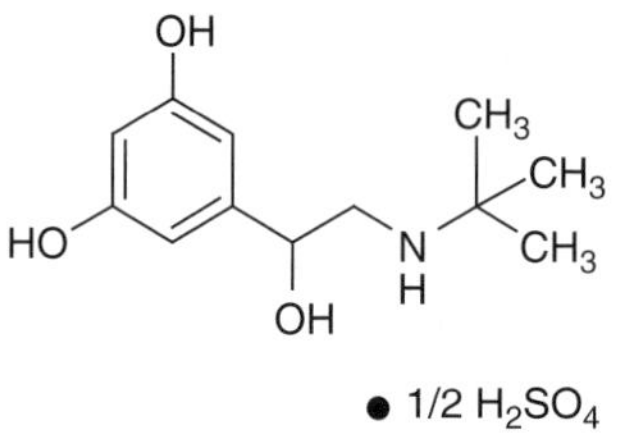

FIGURE 4 Structure of terbutaline sulfate.

TABLE 1 Impurity Results from Three Validation Batches Using the New Suppliers' Drug Substance. Specifications Are <0.1%

Drug product lot #	Lot #1		Lot #2		Lot #3	
Impurity	Peak 2	Peak 4	Peak 2	Peak 4	Peak 2	Peak 4
%	0.2	0.19	0.14	0.14	0.11	0.10

The drug substance in question was known to degrade via an oxidative pathway that in solution is enhanced by low pH and levels of heavy metals in the presence of oxygen. Two specific impurities 3,5-dihydroxybenzaldhyde and 2-*tert*-butyl 4,6,8-trihydroxytetrahydroisoquinoline were known to increase under these conditions (Figure 5) and these were the same two that caused the validation to fail.

As heavy metals were known to influence the degradation pathway, experiments were carried out to determine if excess levels of metals could be the root cause of the degradation. EDTA (ethylenediaminetetraacetic acid) is a known metal chelator and is often used in ampoule formulations. The addition of EDTA to samples of ampoule solutions prepared in the lab was shown to halt the degradation (Table 2).

Upon obtaining this information a number of heavy metal tests were carried out but the levels were found to be insignificant. Further reflection on the problem led to the idea that EDTA is itself a base and the degradation was also known to be enhanced via low pH. Accordingly, experiments were carried out varying the bases used to determine if the culprit of the degradation problem was excess acid in the drug substance (Table 3).

Surprisingly, both the addition of potassium carbonate and sodium hydroxide halted the degradation. From the results obtained it is clear that sodium hydroxide had a larger effect which is to be expected as it is a stronger base.

The consequence for this material was that in order for it to be used in the ampoule formulation, EDTA needed to be included into the

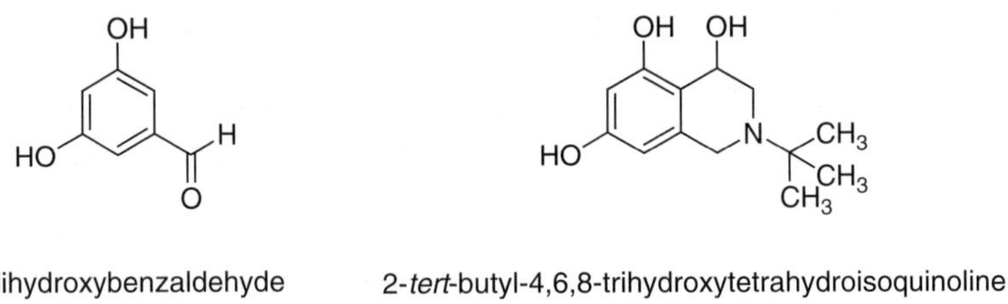

FIGURE 5 Degradation products of terbutaline sulfate.

TABLE 2 Impurity Results with and Without Addition of EDTA

	% Impurities			
	Peak 2		Peak 4	
Batch	**Before sterilization**	**After sterilization**	**Before sterilization**	**After sterilization**
Lot #1	0	1.328	0	0.769
Lot #1 + EDTA	0.0006	0.027	0	0.034

TABLE 3 Impurity Results Varying the Base Added to the Formulation

Sample	Result	
	Dihydroxybenzaldehyde	Isoquinoline
Lot #1	0.508	0.589
Lot #1 + EDTA	0	0
Lot #1 + Pot Carb	0.427	0.516
Lot #1 + NaOH	0.153	0.335

formulation. In this case, the analytical methods used were valid for their purpose. This presents a good example of a situation where the methods used were developed for a particular process. The new material made their product by a slightly different route and this in turn caused problems during drug product production.

The second example comes from a process change to the manufacture of the penultimate intermediate (last intermediate prior to the drug substance). This particular intermediate, which in structure was a dimer, was produced via an electrochemical reaction. Changes were made to the process for safety reasons; however, the intermediate met all release specifications. The impurity profile of a batch made using the original process and that met specifications is shown in Figure 6a. Conversion to the drug substance gives the impurity profile shown in Figure 6b. Batches manufactured using the new process produced slightly different impurity profiles as shown in Figure 7a. Although unknown impurities were observed in the 8–10 min retention time range, these batches still met the specifications for impurities. Upon conversion to the drug substance a new impurity was observed at a failing level as shown in Figure 7b. Using LC-MS the new impurity in the drug substance batches was identified as a trimer (one additional monomer) which was formed at the intermediate step of the process. As the intermediate process contains minimal purification the trimer reacted along with the intermediate at the drug substance step. The purification during the drug substance process was not sufficient to remove these trimer impurities. A new HPLC method was developed for the intermediate during the investigation including the trimer in the chromatogram. The new method was also capable of separating the trimer peak into all of the structural isomers as shown in Figure 8. Specifications were then set reflecting appropriate levels of the trimer which would guarantee drug substances that would meet the release specifications.

These two examples demonstrate how analytical methods provide useful information about the manufacturing processes. Analytical methods are developed to provide information about a particular process. Accordingly they are developed concurrently with the manufacturing process and reflect that process. If they are to be used to provide information about drug

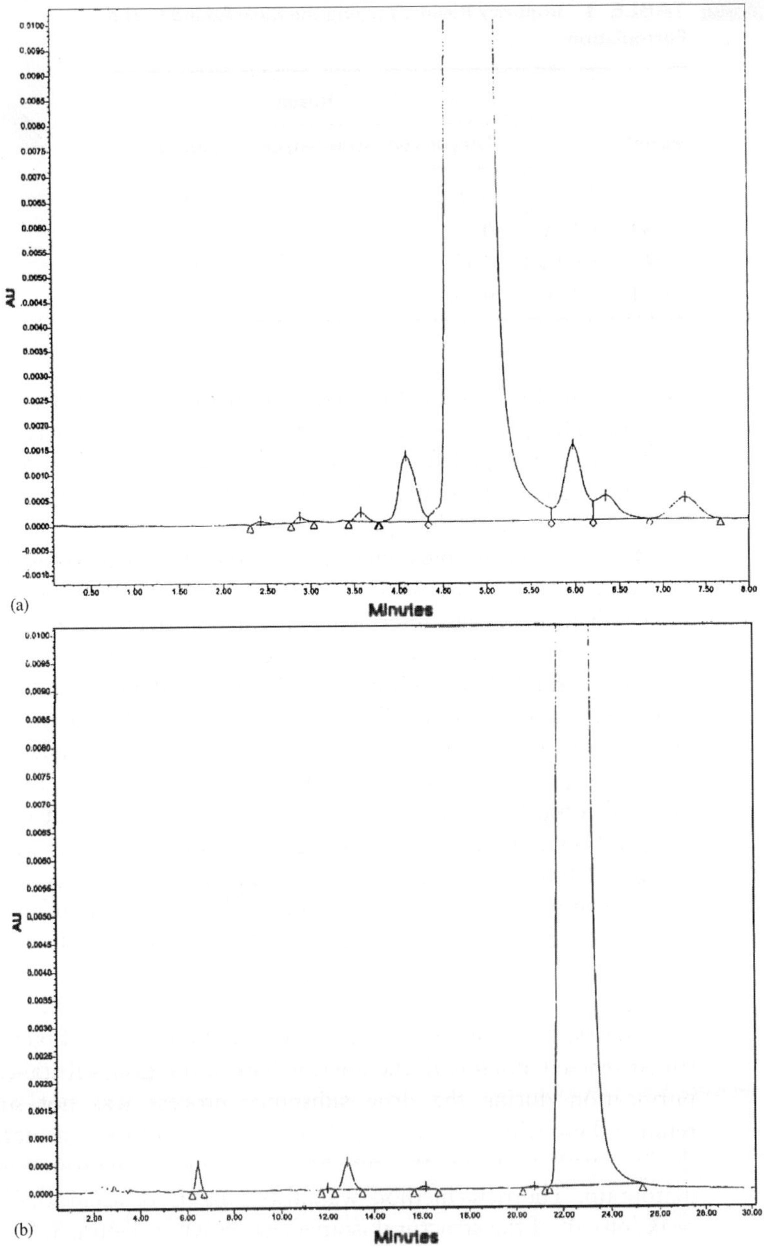

FIGURE 6 (a) Intermediate prepared via the original process. Impurities meet specifications. No impurities elute past 8 min retention time. (b) Impurity profile of the drug substance produced via the original process. No impurities were present after the drug substance which appears at ~22 min. Known impurities are found between 6 and 22 min.

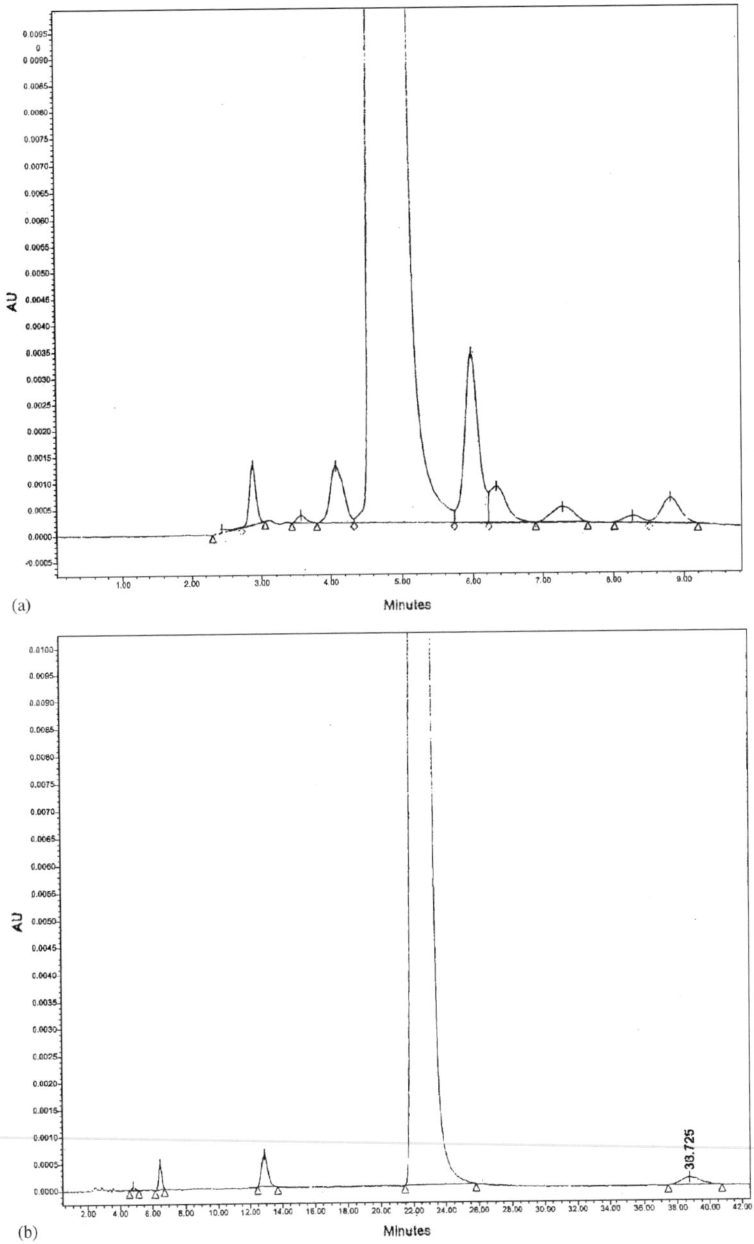

FIGURE 7 **(a) Impurity profile of Intermediate produced by the new process. Peaks after 8 min were impurities that were not previously observed when the Intermediate was produced via the old process. (b) Chromatogram of the drug substance made from intermediate of poor quality. Peak at 22 min retention time is the drug substance; Impurity at 36 min retention time was a new impurity and failed specifications.**

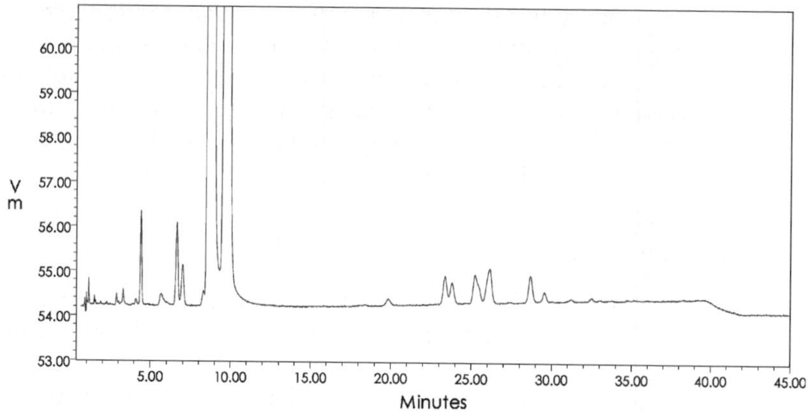

FIGURE 8 Impurity profile using a new method. Intermediate peaks are at 10 min while those in the range of 23–30 min are the trimer peaks which were not separated using the previous method. Peaks prior to 10 min are the solvent front and other known impurities.

substances or drug products made by an alternate route or by a different manufacturer a reevaluation of the method is prudent and may avoid future problems.

IV. IMPURITIES

Major analytical tests recognized by health authorities include impurity tests. The term "impurities" encompasses both process impurities and degradation products. Process impurities, which are indicative of the process, could be starting materials, impurities of starting materials, reagents, or products of side reactions. Degradation products are formed from the interactions of the drug substance and related compounds under various conditions. The general term 'impurities' usually encompasses both process impurities and degradation products and unless otherwise specified can be taken to include both. Impurities fall into three general categories: organic, inorganic, and residual solvents. Organic impurities, in turn, can be classified in a number of ways based on whether they are identified or unidentified, and volatile or non-voltatile.

A. Process-Related Impurities

As mentioned previously, process related impurities are typically starting materials, by-products of side reactions, intermediates, or reagents. Starting materials are easy to identify as their structure is known, and their retention times can be quickly compared to known standards during method development. Reagents that are organic in nature could fall into this category as well. Inorganic reagents may need to be tested by alternate methods depending on whether they are metallic in nature or not. In some cases, compendial methods may apply. By-products would arise from side reactions occurring in conjunction with the primary synthetic pathway. Process chemistry plays an important role in developing routes which minimize unwanted side reactions.

B. Degradation Products

Impurities that form from the degradation of the drug substance are called degradation products. They are identified using degradation or stress studies which are performed during the early phases of drug substance development. A well-designed degradation study involves solid and solution phase challenges which mimic potential storage conditions and include acid/base hydrolysis, thermal/humidity, oxidation, and light exposure. All of these methods of stressing a compound are governed by ICH guidelines. First we will briefly review the different types of stress testing.

Acid/base stress testing is carried out to force the drug substance to degrade to its primary degradants by exposing it to acids and bases. Prior to carrying out such a study, solubility screening will be conducted and a solubility of at least 1 mg/mL can be viewed as a standard although lower concentrations have been used. Initial experiments will be carried out in the absence of light and heat. The temperature can be increased to induce degradation if necessary. To gain an understanding of the degradation pathway samples will be taken and analyzed. Typical experiments will last approximately 1 week and the frequency of the sampling will depend on the rate of degradation observed. It is also important that appropriate controls be included in the study.

Thermal/humidity stress testing involves exposing samples to various conditions of temperature and humidity. Once the conditions are chosen, the samples are placed in either ovens which control the conditions or saturated solutions of salts in desiccators. The physical appearance of the sample is important and should be monitored throughout the study. In addition, any loss or gain of water should be noted as well.

Oxidation. Oxidative studies are designed to reveal the primary oxidative degradation products due to oxidation. Oxidative degradants in drug substances and drug products normally are the result of reaction with molecular oxygen. As in the case of acid/base hydrolysis, appropriate controls are necessary.

Forcing the sample to degrade as a result of exposure to UV conditions tests its photostability. Studies are carried out to gain information about the degradation and also to determine if the compound needs to be protected from light. For UV studies ICH guidelines specify an exposure of $200\,\mathrm{W\,h\,m^{-2}}$ and 1.2×10^6 lux hours for fluorescence confirmatory studies. Again samples should be taken at time points indicative of the degradation pathway.

The culmination of degradation studies is the preparation of a degradation pathway. A degradation pathway will show all the possible degradation products beginning with the drug substance. An example is shown in Figure 9.

C. Drug Product Degradation

The stability of drug products is different to that of drug substances, and separate studies are required. Depending on the formulation the drug product may contain excipients (non-active ingredients) which can react with the active ingredient. Degradation studies on dry products are carried out to understand the physical and chemical compatibility of the drug substance with the excipients when exposed to heat, light, and humidity. For drug products that are solutions or suspensions, acid/base, oxidation, and photostability studies will also become important.

D. Identification of Impurities and Degradation Products

Prior to identifying impurities present in drug substances/drug products the levels of the impurity must be assessed using the analytical

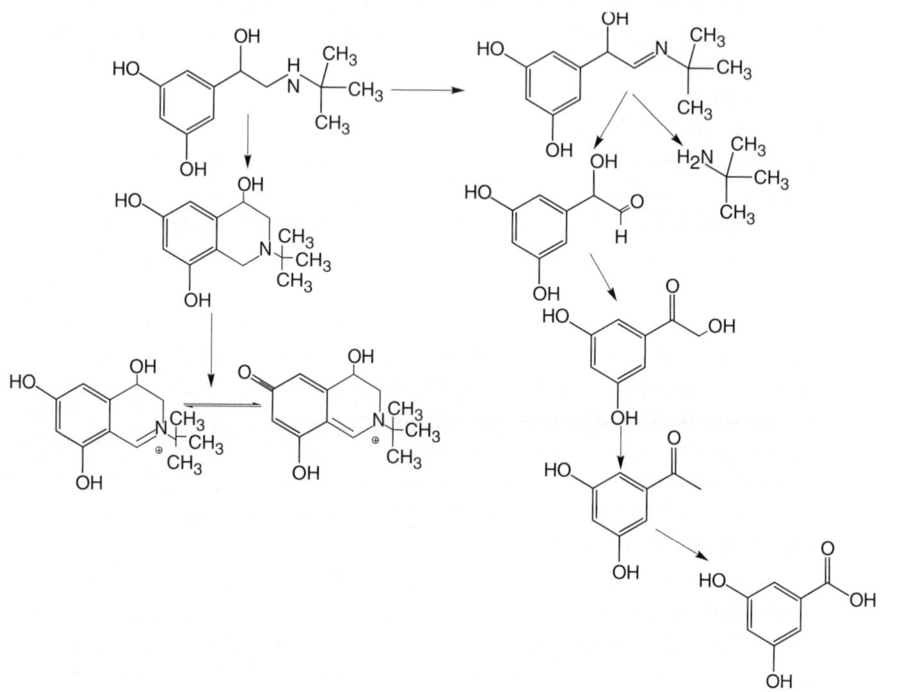

FIGURE 9 Degradation pathway of terbutaline sulfate.

method. Impurities below 0.1% are generally not identified unless they are expected to be potent or toxic. Examples of potent or toxic impurities are methylating agents which may be used in the synthesis and will be checked for at the intermediate level, and residual solvents such as benzene, which is an impurity of toluene.

Once an assessment on a particular impurity has been made all process-related compounds will be examined to confirm that the impurity of interest is indeed an unknown. An easy way of doing this is to compare the retention times of known process-related compounds to that in question. If this analysis confirms that the compound is an unknown, the next step would be to obtain an LC-MS on the compound. Mass spectrometry provides structural information which aids in determining structure. In some cases, mass spectrometry will be enough to identify the compound. In other cases, more complicated methods like LC-NMR are needed or the impurity will need to be isolated in order to obtain additional information. Compounds that are not purified often contain high levels of by-products and can be used for this purpose. Alternatively, mother liquors from crystallizations also contain levels of by-products. Other ways of obtaining larger quantities of impurities include flash chromatography which is typically used for normal phase separations or preparative HPLC which is more common for reversed phase methods. Once a suitable quantity of the compound in question has been obtained a full characterization can be carried out to identify it.

E. Impurities in Drug Substances

For identification and setting specifications for impurities in drug substances, impurities can be addressed using a two-pronged approach in which both chemistry and safety aspects are important. The chemistry side includes classification and identification of impurities, reporting and setting specifications, and the analytical methods that will be used to identify them. The safety aspects include specific guidance on how impurities that were not present (or were present at substantially lower levels in previous batches) will be handled. Oftentimes, during the lifetime of a drug, substance impurities that have not been previously identified will appear as a result of new sources of raw materials or changes in processing during manufacturing. Alternatively, deviations to processing do occur from time to time and can result in increases in impurity levels or in new impurities becoming evident.

For organic impurities summaries of all experiments used to detect impurities should be available and should include comparisons between batches produced during development, those from the proposed commercial process, and results from any stress testing. Any differences between these should be discussed. Studies used to identify and qualify impurities should also be discussed. The ICH has published guidelines depending on

TABLE 4 Regulatory Agencies and Their Associated Web Sites

Regulatory agency	Website
United States Pharmacopoeia	www.uspnf.com
Japanese Pharmacopoeia	www.jpdb.nihs.go.jp/jp14e/
British Pharmacopoeia	www.pharmacopoeia.org.uk
European Pharmacopoeia	www.edqm.org
International Conference on Harmonisation	www.ich.org
US Food and Drug Administration	www.fda.gov
European Agency for the Evaluation of Medicinal Products	www.emea.eu.int

the maximum daily dose which describe levels for reporting, identifying, and qualifying impurities. In all cases these studies should be carried out and in situations where impurities cannot be identified all attempts to do so should be described.

Inorganic impurities are normally followed by pharmacopoeial methods and the acceptance criteria are usually based on these standards.

V. OVERVIEW OF AVAILABLE GUIDANCE DOCUMENTS

There are numerous guidelines which outline the regulatory aspects of HPLC analysis available on the internet. Table 4 represents some of the regulatory authorities and their associated websites.

VI. CONCLUSIONS

HPLC testing methods have come to dominate analytical testing in the pharmaceutical industry. In recent years more and more tests are being carried out by HPLC. Guidelines which govern this testing were discussed. Throughout this chapter the current thinking which governs HPLC testing has been reviewed and several case studies have been presented.

VII. SUMMARY

HPLC has become the most important separation technique for drug substance and drug product testing. Due to the advances in chemistry and pharmacy, drug substances have become more complex and new drug product dosage forms have evolved and HPLC has advanced to meet the separation challenges. Information can be obtained about stability, degradation, assay. Impurity profiles are also readily obtained. This chapter reviews the current thinking and guidelines which regulate the use of HPLC in the Pharmaceutical Industry.

11 HPLC SYSTEM CALIBRATION FOR GMP COMPLIANCE

MICHAEL W. DONG

Purdue Pharma, Ardsley, NY

ABSTRACT

This chapter reviews the principles and strategies used for HPLC system calibration that includes the pump, the detector, the autosampler, and the column oven. A case study is used to illustrate the development of the calibration procedures for all system modules and the rationale of setting up acceptance criteria that balance productivity and compliance.

I. INTRODUCTION

"Calibration," in current Good Manufacturing Practices (cGMP) terminology, refers to instrument "qualification" or performance verification of the HPLC. Note that neither any internal instrumental adjustment nor detector response curve for quantitation is intended here as in the common usage of the terminology. In most pharmaceutical laboratories,

Handbook of Pharmaceutical Analysis by HPLC,
S. Ahuja and M.W. Dong, editors.

calibration is performed every 6–12 months. If results are acceptable according to some predetermined criteria, a sticker is placed on the instrument to indicate its readiness for GMP work. This periodic calibration, coupled with initial system qualification (installation qualification and operation qualification or IQ/OQ) and daily system suitability testing, is of an overall system validation program to ensure data validity and regulatory compliance (Figure 1).[1–6] A company-wide system calibration program also facilitates method transfer by minimizing system-to-system variability between different testing sites.

In this chapter, the principles and strategies behind HPLC calibration are described and illustrated with a case study of a company's revised calibration procedures and acceptance criteria.

II. OVERALL CALIBRATION STRATEGIES

Calibration procedures are often similar to those used in the initial operational qualification of each module of the HPLC system. While each company's procedure might differ in the details, most share these common strategies.

A. By Whom

An analyst, a metrologist, or a qualified contractor can perform the calibration, though all must follow the company's prescribed standard operating procedure (SOP) and acceptance criteria. The cost effectiveness of using outside contractors or an internal metrology department is dependent on company size and the number of HPLC systems in the laboratory.

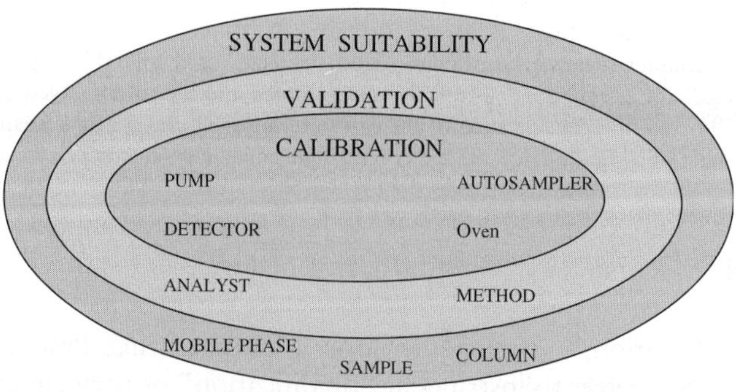

FIGURE 1 Schematic diagram showing the overall validation strategy, which includes calibration, system validation, and system suitability testing. Figure adopted with ideas from Reference 1.

B. When

Most HPLC systems in pharmaceutical laboratories are calibrated every 6–12 months. Periods longer than 12 months are not recommended while periods shorter than 3 months are deemed unnecessary because each HPLC system is also subjected to a daily system suitability check to ensure sufficiency for the application. Ideally, the frequency of calibration should be dictated by the historical data of calibration failures and the manufacturer's recommendation. In practice, 6 months appears to be the norm adopted by most laboratories. Calibration is also required after annual preventive maintenance or major repairs though only the affected modules, and not the entire system, need to be recalibrated.

C. How

Here is a list of common strategies in selecting calibration procedure and adopting acceptance criteria:

- Annual preventive maintenance, in which most wearable items such as pistons, seals, lamps, and filters are replaced, is to be scheduled before calibration.
- Before calibration, each module is shutdown and powered-up to evoke built-in diagnostics for detecting problem situations.
- The calibration order is: detector ⇒ pump ⇒ autosampler (as the detector is often used to calibrate other modules).
- Common performance characteristics to be verified for each module are:
  - UV detector: wavelength accuracy, absorbance linearity, and sensitivity;
  - pump: flow rate accuracy and precision, and compositional accuracy;
  - autosampler: sampling precision and accuracy;
  - oven: temperature accuracy;
  - overall system: system dwell volume and instrumental bandwidth.
- Acceptance criteria generally mirror the manufacturer's specifications though many are necessarily relaxed to accommodate diversified models and aging components.
- The calibration standard employed should be National Institute of Standards and Technology (NIST) traceable and easily obtainable.

III. CALIBRATION PROCEDURE: A CASE STUDY

Since calibration is time consuming and is often viewed as nonproductive work, the decisions in developing a calibration program often walk a delicate balance between maintaining laboratory productivity, scientific thoroughness, and regulatory compliance. The case study of one company's calibration procedure[6] is used here to illustrate the rationale behind these decisions. Table 1 summarizes the procedures and acceptance criteria for each module.

TABLE 1 Summary of Calibration Tests and Acceptance Criteria

HPLC module	Test	Procedure	Acceptance criteria
Detector PDA or UV/Vis	Wavelength accuracy	Measure λ_{max} or maximum absorbance of anthracene solution (1 μg/mL).	251±3 nm 340±3 nm[2]
Pump	Flow accuracy[1]	Run pump at 0.3 and 1.5 mL/min (65% acetonitrile/water) and collect 5 mL from detector into a volumetric flask. Measure time.	<±5%
	Compositional accuracy	Test all solvent lines at 2 mL/min with 0.1% acetone/water, step gradients at 0%, 10%, 50%, 90%, and 100%. Measure peak heights ofrespective step relative to 100% step.	±1% absolute
Auto-sampler	Precision[1]	Determine the peak area RSD of 10 10-μL injection of ethylparaben (20 μg/mL)	RSD <±0.5%
	Linearity[1]	Determine the coefficient of linear correlation of one injection of 5-, 10-, 40- and 80-μL of ethylparaben solution	R > 0.999
	Carryover[1]	Determine the carryover of peak area from injecting 80-μL of mobile phase following 80-μL injection of ethylparaben	<1%
	Sampling accuracy	Determine gravimetrically the average volume of water withdrawn from a tared vial filled with water after six 50-μL injections	50±2 μL
Column oven	Temperature accuracy	Check actual column oven temperature with validated thermal probe	35±2°C

[1]Waters Symmetry C18 (150×3.9 mm i.d., 5 μm) used in most tests except wavelength accuracy and compositional accuracy tests.
[2]Note some UV/Vis detector might required higher conc. (e.g., 2 μg/mL) for the 340 nm verification.

A. Photodiode Array (PDA) or Variable UV/Vis Detectors

A number of reference chemicals with well-defined UV spectra have been used for detector wavelength calibration, including uracil, erbium perchlorate, holmium oxide, caffeine, and anthracene. This procedure uses an anthracene because of its sharp absorbance bands and good solution stability. Wavelength accuracy of the detector is verified by measuring the absorption bands at 251 and 340 nm. In photodiode detectors, a spectrum annotated with λ_{max} is used (Figure 2). In UV/Vis detectors, two scans by stepping up the wavelength program at a 1-nm increment from 247 to 255 nm and 336 to 344 nm are used to locate the actual λ_{max} (Figure 3). Detector absorbance linearity is tested simultaneously with autosampler linearity. Detector noise is a common verification parameter; however, it is not tested in this procedure because the calibration committee believed that detector sensitivity could be covered by system suitability tests in impurity methods, if required. Other detectors (e.g., electrochemical, refractive index, and evaporative light scattering) can be covered by separate equipment calibration procedures using the same principles.

B. Pumps

Flow accuracy is checked at 0.3 and 1.5 mL/min with the column in place by measuring the time required to fill a 5-mL volumetric flask from

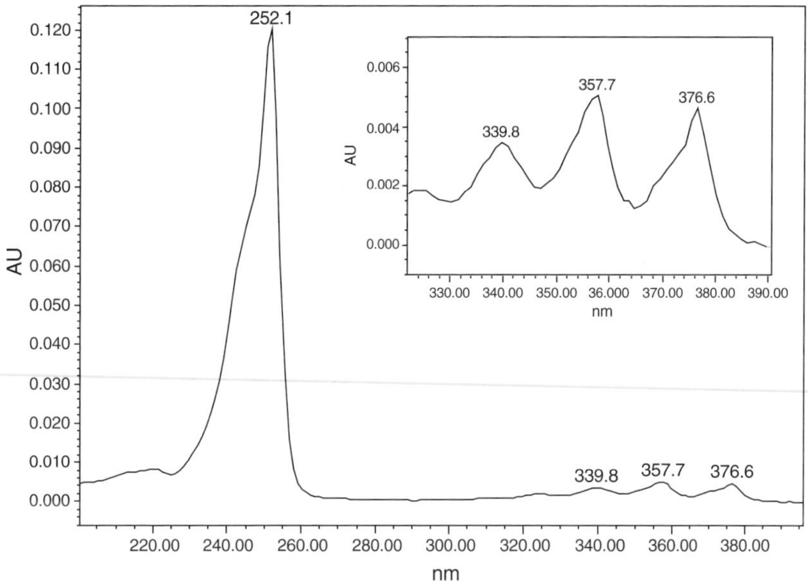

FIGURE 2 **Spectrum of anthracene solution (1 μg/mL in acetonitrile) from Waters 996 PDA detector showing the annotations of λ_{max}. The inset shows an expanded view of the 340-nm band.**

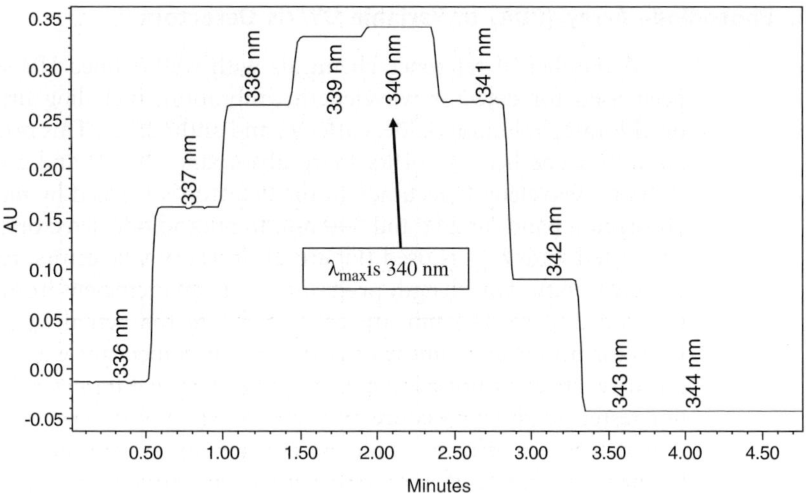

FIGURE 3 **Calibrations of a UV/Vis detector (Water 2487) near 340 nm by incremental scanning from 336 to 344 nm. The λ_{max} of anthracene is determined to be at 339 nm for this detector. The "autozero on wavelength change function" in the detector must be deactivated for this test.**

the detector outlet. This was upgraded from the original procedure requiring only verification at one flow rate of 1.0 mL/min. Other flow rates can be tested if the lab routinely performs analyses using narrow-bore or microbore columns. Compositional accuracy is determined by making 5-min step gradients from water in one solvent line to 0.1% acetone in water in a second solvent line. The absorbance of each step against that of the 100% step is measured. A flow rate of 2 mL/min is used here to sharpen the step definition and to increase test robustness. All four solvent lines are checked in a 1-h-long solvent program (see Figure 4). Calculations of absorbance ratio based on the heights of each step are reported automatically by the data system (see Figure 5 showing the report in Waters Millennium[32]). A stringent criterion of ±1% absolute is required to ensure the blending accuracy of each pump used in method development.

C. Autosampler

For autosampler precision, 10 consecutive 10-µL injections of an ethylparaben solution (20 µg/mL) are used (Figure 6). A Waters Symmetry column packed with 5-µm particles is used. The manufacturer's specification for peak area precision at 0.5% RSD is adopted as the acceptance criterion. This stringent precision criterion is required for precise assay testing of drug substances typically specified at 98–102% purity. The linearity test is performed by single injections of 5, 10, 40, and 80 µL of the

Time (min.)	Flow (mL/min.)	A water (%)	B water (%)	C 0.1% acetone (%)	D 0.1% acetone (%)	Curve
0.0	2.0	100	0	0	0	11
5.0	2.0	90	0	10	0	11
10.0	2.0	50	0	50	0	11
15.0	2.0	10	0	90	0	11
20.0	2.0	0	0	100	0	11
25.0	2.0	0	100	0	0	11
35.0	2.0	0	90	0	10	11
40.0	2.0	0	50	0	50	11
45.0	2.0	0	10	0	90	11
50.0	2.0	0	0	0	100	11
55.0	2.0	0	100	0	0	11

FIGURE 4 **Table showing the solvent programming steps used for verifying pump compositional accuracy. This program is used specifically for the Waters Alliance 2695 system.**

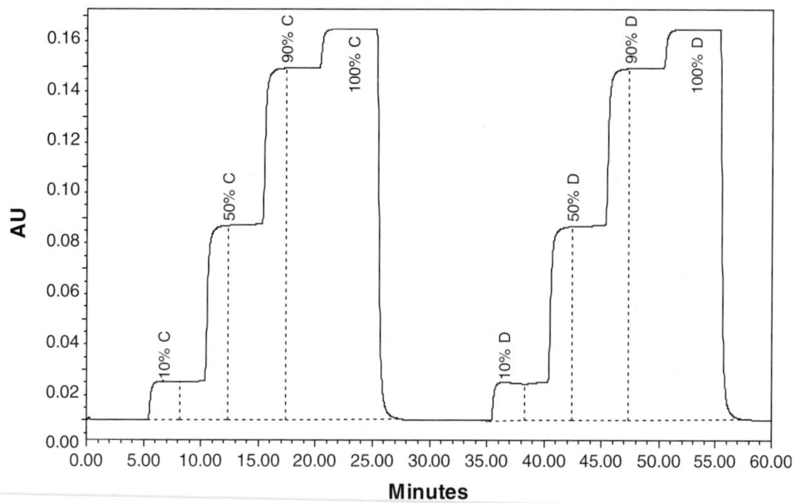

FIGURE 5 **Typical step gradient profile of a four-solvent pump (Alliance 2695 module) using a solvent program at 2 mL/min with water and 0.1% acetone in water. Absorbance ratio of the each step against the 100% step is used to determine compositional accuracy. All four solvent lines are tested in a 1-h-long gradient program.**

ethylparaben solution (Figure 7). The largest volume is set at 80 μL since the default sample loop is 100 μL in many autosamplers. This test actually checks the combined linearity of the injector, the detector, and the data system. Carryover is determined by measuring the carryover peak

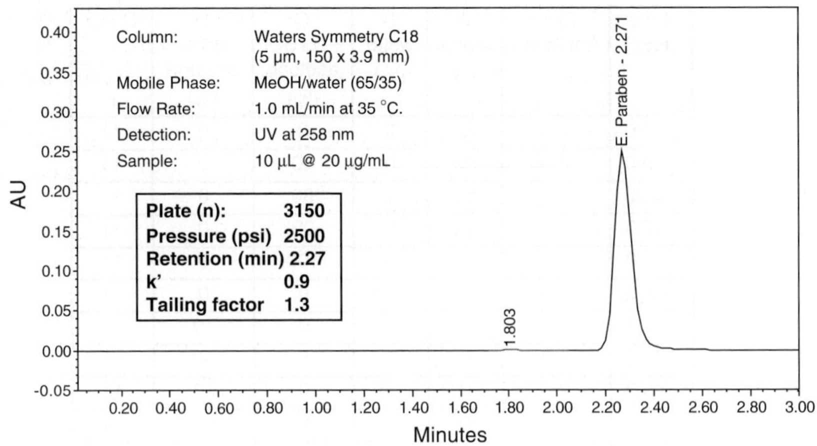

FIGURE 6 HPLC chromatogram of 10 µL injection of ethylparaben used for the verification of autosampler precision.

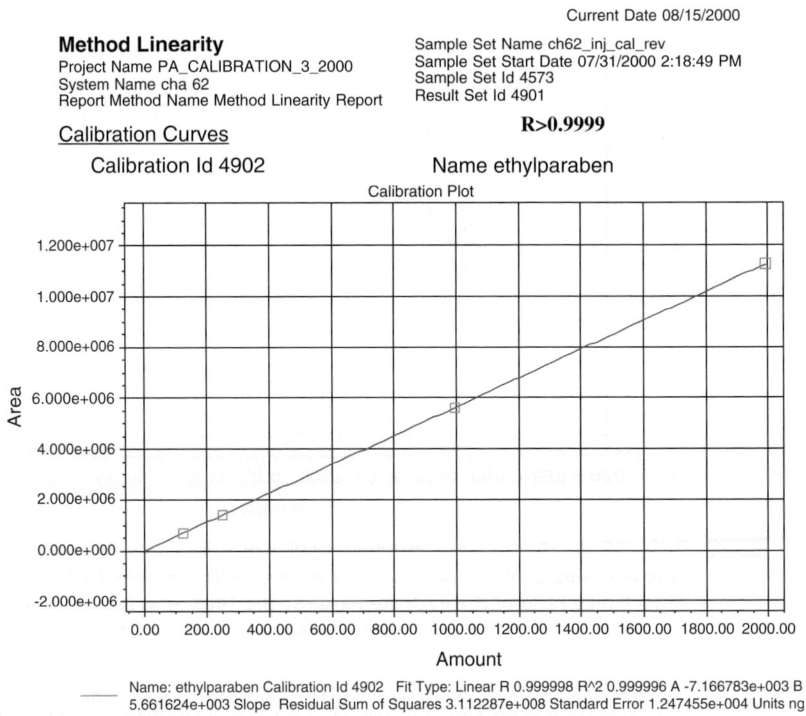

FIGURE 7 Waters Millennium report showing the linearity verification for the autosampler. This test also verifies the combined linearity of the system including the autosampler, detector, and the data system.

area of an 80-μL injection of a mobile phase blank sample immediately after an 80-μL injection of the ethylparaben standard (Figure 8).

The autosampler injection accuracy test involves the gravimetric determination of the average volume of water withdrawn from a tared vial filled with water after six 50-μL injections. The procedure, adopted from a manufacturer's OQ, takes less than 10 min and has an acceptance criterion of $50 \pm 2\,\mu L$ (Figure 9).

D. Column Oven

A temperature accuracy test of the column oven measured with a calibrated thermal probe is used. An acceptance criterion of $35 \pm 2°C$ is adopted.

IV. OVERALL PROCEDURES AND DOCUMENTATION

Significant time is saved in this calibration procedure by minimizing sample preparation and the number of calibrated test apparatus wherein only a balance, a 5-mL volumetric flask, a stopwatch, a thermal probe, two standards, one HPLC column, and one mobile phase are used. The use of MS Word template forms (see Figure 10) saves considerable time by eliminating any notebook entries and also improves the consistency of calibration records. Two important system parameters, dwell volume, and

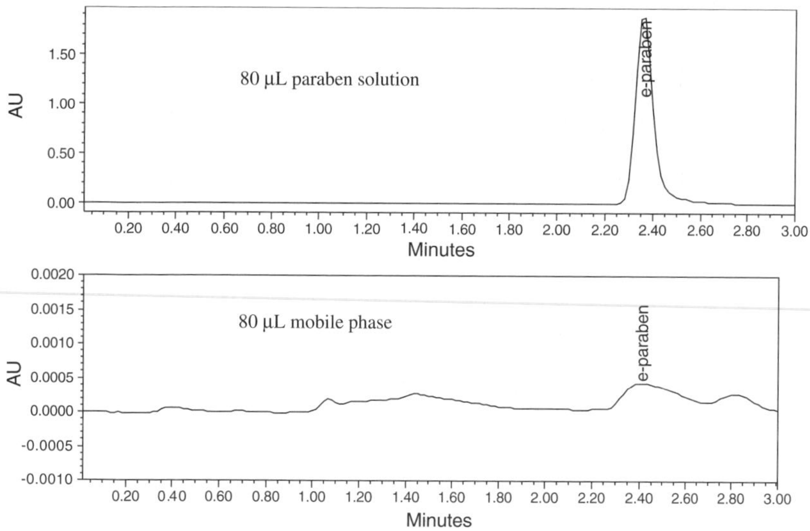

FIGURE 8 HPLC chromatograms used for the evaluation of the autosampler. Upper: 80 μL injection of ethylparaben. Lower: 80 μL injection of the mobile phase.

Sample Data Table for evaluating autosampler sampling accuracy

	Run 1	Run 2
Weight of filled vial in g: W1	2.98872 g	2.93152 g
Weight of vial after six injections in g: W2	2.68380 g	2.63300 g
Average injection volume (µL)= (W1-W2)/6x1000	50.89 µL	49.75 µL
Specification	50 µL ± 2 µL	50 µL ± 2 µL

FIGURE 9 **Gravimetric data table showing the evaluation of autosampler sampling accuracy.**

Autosampler Template

13. AUTOSAMPLER CALIBRATION

Autosampler manufacturer:____________________ Autosampler ID: ______

Flow rate: 1.0 mL/minute Wavelength: 258 nm

Next calibration due: ________________

13.1. Reproducibility of Autosampler

RSD of peak areas:_______________% (Specification: = 0.5%)

Millennium result ID: ______________ Reproducibility: ☐ Pass ☐ Fail

13.2. Linearity of Autosampler

Correlation coefficient (R):______________ (Specification: > 0.999)

Millennium result ID: ________________ Linearity: ☐ Pass ☐ Fail

13.3. Carryover of Autosampler

Carryover (%): __________________ (Specification: < 1.0%)

Millennium result ID: Carryover: ☐ Pass ☐ Fail

FIGURE 10 **Part of the MS Word template form to document the calibration of the autosampler. A 4-page template was used to document the results obtained for HPLC calibration so any laboratory notebook entries are not needed. Reprinted with permission from Reference 6.**

instrumental bandwidth, important for HPLC methods using small bore columns, are considered for future revisions of the calibration procedure.

V. SUMMARY AND CONCLUSION

In summary, this chapter reviews how to develop expedited calibration procedures and acceptance criteria for all the common modules in an HPLC system.

ACKNOWLEDGMENTS

The author acknowledges helpful ideas and suggestions from R. Ornaf, D. Barker, J. Bonilla, and C. Choi of Purdue Pharma. This chapter has been presented in parts in a symposium "Regulatory compliance issues in the pharmaceutical industry" at the Pittsburgh Conference in Orlando, Florida, March 13, 2003.

REFERENCES

1. Wiggins, D. E., *J. Liq. Chromatogr.*, 14(16):3045–3060, 1991.
2. Furman, W. B., Layloff, T. P. and Tetzlaff, R. F. *J. AOAC Int.*, 77(5):1314, 1994.
3. Huber, L., *Validation and Qualification in Analytical Laboratories,* Interpharm Press, Englewood, CO., 1998.
4. Dong, M. W., Paul, R. and Gershanov, L., *Today's Chemist Work*, 10(9):38–42, 2001, http://pubs.acs.org/subscribe/journals/tcaw/10/i09/html/09dong.html.
5. Parriott, D., *LC.GC*, 12(2):132, 1994.
6. Dong, M. W., Paul, R. and Roos, D., *Today's Chemist at Work*, 10(2):42–48, 2001, http://pubs.acs.org/subscribe/journals/tcaw/10/i02/html/02dong.html.

12

SYSTEM VALIDATION: PRE-INSTALLATION, IQ, OQ, AND PQ

WILHAD M. REUTER

PerkinElmer LAS, Shelton, CT

ABSTRACT

This chapter defines the terms, responsibilities, requirements and recommended procedures involved in pre-installation, installation qualification (IQ), operational qualification (OQ) and performance qualification (PQ), which are all part of a typical HPLC system validation process. As the FDA does not publish a definitive reference or "cookbook" for these procedures, the suggestions herein are only recommendations. These have been successfully incorporated into formal standard operating procedures (SOPs) that have been implemented at a number of larger pharmaceutical establishments. As long as proper SOPs

Handbook of Pharmaceutical Analysis by HPLC,
S. Ahuja and M.W. Dong, editors.

are drafted, the SOPs are followed and all is well documented, the following recommendations should greatly improve a company's chances of success in passing FDA inspections.

I. INTRODUCTION: WHAT IS VALIDATION, AND WHY VALIDATE?

Validation itself is considered a required part of good laboratory practice (GLP) and good manufacturing practice (GMP) within any pharmaceutical laboratory and has been defined by the FDA as:

> Establishing documented evidence which provides a high degree of assurance that a specific process will consistently produce a product meeting its pre-determined specifications and quality attributes[1]—FDA Guidelines on General Principles of Process Validation, May 1987.

Although this definition refers specifically to a process, it can also be easily applied to automated systems and software. One could read this as "Establishing documented evidence which provides a high degree of assurance that a specific automated system and/or software product will consistently operate and perform to its pre-determined specifications and quality attributes".

In a pharmaceutical laboratory environment, system validation for analytical instrumentation is certainly both a value and a burden. Albeit a valuable procedure, it is a process that requires considerable time and resource expenditure. Part of the latter concern is exacerbated by the fact that many laboratories are not quite sure about what exactly is required. In this chapter, the system validation terms will be defined, the responsibilities and requirements will be brought into better focus and recommended procedures will be described. This is intended to help relieve the actual burden of the process for those on-site individuals who are directly involved with the drafting and implementation of the validation procedures. The recommendations herein are provided as guidance to help streamline the actual procedures in an effort to keep both the time and financial expenditures to a minimum.

II. TIMELINE AND DEFINITIONS

System validation is actually a process that includes a number of tasks along a specific timeline. This is best shown via the flow chart in Figure 1.

The validation procedure can be subdivided, as per timeline, into three discrete task groups: *pre-installation*, *upon installation* and *post-installation*. The subsequent sections in this chapter will focus on each of these, defining the terms and providing a chronological progression of

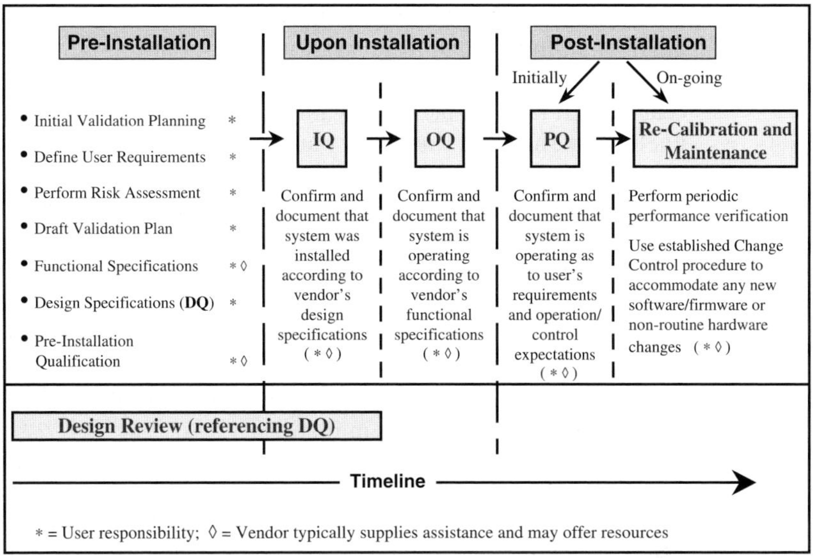

FIGURE 1 Validation timeline.

the series of tasks, responsibilities and documentation requirements. Examples are included to provide guidance as to the suggested content and/or format for some of the documents.

Together, the defined/written series of tasks, test methods, requirements and documentation required to carry out the validation procedure for a system can be described as the *validation protocol*. The "instruction manual" for all of this is the *validation plan*, developed by the validation team.

III. PRE-INSTALLATION

A. Preliminary Planning

If it has not been already completed, the first recommended task is to establish/designate a validation team within the company—perhaps one such team at each of multiple sites. If one elects to have a separate team at multiple sites, these teams must make certain that they carry out the validation processes consistently across all sites. In larger corporations, this is often accomplished via a central validation committee managing the individual groups at the various sites. For smaller companies a "team" may consist of only one individual. The goal of the team(s)/committee is to draft comprehensive validation plans and protocols across all the instruments, systems and processes that require validation. They must then diligently execute, document and verify each validation procedure, making certain that all results and documentation are properly secured. Early on in the validation process, it is important to establish a clear line of responsibility for all the required tasks that must be undertaken, procedures that must be performed and documents that must be written. This line of responsibility should be put into writing and documented.

Corporations are urged to draft specific *SOPs* in this regard. While many end users frown on having excessive SOPs on-site, SOPs are indeed a very important part of everyday operation in that they clearly define what, when, where and how things need to be performed. A well-designed SOP is always valued and appreciated. Unfortunately, SOPs are not always well written, especially some of the older versions. A saying that one comes across, which certainly still applies is: "the one thing worse than having no SOP is a poorly written SOP that isn't followed." This should always be kept in mind.

The *change control* SOP warrants particular attention. This procedure should clearly spell out how any deviations in instrument operation/performance, and hardware/firmware/software updates are to be handled and documented. It should also make clear what distinguishes routine maintenance, minor vs. major changes and what the procedural differences should be. The change control procedure may be a specific SOP, as part of an HPLC system's *validation plan*, or as part of the corporate *validation master plan (VMP)* that is established within each company.

Consider the VMP as the corporate doctrine of a corporation's general validation, testing, recording and archival practices. Typically, the SOPs within the VPM are too general and, therefore, it is advised that one create specific SOPs for a particular system to be validated. These would be part of an HPLC system's validation plan. Note that the VMP should be thought out and drafted well in advance of developing a specific system Validation Plan.

All individuals involved with the validation process should be thoroughly acquainted with the following documents, as these can all be very useful:

- GAMP4, IPEC, International Pharmaceutical Excipients Council, 2002 (http://www.ipec.org).
- Guidelines on General Principles of Process Validation, FDA, May 1987 (http://www.fda.gov).
- Guidance for Industry: Analytical Procedures and Methods Validation, FDA, November 1994 (http://www.fda.gov).
- Furman, W. B. Layloff, T. P. and Tetzlaff, R. F., Validation of Computerized Liquid Chromatographic Systems, *J. AOAC Int.*, 77 (#5):1314, 1994.
- Guidelines on General Principles of Software Validation; Final Guidance for Industry and FDA Staff, FDA, January 11, 2002 (http://www.fda.gov).

- Title 21 Code of Federal Regulations (21 CFR Part 11), Electronic Records; Electronic Signatures, FDA, 2003 (http://www.fda.gov).
- Validation/compliance documents found under the Parenteral Drug Association (PDA) (http://www.pda.org).

B. User Requirements

Moving forward, one needs to establish what the user requirements should be for an HPLC system. These requirements are also referred to as *user requirement specifications (URS—ISPE/GAMP)*;[2] they will be subsequently referred to as URS throughout this chapter. Things to consider/include within the URS are:

- spatial limitations
- solvents to be used
- environmental impact
- safety considerations
- solvent flow range
- number of solvents
- temperature requirements
- sampling requirements
- electrical/gas hookup considerations
- operating system requirements
- computer/network interfacing needs
- computer/network security needs
- basic software requirements
- data storage/archival strategy

C. Risk Assessment

Another suggested task during pre-installation is to perform some type of risk assessment. An important part of this task is to establish the relative importance of various functions and hardware/software operations on the HPLC system relating to how it will be used, or not be used, during routine operation. Only those functions that will be used during the expected operational range of the system need to be tested. For example, if there are certain calibration routines or reporting functions offered by the system that clearly will not be used during operation then they do not have to be tested. If a UV-Vis detector supports a tungsten lamp which will not be used then it also need not be tested. These decisions should be made early on so as to minimize the overall validation load and implementation time.

As a part of risk assessment, each validation team should be well versed with the suggestions and stipulations provided in the FDA's *Title*

21 Code of Federal Regulations (21 CFR Part 11), Electronic Records; Electronic Signatures (see documents list above).[3,4] In essence, these rulings/regulations address how any electronic data and associated records generated within a regulated environment are to be properly stored, secured, archived, validated and maintained. Since these rulings/regulations are being continually revised to accommodate the latest needs in security and the changes in existing technology, it is advised that one check the FDA's web site for the latest updates. Part of the validation team/committee's responsibility should be to establish, well in advance, what needs to be 21 CFR Part 11-compliant at each site and how the process should be implemented and documented. Here again, a well-structured and clearly worded SOP is advised.

D. Validation Plan

Following risk assessment, the next step is to draft a formal validation plan. This is a written plan that includes all the specific validation procedures, installation tasks, acceptance testing, documentation requirements, reviews and verification tasks that need to be followed for proper system validation. The plan should also define individual responsibilities for these tasks and include an expected timeline. The plan should be designed around the URS and take into account the risk assessment determinations performed earlier.

E. Functional Specifications

Next, one should develop the *functional specifications* for a system. The functional specifications document should establish the HPLC system's actual capabilities as they apply to the full scope of the URS. Some of the more specific URS requirements may be included in the functional specifications as well. This document is often developed with significant vendor interaction and input.

F. Design Qualification

For a particular system, many of the above requirements, specifications and considerations are often outlined as part of a *design qualification (DQ)*. This document should also include information about the general architecture, components and interfaces of the system. The DQ document is then used in conjunction with the *design review*.[2] This review is a continuous process, verifying the established DQ design. The DQ design is subsequently evaluated and cross-checked during the actual pre-installation and both the IQ and OQ procedures. The design review should use the various metrics contained in the DQ to evaluate the level of success in implementing the validation procedure and meeting the URS. It should be noted that, in some publications, the DQ and design review are considered synonymous.

G. Pre-installation Qualification

The final task of the pre-installation process should be a *pre-installation qualification.* This is a site-readiness evaluation indicating whether the location into which the HPLC system is to be installed is actually ready to accommodate this system. This should include the following utility/spatial/environmental and safety considerations:

- space availability
- safety requirements (ventilation; solvent collection/disposal strategy)
- electrical/power requirements (number of outlets; voltage; amperage)
- availability of gas supplies
- environmental requirements (temperature; humidity)
- computer/network requirements (operating system; network tie-ins, terminal servers/IP addresses, if required; serial ports; cables, type and length)

IV. UPON INSTALLATION

Following successful completion of the pre-installation process, it is time for the system to be installed and to implement the core of the validation protocol—the IQ/OQ/PQ. Based on the overall timeline, these procedures have been grouped into two task groups: *upon installation* and *pre-installation.* Focusing on the *upon installation* tasks, these include both the IQ and OQ protocol, implemented in respective order. They can be generally defined as follows:

Installation Qualification—The IQ is the protocol that verifies that the installation of the system has followed the guidelines established within the validation plan and is in accordance with the vendor's installation requirements.

Operational Qualification—The OQ is the protocol that verifies that the operation of the system meets the written and pre-approved performance and functional specifications established by the vendor within its required operating range.

For the hardware installation and IQ implementation on a full HPLC system, one should expect to allocate about 2 days. For the OQ, it is recommended that about 1 day be allocated.

Note: As is often the case, the HPLC system will be under computer control, which is likely to be part of a data-handling system. Since the data generated from the OQ hardware tests typically require chromatographic data handling, the data-handling software should be validated beforehand. The data-handling/LC control software installation and IQ/OQ implementation, which are not addressed in this chapter, may take a considerable amount of time. This is often the case since this process typically involves an initial client/server implementation.

The LC control software, either stand-alone or as part of an overall data-handling system, should be tested by means of a separate OQ protocol. This protocol only needs to address the communications/control integrity of the hardware (e.g., setting up a run/sequence with the proper instrument parameters, the ability to start and stop the pump, etc.). It should cover all the required instrument control functions listed as part of the protocol's functional specifications. It does not need to include specific hardware performance testing, such as linearity or flow rate. The latter tests are performed separately, as part of the individual hardware validation described below.

The HPLC hardware IQ and OQ protocols will now be more fully defined.

A. Installation Qualification (IQ)

The IQ protocol for a system is often partly, if not totally, developed by the vendor of that system. This is understandable, considering that, typically, the vendor has the best understanding of how to properly install the system and would also know how best to validate the installation process.

A system is typically comprised of multiple instrument components. Therefore, there is usually an individual IQ for each of these instruments and for any corresponding instrument control/data-handling software. The typical instrument components making up an HPLC system include: a binary or quaternary HPLC pump, an autosampler supporting multiple vials or microtiter plates (autosamplers often include cooled Peltier trays for sample stability), a column oven, and a UV-Vis or photodiode array (PDA) detector.

This chapter will address the validation of the hardware and will not specifically address the validation of the data-handling software as this is beyond the intended scope.

1. IQ Protocol Guidelines

It is recommended that the *IQ* document contain the following:

(a) Title page providing the name of the instrument component being installed, company location/address, protocol number, author and date of document, and may include the protocol approval sign-offs.

(b) Table of contents and any appendices pertaining to such items as SOPs, manufacturer's certificates, service engineer qualifications.

(c) *Scope*
The scope generally defines the purpose of the IQ protocol. It should describe what component/software will be installed and what will be validated/evaluated during installation. The scope should also provide a basic outline as to implementation tasks, and define the overall documentation, validation and installer's qualification requirements.

(d) *Instrument description*
This is the general description of the instrument which should include identifiers of the instrument, such as model number, serial number and, possibly, an internal inventory number. It should also include the component's overall capabilities and electrical, temperature and humidity requirements/specifications.

(e) *Description of intended use*
Description of the intended purpose of the instrument. This should relate to its use as part of an overall HPLC system in the laboratory.

(f) *Materials contact*
Description of solvents, chemicals and materials that will come into contact with the instrument.

(g) *Software and/or firmware verification*
One should verify the version number and date of all associated software/firmware.

Note: Any subsequent updates should to be handled as part of a well-designed, clearly defined *change control* SOP, specific for each software or firmware type.

(h) *Hazards and safety precautions*
One must document any potential hazards and safety precautions, including any possible electrical/line interference. This may also include suggested safety checks.

(i) *Consumable requirements*
Include any related consumables that will be required for the operation of the module.

(j) *Documentation/manual requirements*
This should include all the additional documentation that is required and any manufacturer's certificates that will be used as

part of the validation process. One should also include all user/service manuals that may be referenced by the user during routine operation.

(k) *Installation checklist*
A tabular checklist is recommended to verify that all installation tasks have been successfully completed.

(l) *Installation verification and certificate*
The installation should now be formally verified. A tabular example of such verification is provided in Figure 2.[5] This should be followed by a verification statement that the instrument was successfully installed in accordance with the *Installation Checklist*, above, and signed by the installer.

(m) *Installation certificate*
Finally, a certificate page should be included, verifying that the entire IQ protocol was completed successfully for a particular instrument. This should include the instrument's serial number, protocol approval sign-offs and any remarks.

{Name of Instrument}			
Serial Number:			
	Description	**Yes**	**No**
1	Instrument properly identified		
2	Verified that all accessories were delivered as per the purchase order on the packing list		
3	Checked installation kit(s) for completeness		
4	Adequate room was available for installing and servicing instrument		
5	Adequate electrical supplies were available		
6	Installed correct fuses for instrument's voltage configuration		
7	Firmware verified		
8	Software verified		
9	Start-up diagnostics run to completion		
10	Located all operator manuals and operation guidelines and placed them in the proper location(s), as specified in this protocol		
11	Located {additional documents} and manufacturer's certificates and placed them in the proper location(s), as specified in this protocol		
12	All safety precautions and hazards were properly observed		

FIGURE 2 Example of installation verification.

B. Operational Qualification

The IQ is typically followed in close succession by the OQ protocol. The OQ confirms and documents that an instrument operates in accordance with the vendor's functional specifications. Again, similar to that for the IQ, the OQ protocol is often partly, if not totally, developed by the vendor of the instrument. In some laboratories, the internal calibration tests may be incorporated as part of the OQ protocol. In most cases the protocol is typically carried out by a vendor's qualified service engineer. Since the same service engineer often performs both the IQ and the OQ, the IQ and PQ are often combined as part of an overall IQ/OQ protocol.

Since the vendor's functional specifications are usually defined for a specific instrument component, the OQ is usually also performed on an individual component basis. This is one of the things that sets the OQ apart from the PQ, which is described later.

1. OQ Protocol Guidelines

It is recommended that the OQ protocol contain the following:

(a) *Table of Contents*
Refer to *PQ Protocol Guidelines* (a) described later in this chapter.

(b) *Scope*
The scope, defining the purpose of the OQ protocol may be combined with the overall scope of an IQ/OQ protocol. It should describe what component/software will be validated as operating according to the functional specifications. Again, as described for the IQ, the scope of the OQ should provide a basic outline as to implementation tasks, and define the overall documentation, validation and installer's qualification requirements.

(c) System information, including location, internal ID numbers, model numbers, serial numbers and the recommended re-certification period. *Note*: If the PQ is combined with the IQ document, then these entries are not necessary here.

(d) A list of all test standards, labware and equipment to be used during the validation process. This should include any associated part numbers, lot numbers and expiration dates for calibration and/or test standards. All electronic measuring devices and stop watches must be calibrated and certified beforehand. An example of a representative list is shown in Figure 3.[6] *Note*: If the PQ is combined with an IQ document, then, this information may be included after the *Scope* of the IQ/OQ document.

(e) A list of *SOPs*
This is a list, by title and location, of all internal and manufacturers' SOPs pertaining to the instrument being validated. This should also include any change control procedures.

Test Standards/Equipment*	***Part #/Certificate#***	***Lot #***	***Expiration Date***
Output device (*e.g.*, any accessible printer)			
Autosampler Vials, standard ~1-4mL vials			
Autosampler Vial Caps			
500mL Graduated Cylinder			
2 10mL Graduated Cylinders			
1-liter Class-A Volumetric Flask			
Luer-Lock Adapter			
30mL Solvent Syringe	###		
5ft 0.007" ID PEEK Tubing	###		
1000psi Flow Restrictor Coil	###		
1000psi Back-pressure Device (Holder/Cartridge)	###		
PEEK Nuts for back-pressure device (x2)	###		
PEEK Ferrules for PEEK Nut, above (x2)	###		
Universal PEEK Nut/Ferrule (x4)	###		
Electronic Thermometer (± 0.5 °C accuracy)	###		
Column Holder 150x4.6mm	###		
Validated C18 Column, 100x4.6mm, 5u	###	______	
Test Standard	###	______	______
10mg vial of Caffeine	###	______	______
Acetone	______	______	
Methanol (MeOH) - HPLC Grade	______	______	
Water - HPLC Grade	______	______	
Calibrated Stop Watch	______		______

***** Management and/or designated personnel have approved these materials which will be used to qualify the performance and functionality of the specified instruments.***

***User Verification*______________________ *Date*__________

FIGURE 3 Example of test materials/equipment/standard(s), included as part of OQ and PQ protocols. Courtesy of PerkinElmer LAS, Shelton, CT.

(f) Manufacturers' warranty information
The expected warranty coverage for an instrument should be clearly documented.

(g) Manufacturers' pre-shipping calibration verification
Typically, the manufacturer of each instrument supplies a "calibration verification" document. This document should verify that the instrument had performed within the functional specifications before it was shipped from the manufacturers' site. The name and location of such a document should be provided.

(h) Routine maintenance procedures
These may be listed as an SOP (see (b) above).

(i) OQ tester's qualification verification
A qualification verification certificate should be referenced, verifying that all individuals performing the OQ protocol are qualified to perform this protocol. This may be included as part of an overall IQ/OQ protocol.

(j) Operational test parameters
Operational testing is the main body of an OQ. It should include those tests and corresponding instructions that will help validate that each instrument component in an HPLC system is operating properly according to the vendor's functional specifications.

These tests will be different for each HPLC component. The recommended tests are broken down into individual components below and are presented in a recommended validation order. This is advised since the validation of one component requires that all other components used to validate that particular component have been previously validated. The prescribed order is one that takes this into account.

Recommended characteristics that are commonly tested by hardware:

- All instrument components
  Test for proper power-up and automated internal initialization procedures. Make sure each instrument will go to a "Ready" status.

- **Pump**

  Flow accuracy

  This is typically tested for at least two flow rates. Typical flow rates that are chosen are within a range from 0.2 to 3.0 mL/min. The specific flow rates tested will depend on the operating range of the pump and may be influenced by the user's expected operating range.
  The test procedure can be performed by pumping HPLC grade water at a specified flow rate under a controlled back-pressure, using a commercial back-pressure device connected in-line just after the pump. A 1000 psi back-pressure device is recommended. It is also advised that about 1% MeOH be added to the water in order to minimize biological growth. While timing the process with a calibrated stop watch, the pump effluent should then be collected in a volumetric flask. From this, one can calculate the actual solvent flow at each flow rate.

  Pressure pulsations

  Using a 1000 psi flow restrictor coil (length of 0.007″ i.d. tubing), again pump the water at 1 mL/min. After the flow has stabilized, record the high and low pressure points for a period of time (typically, 1 min). The high and low points should vary by less than a specified amount (e.g., ≤30 psi).

  Note: Do not use a commercial back-pressure device for this test, as this may dampen the true pressure pulsations arising from the pump.

  Compositional accuracy [not applicable for single-solvent (isocratic) pumps]

This can be tested by running the pump at 1mL/min through a programmed series of mobile phase steps from 100% reservoir A (water) to 100% reservoir B (water containing 0.1–0.5% acetone), with a short hold time at the end. A UV-Vis or PDA detector is also required. Before starting, one should disconnect the flow restrictor and reconnect the 1000 psi back-pressure device between the pump and the detector. One should then start the pump with solvent A and let the pump/detector stabilize for at least 20 min before continuing. Using the data generated by capturing the detector output on a data-handling system, the incremental step heights are then calculated as a % of the total height of all the steps. Figure 4 shows an example output of such a test.[6] An alternate example is provided in Figure 5 of Chapter 12 of this handbook (HPLC System Calibration).[7] The calculated value for each step needs to be within a certain tolerance range around the actual percent solvent increment that was programmed for that step. This process is often automated using the data-handling system. For quaternary pumps, one would also have to test both reservoirs C and D in the same manner.

Pump repeatability

It should be noted that this test can only be performed once the complete HPLC system is in place and, therefore, is sometimes only performed as part of the PQ protocol.

One would isocratically run at least six replicate injections of a *certified* test standard containing a component used to check

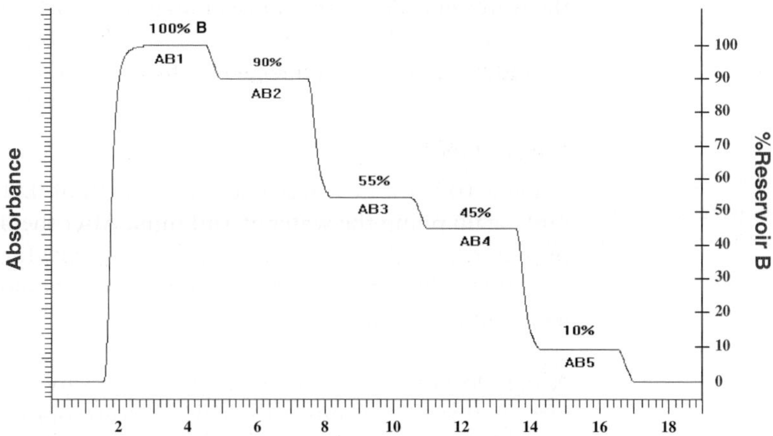

FIGURE 4 An example of a pump step profile as part of the compositional accuracy test for reservoirs A and B. Reservoir A contains 1% methanol/water and reservoir B contains solvent A with 0.5% acetone added. Courtesy of PerkinElmer LAS, Shelton, CT.

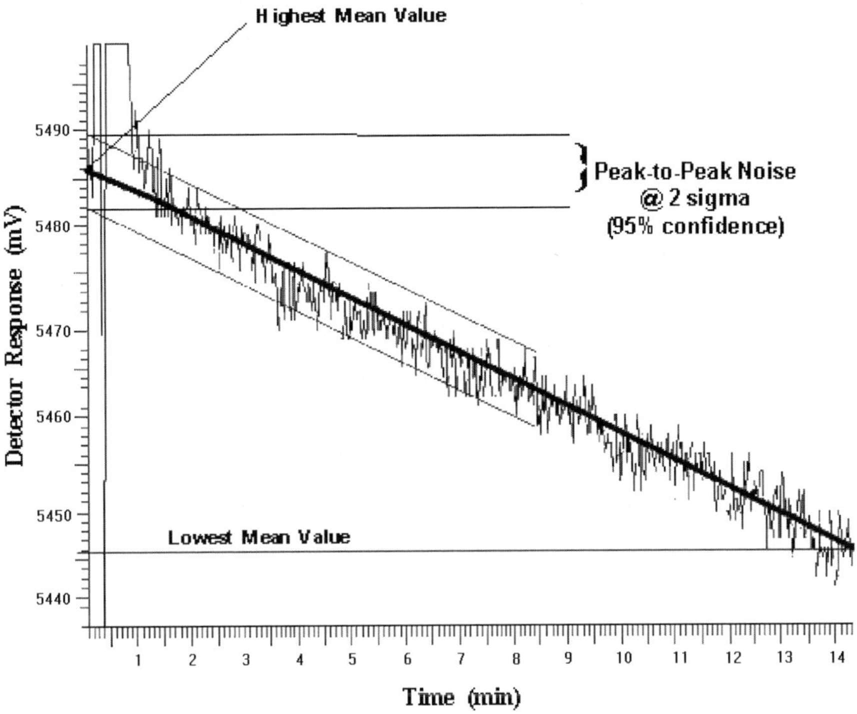

Example Noise and Drift Calculations:

$$\text{Noise (Peak-to-Peak)} = (5.490\ \text{mV} - 5.482\ \text{mV}) \times \frac{1\ \text{V}}{1000\ \text{mV}} \times \frac{1\ \text{AU}}{1\ \text{V}} = 8.0\text{x}10^{-8}\ \text{AU}$$

$$\text{Drift} = \frac{(\text{Highest Mean Value (mV)} - \text{Lowest Mean Value (mV)})}{\text{Evaluation Period (min.)}} \times \frac{60\ \text{min}}{1\ \text{hr}} \times \frac{1\ \text{V}}{1000\ \text{mV}} \times \frac{1\ \text{AU}}{1\ \text{V}}$$

$$= \frac{(5.486\ \text{mV} - 5.446\ \text{mV})}{15\ \text{min.}} \times \frac{60\ \text{min}}{1\ \text{hr}} \times \frac{1\ \text{V}}{1000\ \text{mV}} \times \frac{1\ \text{AU}}{1\ \text{V}} = 1.6\text{x}10^{-4}\ \text{AU/hr}$$

FIGURE 5 Example of how *noise/drift* calculations are obtained. The calculations are performed via a validated data system or certified calculator. Courtesy of PerkinElmer LAS, Shelton, CT.

retention time repeatability (see Note below). As long as both the component and the chromatography are stable, the %RSD of the retention time can then be used as a check for pump repeatability performance. In order to help streamline the OQ process, this test is best performed using the *validated* C18 column described under "*Autosampler*"—*Injection precision*, later in this section.

Note: A "*certified*" standard essentially means that the source of a component in a test standard is traceable to it's origin, that the component's identification is assured and, when applicable, that the component's quantity or concentration in the standard, as obtained, is documented. The formal document certifying the above is often called a "certificate of analysis."

- **Column oven** (if part of system)

Temperature accuracy

To test for temperature accuracy of a column oven, one should place the temperature probe/sensor of a calibrated/verified electronic thermometer (with at least 0.5°C precision) into the oven. The probe/sensor should not make contact with anything inside the oven. With the oven door closed, allow the temperature to stabilize for at least 20 min at each tested temperature (e.g., 30, 45 and 60°C). The thermometer's temperature should then be recorded at each temperature. The difference between the actual and set temperature should typically be within ±1°C.

Temperature stability

Using the same setup as above, one should select a representative temperature between 35 and 45°C. Once the oven has stabilized, record the thermometer's temperature at certain timed intervals (e.g., 15-min intervals) for at least one hour. Again the difference between the actual and set temperature should typically be within ±1°C throughout the test period.

- **Detector**

Noise and drift

All detectors need to be verified for meeting their noise and drift specifications. These tests are often performed with a "dry" cell that is sealed and has no actual solvent flow going through it. Some detectors have on-board diagnostics for these tests, with digital performance readouts.

To independently verify the noise and drift specifications, one would capture the detector's analog signal on a data-handling system or integrator. Again, this is often performed with a dry cell in place. If a dry cell is unavailable or not feasible, then one should pump water at 1 mL/min through a 1000 psi back-pressure device and into the detector. It is recommended that one adds 1% MeOH to the water to minimize biological growth. Once the flow and detector have stabilized, one then measures the peak-to-peak noise (typically, mV or mAU) over a specified time period (e.g., 5 or 15 min) and calculates the signal drift over a period of at least one hour. The actual specifications will vary for each type of detector. Figure 5 shows how the noise and drift is calculated for a representative baseline.[6] Typically, in practice, this is not done manually; rather, the calculations are performed using a validated data system or calculator.

Wavelength accuracy

For UV-Vis and PDA detectors, one should test the detector for wavelength accuracy. Typically, 2 or 3 wavelengths are checked, spanning the detector's wavelength range. This is tested using a test mix containing compounds that are stable and known to have primary/secondary absorbance maxima at various wavelengths throughout the wavelength range to be tested. Calibration standards containing caffeine, anthracene or holmium oxide have all been used for this purpose, depending on the instrument vendor. For caffeine, the wavelengths that can be checked are 205 and 273 nm. For anthracene, 251.5, 340 and 357 nm can be checked and, for holmium oxide, 241, 361, 416 and 536.5 nm can be checked. In all cases, the wavelength accuracy expectation is typically ±1 nm.

The actual testing can be performed in a number of ways:

(i) For wavelength programmable UV-Vis detectors, one can pump a certified caffeine or anthracene standard solution from one of the solvent reservoirs at about 1 mL/min. To make up the solution, a test standard containing the appropriate component is typically first diluted with degassed 75:25 acetonitrile HPLC-grade water by the required amount in an appropriate class-A volumetric flask. While running a blank water injection and collecting the data, the detector should ramp the wavelength in appropriate steps (e.g., 1 minute per step) across the wavelength being tested. This can usually be programmed as part of the chromatographic method. For caffeine, one can combine the wavelength check for 205 and 273 into one analysis; an example of the chromatographic wavelength stepping result is provided in Figure 6.[6] Another example, for anthracene's 340 nm secondary absorbance maximum, is shown in Figure 3 located in Chapter 12 of this handbook.[7]

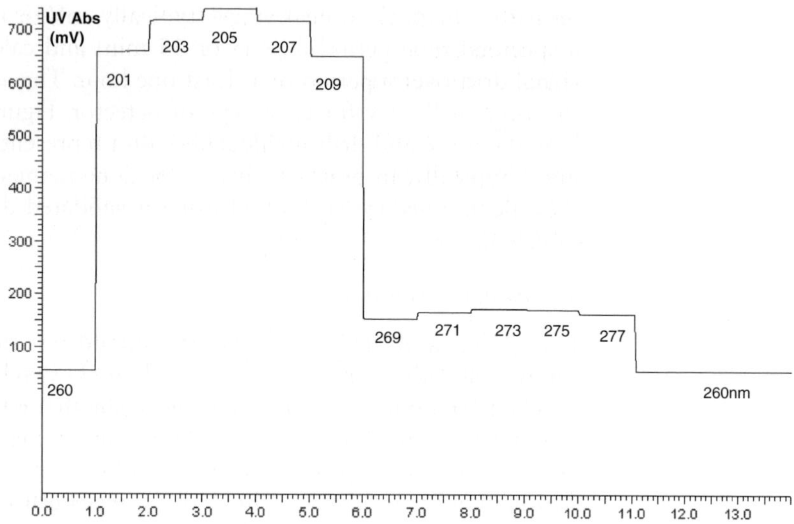

FIGURE 6 Automated 205/273 nm wavelength stepping results for wavelength accuracy test, using caffeine standard solution. Courtesy of PerkinElmer LAS, Shelton, CT.

One would then look at the chromatographic result and note at which wavelength step the highest absorbance was reached. This wavelength should then be compared to the expected wavelength maximum and the difference recorded.

(ii) On all systems with a UV-Vis detector, one can also inject a test standard containing either caffeine or anthracene and monitor successive wavelength steps from one injection to the next. The wavelength maximum is indicated from the corresponding chromatogram providing the highest peak for the compound tested. This wavelength should then be compared to the expected wavelength maximum and the difference recorded.

(iii) On HPLC systems with a PDA detector, one can also inject test standards and then use the spectral output to confirm the wavelength maxima.

On some detectors, a holmium oxide filter or special holmium oxide cell is integrated into the detector; these can then be moved into the optical path during calibration mode. The detector itself or associated control software may then be able to check for wavelength accuracy. *Note*: If the filter or cell is made of glass, and not quartz, 241 nm may not be used.

On a PDA detector, using the spectral energy output from the deuterium lamp, one can also use the sharp calibration lines at 486 and 656 nm (Balmer line).

Note: The actual wavelength maxima may vary slightly depending on the matrix/solution containing the calibration component. Therefore, a calibration solution should always be cross-checked with a calibrated stand-alone UV-Vis instrument.

Linearity

As part of the OQ protocol, testing for detector linearity has always been a topic of some debate, as linearity can be affected by many factors, including column stability, injector performance, sample stability, component concentration, mobile phase, temperature stability, etc. One must certainly minimize the effects of these factors in order to properly test a detector's inherent linearity. If a UV-Vis or PDA detector is being tested and one has a binary or quaternary pump, one can use the data collected during the pump's compositional accuracy tests to test for linearity. One would need to be sure that the last pump step, containing 100% water with acetone, provides enough response to provide a sufficiently high mV or AU signal. This level depends on the linearity range to be tested, typically ranging from 1 to 2 AU, and can be adjusted by using specific amounts of acetone added to the water. The amount of acetone added will typically be around 0.1% to 0.5%, depending on the detector.

Regarding OQ validation, if one has only an isocratic pump available, it is recommended that one does not perform a detector linearity test at this time. However, this test can be subsequently performed as part of either the PQ validation or individual method validation, both of which typically test the performance of the system as a whole (holistically).

- **Autosampler**

Injection precision

This test uses the entire HPLC system with a specific chromatographic method and validated C18 column that should be ≤10 cm and commercially available. "*Validated*" means that the individual column was performance tested before being shipped and that a certified test chromatogram is included with each column. Injection precision testing is typically performed by replicate injections of a test standard (at least six replicates are suggested). One then calculates the peak area %RSD of a stable component in the test standard. Most LC

vendors have a specific lot-numbered standard for this, some including a *certificate of analysis* (see Note on "*certified*" under "*Pump*", earlier in this section). The calculations must be made on either a validated data system or validated calculator.
This test is then repeated at various injection volumes, typically 3–4 volumes in all. Common volumes that are chosen are 5, 10, 25 and 50 µL, but will depend on the range of the particular autosampler, capacity of the test column and the user's intended injection range. Within the 5–20 µL range, the typical %RSD specification is <0.5%.

Carryover

Carryover testing is often performed immediately after injecting the highest volumes during the autosampler injection precision testing. One follows these injections with a mobile phase blank and measures the height of any carryover by the test standard component being evaluated. This height is then divided by the average peak height from the injection precision results for the highest injected volume and the resulting value is then multiplied by 100. The final result is a percent, as compared to the average high-end peak height of the component, and should be less than the vendor's carryover specification.

(k) Results documentation/verification
For each instrument component that is tested, the OQ should contain a corresponding section to document the results. This section should include the actual test results, calculated performance parameters, pass/fail status for each test, as well as any comments. Two examples of such documentation, as part of either an OQ or PQ protocol, are shown: one for a column oven, Figure 7, and another for an HPLC pump, Figure 8.[6]
Immediately following each of these sections, one should include a validation certificate for the corresponding instrument being validated, verifying that the instrument passed all required tests and /or met all the required functional specifications. This must include all the required sign-offs by those individuals performing the validation and authorized on-site managers or users.

(l) Operational testing certificate
As was required for the IQ, a certificate page should be included, verifying that the entire PQ protocol was completed successfully for a particular instrument. This should include the instrument's serial number, protocol approval sign-offs and any remarks.

Column Oven Certificate of Conformity

Model	Serial Number	Date Tested
Thermometer Used (electronic)	**Calibration Number**	**Expiration Date**
	Specification	Test Result
Temperature Accuracy:	30 °C ± 1 °C (≤1 °C)	
	60 °C ± 1 °C (≤1 °C)	
	45 °C ± 1 °C (≤1 °C)	
Temperature Stability @ 45 °C:		
15 min. Interval Data:	Temperature (°C) (using digital thermometer):	
First Interval		
Second Interval		
Third Interval		
Fourth Interval		
	Specification	Test Result
Temperature Stability Verification	$Max_{(reading)} - Min_{(reading)} \leq 1$ °C	

FIGURE 7 Documentation example of column oven temperature verification within OQ or holistic PQ protocol. Courtesy of PerkinElmer LAS, Shelton, CT.

V. POST-INSTALLATION

Following the successful completion of the IQ and OQ protocols, one can then proceed to the *post-installation* tasks. These tasks include both performance qualification, as well as re-validation, re-calibration and maintenance.

A. Performance Qualification (PQ)

PQ can be defined as verifying "that a system is capable of performing or controlling the activities of the processes it is required to perform or control, according to (the user's) written and pre-approved specifications (URS), while operating in its specified operating environment"[2]—ISPE/GAMP4.

While both the IQ and OQ are typically carried out on an individual instrument basis, it is now common for the PQ to be designed and

PerkinElmer
Instruments

Pump Certificate of Conformity

1. **Power up/ Initialization Diagnostics** — *Specification/Requirement: Pass* — ☐ Pass ☐ Fail

2. **Flow Accuracy: 0.5 mL/min** — *Specification: 0.49 to 0.51 mL/min* — ☐ Pass ☐ Fail

RESULTS:

5 mL / (Minutes to collect) = 5 mL / (   )min = (   )mL/min;

3. **Flow Accuracy: 1 mL/min** — *Specification: 0.98 to 1.02 mL/min* — ☐ Pass ☐ Fail

RESULTS:

5 mL / (Minutes to collect) = 5 mL / (   )min = (   )mL/min;

4. **Flow Accuracy: 3 mL/min** — *Specification: 2.95 to 3.05 mL/min* — ☐ Pass ☐ Fail

RESULTS:

9 mL / (Minutes to collect) = 9 mL / (   )min = (   )mL/min;

5. **Pressure Pulsation** — *Specification:* ≤30psi over 1 min. period — ☐ Pass ☐ Fail

RESULTS:

Maximum pressure: ________ - Minimum pressure: ________ = ________ psi

FIGURE 8 Example of HPCL pump power up, flow accuracy, and pressure pulsation verification document, as part of either an OQ or PQ protocol. Courtesy of PerkinElmer LAS, Shelton, CT.

performed *holistically. Holistic validation* refers to the validation of a system as a whole. This makes considerable sense, since this is how an HPLC will be used during routine operation. One typically designs the PQ to run a series of standards under operating conditions that represent those that will be routinely used. Although not covered here, this can include running one or two well-designed chromatographic methods, developed by the user, to fully test the user's requirements (URS). In comparison to the traditional modular instrument approach, the holistic PQ strategy will certainly save considerable time, effort and money.

Since the user knows best what the typical operating condition will be like in the laboratory, it is the user who should have the greatest input in formulating an acceptable PQ protocol. Although the vendor can certainly help provide resources, it is primarily the user's responsibility to finalize a PQ protocol that meets his/her needs and is acceptable to the company.

There may be a number of similarities between the OQ and PQ, but, whereas the OQ testing is based on the vendor's specifications, the PQ testing is based on the user's specifications/requirements. Therefore, the acceptance criteria may be somewhat different or customized, but they must always fall within the vendor's published specification limits. Also, in some laboratories, the system suitability testing (SST) protocol is used as a template for the PQ.[3]

I. PQ Protocol Guidelines

The following is provided as an example of a holistic PQ validation protocol and is recommended as a guideline:

(a) *Table of Contents*. An example of what this might look like is shown in Figure 9.[6] It would be rather similar to the *Table of Contents* in an OQ protocol, described earlier, except that the PQ version may include user customized tests.

(b) *Scope* and documentation layout/procedures

Table of Contents	Page
System Information	3
System Configuration	3
Scope	4
Documentation	4
Performance Testing:	
Testing Procedures	5
Power-Up Procedure	5
Initial Startup	5
Column Oven Temperature Accuracy Test	6
Column Oven Temperature Stability Test	6
Pump Flow Accuracy and Pressure Pulsation Tests	7
Detector Noise and Drift Tests	8
Pump Composition Accuracy Test	9
Detector Linearity Test	9
Wavelength Accuracy Test	10
Autosampler Injection Precision Test	11
Retention Time Test for Pump Repeatability	11
Autosampler Carryover Test	12
System Linearity Test	12
Column Oven Certificate of Conformity	13
Column Oven Performance Qualification Certificate	14
Pump Certificate of Conformity	15
Pump Performance Qualification Certificate	17
Detector Certificate of Conformity	18
Detector Performance Qualification Certificate	20
Autosampler Certificate of Conformity	21
Autosampler Performance Qualification Certificate	24

FIGURE 9 **Example of *Table of Contents* for holostic PQ protocol. Courtesy of PerkinElmer LAS, Shelton, CT.**

The scope, defining the purpose of the PQ protocol, should describe what component/software will be validated as operating according to the functional specifications. Again, as described for the IQ and OQ, the scope of the PQ should provide a basic outline of implementation tasks, and define the overall documentation, validation and installer's qualification requirements.

(c) System information, including location, internal ID numbers, model numbers, serial numbers and the recommended re-certification period.

(d) A list of all test standards, labware and equipment to be used during the validation process. This should include any associated part numbers, lot numbers and expiration dates for calibration and/or test standards. All electronic measuring devices and stop watches must be calibrated and certified beforehand.

(e) PQ tester's qualification verification
A qualification verification certificate should be referenced verifying that all individuals performing the PQ protocol are qualified to perform this protocol.

(f) Performance test procedures
As stated earlier, these tests should be designed to test that the HPLC system, in its entirety, is performing in accordance with the pre-approved user requirements (URS) for routine operation. Ultimately, the best way to do this is to test the system as holistically as possible.

The following is a recommended sequential outline for this section:

- General description of overall test procedure
- Verification of power-up and initialization by all instrument components in the system
- Definition of any initial preparation procedures involving instrument connections, HPLC plumbing requirements or sample/test mix preparation
- **Column Oven Testing** (if part of the system)

  One should perform both the temperature accuracy and stability tests as described in the OQ section. These may be modified to meet the user's requirements, within the functional specifications of the oven.
  It would be quite difficult to incorporate this test holistically. One would need to perform the test by using a very consistent

column (lot to lot) and, using a test standard containing a component whose retention time is rather temperature sensitive. This would not be easy to perform consistently.

- **Autosampler Testing**

  Injection accuracy test

  This test is not always performed nor is it required. If included, it is best performed gravimetrically. Off-line, the autosampler would be programmed to inject 3–6 replicates of a volume of water from a tared vial containing the water. Immediately afterwards, one would re-weigh the vial (in grams), take the before and after difference and then compare the resulting weight to the [intended sample volume per injection (in mL) x replicates], remembering that 1 g equals 1 mL water. For the test to pass, the percent difference between the actual vs. expected results would have to be under a predetermined limit (e.g., <5%).

- **Pump Testing**

  Flow accuracy

  Typically, the flow accuracy test is similar to the procedure used for OQ validation (described earlier in this chapter). The only thing that may differ is that the user may have specific flow requirements.
  This test could conceivably be performed holistically, as part of the injection reproducibility test (defined later). Using retention time as an indicator, such an approach would work provided the retention time for a test component was very stable from week to week. However, this is often not that easy to control consistently, since one often encounters column/mobile phase variations and other variables. Therefore, flow accuracy is typically not tested in this manner.

- **Holistic Testing**

  After completing the above tests, one would then set up a sequence of automated tests which may include the experiments shown in Figure 10.[6] This series of experiments assumes the presence of either a programmable UV-Vis or PDA detector as part of the HPLC system. If the UV-Vis detector is not programmable, one is required to manually step through the various wavelength steps while the data are being collected during experiment 4.

  Note: Experiments 1–4 are typically performed with a 1000 psi back-pressure device instead of a column. These tests should

Experiment # (Row)	Type	Sample Name	Injection Volume	Vial #	Method	Calibration Level
1	Blank	Noise/Drift – Water Blank	1	1	NoiseDrift	
2	Blank	Pump Accuracy and Detector Linearity – Water Blank	1	1	Pump Composition 1	
3	Blank	Detector Linearity – Water Blank	1	1	Pump Composition 2	
4	Blank	Wavelength Check – Caffeine 205/273	1	1	WavelengthSteps 203-209 269-277	
5	Sample	Test Std w/ anthracene 353nm	10	2	LCUV353	
6	Sample	Test Std w/ anthracene 355nm	10		LCUV355	
7	Sample	Test Std w/ anthracene 357nm	10	2	LCUV357	
8	Sample	Test Std w/ anthracene 359nm	10		LCUV359	
9	Sample	Test Std w/ anthracene 361nm	10	2	LCUV361	
10	Cal: Replace	5-μL Test Std	5	5	Prec-Repeat-Linear	L1
11	Cal: Average	5-μL Test Std	5	5	Prec-Repeat-Linear	L1
12	Cal: Average	5-μL Test Std	5	5	Prec-Repeat-Linear	L1
13	Cal: Average	5-μL Test Std	5	5	Prec-Repeat-Linear	L1
14	Cal: Average	5-μL Test Std	5	5	Prec-Repeat-Linear	L1
15	Cal: Average	5-μL Test Std	5	5	Prec-Repeat-Linear	L1
16	Cal: Replace	25-μL Test Std	25	25	Prec-Repeat-Linear	L2
17	Cal: Average	25-μL Test Std	25	25	Prec-Repeat-Linear	L2
18	Cal: Average	25-μL Test Std	25	25	Prec-Repeat-Linear	L2
19	Cal: Average	25-μL Test Std	25	25	Prec-Repeat-Linear	L2
20	Cal: Average	25-μL Test Std	25	25	Prec-Repeat-Linear	L2
21	Cal: Average	25-μL Test Std	25	25	Prec-Repeat-Linear	L2
22	Cal: Replace	100-μL Test Std	50	45	Prec-Repeat-Linear	L3
23	Cal: Average	100-μL Test Std	50	45	Prec-Repeat-Linear	L3
24	Cal: Average	100-μL Test Std	50	45	Prec-Repeat-Linear	L3
25	Cal: Average	100-μL Test Std	50	100	Prec-Repeat-Linear	L3
26	Cal: Average	100-μL Test Std	50	100	Prec-Repeat-Linear	L3
27	Cal: Average	100-μL Test Std	50	100	Prec-Repeat-Linear	L3
28	Blank	Carryover Test	50	91	Prec-Repeat-Linear	

FIGURE 10 Example of an experimental sequence for holistic PQ validation on an HPLC system.

not be performed with a column in place since one does not want to have any retention of the caffeine peak in experiment 4.

The first experiment of this set (row 1) is used to establish the noise and drift characteristics of the detector. 1 mL/min water (with 1% methanol) is pumped through the system. The system is allowed to equilibrate; then, a 1 μL water blank is injected and the data are collected for about 15–30 min. The noise and drift is then calculated as in the procedure described earlier under "Detector" in section (j) of the OQ guidelines.

Experiments 2–3 are used for testing the pump's compositional accuracy. This test should be performed as in the procedure described earlier under "Pump" in section (j) of the OQ guidelines. This time the procedure is programmed as is part of a series of automated experiments with the entire HPLC system.

Experiment 4 is intended to check the detector's wavelength accuracy at 205 nm and 273 nm. Using the same caffeine solution as for the OQ test, this test is performed and calculated as in the procedure described earlier under subsection (i) of "Detector" in section (j) of the OQ guidelines.

Note: After Experiment 4, one should switch the 1000 psi back-pressure device with a validated C18 column and adequately equilibrate the column with the appropriate mobile phase required for the subsequent experiment. The mobile phase solvents are usually percentages of methanol and water, depending on the chromatographic requirements for eluting the standard components that are used.

Experiments 5–9 are performed to check the wavelength at 357 nm by injecting an anthracene standard. This test is only required if the user intends to run the detector above 300 nm. For a description of the procedure, refer to subsection (ii) of "Detector" in section (j) of the OQ guidelines.

Experiments 10–27 are designed to check the autosampler injection precision, pump repeatability and detector/system linearity. One programs the system to automatically inject multiple replicate volumes of a certified test standard. One typically injects 6–10 replicates per volume. The standard component's peak areas are used for calculated injection precision (reproducibility) and system linearity; whereas, the retention times are used to calculate pump repeatability.

For injection precision test details and calculations, refer to the procedure described earlier under "Autosampler" in section (j) of the OQ guidelines.

For pump repeatability test details and calculations, refer to the procedure described earlier under "Pump" in section (j) of the OQ guidelines.

Using a validated data system or validated calculator, linearity is calculated using the averaged standard component areas over all the injected volumes. This resulting linear correlation coefficient (r^2) must be higher than the user's URS (typically >0.999). The calculated linearity is actually that for the entire system, including the detector. The only downside with such a test is that, if it fails, one cannot be certain, from this test alone, which instrument component is at fault. It could be due to any combination of the detector, autosampler, column or even a leak that developed after the autosampler.

Two side notes are provided below:

(1) As shown in the "Level" column of Figure 10, the various injected volumes can be designated as individual component calibration levels. By doing this, experiments 10–27 can also be used to help validate multilevel, multi-replicate calibration algorithms on a chromatographic data system.

(2) To test the autosampler's performance independence of vial location, it would be recommended that standards are injected out of various locations in the autosampler tray. This is shown in the "Vial #" column for experiments 10–27 in Figure 10. It should be noted that the higher vial positions may have to be tailored to the capacity of the actual autosampler tray.

Finally, Experiment 28 is used to check for any autosampler/system sample carryover. A blank water sample is injected immediately following the highest injected volume of the preceding precision/repeatability/linearity test. Any carryover is then calculated as in the procedure described earlier under "Autosampler" in section (j) of the OQ guidelines.

(g) Results documentation/verification
For each instrument component that is tested as part of the overall HPLC system, the PQ should contain a corresponding section to document the results. This section should include the actual test results, calculated performance parameters, pass/fail status for each test, as well as any comments.
Immediately following each of these sections, one should include a validation certificate for the corresponding instrument being validated, verifying that the instrument has passed all required tests and met all the required functional specifications. This must include all the required sign-offs by those individuals performing the validation and authorized on-site managers or users. An example of such a certificate for an HPLC pump is shown in Figure 11.[6]

PUMP PERFORMANCE QUALIFICATION CERTIFICATE

Model: ____________________ *Serial #:* ____________________

This is to certify that the tests have been performed and the configuration tested

(Check one and initial)

Meets specification/requirement listed for each test: ☐ ______ Initial

Does not pass one or all documented tests: ☐ ______ Initial

This certificate does not modify <*vendor's*> standard terms and conditions of sale, including warranty terms.

<*Vendor*> assumes no liability for test results.

Authorized <*Vendor*> Representative: ____________________ Signature

Employee number: ____________________

Date: ____________________

User Name: ____________________

User Signature/Date: ____________________

FIGURE 11 Example of pump performance verification document, as part of PQ protocol. Courtesy of PerkinElmer LAS, Shelton, CT.

(h) Performance qualification certificate
As was required for the IQ, a certificate page should be included, verifying that the entire PQ protocol was completed successfully for a particular instrument. This should include the instrument's serial number, protocol approval sign-offs and any remarks.

B. Revalidation/Calibration and Routine Maintenance

Periodic revalidation of a system can be accomplished by performing another PQ for that system; however, a simplified version is often used. This task has been alternately described as re-calibration, recertification or performance verification. Regardless of what it is called, the accepted strategy should be defined in an SOP and the specific procedure should include a formal testing script, verified results and corresponding documentation. The frequency for this procedure is typically once every year, although some companies insist on performing this once every half-year. Routine re-calibration procedures are covered elsewhere in this book, in the chapter on HPLC system calibration.

Routine maintenance procedures should be included as part of an HPLC's basic operating SOP. This will not be covered here.

VI. SUMMARY

It should be noted that the information in this chapter should be considered as a guideline and not an inflexible mandate. There are not many specific rules that are actually set forth by GAMP4 (ISPE) or mandated by any governing body such as the FDA. The important thing is that one takes reasonable, verifiable action, armed with well-thought-out SOPs, protocols, security and documentation procedures, to verify that an HPLC system used in the development or production of pharmaceutical products is performing both in accordance with the functional specifications set forth by the vendor and the URS set forth by the user. Again, we may recall, the only thing worse than having no validation protocol, is to have one that is poorly designed or one that is not followed.

There are various stages to the validation process and all have to be thought out as part of an overall *timeline*. The validation process starts with preliminary planning, selecting/training those qualified in defining and designing the various procedures, experiments and pertinent documents. This leads to the drafting of specific SOPs and IQ/OQ/PQ protocols, then moves on to carrying out the pre-installation procedures and finally culminates in implementing the IO, OQ and PQ procedures, in that order. One must also have plans/procedures for continued maintenance and re-calibration. All of this is ultimately the responsibility of the user but, especially for the IQ and OQ, the vendor can be of invaluable help, both in supplying needed documentation and in implementing the various tasks. Vendors are now also offering considerable help regarding PQ procedures.

Following pre-installation qualification and the actual installation of an HPLC system, both the IQ and the OQ protocols should be implemented, back to back, soon after the installation. Again, the IQ is used to verify that the installation of the system was successful, with all instrument components powering-up properly. The OQ follows, verifying that the system components perform as they were functionally specified by the vendor. Finally, the PQ protocol serves to verify that the system as a whole performs to the URS established by the user and within the functional limitations of the system as a whole. As part of the PQ, it is recommended to test the system as a whole, called *holistic* validation. This serves to most efficiently test the system as it will be operated by the user, while also expediting the entire validation process.

It is recommended that OQ test the following on an HPLC system: flow accuracy, pump compositional accuracy, pressure pulsations, column oven temperature accuracy/stability, detector noise/drift and wavelength accuracy, autosampler injection precision and carryover.

For PQ testing of the system, besides the OQ tests above, the following additional tests are recommended: pump repeatability and detector/system linearity. Some of the overall PQ tests can be combined into a select series of automated experiments, testing multiple performance criteria.

Though different performance criteria may be defined, many of the OQ and PQ tests will overlap; however, one should try to keep such redundancy to a minimum. Exactly what is tested and what the particular validation criteria are is dependent on both the capabilities of the system and the operating intentions of the user. Throughout, for the entire validation process to be implemented most successfully, it is important for there to be clear communication between the user and the vendor, to know what is expected, who is responsible for each task and to be committed to the details of the process.

ACKNOWLEDGMENTS

The author would like to thank Al Barckhoff, PerkinElmer LAS, for helping clarify some of the nuances in the ever-evolving system validation process and for providing valued expertise on this topic. Michael Dong of Purdue Pharma, should also be thanked for helping maintain a clear focus on the subject matter. Finally, the author would like to thank PerkinElmer LAS for making available the opportunity to write this chapter.

REFERENCES

1. Guidelines on General Principles of Process Validation, FDA, May 1987, (http://www.fda.gov).
2. GAMP4, IPEC (International Pharmaceutical Excipients Council), 2002.
3. Title 21 Code of Federal Regulations (21 CFR Part 11), Electronic Records; Electronic Signatures, FDA 2003 (http://www.fda.gov).
4. Guidelines on General Principles of Software Validation; Final Guidance for Industry and FDA Staff, FDA, January 11, 2002 (http://www.fda.gov).
5. Liquid Chromatography—IO/OQ—Series 200 LC Autosampler, PerkinElmer LAS, 2002.
6. Performance Qualification for Holistic Validation, PerkinElmer LAS, 2002.
7. Dong, M. W., Paul, R. and Roos, D., *Today's Chemist Work,* 10(2):42–48, 2001, http://pubs.acs.org/subscribe/journals/tcaw/10/i02/html/02dong.html.

13 ASSAY AND STABILITY TESTING

ANISSA W. WONG AND ARUNA DATLA

ALZA Corporation, Mountain View, CA

ABSTRACT

This chapter introduces the general concept of drug stability and the factors affecting drug substance and drug product stability. It discusses the guidelines from the International Conference on Harmonisation (ICH) for stability testing on drug substances and products, and the requirements at different phases of the drug development process in the pharmaceutical industry. The role of high-performance liquid chromatography (HPLC) in stability testing along with the chromatographic techniques and the procedures involved in developing a stability-indicating method are also described. Case studies of challenging HPLC method development on dual drug systems in solid dosage and surfactants with different molecular weights in liquid formulation are presented.

Handbook of Pharmaceutical Analysis by HPLC,
S. Ahuja and M.W. Dong, editors.

I. INTRODUCTION

A. Scope

1. What Is Drug Stability?

The term drug stability refers to the extent to which a drug substance or product retains, within specified limits and throughout its period of storage and use, the same properties and characteristics that it possessed at the time of its manufacture. The type of stability is generally divided into chemical, physical, microbiological, therapeutic, and toxicological.[1] Drug stability can be categorized as pre-market and commercial (marketed product) stability. Pre-market stability, which supports the clinical trial where drug products are stored under different conditions for safety and efficacy evaluation, is usually conducted throughout the clinical trial and during the filing period. Commercial stability is continuous assurance on the post-approval batches for long-term stability monitoring on the drug product. Drug stability assessment generally involves the testing of the drug substance or drug product using a stability-indicating method in order to establish the retest period (for pre-market stability) and shelf life (for commercial stability).

Drug substance (also called the active pharmaceutical ingredient [API] as per USP-NF definition[1]) is the material that is used to manufacture, usually with excipients, the drug product. Drug substances can be derived from chemical synthesis, plant or animal sources, or biological or recombinant technology. In addition to the API, the drug substance can contain product- and process-related substances or impurities. From the earliest stage of drug product development, information on drug substance stability has been an integral part of drug development. Data on the physical and chemical characteristics and other properties of the drug substance are helpful for designing methods that indicate the drug product's stability and are also helpful in designing formal stability studies.

Drug product (also called the dosage form or finished product per USP-NF definition[1]) contains one or more drug substances, usually with excipients, in the final packaging intended for marketing. Stability studies on the drug product serve three purposes: (1) to support the stability of the drug product used in clinical/non-clinical studies, (2) to establish commercial expiry dating, and (3) to determine levels for certain specifications (API, preservatives, etc.) and set the control limits for lot release.

2. What Is a Stability-Indicating Method?

A stability-indicating method is a validated quantitative analytical procedure that can detect the changes with time in the pertinent properties of the drug substance and drug product under defined storage condition. A stability-indicating assay method accurately measures the active ingredient(s) without interference from other peaks and is sensitive enough to detect and quantify the degradation products/impurities.[2] To develop a stability-indicating method, stress testing, in the form of forced degradation and a photostability study, should be carried out at an early stage so that impurities and degradation products can be identified and characterized. Stability-indicating assay analytical methods must be discriminating and validated to ensure the accuracy of the long-term stability study trending.

3. Factors Affecting Drug Substance and Product Stability

The stability of drug substances and products can be influenced by environmental factors such as temperature, humidity, light, and oxygen. The major factors that influence drug stability in the solid dosage or liquid form include particle size, pH, solvent system composition, solution ionic strength, cations and anions/excipients compatibility, chemical additives, and primary container and storage conditions.[2] Information obtained from stability studies in different conditions will aid in establishing the retest period of the drug substance and shelf life of the drug product. Strict adherence to the storage requirements specified in the product labeling will help ensure product potency and stability through to the manufacturer's labeled expiration date.

B. Objectives

1. General Stability Study

The general purpose of stability testing is to monitor the characteristics and changes of drug substances and drug products under different conditions (temperature, humidity, and light) over time so that the retest period of drug substances or the shelf life of drug products under specified storage conditions can be established. The test conditions should be wide enough to cover the four climatic zones in the world.[2,3] The four climatic zones are defined as (I)—temperate climate (e.g., Northern Europe, Canada, Russia, and the United Kingdom), (II)—subtropical and Mediterranean climate (e.g., United States, Southern Europe, and Japan), (III)—hot, dry climate (e.g., Iran, Iraq, and Sudan), and (IV)—hot, humid climate (e.g., Philippines, Indonesia, and Brazil). With the data obtained from stability studies, optimum packaging that will ensure the quality and integrity of the drug product in different parts of the world can be developed.

The typical attributes evaluated in stability testing include physical, chemical, biological, and microbiological quality characteristics that cover all dosage forms. A stability protocol should be generated for every batch that is evaluated in a long-term stability study. The stability protocol should record the purpose of the stability test, the method used, testing frequency, the storage conditions, the package description, and any additional information that needs to be included.

2. Role of HPLC in Stability Testing

The following methods are used to evaluate the stability and purity of drug substances and drug products: thin-layer chromatography (TLC), HPLC, gas chromatography (GC), capillary electrophoresis (CE), and ultraviolet (UV) spectroscopy. However, reversed-phase HPLC analysis is generally considered the most effective method of identifying most drug substance degradation or drug–excipient interactions. Hence, it is the typical choice for stability-indicating and stability-specific methods for small molecules. Reversed-phase HPLC accounts for more than 85% of stability-indicating methodologies for small molecular chemical entities as it is suitable for release testing, assay, and assessing impurities.[4] With its well-established techniques, efficiency, robustness, and ease of use, HPLC plays a pivotal role in drug stability testing.

II. STABILITY PROGRAM IN PHARMACEUTICAL DEVELOPMENT

A. Guidelines for Stability Testing on Drug Substances and Drug Products

1. Introduction

The primary purpose of stability testing is to provide data and supporting evidence on the stability behavior of chemical or biological entities in different forms. As drug stability can be affected by many factors like temperature, humidity, light, pH, oxidation, or combinations of these parameters, it is critical to establish the retest periods of drug substances and shelf lives of drug products under recommended storage conditions. In this section, the general approach of stability testing on drug substances and drug products will be discussed.[5]

2. Stress Testing

In the early stages of product development, a systematic stability study plan is critical, as it can affect the product life cycle and project planning. Before the drug product is available, stress testing (a forced degradation study) of the drug substance under different conditions can help to identify possible degradation products, the degradation pathway, and the stability of the drug substance.[2,5] The design of the stress test will depend mainly on the nature of the drug substance and the excipient composition.

Stress testing should be carried out on at least a single batch of the drug substance at the beginning, and minimally, include the effects of temperatures (a wide range from 20°C to 60°C), humidity (e.g., 60%RH, 75%RH, and dew point), oxidation, hydrolysis (wide pH range), and light (see photostability testing section). The information on the impurities and degradation pathway will help develop a stability-indicating method for formal stability studies of drug substances and drug products. (See chapters 6 and 7.)

3. Photostability Testing

Photostability testing is another form of stress testing to ensure that if the drug substances and products are exposed to light, the light will not cause unacceptable changes in the drug's potency or safety profile. To simulate a product left on a window, sitting on a shelf, or left in an open area, light chambers that expose the sample to various types and intensities of light are used. Light or photostability studies can also help determine the optimum packaging required to protect the drug product. Drug substances, in conjunction with excipients and drug products, should be exposed to light providing an overall illumination of not less than 1.2 million lux hours and an integrated near ultraviolet energy of not less than 200-W h/m^2 to allow direct comparisons. Testing should progress until the results demonstrate that the drug product is adequately protected from exposure to light.[6]

The drug substances and products should be examined for any changes in physical properties and for assay and degradants by a suitably validated method. If protected samples are used as dark controls, the sample analysis should be performed concomitantly. The test method should be capable of resolving and detecting photolytic degradants that might arise in the confirmatory studies. To assure that the product is within proposed specifications during the shelf life, the results of photostability studies should be evaluated to determine whether any changes due to light exposure are acceptable.

For drug substances, photostability testing should consist of two parts: forced degradation testing and confirmatory testing. Forced degradation testing is to evaluate the overall photosensitivity of the material for method development and degradation pathway elucidation. Confirmatory studies should then be undertaken to provide the information necessary for handling, packaging, and labeling. For drug products, one batch is required for confirmatory testing in the development phase, but at least two additional batches should be tested if the results of the confirmatory study are equivocal.[6]

4. Selection of Batches and Container Closure System

Stability data on at least three primary batches must be provided. For drug substances, the batches should be manufactured to pilot scale by the same synthetic route that simulates the final process to be used for production batches. Stability studies should be carried out on the drug substance where it is packaged in a container closure system that imitates the packaging proposed for storage and distribution. For drug products, the primary batches should be of the same formulation and packaged in the same container as proposed for marketing. Two of the three batches should be at least pilot-scale batches; the third batch can be smaller in size. If possible, the drug product batches should be manufactured using different batches of the drug substance.

5. Storage Conditions and Testing Frequency

The storage conditions and the length of the studies should be sufficient to cover storage, shipment, and subsequent use. Both drug substances and products should be evaluated under storage conditions that test for thermal stability (temperature effect) and moisture sensitivity (humidity tolerance).

The storage conditions for drug substances and products recommended by the International Conference on Harmonisation (ICH) guidelines are summarized in Table 1. In addition to testing for assay, impurities/degradation products, and general attributes, water content determination is recommended for solid dosage drug products. Alternative storage conditions can also be used with data support and justification.[5]

If drug substances and products are intended for storage in the refrigerator or freezer, real-time data obtained at the long-term storage condition must be provided as supporting evidence for the retest period or shelf life determination. Drug substances and products that require storage below –20°C should be evaluated on a case-by-case basis.

The stability study conditions recommended by ICH guidelines are as follows:

Drug substances and products intended for storage in a refrigerator:
Long term: storage condition at 5 ± 3°C for 12 months.
Accelerated: storage condition at 25 ± 2°C/60 ± 5% RH for 6 months.

Drug substances and products intended for storage in a freezer:
Long term: storage condition at –20 ± 5°C for 12 months.

If a significant change occurs at any time during the 6 months' testing at the accelerated condition, additional testing at the intermediate storage condition should be conducted and evaluated.

TABLE 1 Drug Substances and Products for Storage in General Cases

Study	Storage condition	Minimum time period covered by data at submission
Long term[1]	25 ± 2°C/60 ± 5% RH or 30 ± 2°C/65 ± 5% RH	12 months
Intermediate[2]	30 ± 2°C/65 ± 5% RH	6 months
Accelerated	40 ± 2°C/75 ± 5% RH	6 months

[1]Long-term stability studies at 25 ± 2°C/60 ± 5% RH or 30 ± 2°C/65 ± 5% RH are subject to applicants' decision.
[2]No intermediate condition is required if 30 ± 2°C/65 ± 5% RH is the long-term condition.

A significant change for a drug substance or product as per ICH guideline is defined as follows:

- A 5% change in assay from its initial value; or failure to meet the acceptance criteria for potency when using biological or immunological procedures.
- Any degradation product's exceeding its acceptance criterion.
- Failure to meet the acceptance criteria for appearance, physical attributes, and functionality test (e.g., color, phase separation, resuspendability, caking, hardness, and dose delivery per actuation); however, some changes in physical attributes (e.g., softening of suppositories and melting of creams) may be expected under accelerated conditions.
- Failure to meet the acceptance criterion for pH.
- Failure to meet the acceptance criteria for dissolution for 12 dosage units.

The long-term testing should cover at least 12 months' duration on at least three primary batches at the time of submission and should be continued for a period of time sufficient to cover the proposed retest period of drug substance or shelf life of drug product. For drug substances with a proposed retest period or drug products with a proposed shelf life of at least 12 months, the frequency of testing at the long-term storage condition should normally be every 3 months over the first year, every 6 months over the second year, and annually thereafter through the proposed retest period or shelf life.

For accelerated studies, a minimum of three time points (including the initial and final time points, 0 and 6 months) of at least 6 months is recommended. If a significant change is observed at the accelerated storage condition, testing at the intermediate storage condition for a minimum of four time points (including the initial and final time points) from a 12-month study is recommended.

B. Stability Testing at Different Phases of the Drug Development Process

Stability testing is required throughout all stages of drug development, from early feasibility through clinical trials (pre-filing) to NDA (filing) and product post approval (commercial stability). The requirements and the type of study will change during the drug product development life cycle as the impact is different at various stages.

During the early discovery phase, the primary function of stability studies is to determine the stability characteristics of the drug. Knowing these characteristics helps researchers select and design the most satisfactory chemical or molecular entity for the desired pharmaceutical profile and indication. The pharmaceutical profile focus on obtaining optimum drug stability (chemical and physical stability where physical forms can exist in polymorphs, free acid, salt, base, hydrates, or solvates), suitable pre-formulation, and achievable manufacturing process. Pre-formulation testing is the study of physicochemical properties; preliminary information regarding particle size, distribution, crystal shape, mechanical properties, surface area, and other physical information also plays an important role in estimating drug stability, manufacturability, and the performance of the chemical or molecular entity.[2] Informal stability studies can be carried out in-house with proper documentation that contains all the relevant early-stage information. At the beginning of product development, the information on drug substance stability and its interaction with the excipients in the pre-formulation (excipient compatibility) is crucial where excipient compatibility is considered as an early form of stability study. The effect of temperature and hydrolysis across a wide range of pH, moisture, oxidation, and photolysis under long-term and accelerated conditions should be performed so that the rate and mode of degradation pathways (or mechanism, if possible) can be established. The primary aim of the initial stability evaluation is to ensure that a safe and stable formulation can be developed.

Formulations change as the drug goes through its development stages; therefore, stability studies should be conducted on the promising formulations under both accelerated storage conditions: 40°C/75% RH for short-term and normal storage conditions, and 25°C/60% RH or 30°C/65% RH for intermediate to long-term storage conditions. These studies must be initiated before clinical studies begin to ensure that the clinical formulations are likely to remain stable during clinical studies. During the later stages of product development, the goal is to develop a formulation that has a shelf life of 2–3 years for consumer use and optimum packaging that adequately protects the drug product. Formal stability studies with specified attributes under designated storage conditions and packages usually begin in Phase I or Phase II and continue after market approval (Phase IV). Stability testing must be performed to ensure that the drug products are not only chemically stable but also physically stable. All the chemical testing should be performed by stability-indicating methods where the same method is used throughout the stability-testing period for a particular attribute. A new method can be used if equivalency is demonstrated. If changes are involved in formulation, packaging or testing procedures, the methods must be revalidated according to ICH and FDA guidelines.

C. Testing Requirements for Drug Products Filed in the United States and Outside the United States

Regulatory agencies in various geographical regions enforce guidelines to ensure the quality and uniformity of pharmaceutical substances.

The documents, established separately in the United States, Great Britain, Europe, and Japan, provide standards and specifications for all facets of pharmaceutical materials including their testing, packaging, and storage. With the globalization of the pharmaceutical industry, a need arose to harmonize these separate standards, and the ICH was formed. This committee is responsible for harmonizing the requirements of pharmaceutical products among the participating members: The United States, Europe, and Japan. *The European Pharmacopeia (EP), British Pharmacopeia (BP), and Japanese Pharmacopeia (JP)* are the three largest and most influential official compendia outside the United States. The *United States Pharmacopeia* and *National Formulary* (USP/NF) is the largest and most comprehensive of the national compendia. A detailed discussion of all the pharmacopeia is beyond the scope of this chapter.[7] Overall, methods and acceptance criteria must be acceptable to regulatory authorities in each region. The standards and specifications established by the four major pharmaceutical compendia and harmonized by the ICH guidelines assure uniform product quality throughout the world.

D. Design of Stability Program

1. Matrixing and Bracketing

In an ideal full stability study design, samples for every combination of all design factors are tested at all time points. In a reduced design, samples for every factor combination are not all tested at all time points. Due to the many combinations of formulations, dosages, and container types and sizes in the drug products, a full stability study testing program of all samples for every combination at all time points is time-consuming and costly. Hence, a reduced design can serve as an alternative to save time and cost in the drug development stage if the shelf life of the drug product can be adequately predicted. The concept of bracketing (samples are tested only at extreme conditions for any time point) and matrixing (a selected subset of the total number of stability samples is tested at specified time points) can be applied to the reduced design; assumptions and the statistical analysis should be stated clearly in the study plan.[8]

Bracketing is a design of a stability schedule in which only samples on the extreme of design factors are tested at all time points as in a full design. The design assumes that the stability of any intermediate levels is represented by the stability of the extremes tested. Design factors (strength, container size, and fill size) are variables to be evaluated in a study design for their effect on product stability. Bracketing can be applied to studies with multiple strengths of identical or closely related formulations.[2] With justification, bracketing can be applied to studies with multiple strengths where relative amounts of drug substance and excipients change in a formulation. In cases where different excipients are used with different drug strengths, bracketing generally should not be applied. If the stability of the extremes in the bracketed design is shown to be different, the intermediates should be considered no more stable than the least stable extreme.

Matrixing is the design of a stability schedule where only a fraction of the samples are tested at specified time points. The design assumes that the stability of each of the subsets of samples represents the stability of all samples in the study. Matrixing with reduced testing is an alternative for monitoring stability where multiple factors are involved for the same product. The variables such as different batches, different dosage strengths, and different fills in a container of the same formulation can be matrixed, and the stability of the product studied. The different storage conditions and test attributes may not be matrixed, but should be studied in their own matrix design. If matrixing is applied to time points, all the factors in the design should be tested at the initial and final time points, and at two additional time points through the first 12 months. Matrixing design is usually "one-half reduction" and "one-third reduction" referring to the reduction applied to the full study design. A protocol should be written clearly and followed for all testing. The key aspect of matrixing is that it should be a well-balanced design.

The applicability of matrixing is affected by various factors, especially for the first few batches. The knowledge about the drug substance stability and the information obtained from the pre-formulation work will greatly assist in the design. If the drug substance is well characterized with sufficient stability information, and stability is established for the current formulation with less variability, the applicability of matrix design will be better. However, matrixing is not recommended for formulations with poor drug substance stability and highly variable data. Matrixing design in site qualification lots (SQLs) and post approval lots can be determined based on the historical data. Due to the reduced amount of data collected, shorter retest periods or shelf life should be considered. With appropriate justification, a change to full testing or to a less reduced design can be considered during the course of a reduced design study. The use of any reduced design should be justified. Reduced design of a stability study must be carefully planned and well balanced. The decision to use bracketing or matrixing or both must take into full consideration the information needed and the potential risks involved.

2. Statistical Analysis of Stability Data

As the stability data can be affected by many factors like formulation, manufacturing, storage conditions, in-process and GMP controls, analytical methods, and process validation, the biggest challenge is to figure out the source of the variability in the stability results. At least three batches of the drug substance or product are required to establish the acceptance criteria for future production batches as a measurement standard. The degree of variability of individual batches affects the confidence that a future production batch will remain within the acceptance criteria throughout its retest period or shelf life.

Assuming the drug substance does not show any compatibility problem with the excipient in the stress or forced degradation studies at early stage, long-term and accelerated formal stability supporting data should be evaluated to determine the critical factors that could affect the quality and performance of the drug product. For example, the temperature or humidity conditions during the scale-up process and exposure to excessive oxygen during the drying process could change the assay profile.

Statistical analyses like linear regression, poolability tests, and statistical modeling are examples of mathematical tools that can be used as quantitative ways to evaluate the stability data.[9] Although normal manufacturing and analytical variations are to be expected, it is important that the drug product be formulated with the intent to provide 100% of the labeled amount of the drug substance at the time of batch release. If the assay value of the batches used to support the registration application is higher than 100% of the label claim at the time of batch release, after taking into account manufacturing and analytical variations, the shelf life proposed in the application can be overestimated. On the other hand, if the assay value of a batch is lower than 100% of the label claim at the time of batch release, it might fall below the lower acceptance criterion before the end of the proposed shelf life. Hence a careful and systematic approach should be adopted in the presentation and evaluation of the stability information.

An approach for analyzing data of a quantitative attribute that is expected to change with time is to determine the time at which the 95% one-sided confidence limit for the mean curve intersects the acceptance criterion.[2] If analysis shows that the batch-to-batch variability is small, it is advantageous to combine the data into one overall estimate by applying appropriate statistical tests (e.g., p-values for level of significance of rejection of more than 0.25) to the slopes of the regression lines and zero-time intercepts for individual batches. If it is inappropriate to combine data from several batches, the overall shelf life should be based on the minimum time a batch can be expected to remain within the acceptance criteria.

E. Specifications Setting in Stability Testing

"Specifications" refers to a list of tests regarding analytical procedures and appropriate acceptance criteria which are numerical limits, ranges or other criteria provided in an approved application that confirm the quality of the drug substances, drug products, intermediates, raw materials, reagents, and other components.[10] Typically, the specification is a set of numerical values/ranges or characteristics derived from formal/informal stability results and any experimental data to ensure the consistent, high-quality production of drug substances and drug products. Stability testing of drug products should include general attributes like appearance, average assay, organic/inorganic impurities, release assay, and water content so that sufficient data at different time points can be collected to propose acceptance criteria. Often, only a limited amount of data is available at the time of filing, and this can impose many challenges in setting acceptance criteria. Hence, acceptance criteria may need to be revised as more experience is gained throughout the development of the product. The basis of the specification is to focus on the safety and efficacy of the final drug products. An adequate rationale and data for specifications must be provided.

Specifications set for each test attribute should be evaluated at each test point. Specifications can be set for "release" of the product and for "shelf life" as discussed in ICH guidelines Q6A and Q6B. The shelf life of the product is determined by all available stability information. Justifiable differences between shelf life and release acceptance criteria are appropriate.

Specifications remain a binding standard of quality between the regulatory agency and the applicant. In general, a drug product or drug substance conforms to a specification when (1) the article is tested according to the listed analytical procedure, and (2) values obtained are within the listed acceptance criteria.[7] In other words, methods and specifications are related. If the method is modified or optimized for better performance, a change in the specification can be justified and the regulatory agency needs to be informed. An example of a drug product specification is shown in Table 2.

III. STABILITY TESTING BY HPLC

A. HPLC for Stability Testing

Among all the different analytical techniques used in the pharmaceutical industry, chromatography is the most commonly used technique for drug analysis. HPLC is the most popular technique due to its high accuracy, precision, and ease of usage. Planar chromatography, which consists of TLC and paper chromatography (PC), is less commonly used because it is considered to be a less precise technique, but is a feasible approach for quick pre-screening in searching for impurities and determining the purity of a drug substance.[7] Recent technology includes chiral chromatography, which is used to separate racemic mixtures into individual enantiomers that may have different therapeutic effects.[2] Another separation technique that is growing faster and becoming more popular is CE. CE separates compounds by driving the mobile phase with a voltage differential where the voltage differential results in electroosmotic flow.

TABLE 2 Specification Table of a Drug Product (A Typical Example)

Product name: Analgesic-ABC
Product ID: Lot # 5678abc
Dosage strength: 10 mg
Stage: IND Phase I

Test attribute	Method no.	Specification	Result	Pass/fail
Appearance	#5-001	Round, white tablet with "abc" imprint on one side	Conforms to specification	Pass
Average Assay	#1-001	95–105% label claim	94%	Fail
Impurities	#1-002	Individual known impurities <0.2%	0.7%	Fail
		Total impurities <2.0%	2.5%	
Dissolution testing	USP <711>	10 min >25% LC 30 min >75% LC 45 min >90% LC	10 min = 30% LC 30 min = 80% LC 45 min = 99% LC	Pass
Water content	USP <921>	Report value	1.01%	Pass
Residual solvent	#2-001	<2000 ppm	500 ppm	Pass

Average assay and impurities test did not conform to specification. Upon investigation and confirmation of the results obtained, the product failed the specification. The product is at the Phase I stage where the formulation is not optimized, the manufacturing process is not robust enough, and the analytical methods are not rugged, hence a higher percentage of lots could have failed. In addition, the specification was not set appropriately to reflect the stage of the drug development.

In applying HPLC for separation purposes, the solubility of the compound is the primary consideration. As a rule of thumb, HPLC requires solubility of the analytes in the mobile phase. The choice of separation is based on the size, polarity, and ionic nature of the solute. The preliminary information required is the molecular size and solubility of the compound in water to select the appropriate column prior to optimizing other HPLC parameters. A few chromatographic column types are commonly used for separation. Normal phase (NP) is defined as a polar stationary phase with a nonpolar mobile phase typically used for nonpolar analytes. Reversed phase (RP) is defined as a nonpolar stationary phase with a polar mobile phase, which is the reverse of the NP. In the pharmaceutical industry, RP chromatography is the most popular mode for small molecules <1000 Da. Analytes that are soluble between weakly and relatively polar solvents separated by a nonpolar stationary phase (C18, C8, phenyl or cyano columns) would be the best way to separate small molecules. Ion-exchange chromatography (IEC) and ion chromatography (IC) are all separation techniques for charged analytes (inorganic and organic ions). Size exclusion chromatography (SEC) is a technique that separates analytes based on their sizes in solution and can handle high-molecular weight samples in a short time with simple method development.[7]

In developing a stability-indicating assay method, the basic criterion is to separate the API from the major related substances so that the potency data will not be affected. For a drug product with different dosages, the all-strength method is the best option for efficiency purposes. By changing the sample preparation through varying the volumetric flask size or dilution scheme, the assay target concentrations for different dosages can fall in a narrow range where the analysis can be combined using the same calibration curve or a single point standard. For reporting assay, the typical unit is % label claim (%LC) or mg/system; the final choice usually depends on the specification setting.

During early drug development, combining assay and impurity in one single test method is recommended as more information can be extracted in a shorter time. As the drug product enters a later stage, it is better to separate the active assay by isocratic elution and the impurities/degradants by gradient elution for higher capacity and better analysis for dimers. As a general guideline, gradient methods are more suitable for multi-drug formulations.

In stability testing, HPLC is the most frequently used chromatographic technique for assay. HPLC has illustrated the benefits of speed, resolution, accuracy, and sensitivity, and HPLC can interface with many detectors that other analytical techniques are unable to offer.

B. Detector Considerations

To select a good detector, the basic criteria include sensitivity at low concentrations, linearity over a wide range, and tolerance to temperature or solvent composition changes.[11] A stability-indicating assay method must be specific, selective, and must separate the impurities from the active drug well to avoid interference. In the pharmaceutical industry, the majority of small molecules contain chromophores that display good absorbance in the UV region. Hence, HPLC interfaced with a UV detector becomes the natural choice for RP chromatography. The API peak should have fairly high absorbance where the absorbance value of the highest concentration should be between 1 and 1.5; a wide linearity range should also be developed. A mass spectrometry compatible method can be developed for confirmation or to serve as an alternate method.

However, samples that have no UV absorbance, exist in ionic form, or require structural information have to couple with other types of detectors. Refractive index (RI) detector is almost a universal detector in that it responds to almost any solute (UV-absorbing molecules, sugars, polymers, and pharmaceutical excipients). The drawbacks are inadequate sensitivity, low tolerance to temperature and pressure, baseline instability, and incompatibility with gradient elution. Another type of universal detector is the evaporative light-scattering (ELS) detector. The advantage of ELS is that it can be used for gradient elution, especially for impurity analysis. However, ELS is amendable to nonvolatile and semi-volatile analytes. Electrochemical detectors are mainly designed for ions and can be performed in either oxidative or reductive modes. The main disadvantage is a poor signal-to-noise ratio as a result of dissolved oxygen in the solution. Fluorescence (FL) and direct-current amperometry (DCA) detectors both have high sensitivity and selectivity, but the former detection mode always requires sample derivatization, which may cause problems like incomplete reaction or multiple derivatives formation, and hence is less preferred. The combination of LC and MS (mass spectrometer) is a powerful tool as molecular weight and structural information can be obtained from the mass spectra. LC-MS and GC-MS have become increasingly popular in the pharmaceutical industry. The main applications are to monitor the purity of drug substances and identify the structure of new chemical entities or intermediates.[7,12]

Other than selecting the appropriate detector with the correct mode of chromatography, optimizing the LC parameters is crucial in developing a stability-indicating assay method. The lower the injection volume the sharper the peak will be, but the area count for the low concentrations in the calibration curve should give an acceptable percentage of recoveries and reproducibility. Flow rate must be well-adjusted to avoid backpressure exceeding 3000 psi. The mobile phase selected should avoid interference and absorbance from the matrix and excipient. The wavelength selected for the UV detector should provide maximum detection sensitivity and signal-to-noise ratio on top of a wide linearity range. Compounds that do not have UV absorbance are the ultimate challenge in method development. The detector must be sensitive enough to detect low concentrations, which often require alternative ways like increasing injection volume, serial detectors, or sample derivatization to become a UV-absorbing molecule. For example, an isocratic RI detector can detect a non-UV absorbing molecule, but the detection mode needs to switch to electrochemical (conductivity or amperometry) if the impurities contain inorganic compounds. As gradient elution is not possible with an RI detector, a separate HPLC method must be developed for organic impurities that are different in nature from the API.

In combining the optimized chromatography parameters and the appropriate detector, the accuracy, precision, and robustness (including standard and sample stability) of the method must be tested to ensure that it can be defined as a stability-indicating method. (See section V and VI of chapter 3.)

C. Stability-Indicating Methods Development

1. Different Routes of Administration

In recent years, drug delivery technologies have grown extensively that go beyond conventional oral dosage forms. For example, intravenous, transdermal, and implant technologies are widely used for sustained drug delivery over increased periods of time. The type of drug, its characteristics, and the intended indication determine the mode of delivery.

In developing HPLC stability-indicating methods, sample preparation is the most important part of drug method development, regardless of the route of administration. Sample preparation of the planned dosage forms should be optimized based on the drug's characteristics. Solid and liquid dosage forms can readily be extracted or dissolved in appropriate solvents or media, whereas the sample preparation of transdermal patches needs special attention. For example, it is important to consider (1) the effect of the viscosity of polymers involved in controlled-release tablets, and (2) the effect of external excipient glue-like adhesives that come into contact with the drug and/or the semi-solid dosage form preparation. The stability of the product is dictated by the combined effect of the excipient (solid or liquid) mixture and API. Hence, a stability-indicating method must be developed for the excipient, the drug substance, and the drug product. The HPLC conditions can be generalized regardless of the type of dosage form—solid dosage, liquid dosage, or transdermal patch.

2. Single Drug Vs. Dual Drug System

Developing and validating a stability-indicating assay method becomes more challenging when multiple drugs are present in a drug product. Since developing and marketing new chemical entities (NCEs) for multiple indications is a difficult task, pharmaceutical companies are looking into creating products by combining two or more known, compatible APIs to treat multiple diseases and achieve better patient compliance. Method development for two or more compounds and their related impurities becomes very complex if the solubility and the pK_a values vary greatly and the UV profiles are not similar. It generally poses more issues in developing sample preparation to fully extract the drugs and in optimal HPLC methods for analysis. If the solubility and the pK_a values of the APIs involved are similar, where all the active components are totally soluble in water, the method development is much easier.

In addition, the API's dosage strength also plays an important role in developing a stability-indicating method. The value of forced degradation studies conducted in the early stage of development is critical in identifying the related impurities and degradants and their origins. In the case of two APIs with disparate, new impurities seen in stability samples, the samples must be evaluated carefully and the impurities reported to their corresponding origins. Incorrect identification of the origin source and quantitation against the inappropriate API can cause stability issues (e.g., impurities calculated against the wrong dosage API could have over- or underestimated the actual level, and thus could pose regulatory issues as a result of wrong trending and tracking). The confirmed presence of a new, unknown peak in the sample should initiate further investigation as to the source and identification of the structure. The unknown peak should be trended at specified time points and the structure identification and characterization should be performed and quantitated against the correct API.

3. Active Drug Assay and Impurities Method

For routine analysis in a stability program, a stability-indicating method is required for analyzing both the API and impurities. Stability-indicating analysis for an API is crucial since it measures the potency of the drug at an initial time point and the loss of potency during storage. The evaluation of impurities with a good stability-indicating method is also important in measuring the impurities and degradants, which could have toxicological effects when administered to humans. Forced degradation studies in which the product is artificially exposed to high stress conditions can be supportive in developing a stability-indicating assay method.

A well-defined, precise, and validated method will help to determine the drug content accurately, whereas an assay method capable of detecting at low levels can help calculate drug losses during the manufacturing process. Different vendors can supply common APIs. The API characterization accompanied by information from the manufacturer on the synthetic route determines the impurities profile and the method used for the active assay.

Isocratic elution uses the same mobile phase composition throughout the chromatographic analysis and is the preferred choice for API assay analysis. Isocratic separation is relatively simple and rugged, and produces a stable baseline. However, it may not be appropriate for impurities analysis if the API produces impurities/degradants that have widely divergent affinities for the selected column. In this case, gradient elution is required, where the composition of the mobile phase changes from a weak to a strong one throughout the run. Although gradient separation does not produce as good a baseline as isocratic elution, and is not compatible with all detectors, it shortens analysis time and accommodates a greater variety of compounds in a mixture. Therefore, gradient elution during initial method development significantly saves time and provides an informative background against which to ultimately develop an isocratic method, if possible.[7,11]

Other than selecting the column and mobile phase for the correct mode of separation, optimizing different HPLC parameters (injection volume, run time, wavelength, and detector) is equally important for achieving acceptable capacity factor (k'), resolution (R), and tailing factor (T).

In developing a commercially viable method, the stability of samples, standards, and reagents used for the HPLC method must be considered. For the stability of standard solutions and reagents, long-term stability of up to weeks is desirable. For the stability of sample solutions, a minimum of 3 days is ideal. Generally, the reagents for standard and sample preparation should be the same or very similar to the mobile phase composition.

4. Dissolution Method

Any drug that enters the body must first be disintegrated into small aggregates and eventually dissolved. The dissolution measurement can then be correlated with the biological performance of the drug. Dissolution testing is a measurement of drug solubility over time. The purpose of the dissolution test is an in vitro evaluation to compare the in vivo performance of the solid dosage formulation. The commonly used apparatus of dissolution tests are USP Type I (basket method) and Type II (paddle method) in vessel, which is more suitable for immediate release tablets, where sampling typically occurs at 5–15-min-intervals for a total of 1–2 h. USP Type VII allows samples to be released in calibrated tubes of 25–75mL arranged in designated rows (where total row numbers = total time intervals) containing selected medium. The tablets are typically released every 2–3 h for 24 or more hours based on the design of the delivery system; Type VII is primarily designed for controlled-release tablets.[1]

The medium selected for the dissolution test must consider the drug solubility. Aqueous media with a typical pH range between 1 and 7 to mimic the human gastrointestinal tract are preferred over organic solvents. The operating parameters of the dissolution setting should be optimized to ensure complete dissolution.[7]

To develop an HPLC stability-indicating method for Type I or II dissolution, the linearity must be wide enough, in combination with good sensitivity and minimal interference, to accommodate concentrations from low (possibly LOQ) to very high end, as the samples drawn represent the cumulative drug amount dissolved over time. As for an HPLC method that is designed for Type VII dissolution, the linearity should accommodate the lower concentrations since it is a drug measurement of a controlled-release system.

5. Testing of Preservatives in Pharmaceutical Products

A preservative is a substance that prevents or inhibits microbial growth and extends the shelf life of the drug products. In most pharmaceutical drug products, only a few compounds are typically selected as preservatives. For efficiency, a generic method should be developed for the types of preservatives that are more commonly used. For example, butylated hydroxytoluene (BHT) is an antioxidant commonly used in many solid dosage formulations to retard oxidative degradation of the excipients.

Hence, a generic or universal HPLC method interfaced with a UV detector for BHT can be used on any drug as long as the acceptance criteria on accuracy, precision, robustness, and other necessary requirements have been met. Similar to appearance, drug release, assay, and impurities, preservative testing is also required if a certain degree of preservative has to be included in the drug product to ensure an adequate shelf life.

For a liquid or semi-solid pharmaceutical dosage form, it is crucial to include a preservative in the formulation. Commonly used preservatives in these systems include sodium benzoate, EDTA, sorbic acid, and parabens.[13] A generic HPLC method is also recommended for the preservatives used in liquid formulations for routine monitoring to ensure the stability of the preservative itself and it must be validated specific to its use with the dosage form. (See chapters on Sample Preparation and Method Development.)

D. Case Studies

Case study 1: Stability-indicating method development for dual drug system

An oral dosage form was developed that consisted of two APIs, Drug A and Drug B, in a respective dosage ratio of ~20 to 1 in the system. Drug A is relatively hydrophobic and Drug B is hydrophilic. The pK_a value of Drug A is known, but the pK_a value of Drug B is unknown. The two drugs had similar UV profiles in the low UV region, and Drug A had a slightly different profile at the mid-UV region. Combined standards were prepared in a mixture of organic and aqueous media to enhance Drug B's solubility. An HPLC column was selected and the mobile phase was optimized to separate both compounds with a run time of approximately 5 min for optimal routine dissolution samples analysis. The separation was acceptable and resolution was good for both drugs, but sensitivity was an issue. Drug A with high dosage along with very high absorptive nature presented a peak with high area count and the peak height was off scale at the selected wavelength, while Drug B with low concentration and low absorption exhibited poor signals (Figures 1a and b). To accommodate the wide range required to encompass the dissolution sample concentration, various approaches were attempted. One approach was to collect data from the detector in two channels (two different wavelengths) where a slightly weaker signal was chosen for Drug A and a maximum absorptive wavelength was selected for Drug B. The problem was solved only if a photodiode array (PDA) detector was used. If a regular dual wavelength detector was chosen, the baseline became wavy and the peak shapes were distorted due to a different detection mechanism (Figure 2). Since the detectors available with a dual channel are programmable, a wavelength-switching technique was explored using a single channel at a given time and this technique produced promising results. The wavelength

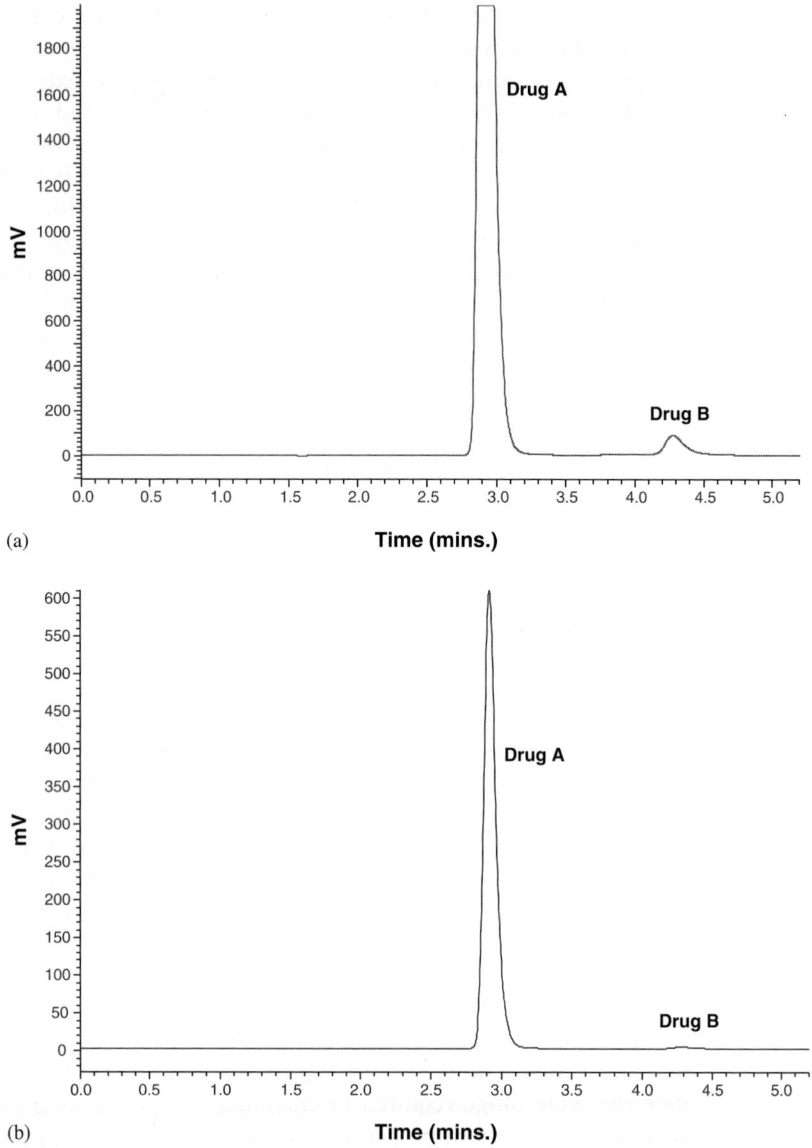

FIGURE 1 **(a) HPLC chromatogram of Drug A and Drug B at single channel (single wavelength: λ_1). (b) HPLC chromatogram of Drug A and Drug B at single channel (single wavelength: λ_2).**

was programmed for the first eluted peak and it was switched immediately afterwards to a second wavelength that was compatible with the second peak (Figure 3). Resolution, capacity factor (k'), tailing factor (T),

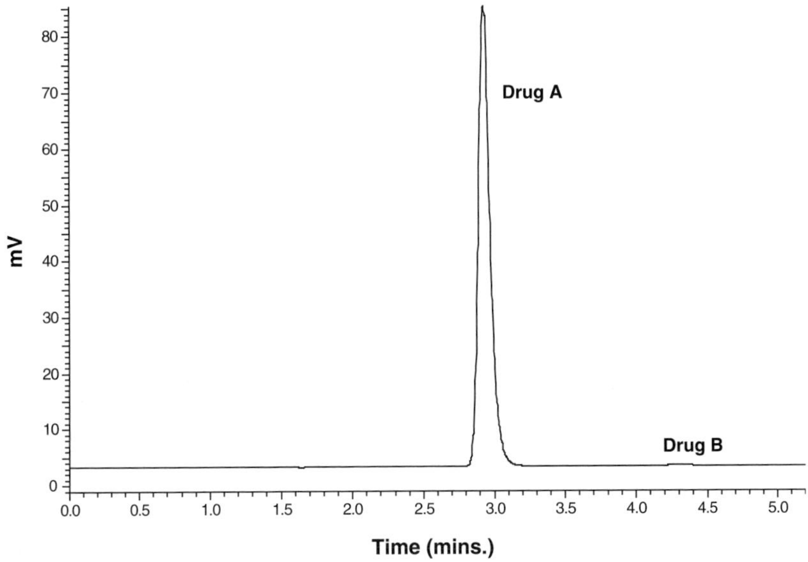

FIGURE 2 HPLC chromatogram of Drug A and Drug B at dual channel (dual wavelength: λ_2 only).

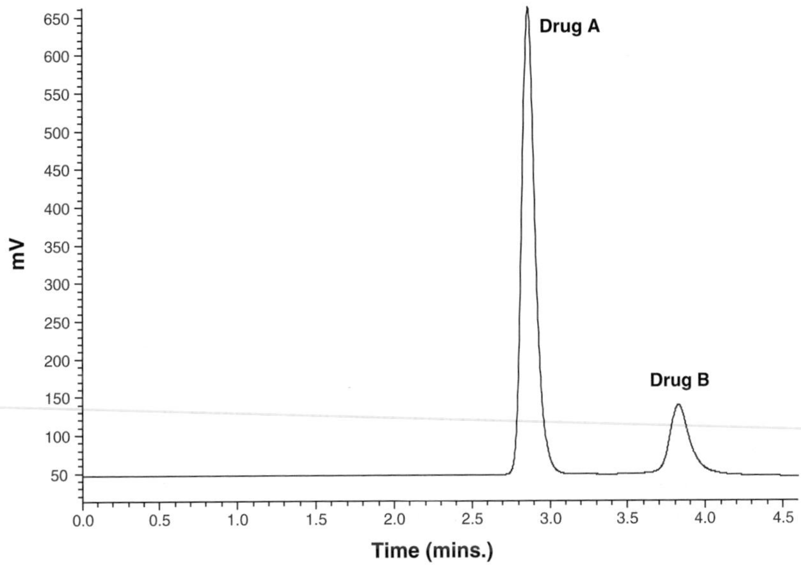

FIGURE 3 HPLC chromatogram of Drug A and Drug B at single channel (dual wavelength with wavelength switch: $\lambda_2 \rightarrow \lambda_1$).

and other suitability parameters were optimized with this new approach. Linearity was developed for the range required for both drugs, and accuracy and repeatability were established. The concern of method development for the dual drug was solved for dissolution and drug content method. Hence, the method was applied to stability studies and to routine formulation development.

The real challenge appeared when a method was to be developed for impurities in the drug product. To add to the complexity, both APIs under study listed at least four individual impurities for each compound. Given the nature of both APIs, detecting all the impurities by an isocratic method was not feasible. A two-step gradient was developed to successfully separate all known impurities with good resolution using dual channel detection (no wavelength switching). The method was successfully validated to Phase-I requirement. Once the stability samples were under testing, new unknown peaks were found in the sample chromatograms. The evaluation as to which API these unknowns belonged to was to be done. As described earlier, as the UV profiles for both drugs are similar in the low UV region and the region was chosen to accommodate the low dosage due to its low sensitivity, all the unknown peaks were also detected at this wavelength. Since the newly discovered impurities were present at this wavelength, they were quantitated against the low dosage form, and some values became alarmingly higher than the ICH threshold. After confirmation of the presence of these peaks at the next stability time point, structural identification and characterization were initiated. The results confirmed that a few unknown peaks seen in the Drug B channel were actually related to Drug A. When quantitated against Drug A, the levels of these impurities were negligible. As the occurrence of new unknown peaks will continue further at future time points, the process of identification and characterization should be an ongoing process for impurities method development and throughout the stability study.

Case study 2: Stability-indicating method development for surfactants

This is a methodology adopted for separation and quantification of two surfactants with a slight difference in molecular weights. A liquid formulation (cleaning solution) was developed incorporating two forms of surfactant with different molecular weights. Size exclusion chromatographic technique did not provide successful results. A RP chromatography was developed for the separation. One molecular form of the surfactant was at a higher concentration than the other. The second surfactant did not provide enough sensitivity with the method. Hence, on-column concentration was attempted. For the test method, two mobile phases with the same composition, but one with higher organic composition and the other with lower organic composition, were prepared. A low-pressure, six-valve, column-switching unit was incorporated into the HPLC system after the autosampler and before the detector. A shorter column (5-cm) C18 column was attached to one of the ports in the column-switching valve. This column was used for concentrating sample in the column. A 15-cm C18 column was attached at another port of the valve, which was the outlet to the detector in the second position (position B) of the valve system. Initially the weaker mobile phase was allowed to run through the system until the HPLC system equilibrated. During this time the switching valve was in the first position (position A), and the mobile phase flowed through the 5-cm column, while the effluent went to waste as set by the valve position. The sample was injected repeatedly and concentrated in the 5-cm column as the weaker mobile phase could not partition the compound from the column. After the accumulation of a set sample size, the valve position was switched to position B, and the pump was programmed to run the stronger mobile phase through the 5-cm column and onto to the 15-cm column for good separation and detection of both compounds. The method was reproducible; accuracy and precision were developed and used for monitoring of the indicated stability studies. The column-switching technique is not a commonly used approach since the reproducibility in a commercial laboratory is difficult. This approach is used as a last resort for difficult applications as described above.

IV. SUMMARY AND CONCLUSION

Stability testing is an integral part of pharmaceutical product development and is an ongoing activity throughout the entire drug development process. Product integrity and shelf life are based on stability testing. The pharmaceutical industry is challenged by the frequently changing regulatory requirements. The need for improvement in analytical techniques poses challenges and opportunities in pharmaceutical stability testing. A wealth of specific information is available in the FDA stability guidance and in ICH stability guidelines. HPLC and other analytical techniques play important roles in stability testing. The precision, ease of use, and ruggedness of HPLC methods are by far preferred over other separation and quantitation techniques. Making continuous efforts to learn updated technology and to remain informed of constantly changing regulations is the way to succeed in the pharmaceutical industry.

ACKNOWLEDGMENTS

The authors acknowledge valuable suggestions from Dr. M. W. Dong from Purdue Pharma L.P.

REFERENCES

1. USP 26-NF 21, 2003.
2. Carstensen, J. T. and Rhodes, C. T., *Drug Stability: Principles and Practices*, 3rd ed., Marcel Dekker, Inc., New York, 2000.
3. ICH Harmonized Tripartite Guideline—Stability Data Package for Registration Applications in Climatic Zones III and IV Q1F, 2003.
4. Higgins, III, J. D. and Rocco, W. L., Pharma Pre-formulation, *Today's Chemist Work*, 12(7):22–26, 2003.
5. ICH Harmonized Tripartite Guideline—Stability Testing of New Drug Substances and Products Q1A (R2), 2003.
6. ICH Harmonized Tripartite Guideline—Stability Testing: Photostability Testing of New Drug Substances and Products Q1B, 1996.
7. Miller, J. M. and Crowther, J. B., *Analytical Chemistry in a GMP Environment: A Practical Guide*, Jossey-Bass, New York, 2000.
8. ICH Harmonized Tripartite Guideline—Bracketing and Matrixing Designs for Stability Testing of New Drug Substance and Products Q1D, 2002.
9. ICH Harmonized Tripartite Guideline—Evaluation for Stability Data Q1E, 2003.
10. ICH Harmonized Tripartite Guideline—Specifications: Test Procedures and Acceptance Criteria for New Drug Substances and New Drug Products: Chemical Substances Q6A, 1999.
11. Harris, D. C., *Quantitative Chemical Analysis*, 3rd ed., W. H. Freeman and Company, California, 1991.
12. Snyder, L. R., Kirkland, J. J. and Glajch, J. L., *Practical HPLC Method Development*, 2nd ed., John Wiley & Sons, Inc., New York, 1997.
13. Gennaro, A. R., *Remington's Pharmaceutical Sciences*, 18th ed., Mach Publishing Company, Pennsylvania, 1990.

INTERNET RESOURCES

http://www.fda.gov
http://www.ich.org
http://www.usp.org
http://www.phrma.org

14
IMPURITY EVALUATIONS

TODD CECIL AND ERIC SHEININ

U.S. Pharmacopeia, Rockville, MD

ABSTRACT

An overview of impurity evaluations of drug substances and products by HPLC is given in this chapter, from regulatory and laboratory viewpoints.

I. INTRODUCTION

Impurities in drug substances and drug products continue to be a source of great concern, discussion, debate,[1] and research.[2–4] These concerns and debates typically center on the potential safety risks associated with impurities due to contamination and the setting of acceptance criteria. However, the bulk of the work being performed in the pharmaceutical industry, with respect to impurities, is focused on the isolation, identification, qualification and quantification of impurities that are found as a result of the manufacturing process or through chemical decomposition. On the

Handbook of Pharmaceutical Analysis by HPLC,
S. Ahuja and M.W. Dong, editors.

surface it appears that there is a divergence between the perceived need for public safety and the industry focus. However, only through a thorough understanding of all known potential impurities (also known as the impurity profile) is it possible to find unknown impurities and contaminants. The most effective means to increase public safety is to create and maintain a public specification. In theory, a public specification defines the boundaries of safety and efficacy of a given drug substance and product, by ensuring the quality attributes of the drug substance and product on which safety and efficacy studies have been performed. The boundaries are set by the regulatory agency that is tasked with this mission (e.g., the United States Food and Drug Administration or the Ministry of Health, Labor and Welfare in Japan.) These boundaries are created by evaluating critical parameters of a drug substance/product with clinical data collected on human subjects. The batches of material that are tested in the clinic are often called the pivotal batches. Once the pivotal lot is manufactured it is evaluated by almost every imaginable means to ensure that the material is well characterized and understood. Through this testing a manufacturer may qualify impurities as nontoxic and can create goalposts to evaluate additional lots, for submission to the appropriate regulatory agency. The tests and procedures used to evaluate and characterize the pivotal batches and the acceptance criteria created to define the goalposts are the basis of a public specification. Public specifications such as those found in the *United States Pharmacopeia* (*USP*) or the *British Pharmacopoeia* (*BP*) provide a benchmark of expected impurities both toxic and nontoxic that are typically found in the manufacture of a drug substance. With this information and a standardized approach, an analyst can ensure that additional lots of a material are equivalent to the pivotal batches. More important in terms of public safety, these public specifications aid in the search for new impurities and contaminants. Therefore, it is vital that the pharmaceutical industry regularly updates these publications for the continued safety of the patient.

The topic of the control and reporting of impurities has been addressed by the International Conference on Harmonisation of Technical Requirements for the Registration of Pharmaceuticals for Human Use (ICH) in the Q3A(R) and Q3B(R) guideline documents.[5,6] These documents were created as guides for the submission of new drug applications to a regulatory agency. These documents include definitions and reporting thresholds that serve to create a common language for the discussion of, and limits for, impurities in the pharmaceutical industry.

ICH Q3A(R) has classified impurities as organic impurities, inorganic impurities, and residual solvents. This classification system provides a useful framework for the discussion of impurities and serves to create fixed and consistent definitions for impurities in pharmaceutical substances.

Organic impurities can arise from the starting materials, the process, the drug substance itself, alone or in the drug product, the interaction of the drug substance with other components or from the degradation products of the other components. They can typically be classified as starting materials, by-products, intermediates, degradation products, reagents, ligands, and catalysts. In essence, anything that is added to the reaction vessel, the reaction products, both intended and unintended, and any breakdown products of the drug define the broad boundaries of expected potential impurities.

Figure 1 is a simplified representation of the manufacturing process of a drug product. Each of the potential classes of impurities that might be present has been identified and numbered. This picture identifies sources and classes that are not specified in the ICH guidelines, but that are clearly potential sources of impurities and interferences.

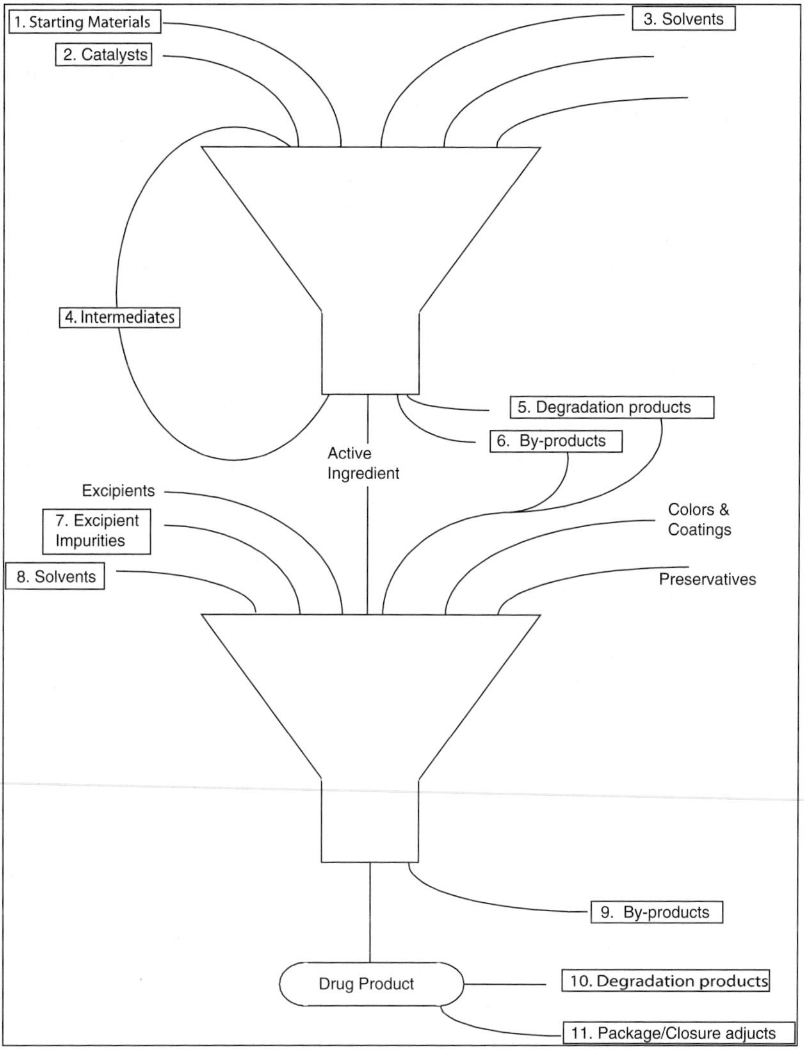

FIGURE 1 Simplified diagram of a pharmaceutical process.

Starting materials can be defined as the raw materials that form the basis of a chemical reaction as a part of the synthesis of an intermediate in the production of a drug substance. Catalysts typically include any material added to a mixture to accelerate, control, or otherwise modify a chemical reaction. Intermediates are those products of a synthesis scheme that will undergo further reaction. By-products are the side-products of a chemical reaction, and may include conjugates, dimers, enantiomers, unintended salts or free-bases, over-substitution, others. These types of impurities are usually considered to be process impurities and are not expected to increase in concentration over time.

Degradation products are those compounds produced through the chemical or thermal decomposition of the analyte. Container/closure adducts are generally the result of an interaction between a drug substance or intermediate with the container/closure material or extractables from these materials.

Starting materials, solvents, intermediates, degradation products, and by-products are often organic impurities found in drug substances and excipients, while excipient impurities, solvents, by-products, degradation products, and container/closure adducts may be found as organic impurities in drug products.

Organic impurities are often evaluated using HPLC analysis. The major part of the chapter will focus on the application of chromatographic methods to evaluate and limit these impurities.

Inorganic impurities are described in ICH Q3A(R) as reagents, ligands, and catalysts, heavy metals, inorganic salts, and other materials (e.g., filter aids, charcoal). Few of these impurities lend themselves to evaluation using liquid chromatography and as such will not be discussed further in this chapter.

Residual solvents are the third general classification of impurities in pharmaceuticals. This class is described as inorganic or organic liquids used during the manufacturing process. Typically, these solvents can only be evaluated by gas chromatography and therefore will not be addressed in this chapter.

A. Impurity Limits

In addition to the discussion of the types of impurities that are typically found in a pharmaceutical ingredient and product, the ICH Q3A guideline also describes how these impurities should be expressed. Impurities may be described as specified or unspecified.

Specified impurities are generally those that are found on stability or nontoxic impurities that cannot be effectively separated from the active ingredient. In these cases there will be specified acceptance criteria for the individual impurities. The acceptance criteria will be supported by the testing of materials produced by the commercial process. These impurities can be either identified or unidentified. In the context of the ICH documents, an identified impurity has an elucidated structure. Unidentified impurities may be reported using a generic description or by citing the relative retention time. The ICH also sets threshold limits for the reporting of unidentified impurities. If the impurity is below a specified level (e.g., 0.1%), then these unidentified impurities can be measured and limited as part of the total impurities limit.

Unspecified impurities are those that are typically found at levels below the reporting threshold. These are typically reported by the summation of all the impurity peaks.

However, the ICH guidelines do not include information about conducting or interpreting impurity studies, except to state that stability studies, chemical development studies, and routine batch analysis should be used to predict the impurities likely to occur in commercial production.

B. Impurity Profiles

An impurity profile is usually defined by the chromatographic procedures used to evaluate impurities in a drug substance. A typical profile includes chromatograms for all specified and unspecified impurities, as well as enantiomers, conjugates, and concomitant species. Once a profile has been developed, it creates a tool to help evaluate trends in the production of future lots of material. One of the principal uses of the impurity profile is to scan materials for new or unexpected peaks. Where these occur, regardless of the peak area, an investigation needs to be conducted. The source of the new impurity will need to be tracked to its source. These impurities may come from unexpected sources including reagents, column artifacts, carryover, and contamination. These profiles are also useful in evaluating the effects of changes to the process, or to identify trends in the data that can indicate issues with the starting materials or the processes.

II. DRUG SUBSTANCE EVALUATION

A. Identification of Impurities

The concurrent identification and quantification of organic impurities is a principal use of liquid chromatography in the pharmaceutical industry. However, the application of liquid chromatography to this task highlights a weakness of this technique when compared to gas chromatography: specifically, the lack of a universal detector. Great strides have been made to create detectors and hyphenated techniques to address these problems. However, multiple detectors and analytical procedures may be necessary to accurately and specifically identify and quantify the impurities in complex systems.

1. A Starting Place

The identification of potential impurities typically begins early in the life cycle of a drug substance candidate. However, the comprehensive evaluation of potential impurities is typically not the first application of HPLC to a new drug substance. Instead, HPLC procedures are first developed to give yield values for the synthetic steps. Liquid chromatography is next implemented in the development of sound, stable procedures for detecting, identifying, and quantifying the drug substance and metabolites at low levels in biological fluids. The very complex nature of these studies as well as the focused effort to quantify only the active ingredient often means that these procedures are inappropriate for the evaluation of impurities in the drug substance. However, these early assay procedures may present a starting point for the determination of impurities and for the development of a stability-indicating assay or set of procedures.

Where this early work does not exist, or proves to be inappropriate, the use of a procedure for a structurally-related species is a good starting place. Procedures for related species can be found in a number of sources including books of standard procedures,[7] compilations of literature methods,[8] the excellent Analytical Profiles series edited by Florey and Brittain[9] and current peer-reviewed journal articles.

Once a starting point has been identified, the analyst should evaluate the appropriateness of the procedure for impurities applications by first examining the purity of the analyte peak using a multidimensional detector such as a photodiode array or mass spectrometer. The time to complete these initial evaluations can be accelerated by working with samples that have not been purified to the level intended for clinical application (also known as dirty samples.) Any impurities that are found should be identified by relative retention time and peak area so that they can undergo additional characterization at a later date.

2. Developing the Impurity Profile

Once the impurities that are found in the drug substance have been identified, then the work of determining an impurity profile can proceed. Initially, the analyst can determine a list of expected impurities by completing a theoretical evaluation of the potential reaction and degradation pathways to the products likely to appear in a sample. Each of these potential impurities may be synthesized, purchased, or created in situ and then evaluated using the tentative analytical procedure. This process is illustrated in Figure 2. This figure includes potential intermediates to produce Drug JA-171, the potential by-products, and degradation products from a purely theoretical approach. Please note that the reactants and reaction conditions have been excluded for purposes of confidentiality. In this example, there are three specific routes for the creation of impurities. These routes include dimerization, hydrolysis to a degradation product, and reactions with the degradation product to form by-products. In this example, structure A represents the final intermediate leading to the desired product (B). However, if this reaction is not carefully monitored, hydrolysis to a degradation product (C) might occur. For the purpose of this chapter, suppose that the neutralization of the reaction mixture prior to the conversion to Compound C will limit the amount of impurities created, but that if an excess of alkali reagent is used in the neutralization, then dimerization will occur (D). Based on this theoretical evaluation the analyst knew to isolate and evaluate compounds A–D to determine their retention behavior. When a mixture of these compounds was evaluated (Figure 3), five peaks were found; however only three peaks appear in the schematic in Figure 2(A–C). The other two peaks were process impurities not included in the scheme for the sake of clarity. Compound D was found to elute much later in the chromatogram (about 60 min). It was not

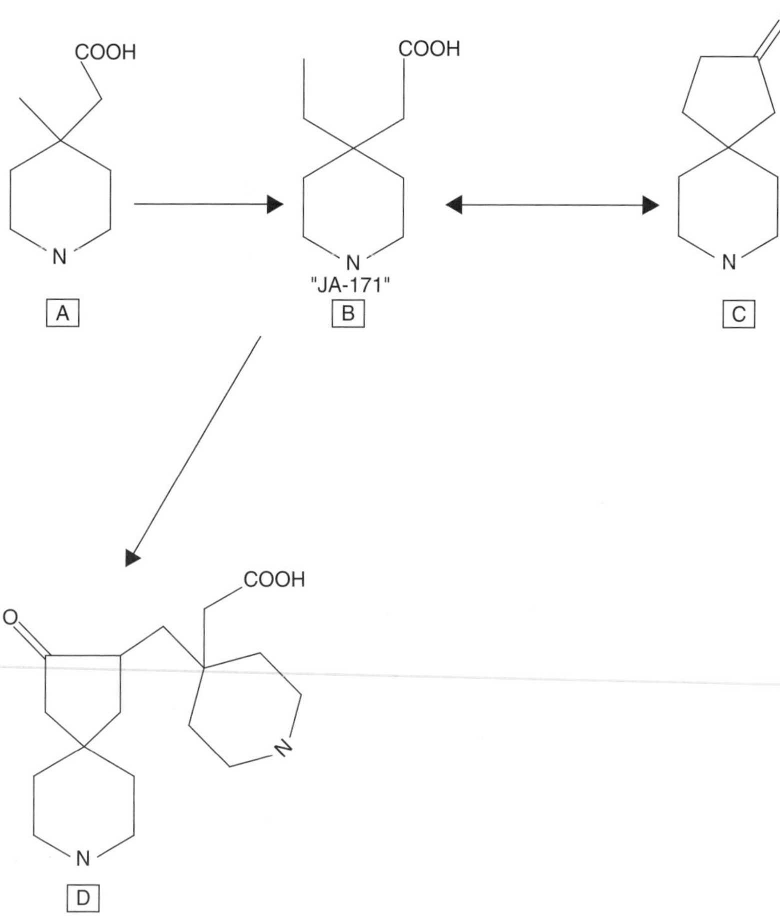

FIGURE 2 **Drug JA-171 synthesis and degradation pathways.**

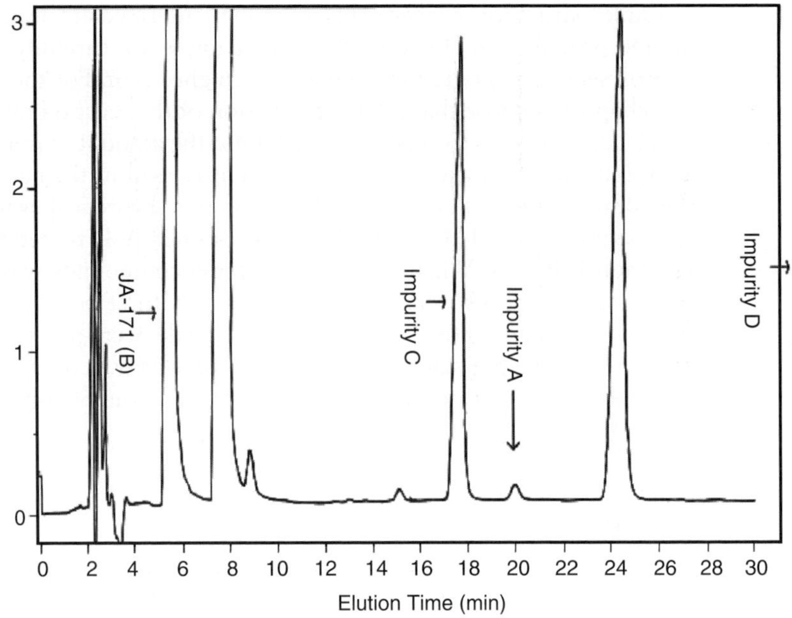

FIGURE 3 **Drug JA-171 impurity mixture—early elution.**

initially found due to a combination of peak broadening and the low molar absorptivity of this compound. These factors had combined to yield a peak below the limit of detection. To address this problem, a second chromatographic system was developed with a much more aggressive mobile phase to make the elution occur within an acceptable time frame. This second system was unable to separate the three other peaks from the solvent front. Therefore, the impurity profile for Drug JA-171 requires two different analytical procedures, (Figures 3 and 4).

Once the procedure(s) needed to evaluate the impurity profile has been developed, it should be challenged through a forced degradation study. This study involves the intentional degradation of a drug substance via several different degradation pathways.[10] The degradation pathways that are employed should be designed around the individual drug substance being evaluated. The most commonly used degradation conditions include acid- and/or base-degradation, light-exposure, heat-degradation, and oxidative degradation. The conditions should be adjusted to obtain about 30% degradation. This level of degradation generally provides a sufficient number and concentration of degradants without the creation of secondary degradants. These degraded samples are appropriately neutralized, and then analyzed using a tentative analytical procedure with a multidimensional detector. The results of a typical forced degradation study are shown in Figure 5. This figure is a portion of a specificity study

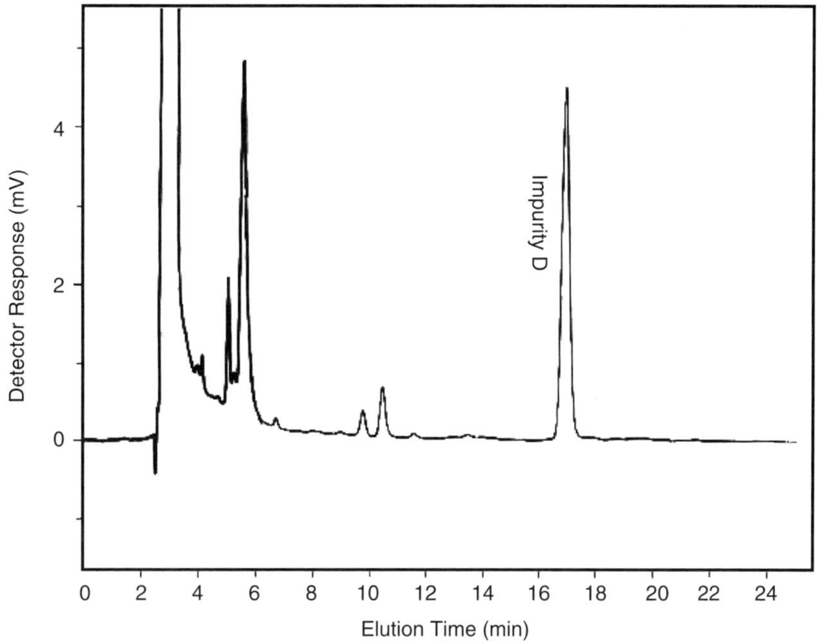

FIGURE 4 **Drug JA-171 impurity mixture—late elution.**

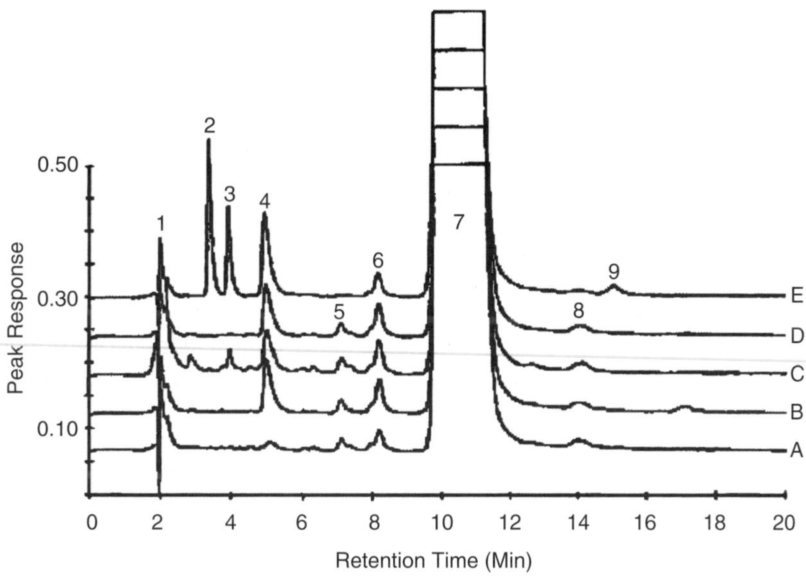

FIGURE 5 **Forced degradation chromatogram.**

using a diode array detector. The chromatograms represented here are compiled from separate data sets using a single wavelength. Chromatogram E is of a standard solution of expected degradation products, and Chromatogram A is the unstressed material. Chromatograms B–D show the results of time and temperature on the drug substance (peak 7). If there were co-elution of the main peak and one of the impurities it would typically be easily seen using the photodiode array detector. However, it should be kept in mind that a small peak is often difficult to find in proximity to a large peak. Therefore, very careful evaluation of the multidimensional chromatogram and the use of an orthogonal detector such as MS are employed at this point solely to avoid surprises later. If co-elution has been detected, the analytical procedure will need to be modified to achieve an acceptable separation of all of the peaks found in the sample. After modifications have been made, then each of the samples (both degradation and mixtures of theoretically expected impurities) will need to be re-evaluated with the modified procedure(s).

3. Selection

After completing this series of experiments and finally optimizing an LC separation, the determination of which impurities to monitor can begin. The primary purpose of this exercise is to determine which impurities are likely to be found in production-scale batches. This process begins with the evaluation of all of the degradation chromatograms to identify common peaks. Where common peaks are found, they should be added to the list of impurities to be characterized and potentially limited.

After completing the impurity profile, the determination of which impurities to monitor can begin. In this step, all of the samples that have undergone forced degradation as described above, are evaluated to identify common products; these are typically compared to accelerated stability samples. The conditions for conducting stability studies are described in the ICH guidelines Q1A(R2)[9] and Q1C.[11] The choice of accelerated conditions is the responsibility of the applicant. Generally speaking, the ICH accelerated conditions of temperature and humidity (40°C/75% RH) can be employed for these studies. The stability studies should give the analyst a clear understanding of the impurities that are most likely to be encountered under typical storage conditions. However, with the information provided by the forced degradation samples, the analyst should be able to detect impurities in the stability samples well below the reporting threshold. This will ensure that these impurities can be evaluated, and specified as necessary. Once this cross-checking between stability results and the impurity profile has been completed, the analyst should be able to recommend which peaks to monitor.

In the Drug JA-171 example in Figures 2–4, suppose Compounds A, C, and D are specified impurities and five additional unspecified impurities were found, but not characterized. Also suppose that during the development of Drug JA-171, it is discovered that the addition of a hydrogen scavenger to the final reaction produces an increased yield and eliminates the need to neutralize the reaction medium, thereby eliminating the production of Compound D. An evaluation of Drug JA-171 using the impurities procedures shows an additional peak. This peak can be identified either by going through the full characterization protocol described below, or the retention time of that peak can be compared to the peaks found in the forced degradation studies. If the retention time corresponds to a peak in one of these spectra, then structural confirmation by LC-mass spectroscopy (MS) is all that would be required to determine the identity of the new impurity. The latter is much more cost-effective and does not preclude the implementation of the former (full characterization).

Although this process is usually very expensive, slow, and labor-intensive, its benefits far outweigh the costs. These benefits include eased regulatory approval through clearly defined and supported release testing and acceptance criteria. Just as important is the ability to create a baseline relative to which changes to the production can be evaluated.

The Drug JA-171 synthesis and degradation diagram (Figure 2) can serve to illustrate this concept. Suppose that a process engineer discovers that the equilibrium of the inter-conversion between Drug JA-171 and its principle degradant can be shifted towards Drug JA-171 by moving to a lower pH range. However, the move to a lower pH gives rise to an increase in the polymerization rate and impurity D. Since these impurities have a larger molar absorptivity than Drug JA-171, a small change in pH is amplified in the chromatogram. The effects of the change are easily identified and can be evaluated when compared to the impurity profile. These peaks can be identified either by going through the full structural characterization protocol, or the retention time of that peak can be compared to the peaks found in the forced degradation studies. If the retention time corresponds to a peak in one of these chromatograms, then only structural confirmation by HPLC-MS would be required. The latter is much more cost-effective and does not preclude the implementation of the former.

4. Relative Response Factors

Although not prohibitive, the cost of implementing an HPLC procedure requiring a mass spectrometric detector in a QA/QC environment is not desirable. Therefore, the analytical procedure will often have to be modified from a gradient procedure with MS-friendly solvents and salts to an isocratic procedure with relatively short retention times using a variable wavelength UV detector. The calculation of relative response factors is most easily accomplished through the use of a photodiode array detector and the following equation:

$$\mathrm{RRf} = C_{\mathrm{i}}R_{\lambda\mathrm{i}}/C_{\mathrm{A}}R_{\lambda\mathrm{A}}$$

where $R_{\lambda\mathrm{i}}$ is the response at the wavelength of maximum absorbance of the impurity, $R_{\lambda\mathrm{A}}$ is the response of the impurity at the analytical wavelength, and C_i and C_A are the concentrations of the impurity and analyte, respectively, in the solution.

B. Characterization of Impurities

The characterization of impurities is an area for which significant guidance from the FDA and the ICH does not exist. Characterization can involve a number of analytical methodologies in addition to liquid chromatography. The number of impurities that are characterized and the degree to which they are characterized is dependent upon a number of factors. These factors include the stage of development of the drug substance, the amount of the impurity that is present, the source and prevalence of the impurity, the peak location with respect to the drug substance and other key peaks, and the perceived need.

Characterization can be defined for this context as the determination of the chemical structure of the impurities. This is a task that can involve research grade equipment and a significant investment of resources. Consequently, a full characterization of individual impurities is not always indicated at the early stages of development of a drug substance. As a drug substance moves towards economic viability the additional expense and effort become increasingly necessary. In the early stages of development it is usually sufficient to identify an impurity by LC-MS retention and molecular weight. However, as the development and understanding of the manufacture of the material progresses, so should the characterization of impurities. By the time the drug substance is ready for clinical trials, the impurity profile should be fixed and well characterized.

C. Quantification of Impurities

Quantification of impurities can be conducted through a number of different methods. The two most often used are the mass balance approach and the standardization approach. The preferred quantification method is standardization, due to its greater accuracy. In this approach, the analyte sample is compared to a sample having a known quantity of a known standard substance. This standard can be either internal or external, or direct or indirect. The mass balance approach adds all of the peaks found in a chromatogram and calculates a percentage of total detected area. This is a much quicker and less costly approach than the standardization approach, but is susceptible to systemic bias, on account of which its accuracy suffers.

1. Internal Standard

In the internal standard procedure, the analyst identifies a region of the chromatogram that is devoid of peaks (quiet region). The analyst then attempts to identify a compound of known purity, that is structurally related to the analyte, and which elutes in the quiet region of the chromatogram. The internal standard should have a relative response factor that is about the same as the analyte. Once this work has been completed and a suitable internal standard has been identified, the sample is prepared to include measured quantities of the internal standard and the drug substance. Due to the complexity of selecting an appropriate internal standard and the potential for obscure unknown impurities, the use of internal standard methods for impurity testing is declining. However, it is still used in specific instances where there is a concern about the reproducibility of the sample injection or manipulation. This concern stems from the lack of reproducibility of the injector, or when a sample needs to be manipulated prior to injection. In both of these cases the internal standard provides the analyst a means to assess the system performance of each analysis.

2. External Standards

The preferred method of analysis for impurity quantification is through the comparison to an external standard. External standard methods rely on a stable chromatographic system and the ability to obtain a reference standard for the impurity under test.

External standards also give the analyst the ability to conduct spiked sample evaluations to evaluate for any procedural biases.

The use of a primary standard gives the analyst the greatest confidence in the accuracy of a measurement. A primary standard has been evaluated for purity using at least two orthogonal methods in a collaborative study using various instruments and reagents. The development of a primary standard is a very costly and time-consuming procedure. Therefore, it is not generally a cost-effective proposition to create a primary standard for each impurity. Many of these standards are available as compendial standards from the USP. These standards are generally of very high purity and have undergone a significant characterization through a collaborative study. Compendial standards are not as expensive as the costs involved in developing and maintaining a primary standard at each site, and are developed for many specified impurities.

The development of in-house standards for impurities is often necessary due to the unavailability of a compendial impurity standard. The process of developing an in-house standard is the topic for another publication. However, there are aspects that should be addressed herein. These include the establishment and maintenance of an impurity standard. The establishment of an impurity standard is a difficult process. The method that has been developed to monitor the impurity level of the drug substance is often inappropriate or insufficient to monitor the purity of the standard. The process described above to develop a procedure to monitor impurities in the drug substance must be re-evaluated using a purified impurity material. Although the procedure is capable of separating the impurity from the drug substance, it has not been evaluated to determine if the system will exhibit acceptable chromatography for the in-house standard. There may be problems with solubility, peak symmetry, solution stability, and linearity that had not been encountered in the drug substance impurity procedure. Once a procedure has been created and validated, the in-house standard can be used in the evaluation of the drug substance. This process will need to be repeated for each specified impurity in the drug substance. Finally, the in-house standard will need to be evaluated periodically to determine its stability. The stability information is important to ensure the accuracy of the impurity evaluation of the drug substance. In cases where a compendial standard exists, its use is recommended.

Where the specific impurity is unavailable or is too costly, the use of composite or degraded samples is possible. This approach involves the use of a "dirty" sample of a drug substance or the creation of a mixture of impurities through the in situ forced degradation method. Both of these approaches are best used for qualitative uses. In each of these mixtures, the impurities are present in unknown quantities. The real benefit of this type of impurity standard is the low cost and the ability to unequivocally identify the peak loci of the impurities. When these mixtures are used in conjunction with a compendial standard and a well-developed set of relative response factors the results will meet most analytical needs.

A mixture of impurities can also be created by combining measured amounts of each impurity. This mixture can be created using well-characterized, but lower purity, materials. Contrary to the statements in the previous paragraph, synthetic mixtures can be used for quantitative uses if properly packaged. Often these mixtures are created in a liquid form, and then dried to create a more stable mixture. When this material is resolvated for use it is vital to ensure complete solvation before use. Since this is not always possible, quantitative use of dried mixtures is not always the ideal approach.

3. Mass Balance and Area Normalization

In the mass balance approach, all impurities are quantified and subtracted from the absolute value of 100%. This approach will result in a purity value that, if all impurities are accounted for, is more accurate than the external or internal standard methods. However, the ability to identify all impurities in a given drug substance may require the use of hyphenated detection techniques and could be extremely costly to complete on a regular basis. Therefore, a related approach, called Area Normalization, is often used where the majority of the impurities can be identified and quantified in a single chromatogram. In the simplest case, all of the impurities would be assumed to have the same relative response factor as the parent drug. The quantification of the individual impurities would be reported as a percentage of the parent drug rather than an absolute value in milligrams. Where the assumption of common relative response factors is not correct, then the analyst will need to calculate the amount of each impurity weighted by the appropriate correction factor.

D. Setting Goalposts

As described at the very beginning of this chapter, there are many sources of impurities. The majority of these are carried over from the production of the drug substance. Others are due to the instability of the drug substance. The analysis of the impurities can give the pharmaceutical scientist information about potential problems in the process and the state of the drug substance. Information can be elucidated by setting different goalposts for differing needs. These goalposts include alert limits, action limits, release limits, and shelf-life limits.

Limits placed upon impurities that are carried along with the production of an API reflect upon the control of the manufacturing process. To ensure that control has been maintained during the process, a manufacturer may set alert limits for specified starting materials found in the drug substance. Alert limits are generally significantly lower than those required by an approved application. These limits are usually so low that they only indicate that the process should be closely monitored.

If alert limits are in place, a manufacturer will typically also have action limits. These are limits that indicate that the manufacturer needs to intercede in some way to bring the process back under control before a lot fails the release specification. These limits are also usually lower than the acceptance criteria required by an approved application.

Release limits (acceptance criteria) are usually those negotiated between the FDA and the manufacturer of a drug substance. Additional information is available in the ICH quality guidelines.

Shelf life or compendial limits are intended to account for all impurities found from the starting materials to the degradation products. These limits determine the societal boundaries of safety and efficacy.

E. System Suitability

A firm understanding of the chromatographic system is expected of the analyst who developed an analytical procedure. However, when this procedure is transferred to another analyst, problems may occur. The problems may stem from differences in the chromatographic system, column variance, temperature fluctuations, mobile phase variability, and other factors. The standard means to ensure that the procedure transfers (technology transfer) successfully is through the use of system suitability parameters. System suitability parameters are performance measures that are typically developed during development and validation of the procedure. The most accurate parameters are usually obtained during the robustness testing. These studies involve the deliberate alteration of procedure parameters to access the effect on the accuracy and precision. From these studies it is possible to identify the critical parameters that must be controlled and those that can be altered.

System suitability parameters are often expressed in terms of resolution between peaks, k′, peak symmetry, and reproducibility. One of the best empirical means to evaluate the system's suitability is the resolution between two closely eluting peaks. However, caution should be exercised when the resolution between peaks of significantly different sizes is used to evaluate system suitability. Caution is needed because the resolution calculation is derived assuming symmetrical peaks of approximately the same size. Generally, the equation is robust enough so that asymmetrical peaks typically found in liquid chromatography do not significantly alter the results. However, differences in peak heights are more problematic. Therefore, resolution should be evaluated using a special solution having at least the two closest eluting components at concentrations differing by less than 25%. The qualitative impurity mixture described previously may also be used. In addition, analysts will often evaluate a standard solution having a known concentration at the expected impurity level to ensure that the transferred system has maintained the limit of detection necessary for the evaluation.

For example, the chromatographic system used to obtain the chromatogram in Figure 6 was:

Column: Lichrospher RP select B
Mobile Phase A (mpA): methanol: KH_2PO_4 buffer (30:70)
Mobile Phase B (mpB): methanol: KH_2PO_4 buffer (70:30)
Column temperature: Ambient
Flow rate: 1.0 mL per minute
Detection: UV at 272 nm
Injection volume: 10 μL
Sample concentration: 0.2 mg/mL each of pentoxifylline and caffeine in diluent
Gradient program:

Time	% mpA	% mpB
0–6 min	85	15
6–13 linear to	10	90
13–30	10	90

Figure 6 shows the effects of altering the mobile phase compositions of each of the gradient components. Although this study shows that the mobile phase modifications have little effect on the retention of the major components, careful inspection will show that there is an impurity

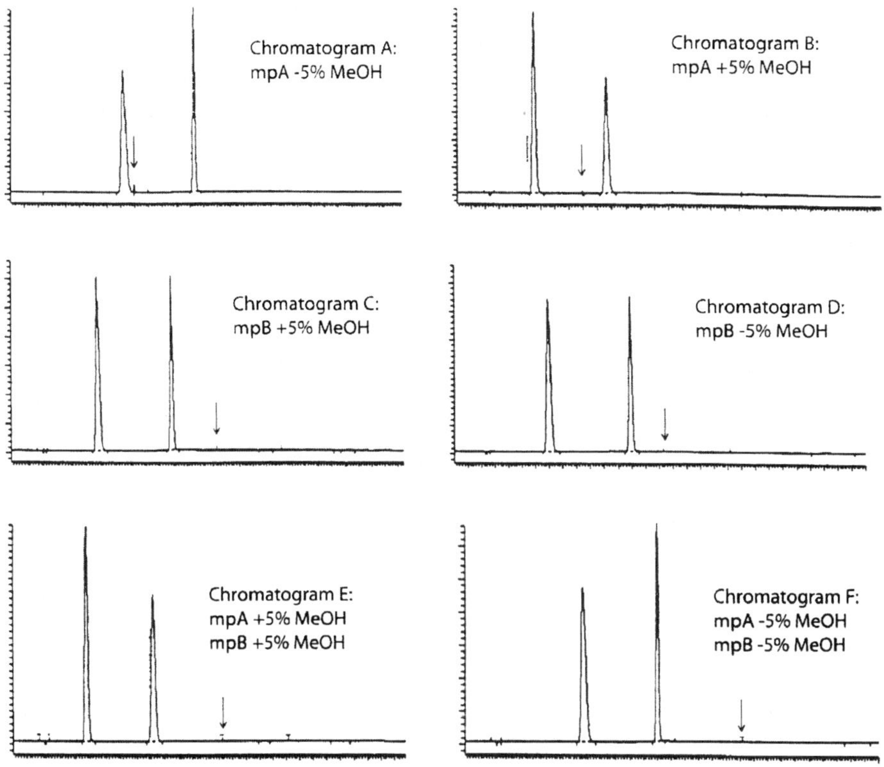

FIGURE 6 **System suitability study.**

that elutes very close to the tail of the caffeine peak, marked with an arrow, in Chromatogram A. Therefore, the system suitability requirement may need to include a limit of resolution between these peaks.

III. IMPURITIES IN DOSAGE FORMS

Whenever energy is added to a mixture of chemicals the potential for reaction increases. Various formulating processes can create the potential for degradation, conjugation, adduct formation, dimerization, polymerization, changes in polymorphic form, or even racemization. Obviously, the energy would have to be greater than the activation energy of the modification, which in most cases is not accomplished in normal production. However, there are areas of increased stress in the process that can lead to impurities. These include high shear areas in the blending process, grinding, granulation, drying, and tableting. Dosage forms other than tablets also have areas of increased stress that can lead to impurity formation.

The generic representation in Figure 1 illustrates the various types of impurities that may arise during the production of a dosage form. It is not all inclusive, as each dosage form has unique sources of impurities, but it includes most of the important ones. The sources of impurities increase with the increase in the number of components and the number of steps in the process. Each drug substance and excipient has its own impurity profile and the potential for interactions and reactions.

A. Formulation-Specific Impurities

Several dosage forms carry an increased risk of degradation or adjunct formation. Products such as injections and aerosols are more likely to interact with volatiles or extractables from packaging and closure systems. Tablets have the potential to form adjuncts with excipients (specifically, lactose has been shown to form adjuncts in tablets). Non-CFC propellants in aerosols have a large number of impurities that typically do not interact with drug substances, but the potential for these interactions does still exist. Creams, ointments, lotions, and other such products will each have specific interactions that should be considered while evaluating the impurity profile of a drug product.

B. Degradation Products

The procedures for measurement and reporting thresholds for drug products are covered in the ICH Q3B guideline. Much of the material presented earlier in this chapter is applicable. One area of difference is the applicability of the ICH guidelines for stability studies. The products of accelerated stability studies represent good sources of samples for the determination of specified degradation products. The ICH guidelines also indicate that only the degradation products need to be monitored in dosage forms. This concept is also espoused by the USP and basically states that since all impurities that are found in the drug substance have been appropriately limited, only those that may increase need to be monitored.

C. Impurity Profiles for Drug Products

The development of an impurity profile for a dosage form follows many of the same principles as in the drug substances; however, instead of identifying all precursors and intermediates, the analyst will identify and place all of the components of the dosage form (excipients, preservatives, and others) and their affiliated impurity profiles within a master profile. Here again additional method development may be needed to separate all of the potential peaks. As in the drug substance, the uses of the profile go beyond quantification to the ability to evaluate changes in incoming materials and the effects on product quality.

IV. SUMMARY AND CONCLUSION

Impurity testing is pivotal in pharmaceutical development to establish drug safety and quality. In this chapter, an overview of impurity evaluations of drug substances and products by HPLC is presented from both the laboratory and regulatory standpoints. Concepts from the development of impurity profiles to the final establishment of public specifications are described. Useful strategies in the identification and quantification of impurities and degradants are summarized with practical examples to illustrate impurity method development.

DISCLAIMER

The views expressed in this chapter are solely those of the authors and do not represent the USP position.

REFERENCES

1. House of Representatives Record of the Subcommittee on Oversight and Investigations—June 7, 2001., energycommerce.house.gov/107/hearings/060/ The_Honorable_Billy_Tauzinprint.html.
2. Geldar, S., *Am. Lab.*, 32(2):32–36, 2000.
3. Evans, E. H., Wolff, J. C. and Eckers, C., *Anal. Chem.*, 73:4722–4728, 2001.
4. Eckers, C., Haskins, N. and Langridge, J., *Rapid Commun. Mass Spectr.*, 11: 1916–1922, 1997.
5. Impurities in New Drug Substances (revised guideline), International Council on Harmonisation, http://www.ich.org/UrlGrpServer.jser?@_ID=363#Q3A(R).
6. Impurities in New Drug Products (revised guideline), International Council on Harmonisation, http://www.ich.org/UrlGrpServer.jser?@_ID=363#Q3B(R).
7. USP 27-NF 22, United States Pharmacopeial Convention, Inc., Rockville, MD, 2004.
8. Lunn, G. and Schmuff, N. R., *HPLC Methods for Pharmaceutical Analysis*, John Wiley & Sons, Inc., New York, NY, 1997.
9. Brittain, H., Ed., *Analytical Profiles of Drug Substances and Excipients*, Vols. 22–27, Academic Press, Inc., San Diego, CA (Vol. 22 in 1993; Vol. 27 in 2001).
10. Stability Testing of New Drugs and Products (Revised guideline), International Council on Harmonisation, http://www.ich.org/UrlGrpServer.jser?@_ID=363#Q1A(R2).
11. Stability Testing for New Dosage Forms, International Conference on Harmonisation, Federal Register 62(90):25634–25635(9 May 1997) http://www.ich.org/UrlGrpServer.jser?@_ID=363#Q1C.

15
HPLC IN DISSOLUTION TESTING

QINGXI WANG[a] AND VIVIAN GRAY[b]

[a]*Pharmaceutical Research & Development, Merck Research Laboratories, West Point, PA*

[b]*V. A. Gray Consulting Inc., Hockessin, DE*

ABSTRACT

This chapter discusses HPLC methodologies used to analyze dissolution samples, the automation of methods, and improved throughput and productivity.

Handbook of Pharmaceutical Analysis by HPLC,
S. Ahuja and M.W. Dong, editors.

I. INTRODUCTION

A. USP Dissolution Testing

Dissolution is one of the critical testing techniques used by the pharmaceutical industry to design formulations and control quality. It is a required performance test for solid dosage forms, transdermal patches, and suspensions, and also the only test that measures the rate of drug release, which can either reflect reproducibility of a product manufacturing process, or in some cases of in vivo drug release. Standard procedures for performing drug dissolution tests constitute an important part of the pharmacopeia of many countries. Dissolution testing is required in most of the drug product monographs published in USP. Each monograph describes the dissolution test procedure, and specifies the volume, medium, timepoints, apparatus, and analytical method. The dissolution test measures the rate of drug release from the drug product matrix in a designated medium. That matrix may include excipients, coatings, and capsule shells. The text, as stated in USP General Chapter Dissolution <711>, describes a vessel containing an aqueous fluid, heated to $37 \pm 0.5°C$ (98°F, body temperature) usually with a water bath.[1] At the beginning of the test, the drug product is dropped into the defined dissolution medium, which is subsequently stirred. The stirring element is typically a paddle or basket that rotates at a specified speed. Sample aliquots are withdrawn at specified time points over a period of time, up to 24 h in the case of some extended-release products. The aliquots are filtered and then analyzed, typically using an ultraviolet light (UV) spectrophotometer or high-pressure liquid chromatography (HPLC). The measurements are compared to a standard preparation and the results are expressed as a percentage dissolved of the label claim (strength) of the product.

B. USP Dissolution Test Modification

Usually, when using HPLC, the standards and samples are prepared in a dissolution medium. However, to resolve compatibility issues between the medium and the mobile phase, the standards and samples may sometimes be diluted with mobile phase or a component of the mobile phase.[2,3]

C. Current Trends in Dissolution—Implications for HPLC Methods

1. Gelatin Coatings

A recent addition to General Chapter Dissolution <711> allows for the addition of enzyme to the dissolution medium.[1] This is justified when gelatin-coated tablets or gelatin-encapsulated products exhibit a slow-down in the dissolution rate attributed to the crosslinking of the gelatin. The addition of enzyme will break up the crosslinking and therefore remove the barrier to the dissolution of the drug. The ramifications of enzyme addition on the analytical method will be discussed later in this chapter.

2. Poorly Soluble Drugs

Increasingly, drugs with low solubility in aqueous media are being developed. Therefore, in dissolution, the concept of "sink conditions" must be considered. Sink conditions are defined in the USP General Chapter <1088> In Vitro and In Vivo Evaluation of Dosage Forms as having a quantity of dissolution medium not less than three times that required for a saturated solution of the drug.[4] The use of surfactants—sodium lauryl sulfate (SLS), polysorbate (Tween), or cetrimide for example—has become commonplace in dissolution testing.[5] Surfactants offer challenges to the analytical method, especially to HPLC, and special considerations of column coating, injector wash, and micelle breakup will be discussed.

3. Special Dosage Forms

More frequently, unique dosage forms that involve liposomes, nanoparticles, micro-emulsions, and other special technologies are being developed. These dosage forms offer challenges to the dissolution test and the analytical method. HPLC methodology is usually required to eliminate the analytical interferences. In addition, dissolution tests are now required for suspensions. Suspension dosage forms usually require HPLC methodology to eliminate interferences from preservatives in the formulation that are often encountered while using direct spectrophotometric methods.

4. Method Development

Along with General Chapter <1225> Validation of Compendial Methods,[6] the proposed USP General Chapter <1092> The Dissolution Test Procedure: Development and Validation, provides a guide to validation parameters unique to dissolution testing, such as automated vs. manual testing.[7] The proposed General Chapter describes in some detail method development for dissolution testing. The choice of media for a dissolution test is a critical component that usually gives the test its discriminatory power. The apparatus and speed also have a major impact on the dissolution rate.

(a) *Apparatus.* There are seven different types of compendial apparatus described in the USP. Apparatus 1 (basket) and Apparatus 2 (paddle) are the most commonly used apparatus in dissolution testing. Apparatus 3 (reciprocating cylinder) is especially useful for beaded products and also has utility for soft gelatin and extended-release products. The dosage unit can be moved to another row of vessels that could be of a different pH, making the apparatus a very good R&D tool for determining pH dependence of the product. The flow-through cell, Apparatus 4, has a number of special uses; the major advantage is that it gives continuous flow over the dosage unit, providing "sink conditions". A number of specially designed cells are available for novel dosage forms, for example, stents, powders, suppositories, and soft gelatin capsules. Apparatus 5, 6, and 7 are for transdermal patches, with Apparatus 7 also having some design variations that will accommodate extended-release products. The apparatus and the speed at which it is operated will have a large impact on the dissolution rate, and the selection of the apparatus and speed will go hand in hand with the media selection to generate meaningful dissolution data. If the test can distinguish between formulations that may have changed in some way either on stability, or through the manufacturing process or in raw materials, then the test is considered suitable.

(b) *Media selection*. When choosing the dissolution media for a product, there are a variety of choices from dilute acids to buffers; the pH range is typically from 1.2 to 6.8. Surfactants can also be used with poorly soluble drugs to increase the dissolution rate. A major factor during media selection is the solubility of the drug substance, but the formulation may also be such that the dissolution rate is enhanced or slowed down by excipients and/or the manufacturing process.

(c) *Method selection*. Dissolution method development is often initiated using the commonly used apparatus, typically Apparatus 1 at 100 rpm or Apparatus 2 at 50 rpm, using a medium with solubility that exceeds sink conditions. Dissolution data is generated on the product of interest and observations are made to see how the product dissolves in the vessel under the chosen conditions. Then, as the results are interpreted, changes to either the media, apparatus, or speed (or combinations) are made until an appropriate dissolution profile is obtained.

5. In Vitro and In Vivo Correlation (IVIVC)

In the effort to demonstrate an IVIVC, the dissolution method may be further refined to show that drug product lots (usually at least 3) of different in vivo performance will demonstrate distinct dissolution profile/rate differences. Guidance to obtaining IVIVC is presented in the USP General Chapter <1088> In Vitro and In Vivo Evaluation and also in the FDA guidance, Extended Release Oral Dosage Forms. An IVIVC is often developed for extended-release products where the formulation is designed especially to give a certain dissolution rate. However, immediate-release products that are poorly soluble, and for which the dissolution is the rate determining step, may also show an IVIVC.

6. Bioequivalence

In the pharmaceutical industry, there are numerous instances where the bioequivalence of a given product vs. some reference products must be demonstrated. A human biostudy is the most important test for bioequivalence and is often required for regulatory purposes. However, the dissolution test can be used as a supplemental tool to show bioequivalence. A statistical tool known as the F2 test compares the two dissolution profiles and is described in several FDA guidances (for example, "Waiver of in vivo bioavailability and bioequivalence studies for immediate-release solid oral dosage forms, based on biopharmaceutics classification system"). Using the F2 calculation, the similarity may be established and if so, bioequivalence is demonstrated in lieu of a human biostudy.

7. Pooled Sampling

In order to reduce the analytical test load, several monographs make reference to "pooled sampling" for dissolution sampling.[8] In this procedure, individual aliquots are taken from each dissolution vessel and these aliquots are pooled into a common sample. This sample is then injected and analyzed. The results are evaluated, using mainly HPLC methodology and the pooled sampling criteria in USP Dissolution General Chapter <711>. Some instrument manufacturers have automated this pooling procedure.

II. UTILITY OF HPLC ANALYSIS IN DISSOLUTION TESTING

In general, the analytical methods used for dissolution testing can be simply classified into two categories: spectrophotometric and chromatographic methods. Since dissolution samples may not require complex sample preparation or sample modification other than filtering (and occasionally, sample dilution), direct UV-VIS spectrophotometric determination of absorbance is still the universal detection method. It is simple, direct, and cost-effective. It is particularly attractive where a simple, single-component determination in immediate-release dissolution is desired. However, within the current environment of the pharmaceutical industry, where multi-component or complex formulations (e.g., controlled-release formulations) and extremely low-potency dosage forms are quite popular, chromatographic methods offer significant advantages.[9]

In comparison with spectrophotometric methods (e.g., UV), HPLC methodology often provides the following features: wide dynamic linear range, improved specificity via separation and increased sensitivity. These features may be used to solve a variety of analytical problems encountered in modern dissolution systems.

A. Dynamic Linear Range

In contrast with UV methods where the linear range is approximately 1–2 orders of magnitude at best, the HPLC method with UV detection typically has a linear range of 3–4 orders of magnitude due to the narrow path length of the detector flow cell. For example, a simple

HPLC method was applied to analyze the dissolution samples of levothyroxine sodium tablets dosage forms of various brands and dosage strengths.[10,11] Levothyroxine present at concentrations ranging from 27 to 333 ng/mL can be detected and accurately quantified by HPLC with UV monitoring at 225 nm. In the early phases of drug development, developmental dosage forms with wide ranges of potency are often used in clinical studies. The dissolution testing of such a wide dose range using direct and derivative UV methods has proven very challenging and often impossible, due to the inadequate sensitivity and narrow dynamic linear range. However, the dissolution testing of such a dose range can be accomplished by HPLC even without any sample treatments.

B. Sensitivity

The potency of a pharmaceutical dosage form is established through clinical trials coupled with safety studies, and may change significantly before the formulation is finalized. Although the volume of the dissolution medium can be somewhat reduced to allow for an increased concentration of the drug, this adjustment may not be enough to compensate for extremely low potencies. However, HPLC can provide superior sensitivity and therefore overcome the challenges resulting from low sample concentrations. With an appropriate detection method, an HPLC method can be used to quantify drug concentrations of the nanogram per milliliter range. For example, the dissolution testing of oral contraceptives containing 1 mg of norethindrone and 0.05 mg ethinyl estradiol was reported using a sensitive HPLC method to analyze sample solutions with concentrations of less than 60 ng/mL.[12] Transdermal dosage forms (patches, creams, and ointments) also pose sensitivity challenges to pharmaceutical researchers in developing dissolution testing procedures. The dissolution samples for these dosage forms are best analyzed by HPLC methods due to their extremely low concentrations.

C. Specificity and Selectivity

One of the most important advantages of HPLC over spectrophotometric methods lies in its specificity and selectivity due to its separation capability. Through chromatographic separations, the analytes of interest can be detected and quantified without interference from the typical matrix that includes excipients, antioxidants, preservatives, and dissolution media. Ion-pair HPLC was used to monitor the dissolution of pentamidine from EVA® sustained-release film where polymeric matrices could create significant bias if a spectrophotometric method were used.[13] Due to their strong UV absorbance, the antioxidants and preservatives (e.g., BHA, BHT, ascorbic acid and propyl gallate) are often the major sources of interference for direct spectrophotometric methods. These excipients are also prevalent in suspensions. On the other hand, dissolution media such as polysorbate solution would saturate the detector below 280 nm if direct UV analysis were used. With the proper selection of columns, mobile phases, and detectors, efficient HPLC methods can be developed to eliminate this interference. Another advantage of HPLC over spectrophotometric analytical methods lies in its ability to concurrently separate and analyze several active components in the presence of excipients and even degradation products that may be present in the dosage form. The simultaneous dissolution monitoring of rifampin, isoniazid and pyrazinamide using a reversed-phase HPLC was reported. Such a method worked well for the analysis of all three drugs in the presence of excipients and degradation products.[14] When two or more active ingredients are included in the dosage form, UV spectrophotometric methods often cannot be used, mainly because of cross-interference of active components. Though recently developed UV-Vis spectrometers can perform multi-component analysis for some combinational dosage forms, improvements in data treatment and reporting are needed for them to be acceptable for routine dissolution testing. The pharmaceutical industry is trending towards formulations that are more complex, lower in dosage strengths, and more sophisticated drug delivery systems. This, in combination with increasing emphasis on product quality control requirements, is making pharmaceutical researchers move away from conventional spectrophotometric techniques and turn to HPLC for dissolution testing and quality control.

III. CONSIDERATIONS OF HPLC METHODS IN DISSOLUTION TESTING

Development of efficient and reliable HPLC methods for dissolution testing can at times be problematic. Several factors need to be taken into consideration during method development.

A. Column Technology

Speed of separation is one of the most desirable characteristics of an HPLC method when it applies to dissolution testing, and the selection of columns and column technologies is crucial to achieve fast separation and sample analysis. The availability of high-speed HPLC for dissolution testing has greatly increased in recent years. Perhaps the main reasons behind such a great increase are as follows: (1) a large number of dissolution samples will be analyzed to support stability programs, clinical release, formulation/process development, scale-up and optimization. As such, achieving high-speed analysis would be justified from the technical, productivity, and economic perspectives; (2) the fast analysis of dissolution samples is desired to synchronize with pre-determined dissolution sampling time points for full automation of dissolution testing; and (3) HPLC methods used in dissolution testing do not require baseline separation of degradates and impurities so long as they are reasonably separated from the active drug of interest. On a case-by-case basis, the interference from impurities or degradates presented at low levels is considered acceptable. Short columns with smaller particle size have the important characteristic of a greater degree of separating power per unit of time, that is, a higher number of theoretical plates per second. Therefore, the separation efficiency provided by short columns with small particle size is highly comparable to traditional reversed-phase columns (typically 25 cm in length), with significantly shorter run time to synchronize with the dissolution sampling timepoints. Typical fast HPLC chromatograms generated from a set of three dissolution samples at three time points are shown in Figure 1.[15] The applications in pharmaceutical analysis (rapid drug screening, content uniformity, dissolution and drug purity assay) and the practical considerations (column lifetime, viscous heating, and analytical precision) of high-speed HPLC using columns with small particle size were discussed by Gant and Dong.[16] Monolithic column technology has been developed recently to allow ultra-fast HPLC methods to operate at high flow rate without a significant increase of backpressure. Major chromatographic features of monolithic columns arise from the large through-pore size/skeleton size ratios and high porosities. The separation efficiency of the monolithic columns has been demonstrated to be similar to or higher than that of the small nonporous particles and much higher than that of conventional columns. The microbore column is another good choice for fast chromatography, and has been predicted to develop a substantial user base in

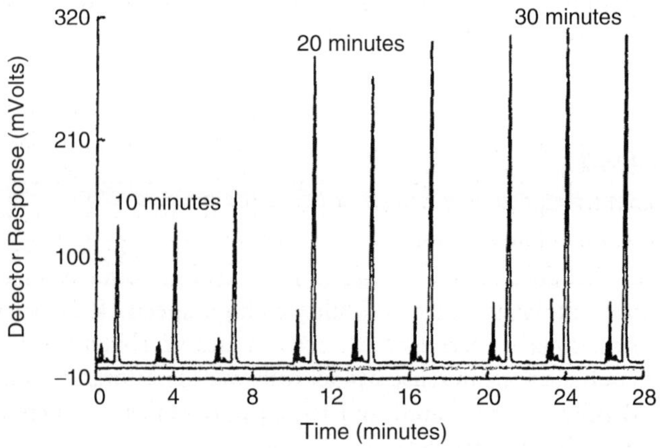

FIGURE 1 Typical fast HPLC chromatograms for a set of three dissolution samples at three time points.

the pharmaceutical industry.[17,18] The detailed illustration of theory and the applications of fast LC are discussed in Chapter 4.

B. Selection of Detection Methods

The selectivity of an HPLC method is commonly augmented by an appropriate detector. UV detection is the most commonly used technique in HPLC methods for dissolution testing. However, other detection methods allow pharmaceutical scientists to use HPLC for dissolution testing of various drug products which otherwise could not be done with UV detection. HPLC interfaced with mass spectrometric detection presents an excellent alternative for high throughput dissolution testing for pharmaceutical dosage forms of both chromophoric and non-chromophoric active ingredients. With tandem mass spectrometric techniques, active ingredients can be analyzed in complex matrix systems with superior efficiency, sensitivity, and specificity.[19] The use of mass spectrometric techniques in the routine dissolution testing for formulation development support and drug product quality control has not been recognized due to the limitations of mass spectrometric instrumentation: (1) HPLC interfaced with mass detector is still relatively expensive for routine formulation development support, stability programs, and quality control. The relatively complicated operation procedures for instrument calibration and performance optimization require extensive training of pharmaceutical chemists; (2) there is still a lack of well-defined analytical procedures that are in compliance with current GMP requirements. Due to the suppression of ionization from the matrix (e.g., excipient, dissolution medium, and sample diluent), quantitation is oftentimes limited to the use of internal standards. Nevertheless, the authors believe mass spectrometric detection will find its use in dissolution testing as instrumentation advances in the future.

Electrochemical detection coupled with LC is the fastest growing technique for biomedical analysis, and also find its use in the dissolution testing of pharmaceutical dosage forms.[20] Electrochemical detection offers the advantages of low detection limits (sub-pico molar range) and selectivity. The advantage of using electrochemical detectors is pronounced when analyzing electrochemically active drugs in highly complex formulations. Thermal conductivity and reflective index detectors are also used for the detection of both chromophoric and non-chromophoric compounds. Though suffering from inadequate sensitivity and method ruggedness, these detection techniques should be able to find applications in the dissolution testing of high potency dosage forms in simple aqueous media (dissolution media containing surfactants can be problematic due to analytical interference).

Overall, a broad choice of detection methods in HPLC has provided pharmaceutical researchers with great flexibility in solving the dissolution problems of pharmaceutical products of various properties and dosage strengths.

C. Chromatographic Considerations Related to Dissolution Media

The adequate specificity of HPLC methods is crucial for method accuracy and robustness. Chromatographic interference with the peaks of interest in dissolution testing can arise from the formulation and dissolution medium. Unlike HPLC methods for composite and content uniformity assay testing, the acceptable media for dissolution testing are limited and often dictated by regulatory requirements. The use of organic solvents is not recommended except under extraordinary circumstances. For aqueous soluble drug substances, water, acids (0.01 or 0.1 N HCl), and buffer (pH ≤ 7.0) are often used as dissolution media. With aqueous insoluble drugs (or those only soluble in basic aqueous media), surfactants are added to enhance solubility or wettability in achieving satisfactory dissolution profiles. The interference of these commonly used dissolution media with HPLC methods is generalized in Table 1. With the recent development of column technology, there are numerous base-deactivated columns that can handle sample solutions with low pHs (≤1) without having a significant impact on column lifetime and chromatography. But occasionally, peak splitting and broadening still occur. With surfactant media, drug compounds can potentially form meta-stable micelles and self-assembly architectures. The impact of such a micro heterogeneity of the sample solution on dissolution method is demonstrated by poor measurement precision and method ruggedness. In this case, the micellar or self-assembly structures can be eliminated by the addition of salts, solvents or mobile phase. As discussed earlier in this chapter, the dissolution testing procedure as defined in USP frequently calls for the sample to be diluted with mobile phase to avoid this incompatibility of the aqueous dissolution medium and mobile phase, and also to break up micelles.

Chromatographic interferences from the surfactants SLS, Tween-80, cetrimonium bromide (CTAB), etc.) are commonly observed, which pose challenges during dissolution method development. The interference of Tween-80 with chromatography has been extensively studied.[21] Typical chromatograms of Tween-80 in aqueous solutions (0.1–3%, w/w) are presented in Figure 2, showing complex impurity profiles which often result in significant interference with active pharmaceutical ingredient(s) of interest. SLS is commonly used in the dissolution media as a wetting and/or solubilizing agent, in lieu of Tween-80, due to its lower interference with chromatography. In general, dissolution media with up to 5% SLS do not pose any significant interference with chromatography at UV detection of greater than 240 nm. Dissolution media with added enzymes are occasionally used for dissolution of gelatin capsule dosage forms

TABLE 1 Compatibility of Commonly Used Dissolution Media with HPLC Methods

	HPLC methods		
Dissolution media	**UV detection**	**Other selective methods (electrochemical, fluorescence, etc.)**	**Mass spectrometric detection**
Water	No interference	No interference	No interference
Buffers (pH 4.5, 6.8)	Potential interference from UV-absorbing organic buffers	Potential interference	Significant interference due to ion suppression
HCl (0.01N, 0.1N)	Occasional incompatibility with mobile phase. Sample dilution may be needed	Potential interference	Significant interference due to ion suppression
Tween-80 (0.1–5%)	Significant chromatographic interference esp. at high concentration. Occasional incompatibility with mobile phase	Sensitivity may be affected. Extensive method development is needed	Significant interference
SLS (0.1–5%)	Occasional incompatibility with mobile phase.Sample dilution may be needed. No analytical interference at >240 nm	Sensitivity may be affected. Extensive method development is needed	Significant interference

when dissolution in commonly used aqueous or surfactant media cannot be accomplished due to capsule crosslinking (pellicle formation during dissolution testing).[22] Similarly, enzymes at certain concentrations create significant interferences during sample analysis by either spectrophotometric or chromatographic methods.

In designing a dissolution method, the absorbance of the active pharmaceutical ingredients needs to be evaluated against that of surfactant or buffer media to determine the appropriate UV detection wavelength. The lowest concentration of surfactants and buffers that affords the satisfactory profiles should always be adopted. This approach also complies with global regulatory guidances/expectations. Another means of countering the analytical interference is to consider the use of HPLC with an appropriate detection method. On a case-by-case basis, other selective detection methods such as fluorescence, electrochemical and mass spectroscopic detection should be considered. Finally, some surfactant media can be viscous. Analytical procedures such as injector washing, decreased injector sampling speed, or sample dilution should be considered and specified in the testing method if needed.

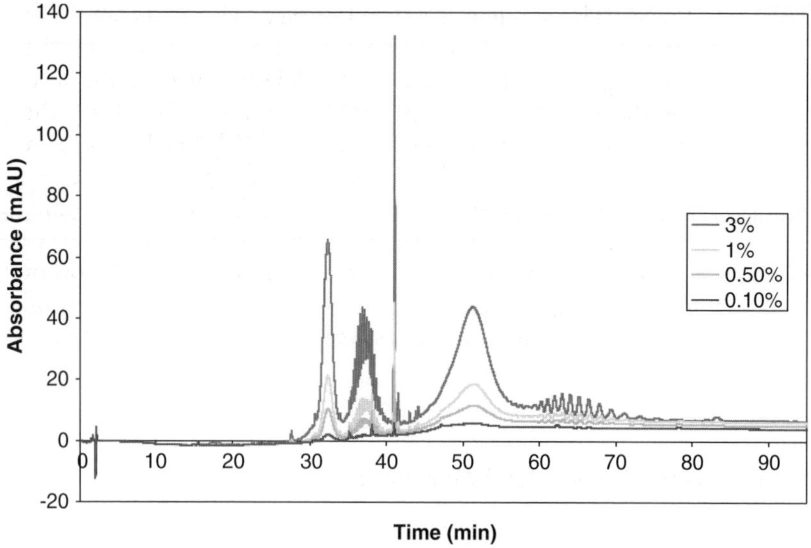

FIGURE 2 Typical chromatograms of Tween-80 in aqueous solutions (0.1–3%, w/w). HPLC method conditions: Waters Symmetry columns (5 μm, 4.6×150 mm) were eluted with a mobile phase containing acetonitrile and 0.1% phosphoric acid (gradient program) with UV detection at 220 nm. Injection volume is 20 μL.

D. Data Collection System and Validation

Data from the chromatographic system may be collected by an electronic integrator or by a chromatographic data system (CDS). Data reduction software may be used to calculate results, typically calculating the percentage dissolved using external standards. When multiple samples are taken from the same vessel (and medium replacement is not employed), it may be necessary to algebraically compensate for media removal during prior sampling points. In some laboratories, data are reported out using a Laboratory Information Management System (LIMS). Chromatographic data may be transferred to LIMS electronically or manually, depending on the capabilities of the system.

As with any data generation or data-handling system, appropriate validation (instrument qualification (IQ)/operational qualification (OQ)/performance qualification (PQ)) is required to demonstrate that the system is accurately and reliably performing in accordance with the user-functional requirements.

IV. HPLC IN AUTOMATED AND ROBOT-ASSISTED DISSOLUTION TESTING

Dissolution is one of the most time-consuming tests in stability testing, clinical release, and process support of pharmaceutical dosage forms. The number of samples to be collected and analyzed can be enormous, and significant labor is often expended in the preparation of reagents and apparatus, the performance of the test, the analysis of the samples and reporting of results. Therefore, the pharmaceutical industry desires fast, accurate, and flexible automation systems. Several automated dissolution testing systems have been developed based on the use of a flow-through cell and UV spectrometric analysis or HPLC.[23] A typical system with on-line HPLC analysis is shown in Figure 3. In general, there are four elements of a dissolution test: dissolution, sample collection, sample preparation, sample analysis/data collection. The automation of each of these elements is illustrated as follows:

A. Automated Dissolution Workstations

Fully automated workstations can automatically fill vessels with dissolution medium, drop the tablets or capsules, collect multiple samples, and clean vessels. Zymark® MultiDose Automated Dissolution Workstation (Zymark Corporation, Hopkinton, MA), Sotax AT70 Smart (Sotax Corporation, Horsham, PA), and AUTO DISS® systems (Pharma Test GmbH, Hainburg, Germany) are several popular automated systems used within the pharmaceutical industry.

B. Sample Collection

The autosampling device is critical to the integrity of automated dissolution testing systems as it serves as the interface between the dissolution apparatus and the sample analysis systems. The autosampler takes

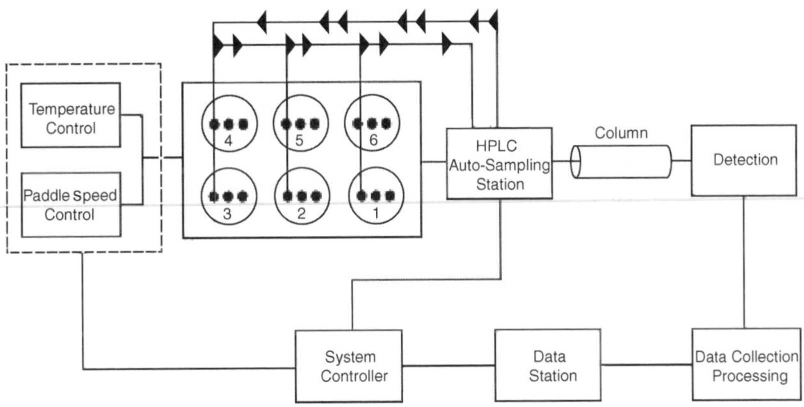

FIGURE 3 The diagram of a typical automated dissolution system with on-line HPLC.

sample aliquots from the dissolution vessels and delivers them to collection tubes or HPLC vials. Compared to manual sampling techniques, which are labor-intensive, and time-consuming, automated sampling systems can be cost-effective and yield consistent and reliable results in less time. Due to the similarity of sample size and mechanics, the automated sampling systems developed for flow-cell UV methods can be applied to HPLC methods with slight modifications. In general, there are three types of autosampling devices: (1) continuous-flow autosamplers that use peristaltic pumps to move fluid continuously from the dissolution vessels through a flow-cell. Systems that use flow-through cuvettes for programmable liquid transport are included in this category. These sampling devices are used only for dissolution with direct UV detection; (2) intermittent-flow samplers, which are modifications of the continuous-flow systems. The equipment withdraws discrete samples separated with an air bubble. Stream-selection valves controlled by microprocessors are a common example of an intermittent-flow sampling device; (3) discrete sampling devices withdraw samples sequentially or simultaneously and either store them in an integrated fraction collector or feed them into a detection system. The advantages and disadvantages of these three autosampling systems are discussed in detail in Hanson's *Handbook of Dissolution Testing*.[24] Discrete sampling devices often serve as the interface between a dissolution apparatus and an HPLC.

C. Sample Preparation

Sample clarification or filtering procedures need to be thoroughly evaluated to ensure the accuracy and robustness of any automated dissolution system. During dissolution testing of pharmaceutical dosage forms, extremely small particles may be generated and the particles will continue to dissolve postsampling if they are not separated from the sample solution in a timely manner. This phenomenon is known as postsampling dissolution. Membrane filters are the best choice to clarify samples for HPLC analysis. PTFE, nylon, glass fiber cellulose ester, PVC, and acrylic-coated filters are commonly used in commercially available dissolution systems. The particle size of the active pharmaceutical ingredient can be used as a guide for filter porosity selection (i.e., the membrane pore size must be smaller than the active pharmaceutical ingredient particle size). However, it is worth noting that filters with smaller pore sizes can present problems to commonly used autosampling devices, which may not be capable of reliably and consistently pulling through filters of less than 10 μm. Automated sampling systems with inert piston pumps capable of delivering up to 100 psi have been developed. These systems have proven applicable with heavy particulate loading and for hydrophobic drugs, and can be used in the dissolution of pharmaceutical dosage forms containing active ingredients with sub-micron particle size to eliminate any issues associated with postsampling dissolution.[25] Automated sampling stations able to use filters with a pore size of as small as 0.2 μm are commercially available. Filters should always be evaluated for clogging potential and absorption of the analytes of interest.

D. Sample Analysis by HPLC

HPLC analysis in automated dissolution systems can be performed off-line or on-line depending on the system configuration and the synergy between sampling time and HPLC run time. Off-line HPLC analysis involves a fraction collector that collects dissolution samples automatically into HPLC vials. The samples are then analyzed off-line by HPLC, sequentially under microprocessor control.[26] Off-line analysis is frequently used for immediate-release dosage forms, where the HPLC method assay time required for each sample does not synchronize with the time specified in sampling intervals for the dissolution test. Automated systems that allow dissolution samples to be directly deposited into automated injection systems at the required time intervals for dissolution sampling have been developed and are commercially available. Such systems allow for completely unattended dissolution/HPLC analysis. The system is computer directed, and the data output follows the dissolution sequence only by the real time necessary to accommodate HPLC assay time vs. sampling time.

Fully automated dissolution systems involving on-line HPLC analysis and data processing are described in numerous reports and have been applied to immediate and controlled-release dosage forms.[27,28] These systems eliminate the need to transfer samples to a fraction collector, because each sample is transferred immediately after sampling directly to the on-line sampling loop valve or to an autosampler carousel for immediate on-line HPLC analysis. The systems are under microprocessor control, which provides system monitoring to improve cGLP and cGMP compliance.

E. Typical HPLC Automated Dissolution Systems

Several automated dissolution systems with HPLC analysis have been developed and are commercially available. The systems from Waters, Anapharm, and Pharma Test GmbH (Auto Diss®) are among those used in the pharmaceutical industry. Based on the Waters 2690D separation module, the Alliance Dissolution System may be the most integrated system (Figure 4). An automated system for the dissolution testing of topical drug delivery systems has also been reported. The system was developed to perform the dissolution testing of topical drug formulations using Franz cells followed by HPLC analysis.[29] Though the

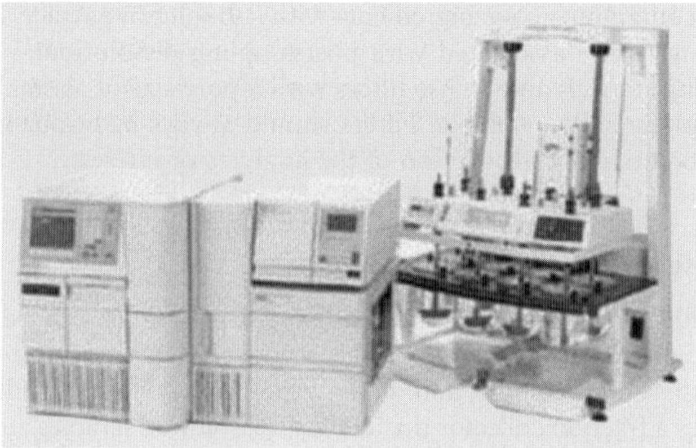

FIGURE 4 **Fully automatic dissolution system based on Waters 2690D separation module.**

fluid path and sample transport can be different, all of these systems allow the continuous analysis of samples by HPLC during drug release, eliminating the lag time between sample collection and analysis. The synergy between dissolution sampling time and HPLC analysis time is critical. This further emphasizes the utility of fast chromatography in automated dissolution testing.[30]

F. Robotics in Automated Dissolution Testing

The application of robotics to the total automation of dissolution testing has been successful in achieving improved productivity and cost-efficiency. Robotic arms are precision mechanical devices controlled by microcomputers to perform physical labor and the results they generate may be more accurate, consistent, and reliable than those produced manually. Although robots require a large initial investment, they can quickly become a sound investment, because they can work unattended for extended periods of time without showing signs of fatigue.

The desire to incorporate robotics into dissolution testing is due to the tedious nature of the preparative workup. During drug dissolution testing, the laboratory robots can perform the following: filling vessels with dissolution medium, deaerating (if necessary) and heating media to the required temperature, dropping dosage forms into vessels, adjusting paddle/basket speed, removing sample aliquots, removing and replacing dissolution medium, and draining, rinsing and drying of dissolution vessels.

Robot-based automated devices have been designed to perform HPLC injections (esp. off-line injections) unattended by pharmaceutical researchers. Each step of an HPLC analysis can be automated either by the robotic arm itself or by the use of a dedicated automated peripheral device with which the robot may interact. In addition, a laboratory robot is an excellent alternative to expensive autosamplers at the sampling end of an HPLC. The autosampler and computer can also work in conjunction with cylinder robots to form a completely integrated HPLC injection system suitable for dissolution testing.

Various robot-assisted, automated dissolution testing procedures have been developed and used in the dissolution testing of pharmaceutical dosage forms.[31–35] To completely eliminate manual intervention for robotic dissolution systems, a complete dissolution robot was developed by combining two robotic arms into a unique system. One arm of this robot performs membrane filtration while the other arm performs UV or HPLC analysis. This dual-armed robotic system provides the versatility to perform totally automated dissolution of tablets and capsules using the USP Apparatus 2 (paddle) method. However, it should be noted that robotic automation involves sequential working steps, and this can result in long processing times. Moreover, due to their limited flexibility, robotic systems are often ill-suited to processing a smaller series of test samples involving frequent change of method, or dissolution testing where short sampling intervals are required.

Automated dissolution systems are attracting increased attention from the pharmaceutical industry. The advantages of these systems can be realized in several ways. First of all, automated dissolution testing can be performed unattended, allowing the analyst to perform other tasks in the lab while the system is running. Secondly, automated systems have the capability to degas and dispense the media, drop the tablets or capsules, record the media temperature, sample the vessels at the programmed time, filter the samples, perform the analysis, and data treatment in consistently sequential order. This eliminates variability due to analyst error. In addition, media evaporation and degradation of active ingredients is minimized. Finally, the increased efficiency and decreased analyst involvement brings great cost saving and increased productivity to the pharmaceutical industry. It was reported that an 80% reduction in expended time can be achieved by using a fully automated system when compared with manual testing.

While offering savings of resources and adding productivity to the pharmaceutical industry, however, automation of dissolution systems can suffer from several drawbacks: (1) automated equipment often requires a long setup time and extensive validation. The analyst must show that the results are accurate and equivalent to the manual method. A significant amount of time must be devoted to training, maintaining logbooks, calibration, and maintenance; (2) errors often occur with some system components. For example, sample lines are often a source of error for a variety of reasons—unequal lengths, crimping, wear beyond limits, disconnection, carryover, mix-ups or crossings, and inadequate cleaning: (3) cleaning time and carryover procedures needs to be evaluated and the volume dispensed, purged, recycled, or discarded needs to be routinely checked; (4) with automation, analysts may develop the tendency to set up the dissolution system and leave the testing area, ignoring valuable visual observations. In an R&D environment, where formulation and dissolution method changes often occur at different stages of drug development, the time needed to develop and validate an automated method is sometimes not cost-effective. Automation is most valuable in research laboratories testing a large number of stability samples and clinical supplies, or in a quality control laboratory where the drug product is tested repeatedly with the same method.

V. REQUIREMENTS OF HPLC METHODS FOR DISSOLUTION TESTING DURING DRUG DEVELOPMENT PROCESS

Dissolution is a qualitative and quantitative tool used during drug development programs, providing valuable information on product quality and sometimes in vivo performance. There are five distinct milestones of the modern drug development process: pre-clinical, phase I, phase II, phase III, and product marketing. Dissolution testing is used from the start of dosage form development and throughout all subsequent phases. It is used for various purposes, as shown in Table 2. The aim of dissolution testing, and the scale and degree of information required from the test may pose different requirements for dissolution methods at each stage of drug development. For example, simplicity and robustness are crucial properties of a quality control method, whereas physiological relevance may overrule these factors if a method is to be used for IVIVC. In general, pharmaceutical researchers often consider five major characteristics when selecting an optimal method for dissolution testing: dynamic range, specificity, extent of automation, efficiency, and robustness. The relative importance of these method characteristics at different stages of drug development is illustrated in Table 3.

TABLE 2 Different Roles of In Vitro Dissolution Testing During Drug Development

To support formulation design and development
To investigate drug release mechanism, especially for extended-release formulations
To probe IVIVC and act as a surrogate for bioequivalence studies
To support process development and validation of manufacturing process
To support probe and long-term stability studies and regulatory filings
To support batch quality control
To support post-approval changes

TABLE 3 Dissolution Method Characteristics and Their Relative Importance in Drug Development

Drug development stage	Method characteristics				
	Dynamic range	Specificity	Extent of automation	Efficiency	Ruggedness
Preclinical—phase II Formulation design Process scale-up Stability	+++	++	+	++	+
Phase III Long-term stability Production of pilot batches IVIVC (if needed)	+	++	+++	+++	++
Product marketing Quality control Post-approval change	+	++	++	+++	+++

Level of importance: + low; ++ medium; +++ high.

At the early stages of drug development (preclinical to Phase II), activities are often focused on formulation design and process scale-up. A wide dose range (multiple doses) is often developed to support dose finding and safety studies. A method with wide dynamic range is strongly preferred to support the development of formulations of wide dose range. Method specificity and efficiency are also considered to support numerous stability programs and formulation/process development. Automation is not considered by most laboratories, due to the lack of experience with the methods or a change of formulations. The HPLC method can be developed based on the stability-indicating assay methods in which all of the potential impurities from drug substances and excipients, and the known degradation products are resolved from the active drug. Since there is no need to monitor degradates and impurities during dissolution testing, chromatography can usually be significantly shortened, either by using a short column of the same type (smaller particle size may be used to maintain the column efficiency), or by adjusting other key method parameters such as the mobile phase and flow rate. At the phase III stage of drug development, the dissolution testing will support long-term stability, production of pilot scale batches and possibly IVIVC or an in vitro and in vivo relationship (IVIVR). Method automation assumes a high priority and may be the most efficient course to take, as the knowledge of drug products and dissolution methods is gained.

The use of automated dissolution has been proven to have a positive impact on the reduction of "time to market" on average for a number of products.[36] Method specificity and ruggedness should be adequately demonstrated, as this method will potentially be transferred to the quality control laboratories. Dynamic range is not as important as the drug development program progresses into phase III, when the market doses are defined. After the drug product is launched into the market, the dissolution method will be transferred to the quality control laboratories. Method ruggedness is the most important factor to consider ensuring that the method can be transferred and used worldwide. The method ruggedness is evaluated by multiple analysts, preferably in multiple laboratories, in research and development and manufacturing sites. Method efficiency is critical for the timely release of drug products. Automation is not strongly desired since the dissolution testing often requires a single sampling point where the specification value, *Q*, is set to make a pass/fail decision. Again, dynamic range is not important post launch. The HPLC method characteristics as they apply to dissolution testing are often governed by the selection of an appropriate detection method. The current available detection methods and their advantages vs. disadvantages are discussed earlier in this chapter.

Reliable quality control in the field of pharmaceutical analysis is based on the use of valid analytical methods. For this reason, any analytical procedures proposed for a particular active pharmaceutical ingredient and its corresponding dosage forms should be validated to demonstrate that they are scientifically sound under the experimental conditions intended to be used. Since dissolution data reflect drug product stability and quality, the HPLC method used in such tests should be validated in terms of accuracy, precision, sensitivity, specificity, ruggedness, and robustness as per ICH guidelines.

VI. CONCLUSION

HPLC instrument and column technologies have progressed to the point that pharmaceutical researchers have tremendous flexibility in selecting an optimal method for any dissolution testing during drug development and routine quality control tests. Novel uses will certainly be discovered as pharmaceutical scientists begin to explore ways of obtaining instant dissolution results with satisfactory accuracy, precision, and reproducibility. As formulation development trends towards more complexity, HPLC methodology will be used more often in dissolution testing, as it can solve unique analytical challenges posed either by the media or dosage form characteristics. Equipped with a variety of detection methods, HPLC can be used to analyze dissolution samples which otherwise cannot be analyzed by UV methods due either to significant interferences or lack of UV absorbance. Automation of dissolution systems with on-line HPLC analysis has been accomplished by the development of fast chromatography which will be increasingly used in the pharmaceutical industry to improve throughput and productivity.

ACKNOWLEDGMENTS

We wish to express our gratitude to Mr. David G. Reed and Mr. Gregory P. Martin for helpful discussions and suggestions during the writing of this chapter. We would also like to thank Drs. Eric W. Tsai, Dominic Ip, and William E. Bowen for reviewing this chapter and providing valuable comments.

REFERENCES

1. <711> Dissolution, *USP,* 27:2303–2304, 2004.
2. Atenolol and Chlorthalidone Tablets, *USP,* 27:185, 2004.
3. Fluoxetine Capsules, *USP,* 27:822, 2004.
4. <1088> In Vitro and In Vivo Evaluation of Dosage Forms, *USP*, 27:2513–2518, 2004.
5. Shah, V. P., Konecny, J. J., Everett, R. L., McCullough, B., Norrizadeh, A. C. and Skelly, J. P., *Pharm. Res.,* 6(7):612–618, 1989.
6. <1092> The Dissolution Procedure: Development and Validation, *Pharmacopeial Forum*, 30(1):351–363, 2004.
7. <1225> Validation of Compendial Methods, *USP*, 27:2622–2625.
8. Acetaminophen and Aspirin Tablets, *USP*, 27:19.
9. Baweja, R. K., *Pharm. Technol.,* 11:28–36, 1987.
10. Rapaka, R. S., Roth, J., Brine, G. A. and Prasad, V. K., *Int. J. Pharm.,* 12(4):285–294, 1982.
11. Bempong, D. K., Adkins, Jr., R. E., Lindauer, R. F., White, P. A. and Mirza, T., *Pharmacopeial Forum*, 25(1):7593–7625, 1999.
12. Swartz, M. and Neitring, I., *Am. Lab.*, 22(40):24–26, 1990.
13. Yang, C.-Z., *Zhongguo Yaoxue Zazhi* (Beijing, China), 27(1):21–22, 1992.
14. Mariappan, T. T., Singh, B. and Singh, S., *Pharmacy Pharmacol. Commun.,* 6(8): 345–349, 2000.
15. Dunkle, J. D., Gleason, D. J. and Hokanson, G. C., *Pharm. Res.,* 9(4):578–581, 1992.
16. Gant, J. R. and Dong, M. W., *Pharm. Technol.*, 11:44–58, 1987.
17. Kucera, P. and Hartwick, R. A., *J. Chromatogr. Library*, 28:179, 1984.
18. Filipi, T. J., In *Small Bore Liquid Chromatography Columns: Their Properties and Uses*, Scott, R. P. N., Ed., John Wiley and Sons, New York, 231pp., 1984.
19. Xie, F.-M., Wong, P., Yoshioka, K., Cooks, R. G. and Kissinger, P. T., *J. Liq. Chromatogr. Related Technol.* 21(9):1273–1282, 1998.
20. Begona Barroso, M., Alonso, R. M. and Jimenez, R. M., *J. Liq. Chromatogr. Related Technol.* 19(2):179–186, 1996.
21. Wuelfing, W. P., Kosuda, K., Templeton, A. C., Mowery, M. D. and Reed, R. A., *J. Pharm. Biomed. Anal.,* to be submitted for publication.
22. Murphy, K. S., Reisch, R. G. J. and Fawzi, M. B., *Pharm. Technol.,* 13:53–58, 1989.
23. Compton, B. J. and Hinsvark, O. N., *Pharm. Technol.*, 10:28–36, 1986.
24. Hanson, W. A., *Handbook of Dissolution Testing*, Aster Publishing Corporation, Eugene, OR, 1991.

25. Hill R. A. and Snider, B. G., *Int. J. Pharm.*, 36:175–183, 1987.
26. Cristina, G., Danzo, L. and Farina, M., *Int. Lab.*, 20(2):52–56, 1990.
27. Mathieu, R. J., Van Den Oever, M., Nicholson, D. G. M. and Kanfer, I., *Am. Lab.*, 24(15):27–30, 1992.
28. Aronson, B., Faber, P. and Raya, M., *Am. Lab.*, 21(2):32–38, 1989.
29. Li, J. B. and Rahn, P. C., *Pharm. Technol.*, 17:44–54, 1993.
30 Dong, M. W. and Hockman, D. C., *Pharm. Technol.*, 11:70–82, 1987.
31. Papas, N. A., Alpert, M. Y., Marchese, S. M. and Fitzgerald, J. W., *Anal. Chem.*, 57:1408–1411, 1985.
32. Kostek, L. J., Brown, B. A., Erhart L. C. and Curley, J. E., *Adv. Lab. Automat. Robot.*, 5:311–328, 1989.
33. Wolfschutz, R., Ganshirt, H. and Tessun, G., *Pharm. Ind.*, 49:393–397, 1987.
34. McCarthy, J. P., *Pharm. Technol.*, 12:72–81, 1988.
35. Curley, J. E., *Lab. Robot. Automat.*, 1(3):183–191, 1989.
36. Sadowitz, J. P., *J. Automat. Methods Manag. Chem.*, 23(6):173–177, 2001.

16 CLEANING VALIDATION USING HPLC FOR ANALYSIS

ANDREW PLASZ

Abbott Laboratories, Pharmaceutical Analytical Research and Development, North Chicago, IL

ABSTRACT

HPLC is the leading Analytical procedure used for the verification of pharmaceutical cleaning validation programs. HPLC provides a linear, sensitive method for quantitating low levels of residues making the chromatographic finish the most reliable part of the cleaning verification.

I. INTRODUCTION

HPLC is the analytical method of choice for most pharmaceutical bulk substances active pharmaceutical ingredients (APIs) and drug product formulations. It is not surprising then, that it is one of the most popular methods for the analytical finish for cleaning validation samples. It is sensitive, specific, and can detect other contaminants besides the analyte in question in the chromatogram. Nothing is more convincing to the validation team, auditor or inspector than a chromatogram with a clean baseline equivalent to the blank or control.

Handbook of Pharmaceutical Analysis by HPLC,
S. Ahuja and M.W. Dong, editors.

The first place at which cleaning becomes an issue is in the synthesis of the bulk drug substance or API. Although many API producers follow their synthesis with HPLC at each step, following the disappearance of starting materials and the formation of the product, they sometimes prefer a UV scan or UV reading at a single wavelength for testing each sample. On the HPLC side, the methodology is specific and can identify the contaminants besides the API (retention time) in the chromatogram. The UV scan, it is agreed, is faster and has merit in the fact that it is not specific and that any difference in absorbance from the blank in the scan or at a specific wavelength indicates that the surfaces are still dirty.

For the drug product, there is no question that HPLC is the preferred analytical finish. As a result it just makes more sense to go with straight HPLC for the whole overall cleaning program. Why have the laboratories develop two separate analytical finishes, double the validation effort, and have to explain to the regulators why there are two methods? Another advantage of HPLC is that it is usually a modification of a previously developed potency method for the API. These methods are stability indicating, and have linearity, specificity, and robustness already built into their validation packages. Method history, solution stability, limits of detection, quantitation, and recovery (from surfaces) may also be established. However, they need to be refined at the much lower levels of analyte concentration required for cleaning validation samples.

II. ROLE OF THE ANALYST

Although the analytical chemist provides only a part of the work necessary for a successful cleaning validation program, the analyst is the heart of the program. The analyst must generate quality method validation packages, and employ impeccable sample analysis techniques and documentation of these analytical results. The analytical results and raw data are the foundation of any defendable cleaning validation program. The analyst brings a high level of science, expertise, and common sense to the cleaning validation program.

It makes good sense to have the same analysts who test the API and formulations throughout the research stage of a molecule's development perform the cleaning validation method development and testing since they are the most familiar with the existing HPLC methodology. Method transfer of the validated analytical cleaning method to other sites and laboratories in the same facility can be accomplished by this team. The analyst must understand the customer and the customer's needs to develop and validate the necessary methodology properly. The analyst interfaces with many other areas of the team and corporation, but all of them have to come to and through the analyst, making the analytical chemist the heart of a cleaning validation program. The analyst has to follow certain chronological steps to achieve the completed methodology. These are listed in Table 1.[1]

TABLE 1 Steps for Analytical Chemist

1.	Get cleaning limit
2.	Determine equipment to be tested
3.	Swab vs. rinse
4.	Recover drug off swab
5.	Stability of swabbing solution
6.	Recover drug off surfaces
7.	Show cleaning removes or destroys drug
8.	Clean and test
9.	Document with data the appropriateness of the cleaning procedure

Data and chromatograms for four antibiotics will be used to help illustrate and characterize representative approaches to real situations. The work on cefadroxil, cefmenoxime, cefsulodin, and clarithromycin are all HPLC assays. The three cephalosporins used a UV finish, while the clarithromycin being a macrolide antibiotic and having a low chromophoric response, required an electrochemical detector for quantitation.

III. LINEARITY, DETECTION AND QUANTITATION

The HPLC method needs to be linear; usually this means extending the linear range at the lower end of the range used for the potency assay. In many laboratories, analysts, realizing that they will have cleaning samples, push the linear range as low as possible in the initial HPLC method development for the API potency assay. Many times during the development cycle, toxicology or clinical data indicate that the cleaning limit must be lowered. It is important for the analyst to know the limits of the existing methodology to ascertain the possibility of accommodating lower cleaning limits. Poor chromatographic response by many of today's molecules above 220–240 nm means that many cleaning assays are forced into the lower end of the UV range. This presents additional challenges for the analyst in both linearity and sample analysis due to solvent and impurity interferences at the low wavelengths. Typical chromatograms of 0.1 μg/mL solutions of the three cephalosporins are shown in Figure 1.[2]

The next step is to determine the practical detection limit (pDL) based on the signal-to-noise ratio at the lowest level at which the analyst can get the HPLC system to function reproducibly on injections of a standard at a known concentration (*S/N* ratio of 3:1 is a rule of thumb). Then the practical quantitation limit (pQL) is determined usually at a level 2–5 times the pDL and the repeatability of the standard at this level is determined. This pQL usually results in analyte concentrations of nanograms or micrograms per milliliter. The repeatability of a 1.0 μg/mL clarithromycin standard preparation is shown in Table 2.[3]

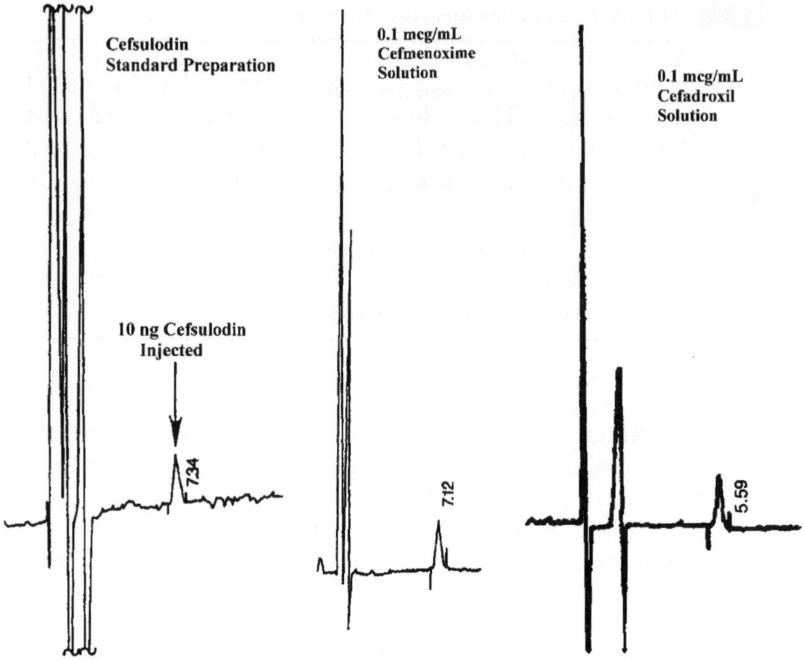

FIGURE 1 Typical chromatogram of a 0.1-μg/mL solution of cephalosporins.

TABLE 2 Repeatability of 1.0 μg/mL Clarithromycin Standard Preparation Injections

	Measured peak height (mm)	Integrated peak height (counts)
	42	845
	44	895
	45	899
	44	883
	48	933
	47	940
Mean	45	899
SD	±2.2	±34.7
RSD	±4.9%	±3.9%

IV. VALIDATION OF RINSE AND SWABBING PROCEDURES

The purpose of the HPLC analysis of cleaning samples is to prove with data that the equipment and cleaning procedures work, and that the surfaces of the equipment are indeed clean. The HPLC chromatographic finish is extremely reproducible and is the easiest part of the analytical procedure to control. To prove that the procedure is accurate, it must be shown that the samples analyzed by the HPLC quantitatively remove any residues of the analyte left behind or incompletely removed by the cleaning procedure. This is done by standard addition and recovery. The analyte is applied to the surface as a solution, allowed to dry, and then either swabbed or rinsed off to prove that if it is there it will be removed by either technique. The rinse samples are pretty straightforward: rinse, possibly dilute, and analyze. With rinse samples, you rinse with a known amount of rinsate and usually collect a portion, but assume the concentration is based on the volume of rinsate that one starts with.

The swabs present several problems. With swabs there are many steps that need to be validated to insure an accurate result. The analyst must be able to quantitatively remove the analyte from the swab. The analyte is added to the swab as a solution, dried, and quantitatively extracted off for analysis. The preferable extraction solution is the swabbing solvent. The HPLC mobile phase or a mobile phase component could be used, but would necessitate a dilution of the swabbing solution. Standard addition and recovery data of clarithromycin added to different lots of polyester fiber are present in Table 3.[3]

The analyst must first determine that the swab material is suitable for use in the analysis. Extractables, impurities, or particulate matter from the swab or its handle are all items that can eliminate a vendor or commodity from use in the analysis. An example of swab material containing interfering chromatographic impurities is shown in Figure 2.[4] This swab material is washed with solvent, dried, and stored in clean bottles before use after the removal of the interfering peaks. Many times it is necessary to wash and dry swabs thoroughly or swab material in large lots and store for future use. All glassware and commodities used, such as caps for storing and transporting the swab or rinse solvents and analytical samples, should also be scrupulously cleaned with swab or rinse solvent and dried before use.

TABLE 3 Standard Addition and Recovery Data of Clarithromycin from Polyester Fiber

Lot of polyester	Clarithromycin added (μg)	Clarithromycin found (μg)	% recovery
1	0.97	0.93	95.9
1	2.43	2.44	100.4
1	9.89	9.55	96.6
2	4.41	3.89	92.4
2	10.51	19.33	98.3
2	21.03	19.63	93.3
3	20.67	18.52	89.6

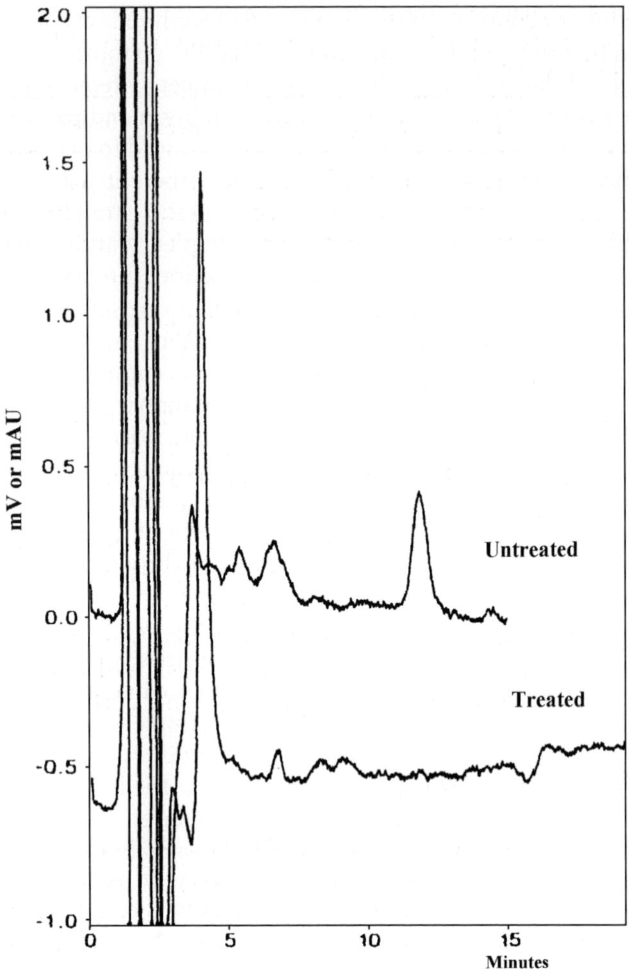

FIGURE 2 **Typical chromatogram of washed and unwashed swab material showing the removal of chromatographic interferences.**

Many times swab samples have to be taken at a remote location where the equipment necessary for the analysis may not be available. Because the samples would be in transit for 24–48 h, an alternative to using test tubes for sample transport was investigated. Screw-cap scintillation vials either with a polyethylene insert or with an aluminum foil liner in the top of the cap were tested. Four polyester swabs were placed into each vial along with 10 mL of 1.00 µg/mL 50:50 ethanol:water solution of clarithromycin or 10 mL of 50:50 ethanol:water. The vials were capped and then shaken to wet the surface of the vials. Half the vials were refrigerated and the other half were left at room temperature on the

TABLE 4 Stability of a 1.00 μg/mL Solution of Clarithromycin in Scintillation Vials Containing Polyester Swabs at Room Temperature (RT) and Under Refrigeration (Refrig.)

Cap insert	RT	Refrig.	RT	Refrig.	RT	Refrig.
Polyethylene	0.88	0.84	0.92	0.92	0.91	0.84
Aluminum foil	1.05	1.00	1.04	1.02	0.94	1.00

laboratory bench. Samples were assayed at 24, 48, and 120 h. The sample blanks showed no peaks that would interfere with the clarithromycin peak. The results are shown in Table 4.[3]

From the data, loss of clarithromycin is obvious in the solutions stored in the scintillation vials with the polyethylene cap inserts at all data points. No loss was seen in the solutions stored in the scintillation vials with the aluminum liners in the caps for 48 h at any of the storage conditions.

When all the commodities and supplies have been selected, the stability of the analyte in the rinse or swab solution must be determined. This should be done at more than one concentration and if analyte determination is noted, refrigeration or freezing of the solutions should be evaluated, to extend the analysts testing window. Degraded out-of-date samples will give lower numbers than true numbers and a false sense of cleaning effectiveness.

V. RECOVERY STUDIES OFF SURFACES

Clean commodities, solution stability, and 100% recovery off of swabs mean we are already to determine the accuracy or recovery of the surfaces or equipment in question. As previously mentioned, the analyte is applied to the surface as a solution usually in a 10 cm × 10 cm square or some other surface whose area is accurately measured and known. The solution is allowed to dry and then the surface wet swabbed, usually with 50 total strokes in the 10 × 10 cm area, 25 in one direction followed by 25 perpendicular strokes. It was noticed in early swabbing studies in the 1980s that many times liquid was still left on the swabbed surfaces. A dry swab was then used to absorb any liquid on the surface and added to the wet swab in the swabbing solution and sent for analysis. The analyst reports the values in a weight/volume mode and the Quality Assurance or manufacturing professionals apply their calculations to determine if the equipment or surface is clean and passes. Results for clarithromycin recovery off of various surfaces are shown in Table 5[3] and for the three cephalosporins in Table 6.[2] Figure 3[3] shows chromatograms of a clarithromycin swab solution blank, a 1.0 μg/mL clarithromycin standard solution and an authentic sample. Figure 4 shows a chromatogram of a clarithromycin high concentration sample undiluted and a 1:1 dilution of that sample.

TABLE 5 **Standard Addition and Recovery Data of Clarithromycin from Surfaces**

		Percentage recovered		
Surface	**µg Added**	**Swab**	**Rinse**	**Mass balance**
Glass	9.0	70.0	31.1	101.1
	103.4	95.9	8.9	104.8
	103.4	92.3	26.2	118.5
Plexiglas	9.0	82.2	11.1	93.3
	103.4	97.7	6.8	104.5
Stainless steel	9.0	96.7	13.3	110.0
	100.1	107.4	6.9	114.3
	100.1	100.3	12.4	112.7

TABLE 6 **Recovery of 10 µg of Cephalosporins Added to Various Surfaces**

	Percentage recovered[1]					
	Cefsulodin		Cefmenoxime		Cefadroxil	
Surface	**Swab**	**Rinse[2]**	**Swab**	**Rinse[2]**	**Swab**	**Rinse[3]**
Glass	60.8	38.2	101.7	<10.7[4]	76.0	14.0
	98.0	<9.8[4]	73.1	16.3	72.0	22.0
Plexiglas	87.2	12.7	85.6	<9.6	95.9	31.6
	—	—	—	—	85.7	9.2
Floor tile	70.3	37.6	73.7	10.5	77.5	8.5
	105.9	9.8	81.0	10.0	55.9	16.7
Stainless steel	59.4	36.6	78.9	18.9	87.6	7.8
Conveyor belt	90.2	<9.8[4]	74.0	14.4	83.3	11.8
	73.5	21.6	85.0	<10.0[4]	—	—
Lab bench	<9.9[4]	<9.9[4]	<10.5[4]	<10.5[4]	<10.1[4]	<10.1[4]
	<9.9[4]	<9.9[4]	<10.5[4]	<10.5[4]	<10.0[4]	<10.0[4]

[1]When more than one value was reported, the experiment was performed on 2 days.
[2]After the surface was swabbed, it was rinsed with 10 mL of 1% pH 6 phosphate buffer.
[3]After the surface was swabbed, it was rinsed with 5 mL of water.
[4]The sample peak height or area was below that obtained for a 0.1-µg/mL standard solution.

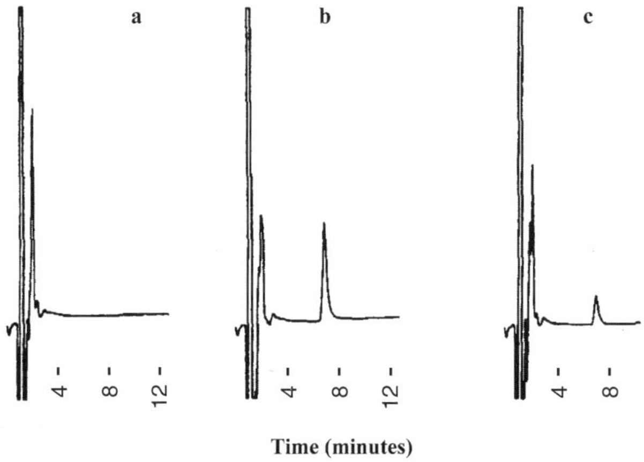

FIGURE 3 Chromatogram of a swab solution blank, 1.0-μg/mL clarithromycin standard solution, and an authentic sample solution: (a) swab blank, (b) 1.0 μg/mL standard, and (c) authentic sample.

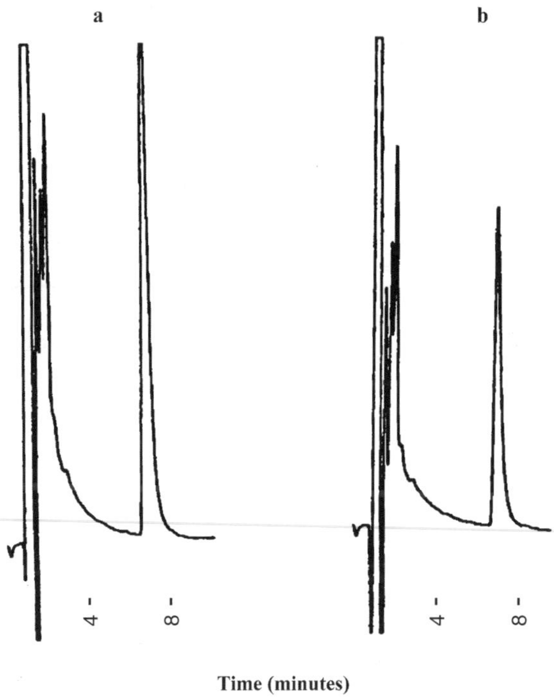

FIGURE 4 Chromatogram of a high-concentration authentic sample, undiluted and then diluted 1:1, concentration of the undiluted determined to be 46 μg/mL: (a) undiluted sample, and (b) 1:1 dilution of the sample.

VI. VERIFICATION OF THE CLEANING PROCEDURE

The last step for the analyst is to show that the selected cleaning agents or procedure actually cleans the surfaces in question. The drug is applied in solution form to the surfaces as in the recovery studies. The surfaces are then cleaned, rinsed, or treated as in the official cleaning procedure. Upon completion of cleaning, the surfaces are swabbed as before and if the cleaning has been successful the analysis will show the surfaces clean and free of analyte residue. The data in Table 7[2] describe the addition of each of the three cephalosporins in 5–10 mg amounts applied to surfaces in separate experiments, cleaned, and tested for residue as described above. It can be seen that this is a necessary step in the procedure in that it shows that the cleaning really works.

VII. VERIFYING THE REMOVAL OF CLEANING AGENTS

The last and final step very often involves the process of showing the complete removal of the cleaning agent. Many of the detergents or cleaners are inorganic but some have a UV chromophore and as a result, HPLC can be used for analysis of the cleaning agent residue. The steps are the same as for the analyte. Some results for Triton X-100, a common non-ionic detergent that is used for cleaning manufacturing equipment surfaces, are shown

TABLE 7 Amount of Cephalosporin Residue on Surfaces Which Were Exposed to Milligram Quantities of Cephalospsorin and Cleaned

		μg/100 cm² found		
Surface	**Detergent[1]**	**Cefsulodin**	**Cefmenoxime**	**Cefadroxil**
Glass	A	<1, <1	<1	1.2
	B	—	—	1.4
Floor tile	A	2.0, <1, <1	<1	<1
	B	—	—	1.3
Stainless steel	A	<1, <1	<1	<1
	B	<1, <1	<1	<1
	C	<1	<1, <1	<1
Plexiglas	A	<1, <1	1.7, <1, <1	<1
	B	—	—	1.5
	C	<1	5.5, 1.0	—
Conveyor belt	C	<1	1.7, 1.0	—

[1](A) One and one-half percent aqueous sodium hypochlorite solution; (B) One part sodium hypochlorite (5.25% min) and three parts 0.5% aqueous detergent; (C) 2% aqueous CONTRAD 70 (pH 11).

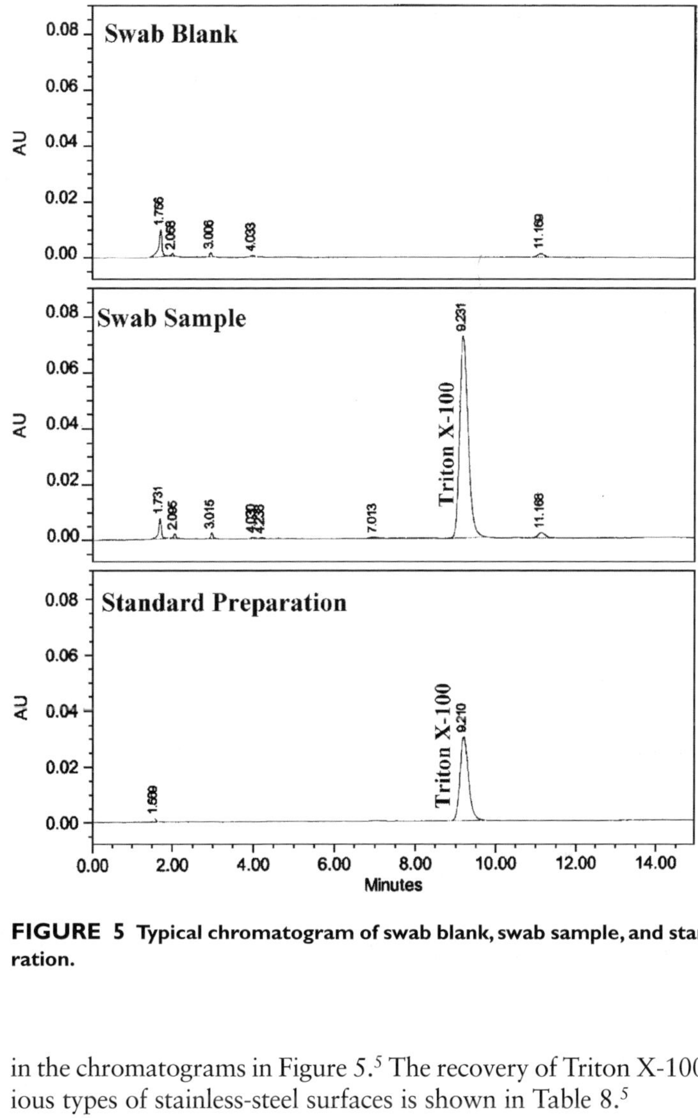

FIGURE 5 **Typical chromatogram of swab blank, swab sample, and standard preparation.**

in the chromatograms in Figure 5.[5] The recovery of Triton X-100 off of various types of stainless-steel surfaces is shown in Table 8.[5]

VIII. CONCLUSION

In summary, HPLC is the method of choice as the analytical finish for pharmaceutical cleaning validation programs. HPLC is versatile, adaptable, available worldwide and provides a reproducible approach as the anchor of any cleaning validation program. The cleaning is only as good as the data that supports it and HPLC has a proven track record with industry and the regulatory agencies.

TABLE 8 Recovery of Triton X-100 from 316, 316L, and 304 Stainless-Steel Surfaces

	Analyst	μg Added	μg Recovered	% Recovery
316 Stainless steel	1	246	238	96.7
	1	246	246	100.0
	1	246	246	100.0
	2	246	249	101.2
	2	246	254	103.3
	2	246	251	102.0
316L Stainless steel	1	246	249	101.2
	1	246	249	101.2
	1	246	250	101.6
	2	246	247	100.4
	2	246	240	97.6
	2	246	230	93.5
304 Stainless steel	1	246	250	101.6
	1	246	249	101.2
	1	246	251	102.0
	2	246	242	98.4
	2	246	248	100.8
	2	246	251	102.0
			Mean	100.3
			% RSD	2.4

ACKNOWLEDGMENTS

I would like to recognize the following contributors who generated all of the work that went into this Chapter: Lin Chang, Jerry Collins, Dave Conine, Phyllis Cugier, Lee Elrod, Ron Gorski, Diane Horgen, Michelle Jensen, Kathy Lenz, Chris Parker, Editha Reisterer, Terry Rotsch, Meredith Spanton, Steve Spanton, Dave Stroz, Eric Volkmann, Landy White, John Yoder.

REFERENCES

1. Analytical Methodology: The Heart of Cleaning Validation, Presented at the 42nd Annual Conference on Pharmaceutical Analysis, July 30, 2002, Merrimac, Wisconsin and 26th Puerto Rico Pharmaceutical Quality Association, January 28, 2003, San Juan, Puerto Rico.
2. Gorski, R. J., Plasz, A. C., Elrod Jr., L., Yoder, J. and White, L. B., Determination of Cefsulodin, Cefmenoxime and Cefadroxil as Residues on Surfaces, *Pharm. Res.*, 8:1525–1527, 1991.
3. Rotsch, T. D., Spanton, M., Cugier, P. and Plasz, A. C., Determination of Clarithromycin as a Contaminant on Surfaces by High Performance Liquid Chromatography Using Electrochemical Detection, *Pharm. Res.*, 8:989–991, 1991.
4. Volkmann, E., Gorski, R. J., Abbott Laboratories, unpublished results.
5. Reisterer, E. and Stroz, D. G., Abbott Laboratories, unpublished results.

17

LC/MS APPLICATION IN HIGH-THROUGHPUT ADME SCREEN

RON KONG

Synaptic Pharmaceutical Corporation, Paramus, NJ

ABSTRACT

This chapter reviews some applications of high-throughput liquid chromatography–mass spectrometry (LC/MS) techniques in absorption, distribution, metabolism, and elimination (ADME) screens in the lead optimization phase of drug discovery. It describes several routinely used techniques in performing these ADME assays with a strong focus on either the instrumental methods or experimental designs that significantly improve the throughput. Selected case studies on both instrumental methods and experimental setups are discussed to illustrate the strategies, trade-offs, and any potential pitfalls of these various techniques. The preparation of biological samples is also presented in great detail for the LC/MS quantitative bioanalysis.

Handbook of Pharmaceutical Analysis by HPLC,
S. Ahuja and M.W. Dong, editors.

I. INTRODUCTION

A. Scope

Liquid chromatography coupled with MS is widely used in drug discovery and development. Mass spectrometric applications are being used in qualitative analysis, quantitative analysis, and preparative high-performance liquid chromatography (HPLC) analysis where the mass spectrometer serves as a detector to trigger fraction collection. This chapter will not attempt to discuss the basic theory of in vitro ADME screen and LC/MS instrumentation method. However, the author will present applications of liquid chromatography–mass spectrometry and tandem mass spectrometry (LC/MS and LC/MS/MS) in in vitro ADME screens and bioanalysis of in vivo biological samples for hit evaluation as well as the lead optimization phase in drug discovery. The author will also make an effort to report the most recent advances in LC/MS software and hardware, analytical instrumentation, and automated sample handling and processing techniques for which significant increase in analysis throughput has been reported.

B. Background

There has been a significant increase in the demand for new drug candidates exhibiting improved efficacy, selectivity, and safety. In response to this demand, drug discovery strategy has dramatically changed during the last two decades.[1–9] Automated and parallel organic synthesis can generate a large number of compounds that can be screened for biological activity and potency against the appropriate targets. These high-throughput screens (HTS) usually identify many hits with potent in vitro pharmacological activity.[10,11]

In order to evaluate the ADME properties of hits from HTS, ADME screens are utilized in parallel with pharmacological screens.[5,12–18] These ADME screens should be relatively inexpensive, have a rapid turnaround and be reproducible and amenable to automation. The need for a high-throughput ADME screen is driven by the necessity not only to accelerate the discovery process identifying new leads rapidly, but also to improve the success rate of clinical development. Therefore, ADME screens for desirable drug-like properties have been developed to help identify promising new chemical entities (NCEs) among libraries in order to optimize the ADME properties and minimize the potential of toxic liabilities. These ADME assays will generate a significant amount of data that can be utilized to categorize and rank-order NCEs according to their aqueous solubility, chemical stability, intestinal permeability, hepatic metabolism, etc. The results from ADME screens can be utilized by medicinal chemists to design and synthesize NCEs with more favorable ADME properties in the lead optimization phase.[1,2,5,9,12–18]

C. Challenges to Analytical Instrumentation

The high-throughput ADME screens have resulted in a significant increase in the number of samples for analysis. The majority of these samples are derived from biological fluids while some of them are derived from other chemical matrices. Determining the compound's bioavailability in animal species and predicting human intrinsic clearance using in vivo pharmacokinetic studies and in vitro ADME assays, respectively, rely on the integrity of the specific quantitative methods for assessing the absolution concentrations of the compounds. Although various analytical techniques are historically employed for the quantitative bioanalysis of in vitro ADME samples and in vivo biological matrices, LC/MS equipped with automated sampling devices and sample preparation instruments has supplanted other analytical techniques for bioanalysis due to its rapid sample handling and excellent selectivity for detection.[19–24] Extensive applications of LC/MS and LC/MS/MS can be found for the various analyses of samples derived from numerous pharmacokinetic studies and in vitro ADME screens.[25,26] In comparison to HTS for pharmacological activity, the throughput of MS-based screens is relatively low. Nevertheless, LC/MS provides a significant means to determine ADME properties of a large number of compounds that include physiochemical properties, permeability, metabolic stability, biotransformation, and in vivo pharmacokinetic profile.[25,26]

II. MEASUREMENT OF PHYSICOCHEMICAL PROPERTIES

Although pharmacological efficacy is a key parameter in nominating a compound for pre-clinical development, the compound's physicochemical properties have significant effects on the success of developing a clinical candidate.[27] Physicochemical properties of a drug candidate, such as solubility, chemical stability, lipophilicity, and membrane permeability, will affect the rate and extent of absorption of an orally administered drug. Unfavorable physiochemical properties may lead to poor pharmacokinetics and therefore poor pharmacodynamics, which will cause an eventual failure of the candidate in clinical development.[27,28] Therefore, screens for favorable physiochemical properties are an extremely important criterion in the drug discovery process. This section provides a brief but systematic overview of the common LC/MS techniques in the characterization of the physicochemical properties of small molecules.

A. Solubility

Determination of equilibrium solubility is a laborious process involving continuously shaking a compound in a selected medium for more than 24 h until the dissolution–precipitation equilibrium is reached.[29]

The concentration of the test compound in the saturated solution is then determined to provide an equilibrium solubility parameter. HPLC with ultraviolet (UV) detection is routinely used to analyze the saturated solution. This traditional shake-flask method with HPLC/UV detection is not suitable for modern high-throughput drug discovery because it requires a relatively large quantity of each compound as well as tedious method development.[29] New high-throughput methods, such as turbidimetric and direct UV assays using 96-well microtiter-plates, have been developed.[29,30] Another method using pH-metric titrations was also reported.[31–33] Neither turbidimetric nor pH-metric titration methods provide a direct measurement of solubility and therefore can only be used to rank-order molecules according to their expected solubility. A direct UV assay provides a fast measurement of thermodynamic solubility; however, this method is limited to pure and stable compounds due to its lack of specificity.[29,30]

Replacing HPLC/UV detection with LC/MS detection for equilibrium solubility studies shows improved throughput, accuracy, and selectivity, especially in studying compounds with low solubilities. Jackson recently developed an LC/UV/MS mechanism to determine the thermodynamic solubility of compounds.[34] In this study, a wide set of pharmaceutical compounds were selected to validate the assay. The standards were prepared in a 96-well plate with known concentrations. Another plate contained samples with varying concentrations to simulate the solubility experiment. Calibration curves were constructed from the standards, and the compounds were quantified to determine the solubility. All samples were quantitatively analyzed by a multiplexed (MUX) LC/UV/MS system as illustrated in Figure 1.[34] The MUX LC/UV/MS system was operated with MassLynx 4.0, which allows four independent LC inlet systems. The parallel system employed four binary gradient pumps (Waters 1525 μ) for parallel analysis of four different compounds simultaneously, resulting in a four-fold increase of throughput in comparison to a single chromatographic system.[34] The mass spectrometer

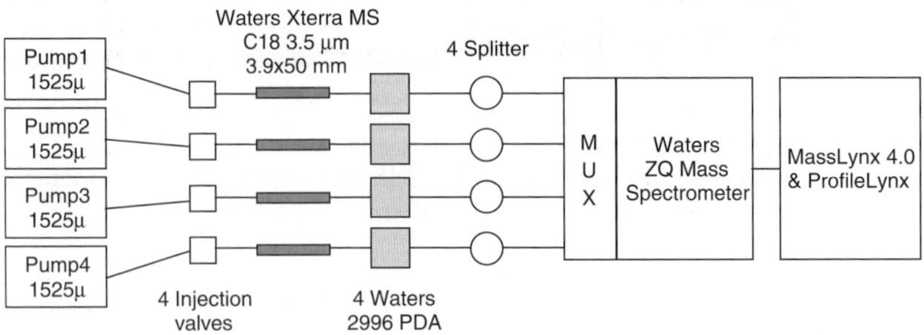

FIGURE 1 Schematic diagram of high-throughput LC/UV/MS analysis for equilibrium solubility assay.[34]

was operated in positive/negative ionization switching mode to maximize the output of mass spectrometric information. A short column (3.9 × 50 mm) with a fast gradient was used to reduce the run time. The data from four-channel parallel LC/UV/MS analyses were managed by ProfileLynx that automatically handled sample management, data processing, and interpreting and formatting results.[34]

Hayward reported a high-throughput ultra-filtration (UF) LC/MS approach to measure the thermodynamic solubility (Figure 2).[35,36] After thermodynamic equilibrium of the test compound in a selected buffer, the saturated solution was filtered through a molecular weight cut-off filter. The concentration was determined by LC/MS relative to a single-point standard solution for maximal throughput. The results from the UF LC/MS method as shown in Table 1 were equivalent to those obtained via other thermodynamic methods and were in accordance with the rank-order provided by the nephelometry method.[35] The success of this UF method relies on the minimal binding of compounds to the filter membrane. In this UF LC/MS method, it was found that total compound accountability was excellent (>90%) and that compound loss to the UF-membrane was minimal (<20%) regardless of the membrane material used.[35] HPLC instrumentation equipped with a gradient pump and a single-quadrupole mass spectrometer equipped with an electrospray ionization (ESI) source will be sufficient for such solubility screens. Other atmospheric pressure ionization (API) methods such as atmospheric pressure chemical ionization (APCI) and atmospheric pressure

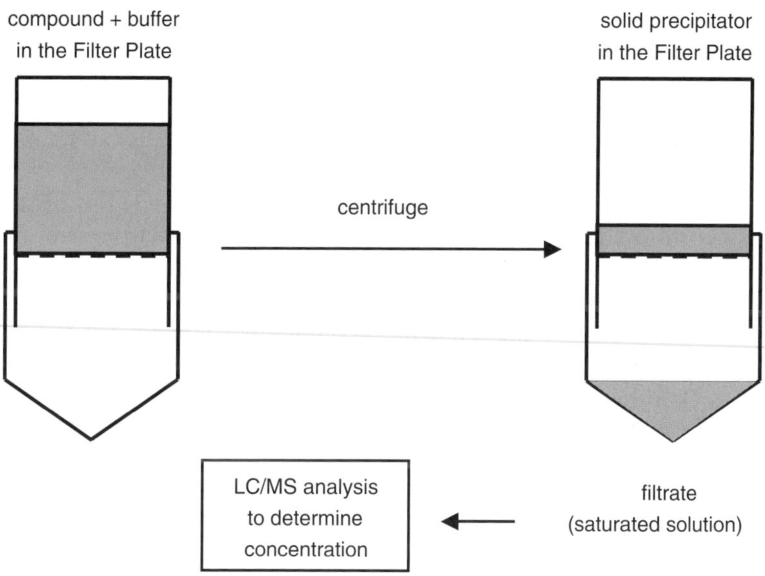

FIGURE 2 Solubility screening by UF LC/MS.[35]

TABLE 1 Comparison of Solubility Measurements by UF-LC/MS, Thermodynamic Approach (TD), and Nephelometry[35]

	Solubility (mg/ml)		
	UF-LC/MS	TD	Nephelometry
Compound	(pH=7.4)	(pH=6.8)	(pH=6.8)
Carbamazepine	0.20 ± 0.01	0.13	0.05
Diclofenac	0.67 ± 0.09	0.38	>0.2
Flurbiprofen	5.3 ± 0.7	5.1	>0.2
Furosemide	5.0 ± 0.3	10	>0.2
Ibuprofen	1.9 ± 0.3	1.1	>0.2
Naproxen	4.6 ± 0.2	4.8	>0.2
Phenytoin	0.020 ± 0.003	0.02	0.02
Terfenadine	0.09 ± 0.02	0.06	0.05

photoionization (APPI) may be needed for the analysis of non-polar low-molecular-weight compounds.[35,36]

Solubility screens using LC/MS detection do not require an ultra-pure sample of the test compound due to the selective detection of the mass spectrometer. Mass spectrometric detection offers high selectivity and low detection limits, which eliminates the need to develop complex chromatographic methods. The LC/MS-based solubility screen surpasses the traditional HPLC/UV-based equilibrium solubility assay with increased throughput, minimal manual intervention, and high sensitivity and selectivity.

B. Lipophilicity

Lipophilicity is a parameter that determines the distribution of a molecule between the lipid and the aqueous environment in a biological system. It is often quoted to give some indication regarding the potential of a drug on protein binding, membrane permeation, and tissue distribution.[27,28,37] Lipophilicity is commonly characterized as Log P, where P stands for partition coefficient. In the drug discovery arena, the partition coefficient is usually referenced as the concentration ratio of a compound (as the neutral molecule) in an immiscible solvent to that in an aqueous phase.[30,37,38] Another frequently used parameter in characterizing the lipophilicity of compound is Log D where D is the distribution coefficient defined as the total concentration (both ionized and unionized forms) ratio of a compound in an immiscible solvent to that in an aqueous phase. Obviously, Log D may vary according to the pH of the aqueous phase. 1-Octanol is often chosen as a simple organic phase to represent the lipid environment in both Log *P* and Log *D* measurements.[30] Several methods are available to measure Log *P*.[30,39–41] The throughput, technical difficulty, and sample requirement amount can vary significantly according to the methods. HPLC using either UV or MS detection is a very useful analytical tool for the determination of both Log *P* and Log *D*.

There are two different mechanisms of measuring Log *P* and Log *D* by using LC/UV/MS. The first approach is a direct employment of the "shake-flask" method using 1-octanol and water.[39,42] The concentrations of a test compound in two immiscible phases are determined using LC/UV/MS detection, followed by calculating the ratio of these two concentrations. This traditional method is time-consuming and does not meet the demand for screening large numbers of compounds in a drug discovery mode. Efforts now have been directed to increasing the throughput by integrating a 96-well microtiter plate with a robotic liquid-handing system. The entire process can be automated and only requires a small amount of sample which can be directly dispensed from a dimethylsulfoxide (DMSO) stock solution.[39] In this 96-well microtiter method, a Beckman Biomek 2000 liquid-handling station was used to automate the sample preparation. After equilibration of the partitioning process, each phase was injected directly from the plate using a custom-made GILSON autosampler, which eliminated the manual sample handling process. Analysis of the two phases was performed by using a reversed-phase HPLC/UV with a fast generic gradient. This method showed a significant increase in the assay throughput with minimal manual intervention.[39] The "shake-flask" approach can be further improved by using HPLC/MS as the detection method in order to eliminate HPLC method development and minimize the need for high purity and large quantity of a compound.[43] In this LC/MS method, a test compound at 10 mM concentration in DMSO was distributed to a 96-well plate containing 1-octanol/pH 7.4 phosphate buffer. The samples were allowed to reach partitioning equilibrium by vigorous shaking for 24 h, followed by centrifugation to remove emulsions. The 1-octanol and phosphate buffer layers were analyzed on an LC/MS system by directly injecting the samples from each well. Adjustment of the needle depth of a GILSON autosampler was required in order to aspirate the samples from two layers. A generic fast gradient was used for LC/MS analysis. Mass spectrometric detection was operated in the positive ionization mode with selected ion monitoring (SIM). The single-channel LC/MS analysis technique was further improved to a parallel technique with four compounds being analyzed simultaneously.[43] A typical run time for four compounds was completed within 2.5 min. Log $D_{pH=7.4}$ was determined by calculating the peak area ratio of the analyte in the 1-octanol sample to that in the aqueous sample. A four-fold increase in the throughput was easily accomplished using a 4-column multiplex LC/MS system.[43]

Another LC/MS approach to determine Log *P* and Log *D* utilizes the chromatographic retention of test compounds. The retention time was converted to a Log *P* scale through a calibration curve obtained from compounds with known Log *P*.[35,36,40,41,44–51] With the aid of highly selective MS detection, Log *P* and Log *D* measurements of multiple compounds were readily carried out on a single LC/MS run. This allowed faster lipophilicity evaluation of a structural series in the parallel synthesis of chemical libraries. Kerns reported a method that was able to analyze up to 20 compounds simultaneously on an Agilent 1100 LC/MSD system using alternating positive/negative electrospray ionization mode.[45] In this LC/MS method, a ballistic 2.5-min gradient with 1.0 mL/min flow rate was applied on a 2 × 50 mm Metachem Polaris C18 column. Variation in the retention times was not observed between individual analyses of each compound and pooling analyses of multiple compounds. Therefore, the pooling strategy to determine Log *P* and Log *D* could reduce the analysis time dramatically. Another strategy to increase analysis throughput was to use a multiplex LC/MS system.[52] This multiplex LC/MS system consisted of a Micromass eight-channel multiplex LCT mass spectrometer equipped with a GILSON 306 HPLC system, a GILSON 215/889 eight-channel injector, and eight UV detectors (Figure 3).[52] LC/UV/MS analysis was performed on a Phenomenex 3 × 50 mm LUNA C18 column with a 50-min isocratic run. The UV detection was essential to obtain the t_0 value that was used to calculate the capacity factor *k*. Log *P* was determined via a correlation to the Log *k* value of each compound. The challenge in this multiplex LC/UV/MS method was to maintain optimal performance of the LC/MS system by careful sample preparation, proper selection of mobile phases, and selection of compatible flow rates.[52] Improper sample cleanup could clog the HPLC

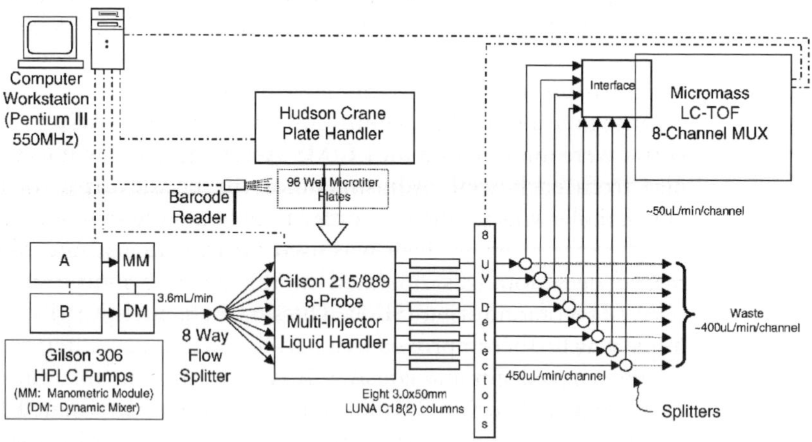

FIGURE 3 A multiplex LC/UV/MS system for Log *P* measurement.[52] (Copyright 2002 © John Wiley & Sons Limited. Reproduced with permission.)

columns, thereby affecting the chromatographic retention of analytes in the remainder of the columns in the eight-channel system. In addition, the change of retention times could throw off either the calibration curve or the correlation between Log *P* and Log *k*. High buffer concentrations in mobile phases and high flow rates could impair the ionization efficiency in the multiplex electrospray ion source, which could lead to unsuccessful detection of test compounds. The multiplex LC/UV/MS methodology was suitable for compounds with a wide range of Log *P* values, with the exception of very basic/acidic or very hydrophilic compounds (Log $P<0.5$).[52]

In general, lipophilicity screening using LC/MS methods is simple and does not require highly pure test compounds. The entire experimental measurement can be automated for a higher throughput when a robotic liquid-handling system is used in the sample preparation.[39] The "shake-flask" method using 1-octanol and water requires a long equilibration period and may not be amenable for HTS.[39,43] The indirect determination of Log *P* and Log *D* through correlation to the retention times offers a quick evaluation of lipophilicity, but the results could be superficial for compounds from different structural series and may require a long run time for compounds with a wide range of Log *P*'s.[40,41,52]

C. Permeability

Gastrointestinal (GI) absorption is one of the key factors governing the bioavailability of an orally administered drug. A drug can be absorbed via passive diffusion through the membranes of the intestinal cells (transcellular mode), paracellular permeation through the tight junctions between cells, or via an active carrier-mediated transport. Since in vivo studies can only evaluate a limited number of compounds, cell-based in vitro models are used to predict the in vivo absorption process.[53–56] In vitro assays using human colon adenocarcinoma (Caco-2) and Madin–Darby canine kidney (MDCK) are extensively used to assess the permeability potential of pharmaceutical compounds.[53–58] When cultured on a permeable porous filter membrane, both Caco-2 and MDCK cell lines differentiate into polarized monolayers with associated tight junctions, mimicking in vivo intestinal epithelial cell barriers. In Caco-2 and MDCK permeation assays, permeability is characterized by the diffusion rate from the apical (A) side to the basolateral (B) side and vice versa. The diffusion rate, the apparent permeability coefficient (P_{app}), is determined by measuring the rate of appearance (concentration) of a test compound in the receiving side under sink conditions (Figure 4).[53–56]

Recently, Caco-2 permeability screening has been optimized to accommodate the increasing demand for permeability assessment.[59–65] Analytical methods like HPLC coupled with UV, evaporative light scattering detection (ELSD), or fluorescence detection were used traditionally

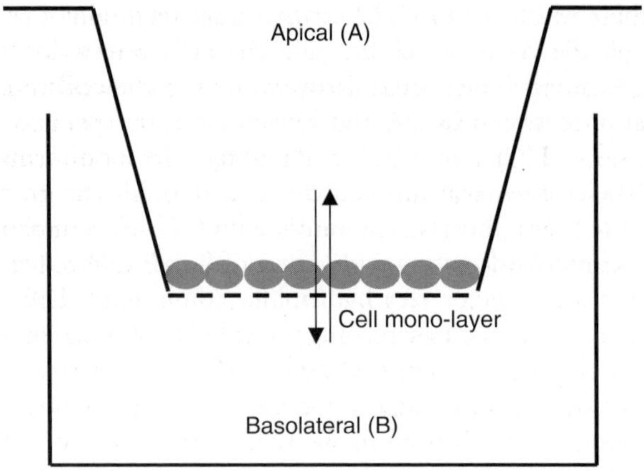

FIGURE 4 Schematic diagram of cell-based permeability assay.[53]

to determine concentrations of a test compound across the Caco-2 cell membrane.[25] However, these detectors are not selective and have limited sensitivity for many compounds. Radioactivity detection can be used, but the availability of radiolabeled materials is a significant limitation in early discovery. All of the above-mentioned HPLC detection systems require extensive method development, which limits the throughput in the permeability screening assays.[25]

The analytical bottleneck can be readily overcome when HPLC is combined with MS. The LC/MS technique has been widely used for quantitative analysis of compounds in Caco-2 permeability screen.[66–74] Its unique selectivity and specificity allow the measuring of small amounts of analytes in complex matrices. Analytes can be separated from matrix components by HPLC, followed by mass spectrometric detection. In order to maximize the throughput, separation is normally conducted using reversed-phase HPLC with a fast gradient.[66–74] To prevent contamination of the mass spectrometer API interface, a switching valve is often placed between the HPLC column and mass spectrometer to divert early eluting salts and hydrophilic matrix components to waste.[66–74] This diverting valve can be automatically controlled through contact closure. Quantitative analysis of drug concentrations from a Caco-2 permeability screen is generally carried out on a triple-quadrupole instrument using tandem MS detection in the multi-channel reaction monitoring (MRM) mode using either APCI or ESI interface.[66,67] Ion trap instruments like the ThermoFinnigan LCQ mass spectrometer can also be used for such quantitative analysis.[68] When the amount of sample supply is low, capillary LC/MS/MS can be used. Stevenson et al. reported a capillary LC/MS/MS method to analyze peptide combinatorial libraries in Caco-2 permeability screening.[69] In this method, chromatography was performed on an HP 1090 HPLC system equipped with an LC Packings capillary column (15 cm × 320 μm ID). The flow rate was 5 μL/min with trifluoroacetic acid (TFA)/acetonitrile/water as the mobile phase. A Sciex API-III was used for mass spectrometric detection operated in total ion chromatogram (TIC) and MS/MS daughter ion monitoring modes. A pool of 2500 peptides was dosed to the apical compartment of the Caco-2 model and LC/MS was used to monitor the transport of peptides across the cell barrier.[69] Although most of the peptides could not be chromatographically resolved even with an 80-min run time, LC/MS analysis was able to detect permeable species in the basolateral compartment. Peptides that were permeable as identified by LC/MS were subsequently characterized by LC/MS/MS to confirm the sequences. The study showed that LC/MS could be used for permeability screening of combinatorial libraries and structural characterization of the resulting permeable compounds.[69] A similar LC/MS method was also reported for the Caco-2 permeability screen of combinatorial libraries.[70]

The above method to assess the permeability potential of a combinatorial library mix can be considered as a cassette dosing mode in Caco-2 screen. This cassette dosing approach was also applied to individual small molecules in Caco-2 permeability screening.[71] The samples were analyzed on a Micromass Quattro II mass spectrometer equipped with a Waters 2690 HPLC system. The separation was performed on a Phenomenex C18 guard column cartridge with a 5-minute run time. Data were recorded in MRM mode. Since MRM detection is so selective, no special effort was made to establish chromatographic separation of compounds. The C18 guard column cartridge played a key role in this instrumentations setup by quickly removing non-volatile matrices at low organic compositions and fast elution of analytes into the mass spectrometer. The use of a guard column cartridge also allowed for a fast gradient that led to the fast equilibration of the column and a short run time.[71] Recently, an HTS for Caco-2 permeability utilizing a multiple sprayer LS/MS/MS system was reported.[72] A four-way multiplexed electrospray interface (MUX) LC/MS system was used to analyze four LC effluents simultaneously with a generic fast gradient HPLC method. The system allowed a throughput of 100 compounds per week for Caco-2 permeability.[72] A recent study indicated a good correlation between the results obtained from the single compound as well as multiple compounds assessments in the Caco-2 permeability screen.[60]

Another in vitro method for permeability screening was parallel artificial membrane permeation assay (PAMPA) initially reported by Kansy.[75] In a PAMPA permeability screen, the Caco-2 cell mono-layer membrane is replaced by an artificially generated membrane. Versions of different artificial membranes that lack active transporter systems and pores have been developed to mimic the in vivo transcellular intestinal epithelial cell barrier.[50,59,76–78] Therefore, the PAMPA screen only measures the intrinsic permeability potential of compounds through a lipid layer. Since the majority of orally administered drugs are absorbed via a transcellular route, PAMPA values can be used as an indicator of in vivo absorption potential.[30,50,59,75–77,79–84]

In order to automate PAMPA for high throughput permeability screening mode, a simple, fast, sensitive, and accurate analytical method is required. LC/MS detection provides an excellent solution for this analytical challenge. The PAMPA screen described herein was carried out on a diverse set of known drugs with known permeability values ranging from low to high.[75–77,85] A solution consisting of lecithin and dodecane mimic lipid membrane was deposited on the filter membrane at the bottom of the donor wells. Compound stock solutions were prepared at 10 μM in 100 mM K_2HPO_4 buffer at pH 7.4 and then were pipetted into donor wells. These donor wells were placed in the acceptor wells that contained 100 mM K_2HPO_4 buffer (pH 7.4). The combined donor–acceptor plate sandwich was then incubated for 4 h at 37°C. Following incubation, aliquots from both donor and acceptor wells of each of the compounds were analyzed sequentially. A typical LC/MS method for high throughput analysis of PAMPA samples included a short and narrow-bored HPLC column, Phenomenex Synergi C18 Hydro-RP 30 × 2 mm 4 μm 80 Å. A fast gradient and a high flow rate were utilized to achieve quick desalting and separation of samples. The LC/MS analysis was implemented using an Agilent 1100 binary pump, a Leap PAL autosampler, and a Micromass Quattro Micro triple-quadrupole mass spectrometer. A typical LC/MS analysis was completed in approximately 5 min. Most compounds showed good chromatographic retention until the gradient started.[75–77,85] This significantly minimized any ion signal suppression effects from the incubation matrices. The high-throughput LC/MS technique allowed potential analyses of hundreds of PAMPA samples per day, thus making the PAMPA assay suitable for high-throughput permeability screening. The throughput of the PAMPA assay can be further improved using a cassette dosing mode in a single well while using currently available MRM mass spectrometric detection.[75–77,85]

LC/MS/MS techniques with selective and sensitive detection methods make it possible to quantitatively analyze samples from Caco-2 cell and PAMPA buffer matrices. A high-throughput permeability screen with robust LC/MS technology can quickly generate information about structure–permeability relationships that are extremely valuable in the lead optimization phase for the selection of pre-clinical candidates with favorable oral bioavailability properties.[79–88]

III. CYTOCHROME *P*450 INHIBITION ASSAY

Both xenobiotics and endogenous substrates are primarily metabolized by a super-family of enzymes, named cytochrome *P*450s (CYP450s,

CYPs, or *P*450s).[89–90] Among the cytochrome *P*450s, CYP1, CYP2, and CYP3 families are the main enzymes involved in the oxidative metabolism of xenobiotics in humans. Five isozymes including CYP3A4, CYP2D6, CYP2C9, CYP2C19, and CYP1A2 have been identified to be primarily responsible for the metabolism of most drugs while CYP3A4 has been implicated for the major metabolizing enzyme responsible for >50% marketed drugs. As a result, the inhibition of CYP450-mediated metabolism may cause potential drug–drug interactions.[89–91] Currently, the inhibition potential of CYP450s is routinely evaluated through an in vitro screen using either cDNA-expressed isozymes or pooled human liver microsomes. The CYP inhibition using cDNA-expressed isozymes is often performed on each of the selected CYP enzymes using 96-well microtiter plates and a fluorescence-based detection method, whereas the pooled microsomal assay uses LC/MS/MS detection methods.[92–104]

The combination of high-throughput biological screens and parallel chemistry synthesis has generated a vast number of pharmacologically active NCEs that need to be evaluated for ADME properties, including their CYP inhibition potential. The in vitro CYP inhibition screen in the liver microsomes that probes for a single CYP isozyme is very laborious and time-consuming. Therefore, a new strategy that uses a cocktail of CYP-specific substrates to probe CYP inhibition of multiple isozymes in pooled human liver microsomes was developed recently.[97–104] When this cocktail incubation approach was integrated with a robust LC/MS/MS analytical technique, it allowed the improved throughput for an in vitro CYP inhibition screen.[97–104]

The cocktail approach uses a mix of well-characterized CYP-specific substrates to assess enzyme activity and inhibition when a test compound is co-incubated in the pooled liver microsomes. The assay is monitored by fast-gradient LC/MS/MS analysis to determine the rate of formation of the corresponding metabolite from each specific substrate. Dierks reported a fast LC/MS/MS method to simultaneously evaluate the activities of seven major human CYP450 isozymes using a cocktail of probe substrates.[98] The CYP-specific substrates consisted of midazolam (CYP3A4), bufuralol (CYP2D6), diclofenac (CYP2C9), ethoxyresorufin (CYP1A2), *S*-mephenytoin (CYP2C19), coumarin (CYP2A6), and paclitaxel (CYP2C8). The inhibition on each isozyme was calculated from the rate of formation of a specific metabolite with respect to the substrate. Separation was performed using a short BDS Hypersil C18 column (20×2 mm, 5 μM) with a fast 4-min HPLC gradient. The LC/MS/MS system for quantitative analyses of metabolites consisted of a TSQ 7000 mass spectrometer operated in APCI positive ion mode and an HP 1090 HPLC system. The samples were analyzed for all specific metabolites from the corresponding substrates respectively, via SRM transitions. Quantification was calculated using Finnigan LCQuan version 1.2. The fast LC/MS/MS analysis of seven metabolites in a single sample allowed rapid assessment of the cytochrome *P*450 inhibitory profile of new chemical entities. The method proved to be robust and readily adaptable to automation by using liquid handlers in a 96-well plate format and using a 96-well plate HPLC autosampler.[98]

A similar cassette probe dosing using a cocktail of CYP-specific substrates was also extensively studied.[99–104] In this approach, the composition of CYP-specific substrates varied slightly from Dierks's method, using tolbutamide, dextromethorphan, and chlorzoxazone as CYP2C9, CYP2D6, and CYP2E1 substrates, respectively. The inhibition for each isozyme was calculated from the rate of formation of the corresponding metabolite from the specific substrate. A novel direct injection/on-line guard cartridge column was used to extract analytes from the microsomal matrix in this method.[99–101] The samples were cleaned and separated on an HP 1090 HPLC equipped with a Phenomenex C18 guard cartridge (4×2 mm). The mass spectrometer, a Micromass Quattro II mass spectrometer, was operated in the ESI mode with positive and negative polarity switching. The guard column HPLC inlet system allowed rapid analyses of supernatant samples that were obtained from the centrifugation of microsomal incubation mixtures. The CYP inhibition assays utilizing cocktail substrates showed comparable results to those assays utilizing individual substrates.[99–104]

LC/MS can also be used for in vivo assessment of CYP inhibition. Pleasance and co-workers reported an LC/MS method to simultaneously analyze Greenford-Ware cocktail substrates and their corresponding metabolites.[97] The substrates containing caffeine, diclofenac, mephenytoin, debrisoquine, chlorzoxazone, and midazolam were administered to humans.[97] Human biological samples, plasma and urine, were cleaned and concentrated with solid phase extraction (SPE) to generate LC/MS/MS samples containing substrates and their corresponding metabolites, and the test compound as well. The challenge was to extract these structurally diverse substrates and metabolites from biological matrices effectively. A custom-built system was used for this generic extraction method.[97] Solid phase extraction experiments were performed on a Zymark robotic system with a 30 mg Oasis HLB extraction 96-well SPE block. The extracts were analyzed on an LC/MS/MS system, a PE Sciex API 365 mass spectrometer equipped with a PE 200 Series HPLC. Separations were performed on a Magellen C18 column (50×2.1 mm) using a 2-min ballistic gradient at a high flow rate (800 μL/min). Analytes were monitored in a positive–negative polarity-switching mode using the SRM detection method. Another challenge in quantifying all substrates and the corresponding metabolites in a single run was to construct good calibration curves for these structurally diversified analytes with concentration range of low ng/mL to low μg/mL. However, the authors successfully demonstrated that LC/MS/MS with SRM could be used to determine all analytes in a single HPLC run, over concentration ranges of 5–3000 ng/mL. The method was demonstrated to be specific, accurate, and precise for all intra- and inter-day analyses.[97]

Recently, Chu et al. reported an ultra-fast LC/MS method for analysis of cytochrome *P*450 3A4 and 2D6 inhibition assays.[105] Testosterone and dextromethorphan were used as the specific substrates for CYP3A4 and CYP 2D6, respectively. LC/MS analyses were performed on a Sciex API 3000 mass spectrometer equipped with a Shimadzu LC-10Advp pump and a PE 200 autosampler. A Phenomenex Luna C18 (4.6×30 mm) column was used along with very steep gradients. Each sample analysis was completed in 0.5 min.

LC/MS/MS with selected reaction monitoring (SRM) offers a fast and simple means to analyze biological matrices, which is a key factor in high-throughput CYP inhibition screens using liver microsomes. Potentially, the LC/MS/MS technique is suitable for analyses of cocktail substrates in other in vitro drug metabolism evaluations such as CYP induction/activation assays, rapid analysis of pooled liver microsomes, rapid reaction phenotyping of tissue (hepatic and extrahepatic) samples, as well as evaluation of hepatocytes/tissue slice CYP activity.[106–111]

IV. STABILITY

Compound stability in a variety of matrices including varying pH buffers, plasma, and liver microsomes is an important consideration related to the success of the compound being developed in the clinic. Low stability profiles in these environments will lead to poor pharmacokinetics of a potential candidate. These stability results can be used by medicinal chemists to optimize the labile portion of the structures to further improve stability.

An orally administered drug will pass through different pH zones in the GI tract before it is completely absorbed into the systemic circulation.[30,112,113] The GI tract pH zones range from pH 1.2 to 8.0.[30,112,113] Assessment of compound stability at different pHs similar to that in the GI tract will be extremely important for the selection of the most stable compounds from a lead series. High-throughput stability screens of compounds in chemical buffers usually contains two automated parts: (1) a robotic sample preparation and (2) rapid LC/MS analysis.[114,115] Sanders et al. reported a similar automated method for determining the chemical stability of test compounds.[115] The compounds were first analyzed by positive/negative polarity-switching in the ESI mode and the data were automatically reviewed to determine whether the results met pre-set criteria. If the pre-set criteria were not met, the samples were re-run in another ionization mode (e.g., APCI) with minimal intervention by the analyst. A custom-made software program allowed the rapid manual review of the data obtained from LC/MS analyses and the analyst was able to provide a summary of the best results for these compounds. This integrated strategy was further applied to LC/MS/MS analysis to generate MS/MS information that could be used for subsequent high–throughput LC/MS/MS quantitative analyses of samples obtained from permeability as well as stability profiling assays. A chemical stability assay using LC/MS requires only a minimal amount of sample and allows direct use of DMSO stock solutions that have been utilized in the pharmacological HTS assay.[114,115]

Compounds showing rapid degradation in plasma usually exhibit poor pharmacokinetic profiles and pose considerable technical challenges in the bioanalysis. Drug instability in plasma will significantly impact the method development and validation of bioanalytical assays in the clinical development stage as well. Therefore, profiling compound stability in plasma is another important parameter in the lead optimization phase. Hsieh and co-workers reported a very simple procedure for measuring drug stability in plasma.[116,117] A Shimadzu HPLC system that consisted of an on-line degasser, an LC-10AD VP pump, an LC-10A VP controller, a LEAP autosampler, and a Phenomenex MF C8 column (4.6×50 mm) was used for extraction and separation. The procedure employed a thermostatic autosampler as the incubator with temperature settings at 4°C, 20°C, and 37°C.[117] Untreated human, monkey, mouse, and rat plasma samples containing the test compound were incubated in the autosampler and were directly injected onto the column that functioned as both an extraction column for on-line protein removal and an analytical column for chromatographic separation. The test compound and its biotransformation products were monitored in SRM mode on a PE Sciex API3000 mass spectrometer with a cycle time of approximately 6 min. The robustness of this semi-automated method was clearly demonstrated through repetitive injections and analyses of the same sample every 24 min. The results showed excellent reproducibility of the retention time and the peak shape. This method eliminated the traditional labor-intensive and time-consuming sample preparation procedures that are typically used for such plasma stability studies.[116,117]

The clearance and bioavailability of orally administered drugs are strongly influenced by their first-pass effect in the liver. Various in vitro assays have been developed to address hepatic metabolic stability in early drug discovery.[118,119] Liver microsomes are widely being used in metabolic stability screen since microsomes have been shown to retain activities of key metabolizing enzymes such as CYP450s, flavin monooxygenases, and glucuronosyltransferases.[122] Hepatocytes that retain a broad spectrum of enzymatic activities, including esterase and amidase activities, are also used for metabolic stability evaluation and other in vitro drug metabolism studies as well.[122] In vitro metabolic stability assays are usually performed by incubating a test compound with liver microsomes or hepatocytes.[118–123]

Liver microsomal metabolic stability screens can be readily automated for high-throughput assays in a 96-well microtiter plate format using a robotic liquid-handling system. The liquid-handling system offers an efficient mechanism to dispense various assay components, mix incubation, and collect assay samples according to a selected time-course.[124] Then, standard sample preparation procedures such as protein precipitation, solid phase extraction, ultra-filtration, and ultra-centrifugation could be used to process the assay samples. The advent of assay automation and robotic sample preparation has created an imminent demand for new analytical instrumentation that can efficiently process a large number of samples and is compatible with the high-throughput 96-well microtiter plate incubation format. Korfmacher et al. reported an automated LC/MS system for the quantitative analysis of liver microsomal incubation samples.[124] In this method, the incubation mixture in a 96-well plate format was treated with cold acetonitrile and centrifuged in the 96-well microtiter plates for collection of the supernatants that were used for LC/MS analyses. This 96-well plate format was designed to accommodate 15 test compounds and one control. The microsomal stability rate was calculated using the percentage of remaining parent compounds at a fixed time point. The LC/MS system consisted of a PE series 200 autosampler, a PE series 200 HPLC system, and a PE Sciex 150EX mass spectrometer. An AppleScript was custom-written to handle the data processing, which proved to bc an effective tool for automatic extraction of MS information, peak area integration, rate determination, and data reporting.[124] Janiszewski et al. reported a dual-injection column-switching LC/MS system to analyze samples obtained from high-throughput metabolic stability screens.[125] In this system, the autosampler—a Gilson 215 Multiprobe equipped with a custom-made dual-injection manifold—allowed parallel loading and injection of samples while the previous samples were being analyzed. A dual-column 10-port valve was used for column switching, where the first column was used to elute the sample into the mass spectrometer while the second column was equilibrating for the next sample to be loaded. An aqueous mobile phase that was directly plumbed into the autosampler was used for loading the samples onto the second column. An organic mobile phase was simultaneously directed through the first column to elute the analytes into the mass spectrometer. A PE Sciex API 150 single quadrupole mass spectrometer or a PE Sciex API 2000 tandem mass spectrometer was operated in SIM or SRM mode for data recording. A short dwell time of 200 ms was used in order to maintain good definition of chromatographic peaks. This high-capacity LC/MS system offered 2000 sample-runs per day and was compatible for analyzing a variety of samples from many ADME screens.[125]

Among the new techniques that have been developed to improve the throughput, turbulent-flow chromatography (TFC) coupled with MS has gained significant popularity in the bioanalysis arena.[126] Lim et al.[127] reported a simultaneous screen for both microsomal stability and metabolite profiling using a direct injection turbulent-laminar flow LC-LC/MS/MS.[127] Sample preparation was reduced through an on-line extraction of both the parent drug and its metabolites from microsomal proteins using turbulent flow chromatography. Samples were then injected onto an extraction column at turbulent flow, and the parent drug and its metabolites were extracted and retained on the column that was backwashed onto a reverse-phase column via on-line column switching. The parent drug and its metabolites were resolved chromatographically at a laminar flow of 2 mL/min. This tandem turbulent-laminar flow chromatographic system that was operated in a total cycle time of 8 min was validated by separating isomeric metabolites of venlafaxine, haloperidol, and adatanserin. This method was further validated to perform both microsomal stability assessment and metabolite profiling in a single analysis.[127] A Finnigan MAT ion-trap mass spectrometer was used to monitor assay samples with SRM recording for microsomal stability and data-dependent MS/MS analysis for metabolite profiling through a single LC analysis. Such an LC-LC/MS/MS approach will dramatically reduce the feedback time for providing metabolism data to drug discovery pioneers.[127]

Metabolic stability is a key assay in the in vitro ADME screening process in the lead optimization phase. High-throughput metabolic stability screens for evaluating and optimizing drug-like properties present a significant analytical challenge. During LC/MS method development, the analyst should pay attention to the analysis speed, system ruggedness, and system compatibility with robotic liquid handling in order to analyze hundreds of samples per day. A successful LC/MS method for high-throughput metabolic stability assays involves robust sample preparation/liquid-handling technique, a sensitive LC/MS detection method for sample analysis, and automated data processing. An automated technique for sample handling must be developed and implemented to minimize the sample plating and liquid handling errors. For sample analysis, the instrumentation must be capable of performing parallel multi-sample injections with associated wash cycles and sample loading. Multiple-column chromatography systems that provide column regeneration, peak stacking, and parallel chromatographic analyses of sample sets will also allow rapid throughput when being used with conventional LC instrumentation. An automated LC/MS/MS quantification method should be implemented to reduce manual data processing.[128–131]

V. METABOLITE PROFILING

The metabolites of a drug candidate are usually characterized after the drug candidate has entered into the developmental stage. However, an understanding of metabolite profiling and the major metabolic pathways of a lead candidate will be extremely valuable during the lead optimization process.[4,12,132] Information generated from metabolite profiling will assist in characterizing structure–activity relationships (SAR) and identifying metabolic labile positions of a lead structural series and thus guide medicinal chemistry efforts to synthesize an optimized structural series with improved metabolic stability and pharmacokinetic profiles.[2] Since metabolite profiling can significantly reduce the cycle time of structural optimization, the demand for metabolite identification to be readily available in the discovery stage has been growing. Such a demand has created a challenge to search for high-throughput metabolite profiling assays. LC/MS/MS has provided drug metabolism scientists a powerful tool with which to rapidly characterize drug metabolites quantitatively and qualitatively.[26,133–135] Rapid metabolite profiling using LC/MS/MS techniques such as neutral loss/gain, parent scan, and product ion scan can be readily implemented in combination with in vitro and in vivo systems and "smart" software for automated data acquisition and interpretation.[136–138]

In vitro systems like liver microsomes and hepatocytes are frequently used to study the biotransformation of drug candidates.[135–138] Fernandez-Metzler et al.[139] developed a set of computer programs to automatically process LC/MS data in search of the presence of metabolites in a biological complex.[139] In this study, generic chromatographic and mass spectrometric methods were employed for both LC/MS and LC/MS/MS analyses that generated fragmentation information used to distinguish drug-related components from endogenous materials. Cross-correlation of MS/MS spectra was also used to identify the relationship between each metabolite and its respective parent drug. In comparison to manual data processing, the automatic data handling combined with rapid LC/MS analysis has significantly increased the level of efficiency in elucidating the metabolic pathways of compounds.[139] Cox and co-workers employed a combination of Q-Tof, LCQ, and triple-quadrupole mass spectrometers for metabolite identification in support of drug discovery.[140–146] The triple-quadrupole mass spectrometer was used to obtain rapid elucidation of metabolites through preliminary LC/MS analysis and to study the correlation between metabolites and their parent compounds by identifying the common fragment ions in MS/MS mode such as neutral loss. Further structural analysis of metabolites was carried out on an ion-trap mass spectrometer that was capable of performing MS^n experiments. The high sensitivity and high-resolution capability of a Q-Tof mass spectrometer allowed accurate mass measurements and collision-induced dissociation (CID) experiments of metabolites present at very low concentrations. Metabolite profiling was further expedited with the aid of commercially-available metabolite ID softwares such as Metabolynx from Waters and Metabolite ID from ThermoFinnigan.[140–146]

With the advent of modern LC/MS instrumentation and sophisticated software, metabolic profiling has gradually become a significant tool to support drug discovery. Early identification of active or toxic metabolites and characterization of metabolic liability of a particular structural series will significantly reduce the cycle time of the lead optimization processes.[25,26,133,134] Rapid LC/MS/MS analysis combined with automatic data processing utilizing very sophisticated software will make metabolic profiling more accessible and affordable.

VI. QUANTITATIVE BIOANALYSIS

Although in vitro assays generate valuable information that can be used to predict in vivo ADME profiles, pharmacokinetic profiling in animals has had a critical impact on the selection of development candidates through utilizing parameters to predict human PK profiles by means of allometric animal scaling.[2] Moreover, medicinal chemists rely heavily on rodent pharmacokinetic data to study structure activity relationship (SAR) with respect to in vivo ADME in the lead optimization process.[12] High-throughput parallel synthesis in the lead optimization phase has generated large numbers of NCEs, among which a subset of NCEs with favorable pharmacological profiles needs to be further evaluated for favorable pharmacokinetic parameters. To meet such a demand, various in vivo screens were developed to characterize exposure (area under curve, AUC) of a large set of compounds quickly.[147–150] Efforts were made to increase the throughput of PK screens through cassette-dosing of multiple compounds into a single animal for reducing the overall cycle time of bioanalysis.[151–153]

Liquid chromatography combined with triple-quadrupole mass spectrometers has been used extensively as the analytical method of choice to determine the plasma concentration of compounds.[25,152,153] With the advent of API sources, LC/MS/MS allows the facile development of quantitative methods that are sensitive, selective, robust, and amenable to the rapid analysis of a majority of small molecules. In order to achieve high-throughput bioanalysis in support of pharmacokinetic studies, many approaches have been reported utilizing automated sample preparation and reducing analysis time by pooling samples, parallel analysis, and fast chromatography.[25,26,152,153]

A. Sample Preparation

Before compounds in biological matrices can be analyzed by LC/MS/MS, the samples must undergo a preparation procedure. There are a variety of techniques available for sample preparation including off-line sample preparation techniques (liquid–liquid extraction, protein precipitation, and solid phase extraction) and on-line sample preparation techniques. These sample preparation methods can also be employed for other in vitro ADME assays as well.

1. Protein Precipitation

Protein precipitation is widely used in preparing LC/MS samples for bioanalysis.[154–156] The plasma samples are usually mixed with 3–5 times their volume of organic solvents such as acetonitrile and methanol or acidified solutions such as diluted trifluoroacetic acid and perchloric acid.[157] Analysts must be aware of the compound stability at low pH before acids can be used for protein precipitation. The mixtures are centrifuged at 3000 rpm or filtered to afford a clear supernatant or filtrate solution. The supernatant or filtrate can be directly injected into an LC/MS system or dried and re-constituted in an HPLC mobile phase to generate final concentrated samples.[157] If evaporation and re-constitution steps are omitted, diluted samples will be obtained, and, as a consequence, the limit of quantitation (LOQ) of the assay may be compromised. Moreover, supernatant and filtrate solutions containing high percentages of organic solvent may not be suitable for direct injection when a fast gradient is being used.[158]

Off-line sample preparation through protein precipitation is often automated with a liquid–handling system that is used to perform sample transferring, mixing, and filtration.[154–156] The final samples can be prepared in a 96-well plate format that is compatible with most LC/MS autosamplers. Protein precipitation does not require very extensive method development and can be implemented as a simple generic method to prepare samples from discovery pharmacokinetic studies.[154–156] One of the caveats of protein precipitation is that the procedure has relatively poor sample cleanup. Therefore, co-eluting components can interfere with the ionization process of analytes by either suppressing or enhancing ion production.[154–158]

2. Liquid–liquid Extraction

Like solid phase extraction, liquid–liquid extraction (LLE) offers a cleanup of matrices and concentration enhancement in preparing LC/MS samples.[157,159–162] LLE is applicable to a variety of biological samples such as plasma and urine. Plasma in a 96-well plate format is mixed with an immiscible organic solvent (3 to 5 times matrix volume) such as hexane or ethyl acetate, and extensively mixed to allow partition equilibration.[157,162] The mixture is then centrifuged and the organic layer is collected, dried, and re-constituted in the HPLC mobile phase to afford LC/MS samples.[157] Automation in LLE cannot be readily implemented due to the utilization of multiple manual steps, like transferring, mixing, centrifugation, and evaporation. LLE methods usually offer a better cleanup of samples and require less method development in comparison to off-line SPE methods.[157,159–162]

3. Off-line SPE

Solid phase extractions (SPEs) are routinely used in the sample preparation for quantifying analytes in biological fluids such as plasma and urine.[163,164] This technique allows both removal of interfering biological matrix components and enhancing the concentrations of analytes in LC/MS samples. SPE is often performed on a liquid handling system using a 96-well plate format. A variety of a 96-well pre-packed SPE blocks with different sorbents are now commercially available.[165,166] In general, the biological samples are loaded onto SPE cartridges and the biological matrices are then washed out of the cartridges. Analytes are often eluted using small aliquots of an organic solvent so that the desired analytes are concentrated enough for analysis without further evaporation and concentration procedures. However, labor-intensive SPE method development as well as optimization is usually required in order to obtain clean and concentrated samples.[167–170]

4. On-line Sample Preparation

In bioanalysis, on-line sample preparation is also frequently used to clean up and extract analytes from biological matrices. Currently, there are two major on-line sample preparation methods available: (1) on-line SPE, and (2) turbulent flow chromatography.[171,172] Both techniques employ short and narrow bore columns that are packed with a large particle size hydrophobic stationary phase. Samples are frequently loaded with weak aqueous phase solvents to extract analytes from biological matrices such as plasma and urine. The analytes are retained on the extraction column. In this extraction procedure, both large molecules such as plasma proteins and small hydrophilic molecules such as inorganic salts will not be retained and will be directly eluted from the extraction column to waste. Once the analytes have been extracted and retained on the extraction column, the mobile phase strength is adjusted to elute the analytes that can be further separated on an analytical HPLC system. Such tandem chromatographic methods allow an automated streamlined process of sample injections, cleanups, separations, and analyses.[171,172]

Application of on-line SPE sample preparation with LC/MS/MS analysis was widely reported in the quantitative analysis of drugs and metabolites in plasma.[172–177] Calderoli et al. described an automated online SPE method in LC/MS/MS quantitation of brostallicin in human plasma.[172] In this automated on-line SPE method, diluted plasma samples containing an internal standard were injected onto an on-line SPE HySphere Resin SH cartridge (10×2 mm ID) and the analytes were back eluted into a Platinum Cyano column (100×4.6 mm) for chromatographic separation. The mobile phase used in this separation was 70/30 acetonitrile/20 mM ammonium formate buffer (pH 3.5) at a flow rate of 1.0 mL/min. LC flow was split so that 300 μL/min of effluent was directed toward the mass spectrometer, an Applied Biosystems/MDS

SCIEX API 365 equipped with a TurboIonSpray interface operated in positive ion mode. The calibration range 0.124–497 ng/mL was readily validated with a negligible carry-over effect from this system. The method offered a total cycle time of 8 min and completely eliminated the manual sample preparation.[172]

Another approach to on-line sample preparation is turbulent flow chromatography (TFC).[178–185] TFC coupled with MS allows the rapid and direct analysis of compounds in biological matrices such as plasma, serum, or urine. High flow rates and large particle size stationary phases are used in TFC, which differs from traditional HPLC in which it incorporates slow and laminar flow technology. Cohesive Technologies has developed and commercialized a suite of TFC instrumentation that consist of single or multiplexing TFC systems.[178,179] A recent article by Herman showed a universal TFC method that eliminated the need for significant method development in bioanalysis during the drug discovery phase.[179] Using TFC/MS, the time utilized for method development was significantly reduced in the quantitative analysis of plasma samples. A Cohesive Technologies 2300 HTLC system with a 2-valve configuration was used while mass spectrometric MRM data were acquired on a Micromass Quattro II.[179] In this method, the plasma sample was mixed with two plasma volumes of acetonitrile containing an internal standard. The mixture was centrifuged to afford a supernatant solution that was separated and injected onto a turbulent flow chromatography system. The analytes were extracted from plasma samples and retained on a Cyclone cleanup column (Figure 5a).[179] The cleanup column was then backwashed and the analytes were transferred from the cleanup column to the analytical column. In this step, the analytes were eluted using 200 μL of a strong organic solvent that was pre-filled in the loop. The organic effluent was merged with an aqueous weak mobile phase from Pump 2 so that the analytes could be retained on the analytical column (Figure 5b).[179] After the analytes were transferred, the cleanup column was washed and the loop was refilled with the strong organic solvent from Pump 1 while the analytes were being analyzed on the analytical column (Figure 5c).[179] Once the loop was filled, Valve 1 was switched and the cleanup column was allowed to be re-equilibrated (Figure 5d).[179] Such a TFC/MS method offered streamlined on-line sample preparation, leading to significantly enhanced bioanalysis productivity with increased sample throughput.

B. High-throughput LC/MS Analysis

After the biological samples have been cleaned up, analytes are introduced to LC/MS for analysis. Approaches to shorten the sample analysis time are well documented.[25,26,152,153] Two major strategies

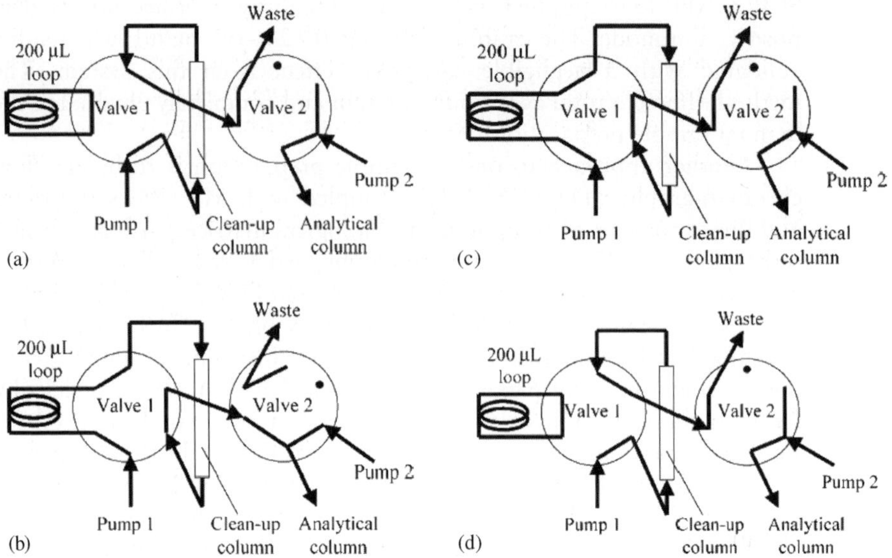

FIGURE 5 **Schematic diagram of valve configuration and flow direction.[179] (Copyright 2002 © John Wiley & Sons Limited. Reproduced with permission.)**

employing pooled sample analysis and multiplexing LC systems are frequently used in combination with a fast HPLC method to maximize analysis throughput.

1. Pooled Sample Analysis

One approach to increase bioanalysis throughput in an in vivo PK screen is to use a cassette analysis mode by pooling the samples acquired from the same time points.[149,186,187] In a cassette analysis method, plasma samples are pooled in order to reduce the number of test samples in an assay. This approach has the advantage of not only reducing the number of samples to be assayed but also of reducing the number of samples to be processed. Thus, an overall turn-around time for an assay study can be significantly reduced by using this methodology. Korfmacher et al. reported a pooled sample analysis strategy to increase the bioanalysis throughput in the evaluation of NCEs in a pharmacokinetic screen termed "cassette-accelerated rapid rat screen" (CARRS).[149] In this assay, NCEs were dosed individually in batches of six compounds per set. A semi-automated protein precipitation procedure was used for sample preparation in a 96-well plate format. Plasma samples were pooled as "cassettes of six" according to time points and were analyzed by liquid chromatography/atmospheric pressure ionization tandem mass spectrometry (LC/API-MS/MS). This assay also employed a mini-calibration curve consisting of three points and a zero. Using this new approach, a three-fold increase in throughput was reported in comparison to the earlier reported "rapid rat screen" methodology.[148,149]

Another strategy in post-dosing cassette analysis is to pool plasma samples of different compounds, leading to the formation of multiple compounds in a plasma sample. Korfmacher further improved the analysis throughput by pooling plasma samples of two different compounds that were obtained from combining samples by the same time point (double pooled analysis).[148,149,188] Although this pooling strategy of different compounds significantly reduces the overall demand for MS analysis, it may compromise the sensitivity due to sample dilution and may introduce the need for excess sample transferring and preparation.[188]

In order to expand the throughput for PK screening, cassette-dosing or N-in-1 was also employed. This cassette dosing method can provide plasma samples containing a mixure of multiple compounds.[147–150,189,190] Typically, fewer than five compounds are cassette-dosed to animals with one of the compounds as the benchmark. The advantage of cassette-dosing is quite obvious; it can process more compounds at a time and reduce the number of animals needed for an assay. However, the PK exposure of testing compounds may be compromised due to the potential of systemic drug–drug interactions.[191,192]

2. Parallel Analysis and Fast LC

While an HPLC analysis is frequently run over the span of a few minutes, the chromatographic peak of interest is usually recorded in less than a minute. As a typical LC/MS system analyzes samples in sequential order, the instrument time is not efficiently used. To overcome this limitation, several approaches have been reported to perform parallel HPLC analyses so that the chromatographic peaks can be staggered and the LC/MS system is constantly used in data recording.[25,26,152,153] Van Pelt et al. presented a novel approach to parallel LC/MS/MS analyses for pharmacokinetic assays.[193] In this parallel method, a conventional LC/MS system was upgraded to perform parallel sample analysis by incorporating three valves and four LC columns. Injections were staggered to allow the mass spectrometer to monitor the chromatographic windows of interest continuously. The parallel chromatography system decreased the overall run time and therefore, increased the overall throughput by a factor of 2.7 in comparison to a conventional LC/MS/MS analytical method.[193] Korfmacher et al. employed a different strategy to improve analytical efficiency by combining the post-column effluents from two parallel HPLC systems, which were then directed into a single MS.[194] The parallel HPLC/APCI-MS/MS system consisted of two Waters 2690 Alliance systems (each one included an HPLC pump and an autosampler) and one Finnigan TSQ 7000 triple quadrupole mass spectrometer. The MS data were later resolved to calculate the concentration results for plasma samples.

Another approach in parallel analysis is to couple parallel chromatography to a mass spectrometer equipped with a multiplexing ESI interface (MUX) that was developed and commercialized by Waters.[195,198] The MUX system allows four or eight flows to be analyzed simultaneously. When an MUX system is coupled with fast LC, the analytical throughput can be dramatically increased for PK screening. Wu and coworkers reported a high-throughput and high-performance bioanalytical system using automated SPE and an HPLC system equipped with monolithic columns.[196,197] This system allowed extraction and analysis of 1152 plasma samples within 10 h. The method involved an automated SPE step that was performed by a Zymark track robot system interfaced with a Tecan Genesis liquid handler. The SPE system could simultaneously process four 96-well plates. The extracted plasma samples were injected onto four parallel monolithic columns for separation via a four-injector autosampler. The total chromatographic cycle time was 2 min on each 100×4.6 mm column using gradient elution. The effluents from four columns were introduced to a triple–quadrupole mass spectrometer equipped with a multiplexing four-probe electrospray ionization source (Micromass MUX interface).[195–197] The entire assay including sample extraction, separation, and detection was carried out in a four-channel parallel format that resulted in an overall bioanalysis time of 30 s per plasma sample. Liquid chromatography using monolithic columns allowed fast and well-resolved separations at a considerably higher flow rate without generating significant column backpressure. The good separation efficiency provided by this system allowed rapid method development for an assay to quantify drug candidates and their metabolites. Despite the successful application of MUX technology, analysts should be aware of the potential carry-over as well as cross talks between four channels. With the advent of new MS software, Masslynx 4.0, the MUX interface on a Micromass mass spectrometer is capable of controlling four HPLC systems simultaneously, eliminating the limitation of flow rates and reproducibility of chromatography observed in the early instrumentation setups.[198]

VII. SUMMARY AND CONCLUSION

During the past few years, bioanalytical instrumentation innovations have changed the existing paradigms used in high-throughput ADME screens. This chapter attempted to review the recent advances of LC/MS/MS as potential tools for bioanalysis in support of ADME screens. In high-throughput bioanalysis practice, attention should be directed to achieve efficient sample preparation and data processing methodologies as well. Consideration should also be given to designing simple assay models, which streamline sample processing, and preparation techniques.

ACKNOWLEDGMENTS

The author acknowledges helpful ideas and suggestions from Prof. Mary C. Whitten from the Chemistry Department at Lock Haven University.

REFERENCES

1. Di, L. and Kerns, E. H., *Curr. Opin. Chem. Biol.,* 7(3):402–408, 2003.
2. White, R. E., *Ann. Rev. Pharmacol. Toxicol.,* 40:133–157, 2000.
3. Ulrich, R. and Friend, S. H., *Nat. Rev. Drug Discov.,* 1:84–88. 2002.
4. Tarbit, M. H. and Berman, J., *Curr. Opin. Chem. Biol.,* 2(3):411–416, 1998.
5. Masimirembwa, C. M., Thompson, R. and Andersson, T. B., *Comb. Chem. High Throughput Screen,* 4(3):245–263, 2001.
6. Dean, J. H. and Olson, H. M., *Biol. Cell,* 77(1):3–8, 1993.
7. Atterwill, C. K. and Wing, M. G., *Toxicol. Lett.,* 127(1–3):143–151, 2002.
8. Amidon, G. L., Lennernas, H., Shah, V. P. and Crison, J. R., *Pharm. Res.,* 12(3):413–420, 1995.
9. Meugge, I., *Med. Res. Rev.,* 23(3):302–321, 2003.
10. Crossley, R., *Modern Drug Discovery* 12(5):18–22, 2002.
11. Bleicher, K. H., Bohm, H. J., Muller, K. and Alanine, A. I., *Nat. Rev. Drug Discov.* 2(5):369–378, 2003.
12. Korfmacher, W., *Curr. Opin. Drug Discov. Devel.,* 6(4):481–485, 2003.
13. Caldwell, G. W., Ritchie, D. M., Masucci, J. A., Hageman, W. and Yan, Z., *Curr. Top. Med. Chem.,* 1(5):353–366, 2001.
14. Venkatakrishnan, K., Von Moltke, L. L. and Greenblatt, D. J., *J. Clin. Pharmacol.,* 41(11):1149–1179, 2001.
15. Venkatakrishnan, K., Von Moltke, L. L., Obach, R. S. and Greenblatt, D. J., *Curr. Drug Metab.,* 4(5):423–459, 2003.
16. Roberts, S. A., *Curr. Opin. Drug Discov. Dev.,* 6(1):66–80, 2003.
17. Ekins, S., Ring, B. J., Grace, J., McRobie-Belle, D. J. and Wrighton, S. A., *J. Pharmacol. Toxicol. Methods,* 44(1):313–324, 2000.
18. Chaturvedi, P. R., Decker, C. J. and Odinecs, A., *Curr. Opin. Chem. Biol.,* 5(4):452–463, 2001.
19. Covey, T. R., Lee, E. D. and Henion, J. D., *Anal. Chem.,* 58(12):2453–2460, 1986.
20. Heinig, K. and Henion, J., *J. Chromatogr. B,* 732(2):445–458, 1999.
21. Zweigenbaum, J., Heinig, K., Steinborner, S., Wachs, T. and Henion, J., *Anal. Chem.,* 71(13):2294–2300, 1999.
22. Zhang, H., Heinig, K. and Henion, J., *J. Mass Spectrom.,* 35(3):423–31, 2000.
23. Hsieh, Y., Merkle, K., Wang, G., Brisson, J. M. and Korfmacher, W. A., *Anal. Chem.,* 75(13):3122–3127, 2003.
24. Hsieh, Y., Merkle, K. and Wang, G., *Rapid Commun. Mass Spectrom.,* 17(15):1775–1780, 2003.
25. Ackermann, B. L., Berna, M. J. and Murphy, A. T., *Curr. Top. Med. Chem.,* 2(1):53–66, 2002.
26. Siegel, M. M., *Curr. Top. Med. Chem.,* 2(1):13–33, 2002.
27. Lipinski, C. A., Lombardo, F., Dominy, B. W. and Feeney, P. J., *Adv. Drug Deliv. Rev.,* 23:3–26, 1997.
28. Lipinski, C. A., *J. Pharmacol. Toxicol. Methods,* 44(1):235–249, 2000.
29. Bevan, C. D. and Lloyd, R. S., *Anal. Chem.,* 72(8):1781–1787, 2000.
30. Avdeef, A., *Curr. Top. Med. Chem.,* 1(4):277–351, 2001.

31. Avdeef, A. and Berger, C. M., *Eur. J. Pharm. Sci.,* 14(4):281–291, 2001.
32. Avdeef, A., *J. Pharm. Sci.,* 82(8):183–190, 1993.
33. Avdeef, A., Berger, C. M. and Brownell C., *Pharm. Res.,* 17(1):85–89, 2001.
34. Jackson, M. R., Balogh, M. P. and Potts, W. B., *An Interactive Physiochemical Property Profiling Software for Early Candidate Analysis in Drug Discovery*, American Society for Mass Spectrometry 2003 Conference Abstract, Montreal, Canada, 2003.
35. Hayward, M. J., Hargiss, L. O., Munson, J. L., Mandiyan, S. P. and Wennogle, L. P., *Validation of Solubility Measurements Using Ultra-Filtration Liquid Chromatography Mass Spectrometry (UF-LC/MS)*, American Society for Mass Spectrometry 2000 Conference Abstract, Long Beach, CA, USA, 2000.
36. Hayward, M. J., Munson, J. L., Mandiyan, S. P., Hargiss, L. O., Wareing, J. R. and Wennogle, L. P., *Fast LC/MS as a Tool for Lead Targeting in Drug Discovery*, American Society for Mass Spectrometry 1999 Conference Abstract, Dallas, TX, USA, 1999.
37. Abraham, M. H., Zhao, Y. H., Le, J., Hersey, A., Luscombe, C. N., Reynolds, D. P., Beck, G., Sherborne, B. and Cooper, I., *Eur. J. Med. Chem.,* 37(7):595–605, 2002.
38. Dearden, J. C., *Environ. Health Perspect.,* 61:203–228, 1985.
39. Hitzel, L., Watt, A. P. and Locker, K. L., *Pharm. Res.,* 17(11):1389–1395, 2000.
40. Valko, K., My Du C., Bevan C., Reynolds, D. P. and Abraham, M. H., *Curr. Med. Chem.,* 8(9):1137–1146, 2001.
41. Kaliszan, R., Haber, P., Baczek, T., Siluk, D. and Valko, K., *J. Chromatogr. A.,* 965(1–2):117–127, 2002.
42. Valko, K., Nunhuck, S., Bevan, C., Abraham, M. H. and Reynolds, D. P., *J Pharm Sci.,* 92(11):2236–2248, 2003.
43. Wang, X., Xu, R., Wilson, D. M., Rourick, R. A. and Kassel, D. B., *High Throughput Log D Determination Using Parallel Liquid Chromatography/Mass Spectrometry (LC/MS)*, American Society for Mass Spectrometry 2000 Conference Abstract, Long Beach, CA, USA, 2000.
44. Gusev, A. I., *Development and Applications of LC/MS for High Throughput Analytical Screening for "Accelerated" Compound Discovery*, American Society for Mass Spectrometry 2001 Conference Abstract, Chicago, ILL, USA, 2001.
45. Kerns, E. H., Petusky, S. L., Kleintop, T., Chlenov, M. and McConnell, O. A., *LC/MS Method for High Throughput log P/log D Profiling*, American Society for Mass Spectrometry 2000 Conference Abstract, Long Beach, CA, USA, 2000.
46. Wilson, D. M., Wang, X., Walsh, E. and Rourick, R. A., *Comb. Chem. High Throughput Screen,* 4(6):511–519, 2001.
47. Valko, K., *Acta Pharm. Hung.,* 73(1):13–22, 2003.
48. Valko, K., Espinosa, S., Du, C. M., Bosch, E., Roses, M., Bevan, C. and Abraham, M. H., *J. Chromatogr. A.,* 933(1–2):73–81, 2001.
49. Canals, I., Valko, K., Bosch, E., Hill, A. P. and Roses, M., *Anal. Chem.,* 73(20):4937–4945, 2001.
50. Valko, K., Du, C. M., Bevan, C. D., Reynolds, D. P. and Abraham, M. H., *J. Pharm. Sci.,* 89(8):1085–96, 2000.
51. Du, C. M., Valko, K., Bevan, C. D., Reynolds, D. P. and Abraham, M. H., *J. Chromatogr. Sci.,* 38(11):503–511, 2000.
52. Zhao, Y., Jona, J., Chow, D. T., Rong, H., Semin, D., Xia, X., Zanon, R., Spancake, C. and Maliski, E., *Rapid Commun. Mass Spectrom.,* 16:1548–1555, 2002.
53. Stenberg, P., Luthman, K. and Artursson, P., *J. Control Release,* 65:231–243, 2000.
54. Hidalgo, I. J., Raub, T. J. and Borchardt, R. T., *Gastroenterology,* 96:736–749, 1989.
55. Tang, A. S., Chikhale, P. J., Shah, P. K. and Borchardt, R. T., *Pharm. Res.,* 10:1620–1626, 1993.
56. Artursson, P. and Borchardt, R. T., *Pharm. Res.,* 14:1655–1658, 1997.
57. Cho, M. J., Adson, A. and Kezdy, F. J., *Pharm. Res.,* 7:325–331, 1990.

165. Albertioni, F., Pettersson, B., Beck, O., Rask, C., Seideman, P. and Peterson, C., *J. Chromatogr. B,* 665:163–170, 1995.
166. Hirai, T., Matsumoto, S. and Kishi, I., *J. Chromatogr. B,* 692:375–388, 1997.
167. Hempenius, J., Steenvoorden, R. J., Lagerwerf, F. M., Wieling, J. and Jonkman, J. H., *J. Pharm. Biomed. Anal.,* 20:889–898, 1999.
168. Hempenius, J., Wieling, J., Brakenhoff, J. P., Maris, F. A. and Jonkman, J. H., *J. Chromatogr.,* 714:361–368, 1998.
169. Penn, L. D., Cohen, L. H., Olson, S. C. and Rossi, D. T., *J. Pharm. Biomed. Anal.,* 25:569–576, 2001.
170. Yang, L., Clement, R. P., Kantesaria, B., Reyderman, L., Beaudry, F., Grandmaison, C., Di Donato, L., Masse, R. and Rudewicz, P. J., *J. Chromatogr. B,* 792:229–240, 2003.
171. Ynddal, L. and Hansen, S. H., *J. Chromatogr. A.,* 1020:59–67, 2003.
172. Calderoli, S., Colombo, E., Frigerio, E., James, C. A. and Sibum, M., *J. Pharm. Biomed. Anal.,* 32:601–607, 2003.
173. Ceccato, A., Klinkenberg, R., Hubert, P. and Streel, B., *J. Pharm. Biomed. Anal.,* 32:619–631, 2003.
174. Zhang, J. Y., Fast, D. M. and Breau, A. P., *J. Pharm. Biomed. Anal.,* 33: 61–72, 2003.
175. Lagana, A., Bacaloni, A., Castellano, M., Curini, R., De Leva, I., Faberi, A. and Materazzi, S., *J. AOAC Int.,* 86:729–736, 2003.
176. Amini, N. and Crescenzi, C., *J. Chromatogr. B,* 795:245–256, 2003.
177. Shen, Y., Moore, R. J., Zhao, R., Blonder, J., Auberry, D. L., Masselon, C., Pasa-Tolic, L., Hixson, K. K., Auberry, K. J. and Smith, R. D., *Anal. Chem.,* 75:3264–3273, 2003.
178. Takino, M., Daishima, S., Yamaguchi, K. and Nakahara, T., *Analyst,* 128:46–50, 2003.
179. Herman, J. L., *Rapid Commun. Mass Spectrom.,* 16:421–426, 2002.
180. Takino, M., Daishima, S., Yamaguchi, K. and Nakahara, T., *Analyst,* 128:46–50, 2003.
181. Ramos, L., Brignol, N., Bakhtiar, R., Ray, T., Mc Mahon, L. M. and Tse, F. L., *Rapid Commun. Mass Spectrom.,* 14:2282–2293, 2000.
182. Wu, J. T., Zeng, H., Qian, M., Brogdon, B. L. and Unger, S. E., *Anal. Chem.,* 72:61–67, 2000.
183. Brignol, N., Bakhtiar, R., Dou, L., Majumdar, T. and Tse, F. L., *Rapid Commun. Mass Spectrom.,* 14:141–149, 2000.
184. Zimmer, D., Pickard, V., Czembor, W. and Muller, C., *J. Chromatogr. A,* 854:23–35, 1999.
185. Ayrton, J., Dear, G. J., Leavens, W. J., Mallett, D. N. and Plumb, R. S., *Rapid Commun. Mass Spectrom.,* 11:1953–1958, 1997.
186. Jonsson, B. A., Malmberg, B., Amilon, A., Helene Garde, A. and Orbaek, P., *J. Chromatogr. B,* 784:63–68.
187. Hop, C. E., Wang, Z., Chen, Q. and Kwei, G., *J. Pharm. Sci.,* 87:901–903, 1998.
188. Hsieh, Y., Brisson, J. M., Wang, G., Ng, K. and Korfmacher, W. A., *J. Pharm. Biomed. Anal.,* 33:251–261, 2003.
189. Berman, J., Halm, K., Adkison, K. and Shaffer, J., *J. Med. Chem.,* 40:827–829, 1997.
190. Olah, T. V., McLoughlin, D. A. and Gilbert, J. D., *Rapid Commun. Mass Spectrom.,* 11:17–23, 1997.
191. Tong, X., Ita, I. E., Wang, J. and Pivnichny, J. V., *J. Pharm. Biomed. Anal.,* 20:773–784, 1999.
192. White, R. E. and Manitpisitkul, P., *Drug Metab. Dispos.* 29:957–966, 2001.
193. Van Pelt, C. K., Corso, T. N., Schultz, G. A., Lowes, S. and Henion, J., *Anal. Chem.,* 73:582–588, 2001.
194. Korfmacher, W. A., Veals, J., Dunn-Meynell, K., Zhang, X., Tucker, G., Cox, K. A. and Lin, C. C., *Rapid Commun. Mass Spectrom.,* 13:1991–1998, 1999.

195. Deng, Y., Zeng, H., Wu, J.-T. and Unger, S. E., *Multiple-Sprayer Tandem Mass Spectrometry with Parallel Turbulent Flow Extraction and Parallel Separation for High-Throughput Quantitation in Biological Fluids*, American Society for Mass Spectrometry 2001 Conference Abstract, Chicago, ILL, USA, 2001.
196. Deng, Y., Wu, J. T., Lloyd, T. L., Chi, C. L., Olah, T. V. and Unger, S. E., *Rapid Commun. Mass Spectrom.*, 16:1116–1123, 2002.
197. Wu, J. T., Zeng, H., Deng, Y. and Unger, S. E., *Rapid Commun. Mass Spectrom.* 15:1113–1119, 2001.
198. Waters Corporation, http://www.waters.com.

18 CHIRAL SEPARATIONS

Y. VANDER HEYDEN[a], D. MANGELINGS[a], N. MATTHIJS[a], AND C. PERRIN[b]

[a]*Vrije Universiteit Brussel – VUB, Dept. Pharmaceutical and Biomedical Analysis, Laarbeeklaan 103, B-1090 Brussels, Belgium*

[b]*Laboratoire de chimie analytique – UMR 5625, Faculté de pharmacie, 15 avenue, Charles Flahault. 34093, Montpellier, France*

ABSTRACT

The chiral separation of drug molecules and of their precursors, in the case of the synthesis of enantiomerically pure drugs, is one of the important application areas of HPLC in pharmaceutical analysis. Besides HPLC, capillary electrophoresis is another technique of choice for chiral separations. In this chapter we give an overview of the different modes (e.g., direct and indirect ones) by which it is possible to obtain a chiral separation in HPLC and CE. The direct approaches, i.e., those where the compound of interest is not derivatized prior to separation, are discussed in more detail since they are the most frequently used

Handbook of Pharmaceutical Analysis by HPLC,
S. Ahuja and M.W. Dong, editors.

approaches nowadays. The latter approaches require the use of so-called chiral selectors to enable enantioselective recognition and enantiomeric separation. Many different molecules have been used as chiral selectors, both in HPLC and CE. They can be classified into three different groups based on their structure: i) low-molecular-weight selectors, ii) macrocyclic selectors, and iii) macromolecular selectors. The different classes are discussed, including their subclasses, while examples of chiral separations obtained with them are given. Finally, some practical guidelines about screening conditions to test the enantioselectivity of a given compound on a limited number of chromatographic systems, and about method optimization when an initial separation has already been obtained, are given briefly.

I. INTRODUCTION

A. General

Chirality has been one of the main interests of the pharmaceutical industry over the last two decades. This is because many drug molecules possess at least one asymmetric center, producing different enantiomers that are often distinguished by biological systems, where receptors and enzyme systems have chiral properties themselves. As a consequence, the enantiomers of chiral pharmaceuticals can behave very differently in the human body. Abderhalde and Meller reported for the first time, in 1908, pharmacological differences between the (−) and (+) enantiomers of epinephrine. Since then, many pharmacological and clinical studies have demonstrated that enantiomers of a given chiral drug frequently differ in their pharmacokinetic, pharmacodynamic and toxicity profiles, and the differences observed can even be dramatic. One enantiomer, the eutomer, expresses the therapeutic effect while the other, the distomer, can be inactive, have a different effect, be responsible for unwanted side effects, or can even be toxic for human beings.[1,2] A historic, commonly known example is thalidomide, where *R*-thalidomide is responsible for the therapeutic sedative effect and *S*-thalidomide has teratogenic properties. Both enantiomers can also have a similar effect but with very different activity ratios. For instance, for most β-blockers, the *S* enantiomer is about 100 times more active than the *R* enantiomer. *S*-propranolol is known to have slower clearance from blood than *R*-propranolol after the oral administration of racemic propranolol to humans.[3] The *S* form of timolol is a more potent β-adrenoreceptor antagonist than the *R*-timolol, but both are equally efficient in reducing intraocular pressure in glaucoma patients.[3] For the anticoagulant warfarin, *S*-warfarin is more highly bound to albumin than *R*-warfarin, while overall *R*-warfarin is more highly protein bound than *S*-warfarin. *S*-warfarin is also 5 times more potent as a blood anticoagulant than the *R*-form.[4] More examples of differences in activity are given in Reference 4. Since one-third of the drugs available on the market possess a chiral center, the systematic study of individual enantiomers and/or the development of pure enantiomeric drugs appears to be essential to improve drug safety.[5] The study of the stereoselectivity of a chiral drug appears to be primordial when its development is envisaged or its safety is evaluated.

However, intensive research in the field has only been possible during the last 20 years, with the emergence of new asymmetric synthesis metho-dologies, and powerful analytical and preparative separation techniques. These technical advances have permitted the production of a wide range of chiral precursors or reagents on a commercial scale. Since then, chirality has become one of the major research areas for the pharmaceutical industry and it has raised increasing expectations mainly due to economic reasons. Today, most new drugs and those under development consist of a single optically active isomer.[5,6] This trend towards the development of single isomer (single enantiomer) drugs is also strongly supported by regulatory authorities who have defined stricter requirements to patent new racemic drugs.[7] Regulatory authorities have imposed strict rules for the development of new drugs possessing an asymmetric center in order to encourage the development of single isomer drugs. The development of a new chiral drug as racemate will have to be highly motivated and the study of the individual enantiomers as well as the racemate mixture will be required during the whole development process.[7] Pharmaceutical industries have also discovered an important economic interest in switching from the racemic form of already marketed chiral drugs to the single isomer form to extend their patent for this drug. In 1992, the US Food and Drug Administration (FDA) published its policy statement for the development of new stereoisomeric drugs: the stereoisomeric composition of a drug with a chiral center should be known as well as the quantitative isomeric composition of the material used in pharmacological, toxicological and clinical studies.[8] Applications for enantiomeric and racemic drug substances should also include a stereochemically specific identity test.

Fast method development and short analysis times are crucial factors in the pharmaceutical industry to enable the increasing number of analyses. This requirement is becoming all the more important with the rapid development of combinatorial chemistry. The synthesis of all sorts of chiral materials created a high demand for fast and highly accurate enantiospecific analyses and has been a strong stimulus for the development of chiral separation methods. Over the last 10 years, the development of chiral separation technologies either to resolve racemic mixtures or to test the purity of synthesized enantiomers has been of particular interest to the pharmaceutical industry. Because of regulations, enantioselective analytical methods are required at every stage of drug development: (i) for the control of the enantiomeric purity of the starting material used in asymmetric synthesis and of the resulting products, (ii) for the pharmacological, pharmacokinetic and toxicological studies on the individual enantiomers, and (iii) for the quality control of the drug product. Thus analytical methods are needed to demonstrate the effectiveness of the manufacturing process and to control the enantiomeric purity of both starting materials and reaction products. At the analytical level, enantioselective separations have been realized in most separation techniques, including gas chromatography (GC),[9–14] liquid chromatography (LC) (see further), thin-layer chromatography (TLC),[15–17] supercritical fluid chromatography (SFC),[18–25] counter current chromatography,[26–28] as well as the electromigration methods, both capillary electrophoresis (CE) (see further) and capillary electrochromatography (CEC).[29–40] LC and CE are currently the most widely used, while SFC is used to a lesser extent. The contribution of TLC in chiral separations is low. The importance of GC, which during the 1960s was the first and most important technique used to separate enantiomers, has been decreasing since the 1970s due initially to the development of HPLC and later of CE methods. Nowadays its contribution to the total number of chiral separations is comparable to that of the SFC methods.[41]

The separation of enantiomers necessitates the presence of a chiral environment. The use of chiral auxiliaries to form selectively diastereomeric species is therefore necessary. Direct separations of the enantiomers are usually achieved by means of the enantioselective interaction of a chiral selector with the enantiomers. This chiral selector is bonded to a stationary phase or added to the mobile phase for chromatographic methods, and is added to the background electrolyte for capillary electrophoretic methods. More than 200 chiral selectors have been developed either for GC, LC, SFC or CE separation techniques, and they should enable, at least theoretically, any enantioseparation. However, the selection of the appropriate selector for a specific pair of enantiomers is a very difficult task. Indeed, the enantioselective recognition mechanisms are complex and often depend on properties that are not well known. Very few chiral recognition models [42,43] have been developed and the prediction of the most suitable analytical conditions to achieve separation is often unreliable. Therefore, the determination of suitable chiral selectors to separate a given chiral molecule often requires a large number of experiments, which can be extremely costly and labor-intensive. Thus, there is a need for generic strategies to facilitate this tedious selection procedure and to avoid to some extent the "trial-and-error" approaches.

The aim of this chapter is to give an overview of chiral separations of pharmaceutical compounds by means of HPLC. Capillary electrophoresis, which is the most popular technique besides HPLC for performing chiral separations at the analytical level, will also be briefly discussed. A second reason to discuss chiral separation in CE in short is the large overlap in the chiral selectors applied in both techniques.

An overview of the principles of chiral separations in CE and HPLC is given in the first part of this chapter. The second part is dedicated to an overview of chiral separation with CE. The popularity of electrophoretic techniques has grown spectacularly in recent years, mainly due to their exceptional performances in the chiral field. Various selectors can easily be added to the background electrolyte (BGE), allowing fast method development. Many chiral selectors have been developed, but cyclodextrins (CDs) are by far the most popular. An overview of them and of the other groups of chiral selectors applied in CE will be provided.

The third part of the chapter is dedicated to chiral separations using LC. Chromatographic techniques are by far the most widely used in the pharmaceutical industry for chiral analysis and many chiral stationary phases have been developed. Among them, the polysaccharide-based ones (containing cellulose and amylose derivatives) have proven to have the widest application range. An overview will be given of those and other chiral stationary phases (CSPs) applied in LC. Their use will be illustrated with an extensive, but not exhaustive, list of references of chiral separations of various drug types, with which we will try to give an overview of the different possibilities of chiral separation in pharmaceutical analysis. The last part of the chapter is devoted to methodologies that can be applied when developing chiral separation in LC. Two aspects will be considered. First, we will focus on the screening strategies for enantioseparation, both in normal and reversed-phase conditions. Secondly, the possibilities of optimizing the separation of a specific enantiomer mixture will be highlighted. The screening strategies are generic approaches, applicable on a large number of different molecules. They give the analyst, for a given chiral mixture, an idea about the enantioselectivity of a given chromatographic system (combination of a CSP and a mobile phase) and this in a limited number of experiments. If required the obtained separation, observed from such a screening approach, can be subjected to a common method optimization procedure.

B. Regulatory Aspects of Chiral Pharmaceuticals

In the last two decades, the regulation of chiral drugs has generated a lot of interest because of their known effects on biological activity, toxicity and specificity. Historically, synthetic chiral drugs were mainly presented as the racemate due to the technical difficulties of either synthesizing the pure enantiomers or separating them. The 1980s introduced new methods for the preparation of single enantiomers accompanied by advances in chiral analytical procedures.

Regulatory requirements for the investigation of chiral drugs have been described for three regions, including Europe, the US and Japan. European Guidelines are available via the European Medicines Evaluation Agency (EMEA) (general website: http://www.emea.eu.int). The FDA is responsible for authorizing human medicinal products in the US through its Center for Drug Evaluation and Research (CDER). The relevant guidance for chiral drugs is the policy statement for the development of new stereoisomeric drugs which can be found on the general FDA website http://www.fda.gov/CDER/guidance/stereo.htm (first published in January 1992 and reviewed in January 1997). The Japanese regulatory authority is located at the Ministry of Health and Welfare (MHW) and the Pharmaceutical and Medical Safety Bureau (PMSB). All three guidances take essentially the same view with respect to the development of chiral drugs but emphasize different aspects. In order to harmonize the regulations, the International Conference on Harmonization of Technical Requirements for Registration of Pharmaceuticals for Human Use (ICH) has been created, which present the duplication of the development work required for the registration of new medicinal products. The guidelines can be obtained from the ICH website which is maintained by the International Federation of Pharmaceutical Manufacturers Associations (IFPMA, http://www.ifpma.org). Some internationally harmonized guidelines regarding specifications and tests, impurities and validation of analytical methods have particular relevance to the development of chiral drugs and are discussed below.

Three important documents (guidances) related to this subject should be mentioned. The first one is the ICH Topic Q6A Specifications on Test Procedures and Acceptance Criteria for New Drug Substances and New Drug Products: Chemical Substances (in operation since May 2000). This ICH guidance provides additional guidance to the recommendations in the regional guidelines described above. Tests for chiral drug substances are included in the category of specific tests and criteria. A decision tree summarizes when chiral identity tests, impurity tests and assays are needed for both new drug substances and new drug products. The differences in testing depend on whether the substance is available as racemate or single enantiomer. Also a possible formation of the opposite enantiomer in single-substance drugs is an important factor for performing a chiral assay or impurity procedure. For instance, for a drug substance developed as a single enantiomer, an identity test should be capable of distinguishing between the enantiomers and the racemate. In a single-enantiomer drug such as this, control of the other enantiomer should be considered in the same manner as that for other impurities. An assay is considered acceptable when performed by either a chiral assay procedure or by the combination of an achiral assay together with appropriate methods of controlling the enantiomeric impurity.

The second and third important guidelines both concern impurities: Topic Q3A makes recommendations on Impurities in New Drug Substances and Topic Q3B on Impurities in New Medicinal Products. In Europe, these guidelines are published as CPMP/ICH/142/95 and as its Annex CPMP/ICH/282/95, respectively. It is stated that enantiomeric impurities are excluded from those guidelines. It is agreed in the main guideline (Topic ICH 6QA) that because of practical difficulties in quantification, the opposite enantiomer (the impurity) is excluded from the qualification and identification thresholds given in the ICH guidelines on impurities, but the principles expressed are expected to apply. The guideline cites thresholds depending on the maximum daily dose of the drug above which qualification studies are required. Different qualification thresholds exist for different classes of drugs. Where the qualification threshold is exceeded, additional safety studies may be required. For example, for substances of which the maximum daily dose ≤ 2 g/day, the identification and qualification thresholds are 0.10% and 0.15%, respectively. A similar qualification of the enantiomeric impurities would be expected, although they are not strictly covered by these guidelines. For the future, it seems possible that the guidelines for chiral drugs can be harmonized because of the recent advances and facilitated progress in the use of chiral HPLC columns, which allow the measurement of trace levels of enantiomeric impurities in drug substances and products.

From a practical point of view, and related to method development for impurity testing, it can be recommended that a method be developed in such a way that the impurity elutes before the main component. It will be easier to obtain a baseline separation and one that reduces quantification problems that might occur when the impurity elutes in the tail of the main peak. Moreover, the peak obtained for the main compound might be rather broad and tailing in chiral chromatography (as will be shown further in this chapter) which favors even more the development of methods with the impurity eluting first.

II. SEPARATION OF ENANTIOMERS OF PHARMACEUTICAL INTEREST USING LC AND CE

A. Principles of Chiral Separation in LC: Direct and Indirect Approaches

Separation of enantiomers can be performed via two different kinds of approaches, *direct* and *indirect* ones.[44,45] In the indirect approach the enantiomers are derivatized prior to their separation, while in the direct approach they are placed in a chiral environment and are not subjected to a chemical reaction.

Some authors distinguish two categories of the indirect approach.[46] In the first, the enantiomers are derivatized using an achiral derivatizing reagent and the derivatives are separated using a chiral stationary phase. The second indirect approach consists in the formation of covalently bonded diastereoisomers by derivatizing the enantiomers with an optically pure chiral agent, i.e., one enantiomer of a chiral derivatization reagent (CDR). The diastereoisomers formed possess different physicochemical properties and can therefore be separated by any analytical method using an achiral separation mechanism, e.g., by reversed-phase chromatography on a C18 achiral stationary phase, which is the one most frequently used as the indirect approach. An example is the determination of warfarin enantiomers after derivatization with (1*S*,2*R*,4*R*)-endo-1,4,5,6,7,7-hexachlorobicyclo[2,2,1]-hept-5-ene-2-carboxylic acid. The derivatized products, *R*-warfarin ester and *S*-warfarin ester, are separated on a reversed-phase C_{18} column with fluorimetric detection.[47] Lists of chiral reagents that are used for the derivatization of enantiomers can be found in References 4, 41, 46. The pharmaceuticals determined with these reagents are also given there.

The indirect approach has been widely applied since many functional groups can be derivatized with various chiral reagents and the covalent diastereoisomers can be separated with inexpensive non-chiral systems. Other advantages of the indirect approach are that method development is rather straightforward and that the detection sensitivity of the enantiomers can be improved by the selection of an appropriate CDR having a strong chromophor or fluorophor.

A general comment about detection in chiral separations, both direct and indirect, is that in general all detectors used for non-chiral separations can also be used. The most frequently applied detector is thus the UV detector.

The indirect separation technique has been performed for many years and is still commonly used,[46,48–51] especially for pharmacokinetic studies and determinations in plasma,[52–54] and for the analysis of one enantiomer as impurity in the presence of the other (enantiomeric purity determination).[55–57] However, one of its disadvantages is that an enantiomeric impurity in the reagent can cause negative effects, i.e., the formation of several diastereomeric products, and therefore a CDR with high optical purity should be used. Furthermore, the derivatization might be time-consuming and give rise to the formation of several reaction products, especially if the enantiomeric solute has different functional groups. The problem of racemization during derivatization must also be taken into account when applying the indirect separation technique, as also the possible variability in the formation rate of the diastereoisomers.[4]

In recent years, for analytical purposes the direct approach has become the most popular. Therefore, only this approach will be discussed in the next sections. With the direct approach, the enantiomers are placed in a chiral environment, since only chiral molecules can distinguish between enantiomers. The separation of the enantiomers is based on the complex formation of labile diastereoisomers between the enantiomers and a chiral auxiliary, the so-called chiral selector. The separation can only be accomplished if the complexes possess different stability constants. The chiral selectors can be either chiral molecules that are bound to the chromatographic sorbent and thus form a CSP, or chiral molecules that are added to the mobile phase, called chiral mobile phase additives (CMPA). The combination of several chiral selectors in the mobile phase, and of chiral mobile and stationary phases is also feasible.[58,59]

1. Direct Chiral Separation Using CMPAs

By adding a chiral molecule to the mobile phase, the direct separation of enantiomers can be obtained on an achiral stationary phase. This approach is also called chiral mobile phase HPLC.[41] The mechanisms of chiral recognition by CMPAs are due to an enantioselective complexation in the mobile phase and a selective distribution of labile diastereomeric complexes to the stationary phase, or because the CMPA is dynamically coated on the stationary phase and forms a CSP potentially showing enantioselectivity. Apparently, a combination of these different separation mechanisms is also possible.[60] Depending on the kind of complexation, ligand-exchange methods, ion-pair methods, inclusion-complexation methods, affinity-interaction methods and host–guest interaction methods can be distinguished.[41] Several applications of the use of CMPAs are described in Reference 41. In the ligand-exchange methods ternary complexes, the so-called ligand–metal complexes are formed. In the applications described, these complexes are usually formed between the enantiomer, a (derivatized) amino acid and a metal ion.[61–67] In the ion-pair method a chiral counter ion is added to the mobile phase. Quinine or quinidine,[68–70] (+) 10-camphorsulphonic acid[71,72] and (+) di-*n*-butyltartrate[73] have been used as counterions. Other chiral phase additives that have been used are CDs,[74,75] which form inclusion complexes, as well as proteins[76] and crown-ethers.[77]

The CMPA technique offers some advantages such as high efficiency and the fact that inexpensive conventional non-chiral columns can be used. Moreover, the chiral selector in the mobile phase can be changed flexibly when exploring new chiral selectors during the development of an enantioselective separation method and a wide range of variables of the chromatographic system can be optimized for the enantioselective separation. CMPAs have been used in studies of qualitative control, determination of pharmacokinetic and pharmacodynamic effects of enantiomers as well as for the isolation of pure enantiomers at a semi-preparative scale.[60]

The limitation of the direct chiral separation using CMPA is that the presence of a chiral additive in the mobile phase may restrict the choice of detection systems and the labile diastereomeric complexes may have a different detector response.[60] Another disadvantage of this approach is the high consumption in the chiral selector. As can be observed from the above references, most of the applications using CMPAs are not very recent. In more recent years most of the chiral additives have been bonded to a stationary phase forming a CSP. A potential interest in the use of CMPAs nowadays might be found in microLC or CEC.[78–80]

2. Direct Chiral Separation Using CSPs

A wide variety of CSPs have been synthesizsed and several of them are commercially available. Their use is nowadays the most favored chromatographic technique used to separate enantiomeric drugs by means of HPLC. In the case of CSPs, the enantiomers that form a stronger association with the chiral selector will be more strongly retained. Interaction of the chiral selector of the system with the solute enantiomers results in the formation of two labile diastereoisomers, provided that at least three active points of the selector participate in the interaction with corresponding sites of the solute molecule. The diastereoisomers differ in their thermodynamic stability. This three-point-interaction rule is generally valid for enantioselective chromatography. An extension to this rule states that one of the required interactions may be mediated by the adsorption of the two components of the interacting pair on the sorbent surface. Numerous reviews and monographs describe the principles of direct chiral separations in LC.[44,81–85] Because of the multiplicity and complexity of the interactions between the enantiomers to be separated and the chiral selector, the sorbent surface and other components of the chromatographic system, their global enantioselectivity can strongly depend on the composition and the pH of the mobile phase, as well as on the temperature. Therefore, it is not possible to predict the enantioselectivity of a given system towards a given drug molecule.

The use of CSPs provides several advantages such as the speed of analysis, the possibility of analyzing or purifying enantiomers in complex mixtures, the reproducibility of the analysis and its flexibility. However, a disadvantage of these CSPs is their relative low stability, i.e., they can be subject to relatively fast column aging.

Different classifications for the chiral CSPs have been described.[4] They are based on the chemical structure of the chiral selectors and on the chiral recognition mechanism involved. In this chapter we will use a classification based mainly on the chemical structure of the selectors. The selectors are classified in three groups: (i) CSPs with low-molecular-weight selectors, such as "Pirkle" type CSPs, ionic and ligand exchange CSPs, (ii) CSPs with macrocyclic selectors, such as CDs, crown-ethers and macrocyclic antibiotics, and (iii) CSPs with macromolecular selectors, such as polysaccharides, synthetic polymers, molecular imprinted polymers and proteins. These different types of CSPs, frequently used for the analysis of chiral pharmaceuticals, are discussed in more detail later.

B. Principles of Chiral Separations with CE

CE has been established as a very efficient technique for the separation of drug enantiomers. Chiral separations in CE are also based on the formation of diastereomeric complexes between the enantiomers and a chiral selector. The main advantages of the technique are its high efficiency, short analysis times, versatility due to the great variety of chiral selectors that can be added to the BGE, short equilibration times required when changing the chiral selector and low consumption of selector.[35,86–90]

However, a separation of the enantiomers will only be achieved if: (i) the complexes have different equilibrium constants so that the two enantiomers have a different average mobility, (ii) the free and the complexed enantiomers have different mobilities and (iii) the exchange between the free and the complexed enantiomers is rapid.[48] As a direct consequence of the second requirement, neutral enantiomers are not resolvable with neutral chiral selectors.

Many chiral selectors have been developed for CE but CDs (natives and their derivatives) are by far the most widely used.[87,91,92] In addition to their wide chiral discrimination abilities, CDs are usually well soluble in aqueous BGE, UV transparent, stable, environment friendly, easily available and relatively cheap, which makes them ideal candidates for CE.

Several CD derivatives (charged and uncharged) are available which should allow the separation of most chiral molecules with at least one of them. However, due to the complexity of chiral recognition mechanisms, the determination of the best selector based on the analyte structure is challenging. Furthermore, separations using CDs are influenced by numerous factors, so that no general rule can be applied for the successful resolution of enantiomers.[91,93]

In most cases, the chiral selector is simply added to the BGE.[94,95] Interactions between the analytes and the chiral selector will determine the stability of the diastereomeric complexes formed. The interactions involved in the chiral recognition process in CE are hydrophobic, electrostatic, Van der Waals and hydrogen bond-type interactions. Several reviews discuss the principles of electrophoretic chiral separations.[86,87]

III. SEPARATION OF ENANTIOMERS BY CAPILLARY ELECTROPHORESIS

A. Chiral Selectors in CE

Many chiral selectors have been used for chiral separations in CE. The most widely used are discussed below. Most of these chiral selectors are also applied in HPLC, as will be discussed later.

1. Cyclodextrins

CDs are by far the most popular chiral selectors in CE and will therefore be discussed in more detail than the others. CDs are torus-shaped, cyclic, α(1,4)-linked oligomers of D(+)-glucopyranose. They contain between six and 12 D(+)-glucopyranose units, but only those with 6 (α-CD), 7 (β-CD) and 8 (γ-CD) units are currently used in chiral separations (Figure 1). The interior of the CD cavity is relatively hydrophobic, while the outside rims are more hydrophilic. The rim of the wider side of the CD contains chiral secondary hydroxyl groups, while achiral primary hydroxyl groups occupy the opposite smaller opening.

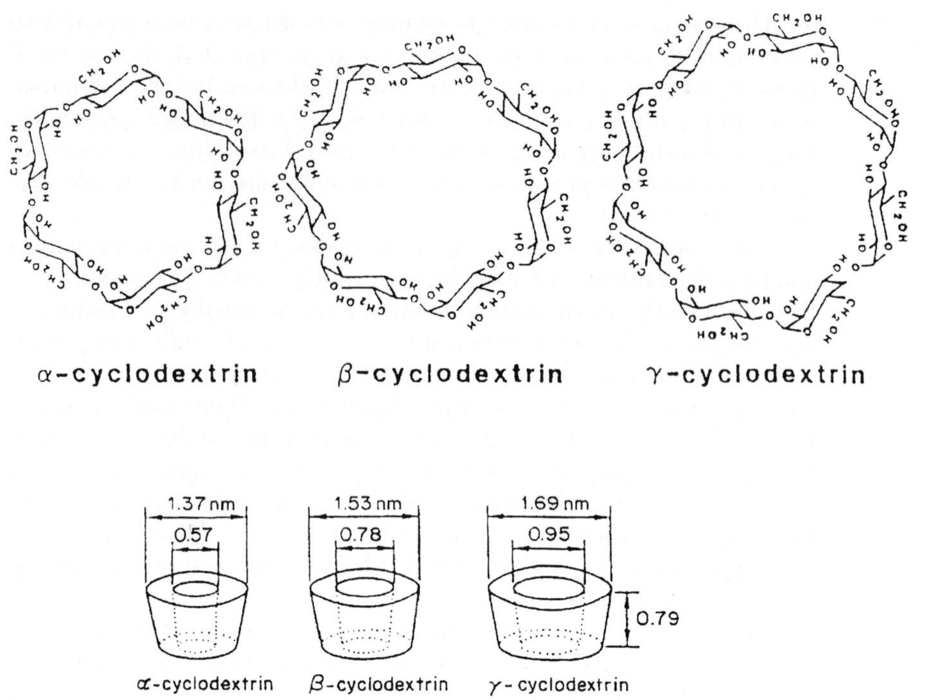

FIGURE 1 Structures, shapes and dimensions of the native *α*, *β* and *γ*-CDs (adapted from Reference 91).

The CD–analyte complex is formed by inclusion of the enantiomer into the cavity of the CD. The complexes are usually formed from the wider opening of the CD (in a few cases, from the narrowest opening). Therefore, the size of the cavity plays an essential role in the chiral recognition mechanism. In general, the *α*-CD can accommodate a single phenyl ring molecule, while *β*-CD and *γ*-CD can accommodate substituted single- and multiple-ring molecules. The inclusion alone is not enough for the chiral recognition. Interactions between substituents on the asymmetric center of the analyte and the hydroxyl groups of the CD rims are responsible for the chiral recognition. The inclusion-complex formation and the size of the analyte binding constant are determined by several interactions. The most important are hydrophobic interactions, which induce the apolar part of the molecule to stay preferably in the CD cavity, and hydrogen bondings between polar segments of the guest molecule and the secondary hydroxyl groups at the mouth of the CD cavity.[86] Other factors which can influence complex formation are Van der Waals interactions, as well as the release of high-energy water from the CD cavity.

Besides the native CDs (Figure 1), a wide range of modified CDs have been synthesized.[96,97] Hydroxyl groups on the CD rim can be substituted by neutral, chargeable or charged groups, which enable more interaction sites. CDs usually have a good solubility in aqueous BGE (except the β-native CD), are non-UV absorbent, easily commercially available, rather inexpensive, stable and environmental friendly. Those properties combined with their wide chiral discrimination abilities make them the most frequently used chiral selectors for chiral separation with capillary electrophoresis.[98–101] The introduction of charged/chargeable groups has several advantages such as the possibility of separating uncharged species, and an increased solubility. Numerous CD derivatives have also been used as chiral selectors, which makes the selection of an appropriate one not so evident. In Reference 97, for instance, 26 different CDs were compared for the separation of 35 different drug compounds. Five CD derivatives, namely, highly-sulfated-β-CD, highly-sulfated-α-CD, hydroxypropyl-β-CD, heptakis-(2,6-O-dimethyl)-β-CD and heptakis(2,3,6-O-trimethyl)-β-CD were identified as most effective.

In general, charged CDs have shown superior discrimination abilities, especially the highly-sulfated (HS-CDs) ones.[102–104] Furthermore, the separation mechanism is altered by the introduction of electrostatic interactions. Finally, the use of chiral selectors carrying a charge opposite to that of the analytes can greatly improve the mobility difference between the two enantiomers. The use of mixtures of CDs in chiral separation is also possible.[105–107]

Charged CDs were used, in a first instance, for the chiral separation of neutral species which was not possible with neutral CDs due to the lack of self-mobility. But they have also shown a higher selectivity towards many basic and acidic compounds than the uncharged CDs.[108,109] The maximum enantioselectivity should be found when analyte and CD have opposite mobilities.[109,110] In such a case, the available separation window is increased as it is limited by the electrophoretic mobility of the analyte and of the CD. A larger separation window provides more opportunities to achieve a baseline separation of weakly interacting or poorly differentiated enantiomers. Furthermore, ion-pairing interactions between analyte and CD of opposite charges can improve selectivity since chiral recognition depends on the strength of the binding and the difference of binding constants between the CD and the two enantiomers.[111] Accordingly, anionic CDs should be tested first for the chiral separation of basic species, and cationic CDs for acidic species. Although good enantioselectivity was achieved with many compounds,[111,112] the use of cationic CDs is not very popular because of their tendency to adsorb to the capillary silica wall leading to non-reproducible experimental results. On the contrary, the use of anionic CDs is very popular, with carboxymethyl CDs and sulphated CDs being the most widely used.[113–119] Many acidic compounds could also be separated using anionic CDs. The electrostatic interactions appear not to be fundamental to achieving enantioselectivity.[113,120]

Numerous applications of chiral separations using CDs can be found in the literature. Some examples, either reviewing chiral separations in CE or presenting the separation of several compounds, can be found in References 97,101,107,121–124. Several studies related to the chiral separation of amino-acid derivatives by CE and micellar electrokinetic capillary chromatography with different types of CDs have also been reported.[102,103,114,125–128]

The main drawback in chiral separation methods using derivatized CDs is that these selectors are mainly available as complex mixtures that contain a large number of isomers differing in their degree of substitution, which may result in batch-to-batch selectivity differences.[129–131] The use of pure single enantiomers or very reproducible mixtures is thus required to obtain reproducible and robust methods.

2. Crown-ethers

Crown-ethers are macrocyclic polyethers capable of forming host–guest complexes, especially with inorganic and organic cations. Modification of the crown-ether by the introduction of four carboxylic groups makes it possible to use this class of compounds as chiral selectors in CE.[132]

Crown-ethers can incorporate protonated primary amine compounds by formation of ion-dipole bonds with the oxygen atoms of the chiral selector. Crown-ethers have been widely used for the separation of several pharmaceuticals both in aqueous and non-aqueous media.[47,133–139]

3. Macrocyclic Antibiotics

Macrocyclic antibiotics, especially ristocetin A, vancomycin, teicoplanin, rifamycins, kanamycin, streptomycin and fradiomycin have proven to be highly selective towards the enantiomers of a broad class of compounds.[140–146] These antibiotics are strongly UV-absorbent and have a tendency to adsorb to the capillary wall. The former property may be a limitation for their use. Usually, however low concentrations of the antibiotics are needed and therefore the detection of the analyte is still feasible. This problem can also be overcome by the use of the partial filling technique.[147,148] The chiral recognition is mainly due to charge–charge interactions, hydrogen bonding, hydrophobic inclusion and π–π interactions.

4. Proteins

Proteins have also been successfully used for chiral separations in CE. One of the characteristics of proteins is their isoprotic point, pI. The protein will mainly be charged positively if pH<pI, and negatively if pH>pI. Therefore, the pH of the BGE is a very important parameter for the optimization of the separations. As with charged-CD derivatives, it is possible to separate both charged and uncharged chiral species with proteins. Among the many proteins that can be used as chiral selectors, bovine serum albumine (BSA), human serum albumine (HSA), α_1-acid glycoprotein, avidin and riboflavin are the most widely applied.[149–153] Haginaka[154] gives an overview of the proteins used as chiral selectors in CE and a rather extensive list of applications describing the chiral separation of drugs. Similar mechanisms as those for macrocyclic antibiotics are involved in the chiral recognition mechanisms.

5. Micelles and Surfactant Pseudostationary Phases

When micelles are used, the CE technique becomes a micellar electrokinetic chromatography (MEKC) one. Natural surfactants, such as bile salts, digitonin and saponins, optically active synthetic surfactants, e.g., amino-acid derived ones, alkylglycoside-, tartaric acid- and steroidal glucoside-based surfactants, and high-molecular mass or polymerized surfactants, have been used as chiral selectors in MEKC.[155–158] In the latter case the surfactants form a pseudostationary phase and the technique is called polymeric surfactant electrokinetic chromatography. A review of the use of this technique is made in Reference 159.

Newly used chiral surfactants often have a low critical micellar concentration, are highly soluble and can be synthesized both in L- and D-forms. This last feature makes it possible to easily change the migration order of the optical isomers, which is very interesting for the determination of the optical purity of drugs, where for quantification purposes it is favorable that the chiral impurity migrates before the main component.

B. Separation Strategies in CE

Separation strategies are interesting for their ability to evaluate the enantioselectivity of a given system towards a chiral molecule. They are of practical use mainly when the system (chiral selector) used shows broad enantiorecognition abilities. For this reason, in the separation of pharmaceuticals, strategies using CDs as chiral selector are most interesting. Several factors (see below) may affect the separation of enantiomers which makes the development of a separation far from evident.

The concentration of the chiral selector, for instance, has considerable influence on the mobility and separation of the enantiomers. Optical resolution varies with the chiral selector concentration and reaches a maximum value at a given optimum concentration. Wren and Rowe[42] proposed a model that describes the influence of the selector concentration on selectivity, and which was extended by Vigh's group[160] by including the pH as a separation parameter for weak acidic enantiomers. The latter model shows that the chiral selectivity is determined by the complex's relative mobility, the CD concentration, the degree of dissociation of the weak acid, and the equilibrium constant of complex formation of both the dissociated and the non-dissociated acid forms.

The chiral recognition of enantiomers can be of three types: (i) desionoselective, (ii) ionoselective, or (iii) duoselective, in which only the non-dissociated, the dissociated or both forms (charged and uncharged), respectively, of the enantiomers selectively interact with the chiral selector. In the case of ionoselective and duoselective interactions, a reversal of the migration order of the enantiomers is theoretically possible by the appropriate selection of CD concentration and the pH of the BGE. The addition of organic modifier to the BGE can also change selectivity by modifying the solubility of the chiral selector and/or of the solute, the complex equilibrium, the conductivity of the BGE and the electroendosmotic flow (EOF) level. Several other factors, such as the temperature, the type and the concentration of the BGE, and the level of the EOF can influence the separation.[96,161]

Separation strategies are therefore valuable since they propose a limited set of experimental conditions, which maximize the probability of separating numerous pharmaceuticals with different molecular structures and properties. A screening strategy, using different types of CD derivatives and experimental designs, for the rapid selection of a suitable selector as well as of acceptable experimental conditions to separate chiral amino acids is described in Reference 114. Three widely used β-CD derivatives, well recognized for their efficiency,[91] were selected in this strategy. They are the hydroxypropyl-β-CD (HP-β-CD), the carboxymethyl-β-CD (CM-β-CD) and the sulfobutylether-β-CD (SBE-β-CD).

A second strategy focuses on the use of highly-sulfated CDs.[102,103] which possess extremely broad enantiorecognition abilities.[88] The latter strategy contains a screening phase to evaluate rapidly the enantioselectivity of the α-, β- and γ-HS-CDs for a large series of drug substances and their precursors or synthesis building blocks. The strategy also contains optimization phases, which allow either the improvement of an incomplete separation obtained in the screening step or the fine-tuning of a baseline separation (e.g., make it faster, improve peak shape).

Although good results can be obtained with CE, specific problems limit its use. First of all, the repeatability and the robustness of separations in CE is somewhat more critical than in chromatography since many experimental parameters, which are potentially critical for the separation, have to be controlled.[162] Furthermore, most chemically derivatized CDs are commercialized as mixtures with an average degree of substitution. The variability in the extent of the derivatization can lead to changes in the degree of selectivity.[129–131] Thus, the reproducibility (batch-to-batch variability) of the chiral selectors is often critical and may dramatically influence the separations.[129–131] The robustness of chiral separations using highly-sulfated CDs was found to be good and is discussed in Reference 104.

IV. SEPARATION OF ENANTIOMERS BY LIQUID CHROMATOGRAPHY ON CHIRAL STATIONARY PHASES

As mentioned earlier, a wide variety of CSPs have been synthesized and are commercially available. CSPs are most commonly used today. Therefore they are discussed in somewhat more detail in the following part of this chapter. Examples of drug substances separated on the different CSPs also are given. For most examples, the stationary phase and the mobile phase composition are given, as also the reference they originate from. Other chromatographic conditions such as flow rates and column temperatures are not specified since they can usually be considered similar to regular chromatographic separations. For some stationary phases, a relatively low flow rate, i.e., below 1 mL/min, can be applied in order to increase the column's lifetime. However, for those specific experimental conditions we refer to the indicated publications.

The most widely used CSPs can be classified into three groups: (i) CSPs with low-molecular-weight selectors, such as Pirkle-type CSPs, ionic- and ligand-exchange CSPs, (ii) CSPs with macrocyclic selectors such as CDs, crown-ethers and macrocyclic antibiotics, and (iii) CSPs with macromolecular selectors such as polysaccharides, synthetic polymers, molecular imprinted polymers and proteins. It can be observed that several of the chiral selectors applied in CE are used in LC. They are then bonded or coated to an inert carrier to form the CSP.

A. Low-Molecular-Weight Selectors

1. Pirkle Type: π-Donor, π-Acceptor Chiral Selectors

Historically, this type of CSP preceded all the others described later.[163,164] Their success was the key point for the rapid development of chiral liquid chromatographic separations using CSPs. The structure of these stationary phases is based on single strands of chiral selectors, connected via amidic linkage on aminopropyl silica.[163,164] The strands possess either π-donor or π-acceptor aromatic fragments as well as hydrogen bonding agents and dipole-stacking inducing structures. The pioneering work of Pirkle's group had such an impact on the field that the whole category of donor–acceptor type stationary phases was named after him. These columns are also called brush-type columns. Many π-donor and acceptor phases have been synthesized. The theoretical basis for separations on those phases is the so-called 'three-point theory'. The columns are designed to give a strong three-point interaction with one compound of an enantiomer pair.

Pirkle-type CSPs do not involve ionic interactions and therefore are almost exclusively operated in the normal-phase mode. The use of subcritical carbon dioxide based mobile phase, i.e., subcritical fluid chromatography, is also a possibility.[165,166] When needed, analytes lacking the complementary functional groups for attractive interactions with the chiral selector can be derivatized to enhance their interactions with the selector. For example, compounds containing an amine or carboxylic groups are derivatized to an amide, ester or anilide. The most commonly used derivatizing reagents contain either a naphthyl, a 3,5-dinitrobenzoyl or a 3,5-dinitrophenyl moiety to ensure maximum π–π interactions between the analyte and the chiral selector.[167] Pirkle-type CSPs can be divided into three classes, the π-electron acceptor CSPs, the π-electron donor CSPs and a third group containing CSPs with both π-acceptor and π-donor substituents. The latter has broad applicability to many compound classes and is able to separate most enantiomers separated on the single π-acceptor or π-donor phases.[168–170] An overview of some CSPs and their chiral selectors is given in Table 1. Some chiral selector units are given in Figure 2.

A π-donor will have a tendency to lose an electron because the resulting positive charge is accommodated by the π-system. Conversely, a π-acceptor can readily stabilize a negative charge and therefore has a tendency to accept an additional electron in its π-system. In this way a π-donor/π-acceptor pair will form a complex when a charge can be transferred from the donor to the acceptor molecule. The most common CSPs are the π-acceptor dinitrobenzoylphenyl (DNBP) derivatives of amino acids. Since the DNPB group is a π-acceptor, solutes should possess a π-donor group such as an aromatic ring with alkyl, ether or amino substituents, in order to be separated. Moreover, solutes should be able to form hydrogen bonds or enter into dipole stacking with the amide group attached to an aromatic system on the CSPs. Also, amino acids such as leucine and phenylglycine are used as π-acceptor chiral phase. Both are available in both D- and L-configurations.

The π-donor phases (typically naphthyl-amino-acid derivatives covalently bonded to silica) require the analyte to contain a π-acceptor group such as the dinitrobenzoyl group. The dinitrobenzoyl group can easily be added to a wide range of compounds such as alcohols, amines and carboxylic acids using dinitrobenzoyl chloride, isocyanate or dinitroanaline, as already mentioned above.

More recent developments in the field of the Pirkle-type CSPs are the mixed π-donor/π-acceptor phases such as the Whelk-O1 and the Whelk-O2 phases.[185,186] The Whelk-O1 is useful for the separation of underivatized enantiomers from a number of families, including amides, epoxides, esters, ureas, carbamates, ethers, aziridines, phosphonates, aldehydes, ketones, carboxylic acids, alcohols and non-steroidal anti-inflammatory drugs.[172,187] It has been used for the separation of warfarin,[181,188] aryl-amides,[187,189] aryl-epoxides[172,187] and aryl-sulphoxides.[187] The phase has broader applicability than the original Pirkle phases. The broad versatility observed on this phase compares with the polysaccharide-derived CSPs

TABLE 1 Overview of Some Pirkle-type CSPs and Their Chiral Selectors

Column type	Chiral selector	References
(i) π-electron acceptor stationary phases		
α-Burke 2	Dimethyl-*N*-3,5-dinitrobenzoyl-α-amino-2,2-dimethyl-4-pentenylphosphonate	171
β-GEM 1	3,5-Dinitrobenzoyl-3-amino-3-phenyl-2-(1,1-dimethylethyl)propanoate	172
ChyRoSine A	*N*-(3,5-dinitrobenzoyl)tyrosine butylamide	173
CHIREX 3010	*N*-(3,5-dinitrobenzoyl)valine urea	173a
Pirkle 1-J	3,5-Dinitrobenzoyl-β-lactam derivate	174
—	Quaternary ammonium derivative of 3,5-dinitrobenzoyl-L-leucine on α-zirconium phosphate	175
—	*N*-(3,5-dinitrobenzoyl)leucine amide	176
—	*N*-(3,5-dinitrobenzoyl)phenylglycine amide	174
(ii) π-electron donor stationary phases		
—	*N*-(1-naphthyl)leucine ester	177
CHIREX 3014	[*N*-1-[(1-naphthyl)ethyl]amido]valine amide	178
CHIREX 3022	[*N*-1-[(1-naphthyl)ethyl]amido]indoline-2-carboxylic acid amide	179
(iii) π-electron acceptor/donor stationary phases		
(*R*,*S*)-Whelk-O 1	(3*R*,4*S*)-4-(3,5-dinitrobenzamido)-1,2,3,4-tetrahydrophenanthrene	180, 181
ULMO	(*R*,*R*)-*N*-3,5-dinitrobenzoyl-1,2-diphenylethane-1,2-diamine or (*R*,*R*)-DNB-DPEDA	182
DACH DNB	*N*-(3,5-dinitrobenzoyl)-1,2-diaminocyclohexane	183
CHIREX 3005	*N*-(3,5-dinitrobenzoyl)-(1-naphthyl)glycine amide	184

—, No specific name available.

(see later). However, in Reference 190, where a comparison was made, it was found that the polysaccharide phases still tend to be more versatile. In Reference 191, a comparison between the two types of columns was also made leading to the conclusion that the chiral recognition processes responsible for enantioseparation on the Whelk-O 1 column were relatively more systematic and easier to manipulate than on the polysaccharide (Chiralpak-AD) column. On the other hand, the performance on the Chiralpak-AD column was superior to that of the Whelk-O1 column.[191]

This phase is also highly loadable and has been used in SFC where some exceptional loadings have been achieved.[166,180,189] In LC conditions, normal-phase solvents, such as hexane-ethanol or isopropanol, can be used.

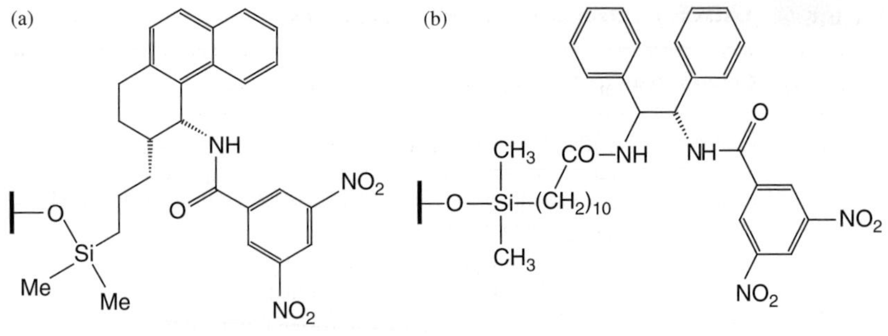

FIGURE 2 Chiral selector units for the (a) 3*R*,4*S*-Whelk-O1 and (b) (*R*,*R*)-DNB-DPEDA columns (see also Table 1).

The Whelk-O2 phase retains the same chiral selector as Whelk-O1 but it incorporates a trifunctional linkage to the silica support (only a single linkage in Whelk-O1). In most cases, the enantioselectivity remains similar to the one obtained with the Whelk-O1. The α-BURKE-II phase was developed specifically to separate β-blockers.[171] The Pirkle 1-J column contains an unusual β-lactam structure, which significantly alters its molecular recognition properties.[192] It is useful for the direct separation of underivatized β-blockers.[192] An important advantage of the Pirkle-type CSPs is the ability to invert elution order by using the same type of CSP, but with the absolute opposite configuration (e.g., *R*,*R* or *S*,*S* Whelk-O1). As a result it is possible to have the trace enantiomer eluted before the major one, which is a desirable feature for enantiomeric purity determinations. Besides a control of the elution order it also allows identification of the stereochemistry of the analyte.[190] For preparative separations it is beneficial to elute the desired component first. Most CSPs of this type are available in both analytical and preparative size.[193]

New brush-type phases (donor–acceptor interactions) are appearing all the time.[194–196] Examples are stationary phases comprising quinine derivatives and trichloro-dicyanophenyl-L-α-amino acids as chiral selectors. Quinine carbamates, which are suitable for the separation of acidic molecules through an ionic interaction with the basic quinine group, are also commonly used but in general they are classified with the anion-exchange type of chiral selectors (see further) because of their interaction mechanism,[194] even though π-donor, π-acceptor properties occur. (Some separations on Pirkle-type CSPs are shown in Table 2.)

2. Chiral Ion-Exchangers

CSPs possessing charged functional groups act as ion-exchangers if the chiral analyte to be separated possesses oppositely charged functional

TABLE 2 Examples of Separations on Pirkle-type CSP

Substance	Stationary phase	Mobile phase	References
Nadolol	(*R*)-*N*-3,5-dinitrobenzoyl-L-leucine, covalently bounded to 5 μm aminopropylsilica	*n*-Hexane/2-propanol/ acetonitrile mixture	176
Metoprolol, bisoprolol	(*R*)-1-naphthylglycine or 3,5-dinitrobenzoic acid	*n*-Hexane/1,2-dichlorethane/ methanol mixture	197
Clenbuterol	Chirex 3005	*n*-Hexane/1,2-dichloroethane/ methanol (54:38:8, v/v/v)	184
Carvedilol	(*S*)-indoline-carboxylic acid and (*R*)-1-(α-naphthyl) ethylamine	*n*-Hexane/dichloromethane/ ethanol (50:35:15, v/v/v) + 0.25% (v/v) trifluoroacetic acid (TFA)	198
An HIV protease inhibitor	*S,S*-Whelk-O1	Ethanol—0.1% acetic acid in hexane (30:70, v/v)	199
15 diverse compounds	Whelk-O1	Isopropanol or ethanol in hexane	190
Acenocoumarol and acenocoumarol	Whelk-O1	*n*-Hexane–ethanol containing 0.5% acetic acid (gradient)	200
10 related anti-varial	*S,S* - Whelk-O1	Various mobile phases	191
uridine analogs warfarin	*R,R* - Whelk-O1	0.5% glacial acetic acid in acetonitrile	188
5-Arylhydantoins	Whelk-O1	2-Propanol–hexane or methanol—(subcritical) carbon dioxide	165
Alcohols and ketones, arylnaphthalenes	Whelk-O 1 and poly Whelk-O (= its polysiloxane borne analog)	Subcritical carbon dioxide-based mobile phase at cryogenic temperatures	166
29 racemic 4-Aryldihydro-pyrimidine-5-carboxylates	3*S*,4*R*-Whelk-O1 and *R*, *R*-*N*-DNP-DPEDA	*n*-Heptane/2-propanol mixtures	201
Aromatic alcohols, tertiary arylalkyl-carbinols and *trans*-2-arylcyclohexanols	*R,R*-*N*-DNP-DPEDA	Isopropanol and *n*-heptane mixtures	182

groups. In general, the ion-exchangers are resins that are polymers with cross-links. The resin has active groups in the form of electrically charged sites. At these sites, ions of opposite charge are attracted but may be replaced by other ions depending on their relative concentrations and affinities for the sites.

Buffered mobile phases are inherently used to adjust and control the adsorption–desorption process. These CSPs are especially useful for the separation of very polar charged analytes, such as sulphonic acids. Chiral anion-exchangers are the most successful CSPs and among them the cinchona alkaloids, quinine and quinidine (Figure 3).[169,194]

Also molecular-imprinted-polymer CSPs (see further) are sometimes classified as ion-exchangers because the similarity in composition between these CSPs and methacrylate-based weak cation-exchange resins suggests that they may be applied in simple cation-exchange separations as well.[202]

Despite the fact that there are huge numbers of polar, basic and acidic drug compounds, only a few papers deal with chiral ion-exchange chromatography. The anion CSPs are mainly used to separate *N*-derivatized amino acids, α-aryloxyalkyl carboxylic acids and α-arylalkyl carboxylic acids like the profens.[194,203,204] Table 3 gives an overview of some separated compounds and the experimental conditions used.

3. Chiral Ligand-Exchangers

Chiral ligand-exchange chromatography (CLEC)[209] separates enantiomers by the formation of diastereomeric metal complexes. In a first instance the technique was mainly used for the separation of amino acids. Impressive results of the first separations gave rise to intensive investigation in the field and numerous publications appeared in the literature, which have been reviewed.[210,211]

Separation is based on the reversible chelate-complex formation between the chiral selector covalently bonded to the chromatographic support, and the chiral solute with transition metal cations. Chelation properties of both the chiral selector and the chiral solute are required. Compounds therefore need to have two polar functional groups in a favorable arrangement to each other, like *α/β*-amino acids, amino alcohols and α-hydroxy acids, which can form rings membered with central chelating metal ions, like Cu(II), Zn(II), Cd(II).[209,212–214] Cyclic amino acids such as L-proline and L-hydroxyproline seem to give the best

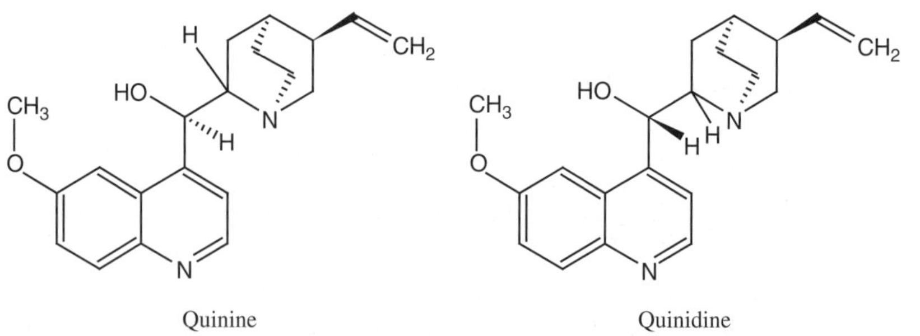

FIGURE 3 Structure and stereochemistry of quinine and quinidine.

TABLE 3 Chiral Compounds Tested for Separation by Chiral Ion-Exchange Chromatography

Substances separated	CSP or selector	Mobile phase	References
Aryl-, aryloxy- and arylthiocarboxylic acids, *N*-derivatized amino acids	Quinine and quinidine carbamates	Methanol—0.1 M ammonium acetate, pH 6.0 (80:20)	194
N-derivatized α-amino acids	*N*-methyl *tert*-butyl carbamoylated quinine	Methanol—0.1 M ammonium acetate, pH 6.0 (80:20)	203
N-acylated and *N*-oxycarbonylated α-amino acids	Quinine, quinidine, epiquinine and epiquinindine *tert*-butylcarbamate	Methanol—0.1 M ammonium acetate, pH 6.0 (80:20)	204
Carboxylic acids, dansyl derivates of amino acids	Aminopropyl derivate of the ergot alkaloid (+)-terguride	Methanol—acetate, pH range 3.8—5.5	205
DNP derivates of amino acids, profens	Quinine carbamate C_9-dimers	Methanol—0.1 M ammonium acetate, pH 6.0 (80:20)	206
Ibuprofen-l-naphthyl-amide, benzoic acid 1-phenylethylamide, 1-(1-naphthyl)ethyl-phenylurea, sulfoxides, propranolol oxazo-lidin-2-one] and others	(*S*)-*N*-(3,5-dinitro-benzoyl) tyrosine derivatives	*n*-Heptane, 2-propanol and diethylamine mixtures	207
Mepivacine, bupivacine, doxapram, quinine, quinidine, atenolol, metoprolol, propranolol, pethidine, methadone	(*S*)-*N*-(3,5-dinitro-benzoyl) tyrosine	Acetonitrile—ammonium acetate (60:40)	208

results.[213,215] Cu(II) has proved to be the most strongly chelating agent.[209–211,216]

A chiral amino-acid/copper complex is bound to a silica- or polymeric stationary phase and copper ions are included in the mobile phase to ensure there is no loss of copper. Amino acids then may be separated by the formation of diastereomeric copper complexes. Water stabilizes the complex by coordinating in an axial position. Steric factors then determine which of the two complexes is more stable. One of the water molecules is usually sterically hindered from coordinating with the copper.[217]

The preparative capacity of CLEC columns is reasonable. However, the presence of copper in the mobile phase is not convenient. Copper can be removed using a iminodiacetic acid-based resin column.[218]

Compared to other chiral separation techniques, the practical importance of CLEC is relatively low because of its complexity. Nevertheless, CLEC is one of the best-investigated techniques from the theoretical point of view.[210,219] Many theoretical concepts developed in CLEC were of general interest for the explanation and prediction of chiral recognition in all chromatographic systems.

A large number of chiral molecules have been separated with ligand-exchange chiral stationary phases. A few examples with commercially available columns are given in Table 4.

B. Macrocyclic Selectors

Another way to achieve chiral discrimination on a stationary phase is the creation of chiral cavities, in which stereoselective guest–host interactions govern the resolution. The first important consideration for retention and chiral recognition in such stationary phases is the proper fit of the molecule to the chiral cavity in terms of size and shape. This category of stationary phases includes CDs and crown-ethers. Another type of CSP, using macrocylic chiral selectors, is based on macrocyclic antibiotics. These antibiotics contain multiple chiral centers surrounding several pockets or cavities. Enantioseparations are achieved due to π–π complexation, hydrogen bonding, hydrophobic inclusion, dipole stacking and steric repulsions.[142,231]

1. CD CSPs

As already discussed for CE, CDs have the shape of a truncated cone with a relatively hydrophobic interior chiral cavity and a hydrophilic exterior surface surrounded by hydroxyl groups. Native CDs and also derivatives of β- and γ-CD are used in CSPs.[167] These CSPs are mainly used in reversed-phase mode[232–236] and polar organic solvents mode,[237–240] but less in normal-phase mode.[234,241] The mechanism of retention is based on inclusion-complexation. This mechanism represents the attraction of the apolar part of the chiral molecule to the apolar cavity. When the molecule possesses an aromatic group, the orientation in the cavity will be stereoselective due to interactions with the glucoside oxygens.[242] It is therefore preferable that the chiral molecule has at least one aromatic ring if a chiral separation is attempted in the reversed-phase mode. Generally, compounds containing a single aromatic ring are best resolved with α-CD, those possessing naphthylrings with β-CD, and where the aromatic system of the compound is larger than a naphthylring,

TABLE 4 Examples of Chiral Separations in Chiral Ligand-Exchange Chromatography

Substances separated	CSP or selector	Mobile phase	References
Threonine, phenylserine	Chiral ProCu	1–5 mM $CuSO_4$ solution	220
Fluoro-substituted aromatic amino acids	Chiral ValCu	0.05 M KH_2PO_4— 1 mM Cu^{2+}	221
α-Trifluoro methyl-α-amino acids	Nucleosil Chiral-1	0.01–0.05 mM $CuSO_4$	212
Mono-, di- and trifluoro-derivates of alanine	Chiral ProCu	1–2.5 mM $CuSO_4$	222
Glycine acylated with 2-phenylpropionic acid	Chiralpak WH ligand exchange	0.3 mM $CuSO_4$	223
HIV anti-infective nucleosides	Nucleosil Chiral-1	CH_3CN/Cu(acetate)$_2$ (0.5 mM), pH 5.75 (5/95)	224
Proline and pipecolic acid in antibiotics	TSK gel enantiomer	0.25 mM $CuSO_4$	225
Pantoic acid and Ca-pantothenate	MCI gel CRS 10W	2 mM $CuSO_4$/ CH_3CN (90/10)	226
Amino acid: 2, 6-diamino-pimelic acid	MCI gel CRS 10W	2 mM $CuSO_4$/ CH_3CN (98/2)	227
Amino acids (serine, proline, leucine), atenolol, norephedrine, homocysteine thiolactone	RP-HPLC column coated with chiral Schiff base	1 mM $CuSO_4$ in water	213
Propranolol, ketamine, phenylalanine, tryptophan, lactic acid	RP-HPLC column coated with chiral Schiff base	2 mM $CuSO_4$ in water —CH_3CN (85:15)	213
1-phenyl-(*p*-tolyl) ethylamine	RP-HPLC column coated with chiral Schiff base	2 mM $CuSO_4$ in water —CH_3CN (80:20)	213
α-Amino acids	Molecular imprints using a monomer, Cu(II)-*N*-(4-vinylbenzyl) iminodiacetic acid	1.5 mM glycine	214
Racemic mixtures of underivatized amino acids	Cu (II)-N^{τ}-*n*-decyl-L-histidine on ODS silica	Water-Cu(II) acetate + organic modifier (85/15)	228
Racemic mixtures of underivatized amino acids	*N*-alkyl- and *N*-aryl-substitued derivates of amino acids (phenyl-alanine and proline) on graphitic carbon	0.001 M Cu(acetate)$_2$ aqueous solution	229, 230

γ-CD performs best.[243] The chiral center of the analyte must be close to the cavity entrance since interactions between the chiral solute and the groups situated at the mouth of the CD are necessary to achieve chiral separation. The high density of secondary hydroxyls at the larger opening of the cone is responsible for the preferential hydrogen bonding.[244,245] Amines and carboxyl groups strongly interact with these hydroxyl groups. The interaction depends on the pK_a of the solute and the pH of the mobile phase.[246,247]

A large number of chiral molecules have been separated with CD-based CSPs. Some examples with commercially available columns are given in Table 5. However, several applications with in-house CSPs are also described.[248–251]

2. Crown-Ethers

Crown-ether CSPs have the ability to include some chiral molecules stereoselectively. These CSPs are well suited for the separation of amino acids and compounds containing a primary amine at or near the stereogenic centre. The most used commercially available crown-ether CSP is Crownpak CR (+), developed by Daicel (Osaka, Japan).[133,134,167,254–259] This CSP is based on a chiral hydrophobic crown ether ((*S*)-2,3:4,5-Bis(1,2,3-phenylnaphtho)-1,6,9,12,15,18-hexaoxacycloeicosa-2,4-diene) that is physically adsorbed onto silica particles. Crownpak CR (−) is also available, but less used. Only a few alterations are allowed in the mobile phase for this CSP. Usually it consists of an aqueous perchloric acid solution with an amount of organic modifier. Varying pH or temperature afterwards can modify the obtained separation.[167,243] Some examples of separations obtained on this CSP are given in Table 6.

However, other separations with a CSP based on (+)-(18-crown-6)-2,3,11,12-tetracarboxylic acid are also described.[260–264] This CSP allows a larger organic modifier content in the mobile phase. Some separations obtained on this CSP are shown in Table 7. This type of CSP was recently commercialized by RSTech (Daejon, Korea) under the name Chirolsil RCA(+). The CSP with the (−) form of this crown ether exists under the name Chirolsil SCA(−).[265]

3. Macrocyclic Antibiotics

Armstrong et al.[140] first introduced chiral stationary phases based on macrocyclic antibiotics. Vancomycin, ristocetin A, teicoplanin, avoparcin, rifamycin B and thiostrepton are used as chiral selectors. They posses a broad enantiorecognition range, similar to protein based CSPs. However, CSPs based on macrocyclic antibiotics show higher stability and capacities.[140] Underivatized amino acids, *N*-derivatized amino-acids, acidic compounds, neutrals, amides, esters and amines can be separated.[266] The first four of the above-mentioned chiral selectors appear to have the largest enantiorecognition range.[267] The selectors can also be derivatized to obtain different enantioselectivities.

TABLE 5 Examples of Separations Obtained on Cyclodextrin-based CSP

Substance	Stationary phase	Mobile phase	References
Warfarin	Cyclobond I	ACN/HOAc/TEA (1000/3/2.5)	237
Ibuprofen	Cyclobond I	ACN/0.02% TEA in water adjusted to pH 4 with HOAc (60/40)	232
Various β-blockers	Cyclobond I	ACN/MeOH/acetic acid/TEA	238, 239
Pindolol	Cyclobond II	ACN/MeOH/acetic acid/TEA (99/1/0.2/0.1)	239
Oxprenolol	2 Cyclobond I	ACN/MeOH/acetic acid/TEA (99/1/0.2/0.16)	239
Chlorpheniramine and metabolites	Cyclobond I 2000	0.25% DEA acetate pH 4.4/ACN/MeOH (85/7.5/7.5)	233
Benzoic acid 1-phenylethylamide, benzoic acid 1-naphthylethylamide, dinitrobenzoic acid 1-phenylethylamide	Cyclobond I SN	Heptane/IPA/DEA (70/30/0.1)	234
Benzoic acid 1-phenylethylamide, dinitrobenzoic acid 1-phenylethylamide, hexobarbital, mephenytoin, dinitrobenzoic acid 1-phenylethylamide, mephenytoin, mephenobarbital, hexobarbital	Cyclobond I SN	0.1 M ammonium acetate pH 7/ACN	234
Norgestrel	Cyclobond II	ACN/1% TEAA pH 4.1 (30/70)	235
Fluoxetine	Cyclobond I	MeOH/1% TEAA pH 4.1 (40/60)	235
Pirlindole	Chiradex	Phosphate buffer pH 6/MeOH (65/35)	236
Oxazepam, lorazepam, temazepam	Cyclobond-I-2000 RSP	ACN/TEAA pH 4.5/H_2O (19/8/73)	252
Methadone	Cyclobond-I-2000 RSP	0.5% TEA phosphate pH 3/ACN (90/10)	253

ACN: acetonitrile, TEA: triethylamine, MeOH: methanol, DEA: Diethylamine, IPA: isopropanol, TEAA: triethylammonium acetate, HOAc: glacial acetic acid.
Cyclobond I: β-CD, cyclobond I 2000: β-CD, cyclobond I SN: *S*-naphthylethylcarbamate-β-CD, cyclobond I 2000 RSP: *R*,*S*-hydroxypropylether-β-CD, cyclobond II: γ-CD, chiradex: β-CD.

Enantioseparations occur by different interactions, such as π–π complexation, hydrogen bonding, hydrophobic inclusion, dipole stacking, steric repulsions, but also due to combinations of them.[140,142]

TABLE 6 Examples of Separations Obtained on the Crownpak CR(+) CSP

Substance	Mobile phase	References
DOPA, tyrosine	0.16 M $HClO_4$ pH 1.05	133
Three GABA analogues	0.11 M $HClO_4$ pH 1.05	133
Baclofen	$HClO_4$ pH 2 or $HClO_4$ pH 2/MeOH (90/10)	254
Phaclofen, saclofen, hydroxysaclofen	$HClO_4$ pH 2/MeOH (90/10)	255
Amphetamine, methamphetamine	Aqueous $HClO_4$ pH 1.8	256
Mexiletine, naphthylethylamine, norephedrine, primaquine	0.1% $HClO_4$ pH 1.9/MeOH (85/15)	134
Clinafloxacin	H_2O/MeOH (88/12) containing 0.1 mM decylamine and set to pH 2 with $HClO_4$	257

$HClO_4$: perchloric acid, DOPA: (−)-3-(3,4-Dihydroxyphenyl-L-alanine, GABA: γ-amino butyric acid. Other abbreviations see Table 5.

TABLE 7 Examples of Separations Obtained on a (+)-(18-Crown-6)-2,3,11,12-Tetracarboxylic Acid-Based CSP

Substance	Mobile phase	References
Arginine, glutamic acid, lysine, methionine, phenylalanine, serine, phenylglycine	10 mM H_2SO_4/MeOH (20/80)	260
Thyroxine, tocainide	H_2O/MeOH(1/4)	261

H_2SO_4: Sulfuric acid, Other abbreviations see Table 5.

They can be used in normal-phase,[140,231,267a] reversed-phase[49,268–271] and polar organic solvents mode[49,167,269,272,273] conditions. The most influential factors in RPLC are pH, type and content of organic modifier and temperature. In normal-phase, mixtures of hexane with ethanol or isopropanol usually give good results. Switching between the different elution modes does not induce irreversible changes in the stationary phases.[167]

Three types of macrocyclic antibiotic CSPs have been commercialized by Astec (Whippany, New York) and show complementary enantioselectivity: Chirobiotic V (selector: vancomycin), Chirobiotic T (teicoplanin)

TABLE 8 Examples of Separations Obtained on Chirobiotic Columns

Substance	Stationary phase	Mobile phase	References
Amino acids and their *N-tert*-butyloxycarbonyl derivatives	Chirobiotic T	1% TEA acetate pH 4.1 with ACN or MeOH	268
Synthesized tyrosine-, phenylalanine-, tetrahydroisoquinoline-, tetralin- and tryptophan analogs	Chirobiotic R	Reversed-phase mode: H_2O/MeOH Polar organic mode: MeOH/HOAc/TEA	269
Bendroflumethiazide, chlorthalidone	Chirobiotic T	MeOH/H_2O (20/80)	270
Citalopram and metabolites	Chirobiotic V	MeOH/acetic acid/TEA (99.9/0.055/0.060)	272
β-amino acids	Chirobiotic T	H_2O with MeOH or EtOH	49
β-amino acids	Chirobiotic R	Reversed-phase mode: H_2O with MeOH or ACN Polar organic mode: MeOH/HOAc/TEA	49
Flurbiprofen, ketoprofen	Chirobiotic V	100 mM ammoniumnitrate pH 5/THF (80/20)	271
1-methyl-2-piperidinoethylesters of 2-, 3 and 4-alkoxy-phenylcarbamic acid	Chirobiotic T	MeOH/ACN (75/25) with 17.5 mM acetic acid and 7.1 mM DEA	273

THF: tetrahydrofuran. Other abbreviations see Table 5.
Chirobiotic T: teicoplanin, chirobiotic R: ristocetin A, chirobiotic V: vancomycin.

and Chirobiotic R (ristocetin A).[142] Table 8 shows some examples of separations that were obtained on these types of CSP.

C. Macromolecular Chiral Selectors

Due to their macromolecular structure and structural heterogeneity, this type of CSPs possess several different binding sites. As a consequence, they usually have a very broad application range.

1. Protein CSPs

The ability of proteins to form enantioselective interactions with a large variety of drugs is used in chiral affinity chromatography. Protein CSPs that are most frequently used for the enantioseparation of pharmaceuticals include bovine serum albumin (BSA), human serum albumin

(HAS), α_1-acid glycoprotein (AGP), ovomucoid (ES-OVM) and α-chymotrypsin.[274]

Enantiorecognition mechanisms involved with protein-based CSPs have not been clearly elucidated. The chiral recognition could be attributed to the specific structure of the protein used and/or to the formation of cavities able to discriminate between enantiomers. Hydrophobic, electrostatic, hydrogen bonding and charge transfer interactions are also involved in chiral recognition. The relative importance of these mechanisms mainly depends on the tertiary structure of the protein that varies with the pH of the mobile phase, and on the nature and amount of organic modifier in the mobile phase. Protein-based CSPs have several charged adsorption sites, which require the use of buffered mobile phases containing a limited percentage of an uncharged organic modifier (1-propanol, ethanol, methanol, acetonitrile, etc.).[275] Occasionally, the use of a charged organic modifier has also led to excellent separations.[167]

The use of protein immobilised to the surface of a silica gel or to another support has been a very successful approach for the chiral separation of various pharmaceuticals. The AGP stationary phase has been shown to have the broadest enantiorecognition abilities while the BSA stationary phase is especially useful for aromatic compounds.[274] Table 9 shows some examples of separations that were obtained on the protein-type of CSPs.

2. Synthetic Polymers and Molecular Imprinted Polymers (MIPs)

Synthetic polymers: This type of chiral selectors (polyacrylamides, polymethacrylamide, etc.) was first developed by the group of Blaschke[286,287] by polymerization of chiral monomers with cross-linking agents to form a three-dimensional polymer network.

The inclusion of enantiomers into the chiral cavities of the network is supposed to be the main chiral recognition mechanism. Moreover, hydrogen bonding between polar groups of the solutes and the amide groups of the polymers are also assumed to participate in the chiral recognition process.[85] Apolar mobile phases such as hexane-dioxane and toluene-dioxane mixtures are therefore commonly used with this type of CSPs.

The main limitation of these CSPs is their limited pressure stability, which makes them not very suitable for HPLC application. However, they have proved to be an excellent tool for the preparative separation of drugs by low-pressure HPLC. To make these CSPs accessible to HPLC, silica gel-based phases were developed.[274] This type of phase is available from Merck (Darmstadt, Germany) under the name Chiraspher.® Polymer phases of different types have been developed by Okamoto's group.[288,289] They are prepared by the asymmetric polymerization of triphenylmethylmethacrylate monomers. The original character of these polymers is that they do not possess any chiral centre and therefore their chirality is only due to their helicity. However, clear mechanisms have not been proposed

TABLE 9 Examples of the Chiral Separation of Drugs Using Proteins as Chiral Selectors

Drug substance	Stationary phase	Mobile phase	References
Ceterizine	AGP	10 mM phosphate buffer pH 7.0/acetonitrile (95/5 v/v)	276
Epibatidine	AGP	10 mM NH_4COOH/NH_3 aq. buffer pH 7.4	277
Fluoxetine	ES-OVM	10 mM potassium phosphate pH 3.5/acetonitrile (98/2 v/v)	278
Ibuprofen	AGP	0.1 M phosphate buffer pH 7.0/2-propanol (100/0.4 v/v)	279
Ketoprofen	AGP	5 mM phosphate buffer pH 7.0 with 1 mM *N,N*-dimethyloctylamine	280
	HSA	0.01 M phosphate buffer/ 2-propanol + 5 mM octanoic acid (94/6 v/v) pH 5.5	281
Methadone	AGP	10 mM ammonium acetate buffer with 0.05% of *N,N*-dimethyloctylamine pH 6.6/2-propanol (85/15 v/v)	282
Nimodipine	ES-OVM	23.5% ethanol in 20 mM phosphate buffer	283
Propranolol	AGP	20 mM ammonium acetate buffer pH 4.1/acetonitrile (98/2 v/v)	284
	ES-OVM	20 mM phosphate buffer pH 6.9/acetonitrile	285

so far. Methanol or hexane-propanol mixtures are usually used as mobile phase. Tetrahydrofuran and chloroform are excluded since they dissolve the polymer. Such chiral phases are commercially available from Daicel under the names Chiralpak OT(+) and OP(+).

Molecular imprinted polymers: MIPs exhibit predetermined enantioselectivity for a specific chiral molecule, which is used as the chiral template during the imprinting process. Most MIPs are obtained by copolymerization from a mixture consisting of a functional mono-unsaturated (vinylic, acrylic, methacrylic) monomer, a di- or tri-unsaturated cross-linker (vinylic, acrylic, methacrylic), a chiral template (print molecule) and a porogenic solvent to create a three-dimensional network. When removing the print molecule, chiral cavities are released within the polymer network. The MIP will memorize the steric and functional binding features of the template molecule. Therefore, inclusion of the enantiomers into the asymmetric cavities of this network can be assumed as the main chiral recognition mechanism. One important advantage of polymers imprinted with chiral templates is the possibility of predicting selectivity.[290] Furthermore, in the elucidation of retention mechanisms, another advantage of using enantiomer templates is that non-specific binding, which affects both enantiomers equally, cancels out. Therefore, the separation factor only reflects the contribution from binding to the enantioselectively imprinted sites.

Several approaches exist to prepare MIPs and they differ in the way the template is linked to the functional monomer and subsequently to the polymeric binding sites. The most widely used technique to synthesize molecular imprinted binding sites is the non-covalent way developed by the group of Mosbach.[291] These polymerizations have the advantage of being relatively robust, allowing polymers to be prepared in high yield using different solvents (aqueous or organic) and at different temperatures. This is necessary in view of the varying solubilities of the template molecules. The most successful non-covalent imprinting systems are based on commodity acrylic or methacrylic monomers, such as methacrylic acid (MAA), cross-linked with ethyleneglycol dimethacrylate (EDMA).[292]

Some limitations of this molecular imprinting technique are obvious: the template must be available in preparative amounts, it must be soluble in the monomer mixture and it must be stable and unreactive under the conditions of the polymerization. The solvent must be chosen considering the stability of the monomer-template assemblies and it should result in the porous structure necessary for rapid kinetics of the template interaction with the binding sites. If these criteria are satisfied, a robust material capable of selectively rebinding the template can be easily prepared and evaluated in a short period of time.

A large number of racemates has been successfully resolved on tailor-made MIPs. Using MAA as functional monomer, good recognition is obtained for templates containing Brönsted-basic or hydrogen containing functional groups close to the stereogenic centre. On the other hand, templates containing acidic functional groups are better imprinted using a basic functional monomer such as vinylpyridine. One important limitation of this approach is that MIPs being made with a specific template can only resolve a restricted number of racemates, meaning that one will nearly need one column per pair of enantiomers to be separated. However, when successful separations are obtained, very high separation factors are usually achieved. For low and moderately polar templates, good recognition is generally seen using either organic solvents or solvent mixtures, such as acetonitrile, chloroform and heptane with or without added acetic acid, where the templates interact mainly electrostatically with the binding sites.[293,294] In fact, MIPs are commonly evaluated in media consisting of the same solvent used as diluent during their synthesis (usually a poorly hydrogen-binding solvent such as chloroform or acetonitrile) together with a polar modifier. On the other hand, templates with protolytic functional groups, i.e., Brönsted basic

TABLE 10 Examples of Chiral Separations of Drugs Using Chiral MIPs as Stationary Phase

Drug substance	Monomer and cross-linker used to prepare the polymer	Mobile phase	References
Ephedrine	MAA-EDMA	Acetonitrile/acetic acid	295
Naproxen	4-Vinylpyridine-EDMA (template (*S*)-naproxen)	THF/heptane/acetic acid (250/250/1 v/v/v)	296
Propranolol[1]	MAA-2,2-bis (hydroxy-methyl)butanol trimethacrylate (template: (*R*)-propranolol)	Acetonitrile/4 M acetate pH 3.0 (80/20 v/v)	297
Timolol	MAA/EDMA	Acetonitrile/acetic acid	298

MAA: methacrylic acid, EDMA: ethyleneglycol dimethacrylate.
[1] Separation was performed in capillary electrochromatography.

or acidic groups, usually perform very well in aqueous mobile phases where the retention is driven by ion-exchange.[202] Table 10 shows some chiral separations that were executed using chiral MIPs as stationary phase.

3. Polysaccharides

Polysaccharides, such as cellulose and amylose, consist of D-glucose units linked by 1–4 glucosidic bonds, forming natural polymers which have a helical structure. When not derivatized, cellulose and amylose have limited enantioselectivity since their very dense structure does not allow inclusion of molecules. Derivatization of the hydroxyl groups on each glucose unit, while preserving the helical structure, forms chiral cavities that are able to include molecules stereoselectively. A wide variety of polysaccharide-based CSPs, mainly cellulose and amylose derivatives, have been developed by Okamoto's group.[299–303] Among all derivatives, the acetate ester, benzoate ester and phenylcarbamate derivatives of glucose have shown the best performance.[299–301]

These polysaccharide-based stationary phases appear to be the most useful in organic, bio-organic and pharmaceutical analysis.[304–308] Of the above-mentioned derivatives three of them, namely cellulose tris-(3,5-dimethylphenylcarbamate), amylose tris-(3,5-dimethylphenylcarbamate) and cellulose tris-(4-methylbenzoate), have very complementary properties and numerous publications have demonstrated that they have been able to achieve the chiral resolution of more than 80% of the drugs currently available on the market.[307–315] These CSPs are known under the commercial names, Chiralcel OD-H®, Chiralpak AD® and Chiralcel OJ®, respectively (Figure 4). Their very broad enantiorecognition range is also the

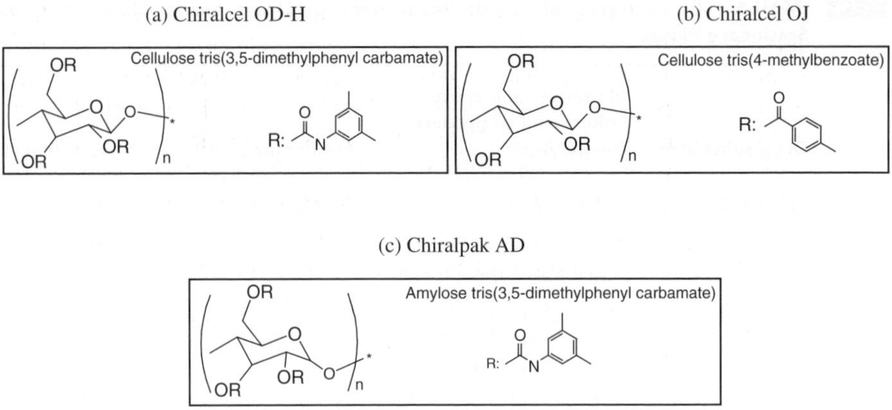

FIGURE 4 Structures of the (a) Chiralcel OD-H®, (b) Chiralcel OJ® and (c) Chiralpak AD® stationary phases.

reason why we did not add a table illustrating their application. Some examples are given later. Polysaccharide-type CSPs have no ionic interaction sites and thus were primarily operated under normal-phase conditions. However, they have recently been shown to perform well under reversed-phase and polar organic mode conditions as well. The mechanisms of chiral recognition are difficult to study due to the complexity of the macromolecules and their multiple binding sites. Several studies have suggested that the enantiomers are discriminated by enantioselective inclusion into chiral cavities resulting in multiple binding sites.[302] The exact geometry of the cavities was found to be influenced by the mode of preparation,[316] but also by the mobile phase used and the temperature conditions. Studies have suggested that both the insertion of an aromatic group and hydrogen bonding stabilize the enantiomers inside the chiral cavity.[317]

For the carbamate derivatives, it is thought that the primary sites of interactions are the polar carbamate groups. Cellulose and amylose carbamates are excellent for the enantioseparation of chiral molecules with hydrogen donor and/or acceptor sites. The aromatic moieties of the chiral molecule may undergo π–π interactions with the aromatic binding sites of the polymer which are located on the exterior of the strand.[318] The discrimination is affected by the steric fit in the cavity. Changing from cellulose (1,4-β-linkage of glucose) to amylose (1,4-α-linkage of glucose) might lead to an inversion of the elution order of the enantiomers. Excellent reviews about the applicability of these CSPs have been published.[316–318]

Cellulose and amylose derivative CSPs are mostly used, in the normal-phase mode, with *n*-hexane-based mobile phases containing some alcohol as modifier. Chromatographic performances, retention and selectivity, are reported to be affected by the composition of the mobile phase and the type of the alcohol present in the mobile phase.[44,45] Isopropanol (IPA) and ethanol (EtOH) are the most commonly used modifiers and, according to the literature and our own experience, should allow the separation of most drug enantiomers on the Chiralcel OD-H, Chiralcel OJ and Chiralpak AD columns. Figure 5 illustrates the differences in selectivity that can be observed on these three columns. Basic and acidic mobile phase additives are often required to improve separations and peak shapes.[299] Diethylamine (DEA) is often added to the mobile phase when the analytes contain an amino basic function in order to reduce peak tailing by masking the residual silanol groups of the CSP.[300] Trifluoroacetic acid (TFA) is usually added to the mobile phase to attenuate the binding of acidic analytes which are often excessively retained under normal-phase conditions with polysaccharide CSPs.[301] Important enhancements in selectivity and resolution by addition of the additives

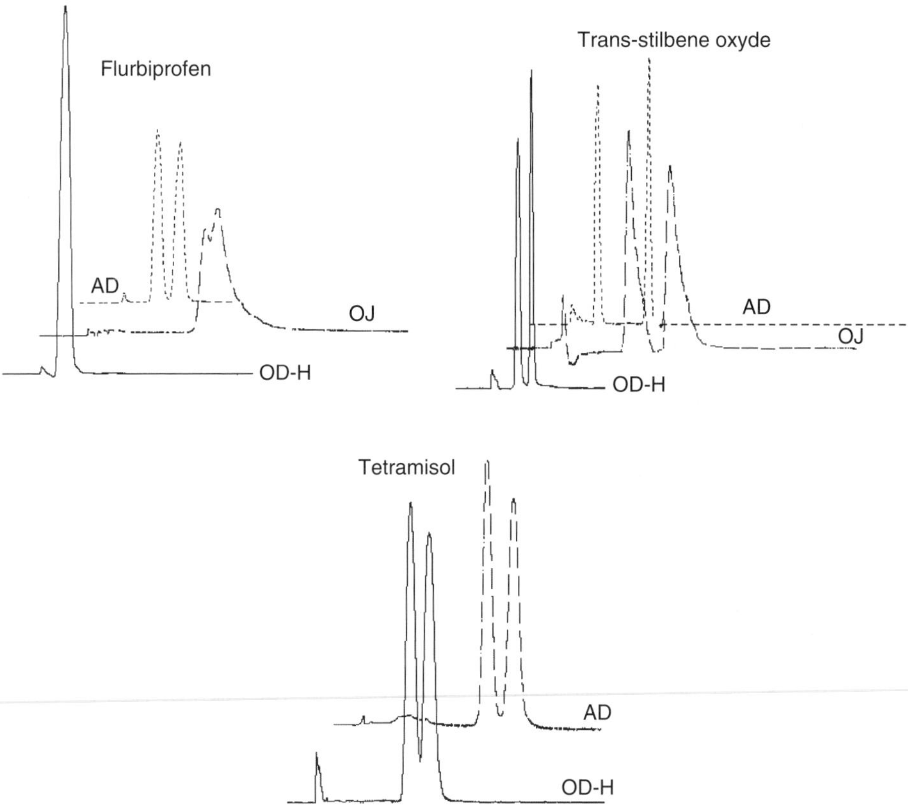

FIGURE 5 Differences in selectivity obtained on Chiralcel OD-H, OJ and Chiralpak AD columns under NPLC conditions. Chromatographic conditions: hexane/ethanol 90/10 (v/v) with 0.1% (v/v) diethylamine or trifluoroacetic acid for basic or acidic compounds, respectively. Flow rate: 1 mL/min.

are reported in the literature.[300–302,316–318] Inversion in the elution order of the enantiomers may occur from one column to another or when changing the type of organic modifier used.[45,319]

In the reversed-phase mode, mixtures of aqueous buffer and acetonitrile are commonly used as mobile phase. Other modifiers are possible, but as shown in Figure 6, acetonitrile often produces the best separations and peak shapes. Several studies have shown the fundamental importance of keeping the chiral analytes neutral when working with polysaccharide stationary phases in the reversed-phase mode.[303,320] Therefore, acidic compounds are preferably analyzed at low pH while basic compounds will be analyzed either in basic media or at low pH in the presence of a chaotropic salt such as sodium perchlorate ($NaClO_4$) or potassium hexafluorophosphate (KPF_6).[321,322] Some illustrations of the effect of the addition of sodium perchlorate on the separation of oxprenolol on different columns and at different pH values are given in Figure 7. It is believed that the perchlorate or the hexafluorophosphate anion forms an ion pair with the positively charged analyte so that the global charge of the analyte is reduced. Furthermore, chaotropic salts have the property to disrupt the water structure by breaking up hydrogen bonds and hydrophobic interactions and therefore, their addition to the mobile phase probably favors interactions between the chiral analytes and the CSPs.

Successful enantioseparations using polysaccharide CSPs in combination with pure polar organic mobile phases have been achieved

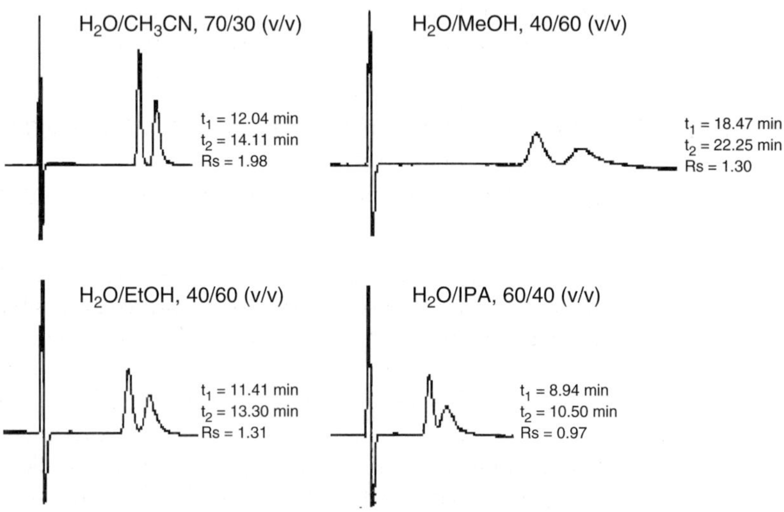

FIGURE 6 **Influence of the type organic modifier on the separation of metoprolol enantiomers (mobile phase: 50 mM phosphate buffer pH 2.0, 500 mM $NaClO_4$/organic modifier).**

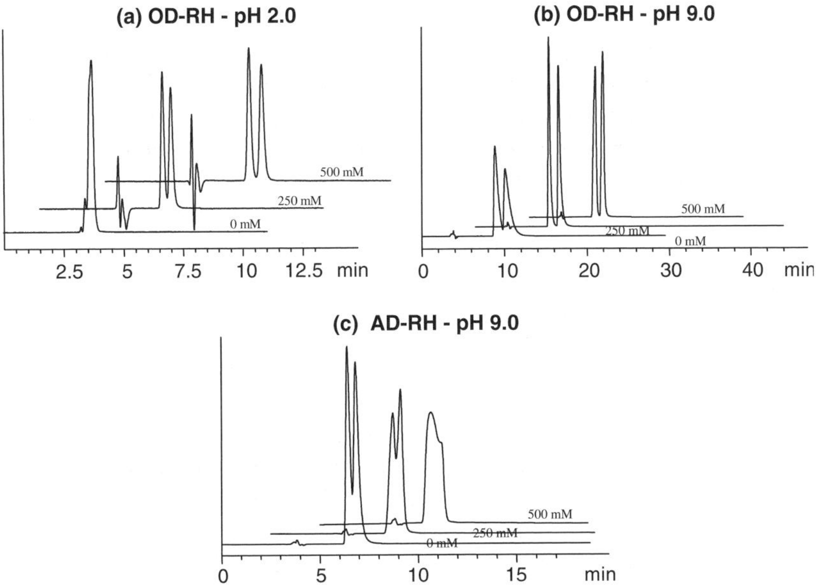

FIGURE 7 **Effect of the addition of sodium perchlorate on the separation of oxprenolol enantiomers (RPLC conditions).**

recently.[323,324] Polar organic mode offers an advantage of higher solubility of analytes as well as alternative chiral recognition mechanisms. A recent publication[325] suggests that the Chiralpak AD column gives the most successful results for a wide variety of compounds.

V. PRACTICAL GUIDELINES TO CHIRAL HPLC SEPARATIONS OF PHARMACEUTICALS

A. Rapid Screening Strategies in Normal- and Reversed-Phase Liquid Chromatography

More than 100 CSPs are commercially available nowadays, which should make the separation of any pair of enantiomers feasible. However, the enantiorecognition mechanisms involved in the chiral recognition between the analytes and the CSPs are complex and therefore the selection of the appropriate CSPs, depending on the structure of the analyte, is a difficult task. A common approach to develop a new enantioseparation is the stepwise trial-and-error approach based on detailed consideration of the enantiorecognition mechanisms between the chiral selector and the analyte, or on the analyst's experience, or on the consultation of literature or databases. However, this approach is time-consuming and often unsuccessful owing to the fact that achieving enantioresolution is often purely empirical and its prediction is thus far from being perfect. Therefore this approach appears not to be practical in an industrial context where a constantly increasing number of samples is submitted for chiral analytical separation, which requires rapid method development.

Consequently, in the attempt to shorten method development time, new screening approaches are continuously emerging.[325–328] The aim of these screening strategies is to be able to analyze quickly series of molecules very diverse in structure and chemical properties, with a minimal set of experimental conditions. The aim at the drug discovery stage of drug development is not to achieve optimal separations as this is only required in later stages, but to evaluate the enantioselectivity of the system towards a given compound and to determine quickly whether an acceptable separation may be achieved with a given CSP. The conditions evaluated could then constitute a good starting point for further optimization. The key point for developing rapid screening schemes is to include a reduced set of chiral selectors with broad application range. Of the commercially available CSPs, the polysaccharides and the macrocyclic antibiotics have appeared to be broad-spectrum chiral selectors and have been privileged for the development of screening strategies in liquid chromatography. Perrin et al.[326,327] have proposed two screening strategies for the enantiomeric separation of drugs using polysaccharide columns in the isocratic normal- and reversed-phase modes. In both modes, three columns, which were found to be complementary and to have broad enantiorecognition abilities for a wide range of pharmaceutical compounds, are used. These columns are the Chiralcel OD-H, Chiralpak AD and Chiralcel OJ for the normal-phase mode and the Chiralcel OD-RH, Chiralpak AD-R and Chiralcel OJ-R for the reversed-phase mode. In the normal-phase strategy, the compounds are screened on each column using two different mobile phases, i.e., *n*-hexane/2-propanol and *n*-hexane/EtOH (both 90/10 v/v). An additive 0.1% (v/v) DEA is added to the mobile phase for the analysis of basic compounds, while the addition of 0.1% (v/v) TFA is required for acidic compounds. The flow rate used is 1 mL/min. This strategy was successfully applied to a set of 36 different drugs. The study has also shown that for basic compounds, the screening on the Chiralcel OD-H and the Chiralpak AD columns is usually sufficient to achieve the separation. Resolution of the enantiomers was observed for 32 compounds on at least one column. Short analysis times (i.e., 20 min or less) were usually achieved. In the reversed-phase strategy, the three columns are screened using two mobiles phases. The compounds are first analyzed with a mobile phase consisting of an aqueous phosphate buffer, pH 2.0, containing 100 mM KPF_6, mixed with acetonitrile (60/40 v/v). The chaotropic agent PF_6^- is added to the mobile phase to achieve the separation of basic analytes at low pH. According to Reference 327, most enantiomers are already separated at this stage. If no or very little enantioselectivity is achieved for some components, they are analyzed on the three columns with a basic mobile phase consisting of an aqueous 20 mM borate buffer, pH 9.0, mixed with acetonitrile (60/40 v/v). Analyses are done with a flow rate of 0.5 mL/min. An example of the obtained separations with the RPLC strategy can be found in Figure 8, while the different separations obtained for one drug substance using either the NPLC or the RPLC strategy are shown in Figure 9. In Reference 327, enantioselectivity on at least one of the columns was achieved for 89% (33 out of 37) of the drugs analyzed. Analysis times were usually less than 30 min.

Another screening strategy using the same type of columns but with normal-phase gradient elution has been proposed by the pharmaceutical group Lilly.[325] In this strategy, each compound is screened on four columns, i.e., Chiralcel OD-H, Chiralpak AD, Chiralpak AS and Chiralcel OJ. An *n*-hexane/2-propanol and an *n*-hexane/EtOH gradient elution system are

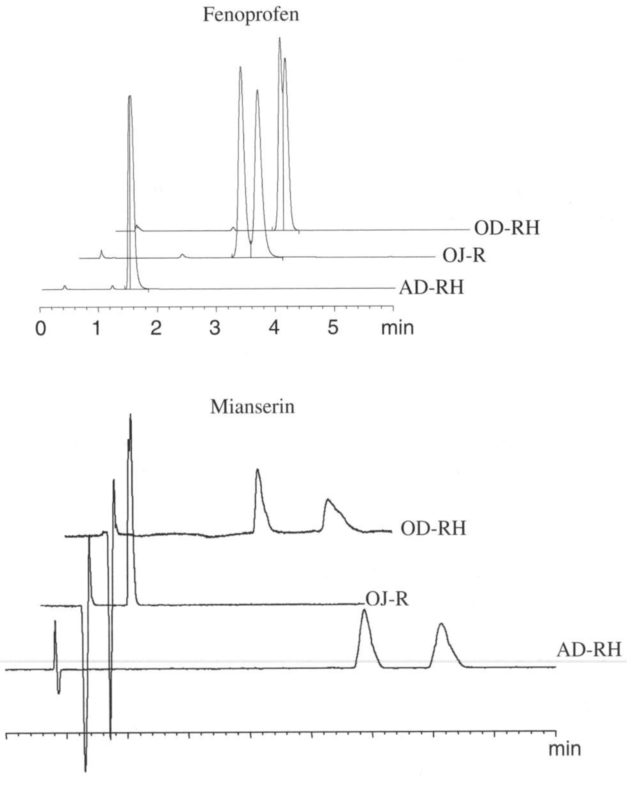

FIGURE 8 **Separation of fenoprofen and mianserin enantiomers on three different polysaccharide CSPs. Mobile phase: 50 mM phosphate buffer pH 2.0 with 100 mM KPF_6/ACN (60/40), except for mianserin on Chiralcel OJ-R: 20 mM borate buffer pH 9.0/ACN (60/40).**

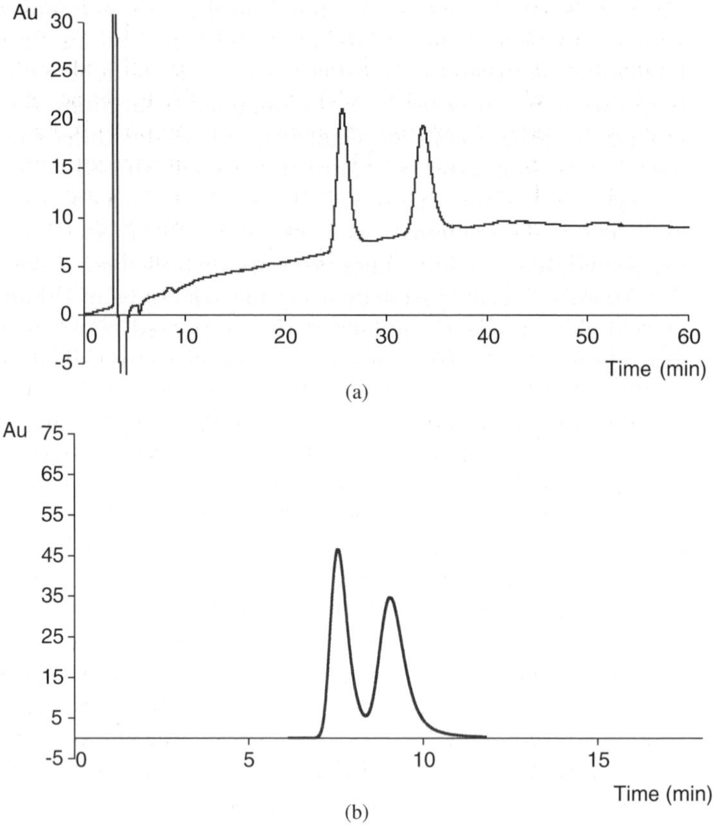

FIGURE 9 Chiral separation of acebutolol using (a) the NPLC and (b) the RPLC strategies. Experimental conditions. NPLC: stationary phase: Chiralpak AD-H, mobile-phase composition: hexane/EtOH/DEA (90/10/0.1), temperature 20°C and flow rate 1.0 mL/min; RPLC: Chiralcel AD-RH, 20 mM borate buffer pH 9.0/CH_3CN (60/40; v/v), room temperature, flow rate = 0.5 mL/min.

used to screen the compounds except for the Chiralpak AD column where the ethanol gradient conditions are discarded. In order to speed-up the analysis, a column-switching device is used. The gradient is run from 20% to 70% alcohol in 20 min with a flow rate of 0.75 mL/min. Baseline resolution was achieved with this strategy for 85% of the substances tested (more than 800 compounds analyzed). When non-baseline separation is achieved, isocratic optimization of the separations is proposed.

Very recently, the pharmaceutical group Astra-Zeneca proposed a screening strategy involving the use of both polysaccharide and macrocyclic glycopeptide columns.[328] It is proposed to screen the compounds on seven different columns: Chiralcel OD-H, Chiralpak AD, Chiralpak AS, Chiralcel

OJ, Chirobiotic R, Chirobiotic V and Chirobiotic T. The polysaccharide columns are tested in the normal-phase and the polar-organic modes. Two mobile phases consisting of iso-hexane/2-propanol and iso-hexane/EtOH (both 85/15 v/v) with 0.1% (v/v) DEA or TFA for basic and acidic compounds, respectively, are used to analyze the compounds. Basic and neutral compounds are also screened in the polar-organic mode with a mixture of MeOH/EtOH (50/50 v/v). A flow rate of 1 mL/min is used for both modes. The macrocyclic columns are evaluated in the reversed-phase and the polar organic modes depending on the basic/acidic properties of the compounds. Basic compounds are analyzed in the aqueous mode with a mobile phase consisting in MeOH/TEA acetate buffer pH 6.0 (25/75 v/v) with a flow rate of 1 mL/min. For neutral compounds, a mixture of MeOH/acetic acid/TEA (100/0.02/0.01 v/v/v) is used in the organic mode with a flow rate of 2 mL/min. Acidic compounds are screened with both mobile phases. This strategy has been applied to a set of 53 chemically diverse compounds. Enantioselectivity was observed for 87% of the test compounds with the polysaccharide columns and for 65% of them with the macrocyclic glycopeptide columns. In total, 96% of the compounds could be resolved into their enantiomers. The use of a switching device is necessary in this strategy due to the number of columns and mobile phases used.

Although earlier a screening strategy already has been proposed with glycopeptide columns connected in series,[329] the authors suggest the evaluation of the columns individually since separations can be missed when columns are connected in series, due to possible elution inversion.

Finally, it may be noted that some chiral column vendors provide free screening or method development support.

B. Optimization of the Separation of a Specific Enantiomer Mixture

When from initial experiments, conditions that indicate the enantioselectivity of the system towards a given enantiomer pair or towards a limited series of substances are known, one might optimize their separation. To obtain optimal conditions, the different chemometric techniques used for method optimization in classic chromatographic or electrophoretic separations can also be applied for the chiral ones.[330] Different experimental design approaches, using both screening and response surface designs can be used.[103,326,331,332] In Reference 331, for instance, a response surface design, i.e., a central composite design,[330] was executed to optimize the chiral separation of some non-steroidal anti-inflammatory drugs (NSAIDs), the 2-arylpropionic acids (e.g., ketoprofen, pirprofen, fenoprofen and tiaprofenic acid). The responses modelled, i.e., those for which a response surface was determined, were the resolution or the Kaiser's peak separation index, indicating the quality of the separation, and the retention time of the last eluting enantiomer peak, indicating analysis time. The optimal conditions are those for which the best compromise exists between the two mutually conflicting responses, i.e., conditions for which the separation is still sufficiently good in the shortest analysis time possible. These compromise conditions can be obtained either from the individual response surfaces for both responses or from applying a multicriteria decision-making (MCDM) approach on the obtained experimental design results. The MCDM methods applied are, for instance, the Pareto-optimality[333,334] method and the method using Derringer's desirability functions.[335–338] The first method is applied for chiral separation optimization in, for instance, Reference 103, while the second is used in References 331 and 332 to optimize the separation of the NSAIDs and the cisapride enantiomers, respectively.

VI. CONCLUSION

As can be observed from the above overview, the possibilities potentially applicable to induce the separation of enantiomers of interest are numerous. Not only can different techniques be applied, but also within each technique large numbers of chiral selectors might be used. This shows that almost any chiral separation can be achieved with at least one of the dozens for the different techniques of commercially available chiral selectors. Difficulties, however, originate from the selection of the appropriate chiral selector for a given separation. Since the chiral recognition mechanisms are complex, the prediction of a suitable chiral selector is very difficult. Frequently, the identification of suitable selectors for a specific pair of enantiomers requires considerable experimentation and might therefore be highly demanding with respect to time, material and labor. There is a clear need for fast method development strategies to avoid to some extent trial-and-error approaches, and recently some attempts in this context have been made.

Among the different CSPs used in chiral HPLC, the polysaccharide chiral stationary phases have proven to be very useful and might be recommended as first-choice columns when one wants to develop chiral separations. These CSPs know many applications, both in normal-phase and reversed-phase mode. As indicated above, the chiral selectors of these CSPs have a broad enantiorecognition range (80% of marketed drugs), which makes them excellent candidates for developing both screening strategies and individually optimized separations. Moreover, the number of different types of stationary phases can be limited to three or four (Chiralcel OD-H and OJ, Chiralpak AD and, occasionally, AS) since both reversed-phase and normal-phase versions exist, while the latter can even be used for chiral SFC.

REFERENCES

1. Williams, R. C., Riley, C. M., Sigvardson, K. W., Fortunak, J., Ma, P., Nicolas, E. C., Unger, S. E., Krahn, D. F. and Bremner, S. L., *J. Pharm. Biomed. Anal.,* 17:917–924, 1998.
2. Krieg, H. M., Breytenbach, J. C. and Keizer, K., *J. Membr. Sci.,* 180:69–80, 2000.
3. Grahame-Smith, D. G. and Aronson, J. K., *Oxford Textbook of Clinical Pharmacology and Drug Therapy,* 2nd ed., Oxford University Press, New York, 1992.
4. Adamovics, J. A. and Fard, D. L., High-Performance Liquid Chromatography, In *Chromatographic Analysis of Pharmaceuticals,* Adamovics, J. A., Ed., 2nd ed., Dekker, New York, pp. 135–207, 1997.
5. Stinson, S. C., *Chem. Eng. News* 78(43):55–78, 2000.
6. Maier, N. M., Franco, P. and Lindner, W., *J. Chromatogr. A,* 906:3–33, 2001.
7. Daniels, J. M., Nestmann, E. R. and Kerr, A. *Drug Inf. J.,* 31:639–646, 1997.
8. FDA Policy, Statement for the Development of New Stereo-isomeric Drugs, Washington, DC, May 1992.
9. Yasuhara, F., Takeda, M., Ochiai, Y., Miyano, S. and Yamaguchi, S., *Chem. Lett.* 2:251–252, 1992.
10. Nolin, T. D. and Frye, R. F., *J. Chromatogr. B,* 783:265–271, 2003.
11. Tao, Q. F. and Zeng, S. J., *Biochem. Biophys. Meth.,* 54:103–113, 2002.
12. Sybilska, D. and Asztemborska, M., *J. Biochem. Biophys. Methods,* 54:187–195, 2002.
13. Juvancz, Z., Markides, K. E., Petersson, P. , Johnson, D. F., Bradshaw, J. S. and Lee, M. L., *J. Chromatogr. A,* 982:119–126, 2002.
14. Kasai, H. F., Tsubuki, M., Takahashi, K., Shirao, M., Matsumoto, Y., Honda, T. and Seyama, Y., *J. Chromatogr. A,* 977:125–134, 2002.
15. Armstrong, D. W., Faulkner, Jr. J. R. and Han, S. M., *J. Chromatogr.* 452:323–330, 1988.
16. Armstrong, D. W., He, F.-Y. and Han, S. M., *J. Chromatogr.*, 448:333–343, 1988.
17. Bhushan, R., Martens, J. and Thuku Thiongo, G., *J. Pharm. Biomed. Anal.* 21:1143–1147, 2000.
18. Gyllenhaal, O. and Karlsson, A. J., *Biochem. Biophys. Meth.*, 54:169–185, 2002.
19. Bernal, J. L., Toribio, L., del Nozal, M. J., Nieto, E. M. and Montequi, M. I., *J. Biochem. Biophys. Methods,* 54:245–254, 2002.
20. del Nozal, M. J., Toribio, L., Bernal, J. L., Nieto, E. M. and Jiménez, J. J., *J. Biochem. Biophys. Methods,* 54:339–345, 2002.
21. Liu, Y., Berthod, A., Mitchell, C. R., Xiao, T. L., Zhang, B. and Armstrong, D. W., *J. Chromatogr. A,* 978:185–204, 2002.
22. Garzotti, M. and Hamdan, M., *J. Chromatogr. B,* 770:53–61, 2002.
23. Johannsen, M., J. *Chromatogr. A,* 937:135–138, 2001.
24. Toribio, L., Bernal, J. L., del Nozal, M. J., Jiménez, J. J. and Nieto, E. M., *J. Chromatogr. A,* 921:305–313, 2001.
25. Terfloth, G., *J. Chromatogr. A,* 906:301–308, 2001.
26. Foucault, A. P., *J. Chromatogr. A,* 906:365–378, 2001.
27. Ma, Y., Ito, Y. and Foucault, A., *J. Chromatogr. A,* 704:75–81, 1995.
28. Ching, C. B., Lim, B. G., Lee, E. J. D. and Ng, S.C., *J. Chromatogr. A,* 634:215–219, 1993.
29. Chankvetadze, B., Kartozia, , I., Yamamoto, C., Okamoto, Y. and Blaschke, G., *J. Pharm. Biomed. Anal.,* 30:1897–1906, 2003.
30. Kawamura, K., Otsuka, K. and Terabe, S., *J. Chromatogr. A,* 924:251–257, 2001.
31. Tanaka, Y. and Terabe, S., *J. Biochem. Biophys. Methods,* 48:103–116, 2001.
32. Fanali, S., Catarcini, P., Blaschke, G. and Chankvetadze, B., *Electrophoresis,* 22:3131–3151, 2001.
33. Blaschke, G. and Chankvetadze, B., *J. Chromatogr. A,* 875:3–25, 2000.

34. Desiderio, C., Aturki, Z. and Fanali, S., *Electrophoresis*, 22:535–543, 2001.
35. Chankvetadze, B. and Blaschke, G., *J. Chromatogr. A*, 906:309–363, 2001.
36. Lämmerhofer, M., Svec, F., Fréchet, J. M. J. and Lindner, W., *TrAC Trends Anal. Chem.*, 19:676–698, 2000.
37. Otsuka, K., Mikami, C. and Terabe, S., *J. Chromatogr. A*, 887:457–463, 2000.
38. Wistuba, D. and Schurig, V., *J. Chromatogr. A*, 875:255–276, 2000.
39. Mayer, S., Briand, X. and Francotte, E., *J. Chromatogr. A*, 875:331–339, 2000.
40. Carter-Finch, A. S. and Smith, N. W., *J. Chromatogr. A*, 848:375–385, 1999.
41. Arai, T., *J. Chromatogr. B*, 717:295–311, 1998.
42. Wren, S. A. C. and Rowe, R. C., *J. Chromatogr.*, 603:235–241, 1992.
43. Wren, S. A. C. and Rowe, R. C., *J. Chromatogr.*, 609:363–367, 1992.
44. Subramanian, G., Ed., *A Practical Approach to Chiral Separations by Liquid Chromatography*, VCH, Weinheim, 1994.
45. Lämmerhofer, M. and Lindner, W., *Recent Developments in Liquid Chromatographic Enantioseparation*, In *Handbook of Analytical Separations*, Valkó, K., Ed. Vol 1, Separation Methods in Drug Synthesis and Purification, Elsevier, Amsterdam. pp. 337–426, 2000.
46. Haginaka, J., *J. Pharm. Biomed. Anal.*, 27:357–372, 2002.
47. Kuhn, R., Erni, F., Bereuter, T. and Haeusler, J., *Anal. Chem.*, 64:2815–2820, 1992.
48. Vékes, E., Török, G., Péter, A., Sápi, J. and Tourwé, D., *J. Chromatogr. A*, 949:125–139, 2002.
49. Péter, A., Lázár, L., Fülöp, F. and Armstrong, D. W., *J. Chromatogr. A*, 926:229–238, 2001.
50. Péter, A., Olajos, E., Casimir, R., Tourwé, D., Broxterman, Q. B., Kaptein, B. and Armstrong, D. W., *J. Chromatogr. A*, 871:105–113, 2000.
51. Péter, M., Péter, A. and Fülöp, F., *J. Chromatogr. A*, 871:115–126, 2000.
52. Lanchote, V. L., Bonato, P. S., Cerqueira, P. M., Pereira, V. A. and Cesarino, E. J., *J. Chromatogr. B*, 738:27–37, 2000.
53. Toyo'oka, T., Toriumi, M. and Ishii, Y., *J. Pharm. Biomed. Anal.*, 15:1467–1476, 1997.
54. Christensen, E. B., Hansen, S. H. and Nørgaard Rasmussen, S., *J. Chromatogr. B*, 670:243–249, 1995.
55. Freimüller, S. and Altorfer, H., *J. Pharm. Biomed. Anal.*, 30:209–218, 2002.
56. Ho Kim, K., Ju Kim, H., Kim, J.-H., Hyun Lee, J. and Cheal Lee, S., *J. Pharm. Biomed. Anal.*, 25:947–956, 2001.
57. Kagawa, M., Machida, Y. and Nishi, H., *J. Chromatogr. A*, 857:127–135, 1999.
58. Sybilska, D., Bielejewska, A., Nowakowsky, R., Duszczyk, K. and Jurczak, J., *J. Chromatogr.*, 625:349–352, 1992.
59. Duff, K. J., Gray, H. L., Gray, R. J. and Bahler, C.C., *Chirality* 5:201–206, 1993.
60. Pettersson, C. and Persson, B., HPLC Applications for Chiral Pharmaceutical Analysis, In *Handbook of HPLC*, Katz, E., Eksteen, R., Schoenmakers P. and Miller, N., Eds., Dekker, New York, pp. 669–693, 1998.
61. Gundlach, G., Sattler, E. L. and Wagenbach, U., *Fresenius Z. Anal. Chem.*, 311:684–686, 1982 (in German).
62. Kurganov, A. A. and Darankov, V. A., *J. Chromatogr.*, 218:559–567, 1981.
63. Forsman, U., *J. Chromatogr.*, 303:217–221, 1984.
64. Benecke, I., *J. Chromatogr.*, 291:155–164, 1984.
65. Horikawa, R., Sakamoto, H. and Tanimura, T., *J. Liq. Chromatogr.*, 9:537–549, 1986.
66. Galaverna, G., Corradini, R., Dossena, A., Marchelli, R. and Dallavalle, F., *Chirality*, 8:189–196, 1996.
67. Galaverna, G., Pantò, F., Dossena, A., Marchelli, R. and Bigi, F., *Chirality*, 7:331–336, 1995.
68. Pettersson, C. and No, K., *J. Chromatogr.*, 282:671–684, 1983.
69. Pettersson, C., *J. Chromatogr.*, 316:553–567, 1984.
70. Karlsson, A. and Pettersson, C., *Chirality*, 4:323–332, 1992.

71. Pettersson, C. and Schill, G., *J. Chromatogr.*, 204:179–183, 1981.
72. Pettersson, C. and Schill, G., *Chromatographia,* 16:192–197, 1982.
73. Pettersson, C. and Stuurman, H. W., *J. Chromatogr. Sci.,* 22:441–443, 1984.
74. Bielejewska, A., Duszczyk, K. and Sybilska, D., *J. Chromatogr. A,* 931:81–93, 2001.
75. Bielejewska, A., Nowakowski, R., Duszczyk, K. and Sybilska, D., *J. Chromatogr. A,* 840:159–170, 1999.
76. Pettersson, C., Arvidsson, T., Karlsson, A.-L. and Marle, I., *J. Pharm. Biomed. Anal.* 4:221–235, 1986.
77. Shinbo, T., Yamaguchi, T., Nishimura, K.and Sugiura, M., *J. Chromatogr. A,* 405:145–153, 1987.
78. Lelièvre, F., Yan, C., Zare, R. N. and Gareil, P., *J. Chromatogr. A,* 723:145–156. 1996.
79. Mangelings, D., Perrin, C., Massart, D. L., Maftouh, M., Eeltink, S., Kok, W. Th., Schoenmakers, P. J. and Vander Heyden, Y., *Anal. Chim. Acta,* 509:11–19, 2004.
80. Mangelings, D., Maftouh, M., Massart, D. L. and Vander Heyden, Y., *J. Pharm. Biomed. Anal.,* 35:789–799, 2004.
81. Krstulovic, A. M., *Chiral Separation by HPLC, Applications to Pharmaceutical Compounds*, Ellis Horwood, New York, pp. 548, 1989.
82. Davankov, V. A., Kurganov, A. A. and Bochkov, A. S., *Adv. Chromatogr.* 22:71–116, 1983.
83. Amstrong, D. W. and Han, S. M., *CRC Crit. Rev. Anal. Chem.* 19:175–224, 1988.
84. Pirkle, W. H. and Pochapsky, T. C., *Chem. Rev.* 89:347–362, 1989.
85. Allenmark, S. G., *Chromatographic Enantioseparation*, 2nd ed., Ellis Horwood, New York, 1991.
86. Gübitz, G. and Schmid, M. G., *J. Chromatogr.* A, 792:179–225, 1997.
87. Vespalec, R. and Boček, P., *Chem. Rev.*, 100:3715–3754, 2000.
88. Verleysen, K. and Sandra, P., *Electrophoresis*, 19:2798–2833, 1998.
89. Chankvetadze, B., *J. Chromatogr. A,* 792:269–295, 1997.
90. Amini, A., *Electrophoresis* 22:3107–3130, 2001
91. Chankvetadze, B., *Capillary Electrophoresis in Chiral Analysis,* John Wiley, Chichester, pp. 71–133, 141–228, 1997.
92. Rizzi, A., *Electrophoresis,* 22:3079–3106, 2001.
93. Altria, K. D., *Analysis of Pharmaceuticals by Capillary Electrophoresis,* Vieweg, Wiesbaden, 1998.
94. Terabe, S., *TrAC Trends Anal. Chem.,* 8:129–134, 1989.
95. Kuhn, R. and Hoffstetter-Kuhn, S., *Chromatographia,* 34:505–512, 1992.
96. Fanali, S., *J. Chromatogr. A,* 792:227–267, 1997.
97. Vescina, M. C., Fermier, A. M. and Guo, Y., *J. Chromatogr. A,* 973:187–196, 2002.
98. Lin, C.-E. and Chen, K.-H., *J. Chromatogr. A,* 930:155–163, 2001.
99. Zhou, L., Thompson, R., Song, S., Ellison, D. and Wyvratt, J. M., *J. Pharm. Biomed. Anal.,* 27:541–553, 2001.
100. Kuo, C.Y., Wu, H.L. and Wu, S. M., *Anal. Chim. Acta,* 471:211–217, 2002.
101. Zukowski, J., De Biasi, V. and Berthod, A., *J. Chromatogr. A,* 948:331–342, 2002.
102. Perrin, C., Vander Heyden, Y., Maftouh, M. and Massart, D. L., *Electrophoresis,* 22:3203–3215, 2001.
103. Matthijs, N., Perrin, C., Maftouh, M., Massart, D. L. and Vander Heyden, Y., *J. Pharm. Biomed. Anal.,* 27:515–529, 2002.
104. Perrin, C., Fabre, H., Maftouh, M., Massart, D. L. and Vander Heyden, Y., *J. Chromatogr. A,* 1007:165–177, 2003.
105. Fillet, M., Fotsing, L. and Crommen, J., *J. Chromatogr. A,* 817:113–119, 1998.
106. Abushoffa, A. M., Fillet, M., Hubert, Ph. and Crommen, J., *J. Chromatogr., A,* 948:321–329, 2002.
107. Fillet, M., Hubert, Ph. and Crommen, J., *J. Chromatogr. A,* 875:123–134, 2000.
108. Schmitt, T. and Engelhardt, H., *Chromatographia,* 37:475–481, 1993.

109. Tait, R. J., Thompson, D. O., Stella, V. J. and Stobaugh, J. F., *Anal. Chem.,* 66: 4013–4018, 1994.
110. Chankvetadze, B., Endresz, G. and Blaschke, G., *Electrophoresis,* 15:804–807, 1994.
111. Schulte, G., Chankvetadze, B. and Blaschke, G., *J. Chromatogr. A,* 771:259–266, 1997.
112. Tanaka, Y. and Terabe, S., *J. Chromatogr. A,* 781:151–160, 1997.
113. Desiderio, C. and Fanali, S., *J. Chromatogr. A,* 716:183–196, 1995.
114. Perrin, C., Vargas, M. G., Vander Heyden, Y., Maftouh, M. and Massart, D. L., *J. Chromatogr. A,* 883:249–265, 2000.
115. Lurie, I. S., Odeneal II, N. G., McKibben, T. D. and Casale, J. F., *Electrophoresis* 19:2918–2925, 1998.
116. Verleysen, K., Van den Bosch, T. and Sandra, P., *Electrophoresis,* 20:2650–2655, 1999.
117. Vincent, J. B., Sokolowski, A. D., Nguyen, T. V. and Vigh, G., *Anal. Chem.,* 69:4226–4233, 1997.
118. Vincent, J. B., Kirby, D. M., Nguyen, T. V. and Vigh, G., *Anal. Chem.,* 69:4419–4428, 1997.
119. Cai, H., Nguyen, T. V. and Vigh, G., *Anal. Chem.,* 70:580–589, 1998.
120. Chankvetadze, B., Endresz, G. and Blaschke, G., *J. Chromatogr. A,* 704:234–237, 1995.
121. Koppenhoefer, B., Zhu, X., Jakob, A., Wuerthner, S. and Lin, B., *J. Chromatogr. A,* 875:135–161, 2000.
122. Christians, T., Diewald, D., Wessler, C., Otte, Y., Lehmann, J. and Holzgrabe, U., *J. Chromatogr. A,* 853:455–460, 1999.
123. Maruszak, W., Trojanowicz, M., Margasińska, M. and Engelhardt, H., *J. Chromatogr. A,* 926:327–336, 2001.
124. Rudzińska, E., Dyżgiel, P., Wieczorek, P. and Kafarski, P., *J. Chromatogr. A,* 979:115–122, 2002.
125. Yowell, G. G., Fazio, S. D. and Vivilecchia, R. V., *J. Chromatogr. A,* 745:73–79, 1996.
126. Nardi, A., Ossicini, L. and Fanali, S., *Chirality,* 4:56–61, 1992.
127. Terabe, S., Miyashita, Y., Ishihama, Y. and Shibata, O., *J. Chromatogr.,* 636:47–55, 1993.
128. Ueda, T., Kitamura, F., Mitchell, R., Metcalf, T., Kuwana, T. and Nakamoto, A., *Anal. Chem.,* 63:2979–2981, 1991.
129. Rickard, E. C. and Bopp, R. J., *J. Chromatogr. A,* 680:609–621, 1994.
130. Valkó, I. E., Billiet, H. A. H., Frank, J. and Luyben, K. Ch. A. M., *J. Chromatogr. A,* 678:139–144, 1994.
131. Liu, L., Osborne, L. M. and Nussbaum, M. A., *J. Chromatogr. A,* 745:45–52, 1996.
132. Tanaka, Y., Otsuka, K. and Terabe, S., *J. Chromatogr. A,* 875:323–330, 2000.
133. Walbroehl, Y. and Wagner, J., *J. Chromatogr. A,* 685:321–329, 1994.
134. Nishi, H., Nakamura, K., Nakai, H. and Sato, T., *J. Chromatogr. A,* 757:225–235, 1997.
135. Kuhn, R., Steinmetz, C., Bereuter, T., Haas, P. and Erni, F., *J. Chromatogr. A,* 666:367–373, 1994.
136. Kuhn, R., Riester, D., Fleckenstein, B. and Wiesmüller, K.-H., *J. Chromatogr. A,* 716:371–379, 1995.
137. Schmid, M. G. and Gübitz, G., *J. Chromatogr. A,* 709:81–88, 1995.
138. Castelnovo, P. and Albanesi, C., *J. Chromatogr. A,* 715:143–149, 1995.
139. Mori, Y., Ueno, K. and Umeda, T., *J. Chromatogr. A,* 757:328–332, 1997.
140. Armstrong, D. W., Tang, Y. B., Chen, S., Zhou, Y. W., Bagwill, C. and Chen, J.-R., *Anal. Chem.,* 66:1473–1484, 1994.
141. Armstrong, D. W., Rundlett, K. L. and Reid, G. L., *Anal. Chem.,* 66:1690–1695, 1994.
142. Ward, T. J. and Farris III, A. B., *J. Chromatogr. A,* 906:73–89, 2001.
143. Desiderio, C. and Fanali, S., *J. Chromatogr. A,* 807:37–56, 1998.
144. Fanali, S., *J. Chromatogr. A,* 735:77–121, 1996.

145. Vespalec, R., Billiet, H. A. H., Frank, J. and Boček, P., *Electrophoresis*, 17:1214–1221, 1996.
146. Gasper, M. P., Berthod, A., Nair, U. B. and Armstrong, D. W., *Anal. Chem.*, 68:2501–2514, 1996.
147. Ward, T., *LC-GC*, 14:886, 1996.
148. Oswald, T. M. and Ward, T. J., *Chirality*, 11:663–668, 1999.
149. Busch, S., Kraak, J. C. and Poppe, H., *J. Chromatogr.*, 635:119–126, 1993.
150. Tanaka, Y. and Terabe, S., *J. Chromatogr. A*, 694:277–284, 1995.
151. Zhang, X.-X., Hong, F., Chang, W.-B., Ci, Y.-X. and Ye, Y.-H., *Anal. Chim. Acta*, 392:175–181, 1999.
152. Altria, K. D., Kelly, M. A. and Clark, B. J., *TrAC Trends Anal. Chem.*, 17:214–225, 1998.
153. De Lorenzi, E. and Massolini, G., *Pharm. Sci. Technol.Today*, 2:352–364, 1999.
154. Haginaka, J., *J. Chromatogr. A*, 875:235–254, 2000.
155. Wan, H. and Blomberg, L. G., *J. Chromatogr. A*, 875:43–88, 2000.
156. Tran, C. D. and Kang, J., *J. Chromatogr. A*, 978:221–230, 2002.
157. Otsuka, K. and Terabe, S., *J. Chromatogr. A*, 875:163–178, 2000.
158. El Rassi, Z., *J. Chromatogr. A*, 875:207–233, 2000.
159. Yarabe, H. H., Billiot, E. and Warner, I. M., *J. Chromatogr. A*, 875:179–206, 2000.
160. Rawyee, Y. Y., Staerk, D. U. and Vigh, G., *J. Chromatogr.*, 635:291–306, 1993.
161. Pálmarsdóttir, S. and Edholm, L.-E., *J. Chromatogr. A*, 666:337–350, 1994.
162. Fabre, H., *J. Pharm. Biomed. Anal.*, 14:1125–1132, 1996.
163. Pirkle, W. H., Hyun, M. H. and Bank, B., *J. Chromatogr.*, 316:585–604, 1984.
164. Pirkle, W. H., Hyun, M. H., Tsipouras, A., Hamper, B. C. and Banks, B., *J. Pharm. Biomed. Anal*, 2:173–181, 1984.
165. Pirkle, W. H. and Gan, K. Z., *J. Chromatogr. A*, 790:65–71, 1997.
166. Wolf, C. and Pirkle, W. H., *J. Chromatogr. A*, 785:173–178, 1997.
167. Ceccato, A., Hubert, Ph. and Crommen, J., *STP Pharma Pratiques*, 9(4):295–310, 1999.
168. Welch, C. J., *J. Chromatogr. A*, 666:3–26, 1994.
169. Gasparrini, F., Misiti, D. and Villani, C., *J. Chromatogr. A*, 906:35–50, 2001.
170. Hyun, M. H., Ryoo, J.-J. and Pirkle, W. H., *J. Chromatogr. A*, 886:47–53, 2000.
171. Pirkle, W. H. and Burke III, J. A., *J.Chromatogr.*, 557:173–185, 1991.
172. Pirkle, W. H., Lee, W. and Welch, C. J., *Enantiomer* , 2:423–431, 1997.
173. Caude, M., Tambuté, A. and Siret, L., *J. Chromatogr.*, 550:357–382, 1991.
173a. Tang, L. Y., Xiang, D. and Blackwell, J. A., *Enantiomer*, 5:345–355, 2000.
174. Pirkle, W. H. and Koscho, M. E., *J. Chromatogr. A*, 840:151–158, 1999.
175. Garcia, M. E., Naffin, J. L., Deng, N. and Mallouk, T. E., *Chem. Mater.*, 7: 1968–1973, 1995.
176. Dyas, A. M., Robinson, M. L. and Fell, A. F., *J. Chromatogr.*, 586:351–355, 1991.
177. Schleimer, M., Pirkle, W. H. and Schurig, V., *J. Chromatogr. A*, 679:23–34, 1994.
178. Li, Z. Y., Zhang, Z. C., Zhou, Q. L., Wang, Q. M., Gao, R. Y. and Wang, Q. S., *J. Assoc. Anal. Commun. Int.* 86:521–528. 2003.
179. Medvedovici, A., Sandra, P., Toribio, L. and David, F., *J. Chromatogr. A*, 785:159–171, 1997.
180. Selditz, U., Copinga, S., Grol, C. J., Franke, J. P., de Zeeuw, R. A., Wikström, H. and Gyllenhaal, O., *Pharmazie*, 54:183–190, 1999.
181. Zheng, J. and Shamsi, S. A., *J. Chromatogr. A*, 1005:177–187, 2003.
182. Maier, N. M. and Uray, G., *J. Chromatogr. A*, 732:215–230, 1996.
183. Gasparrini, F., Maggio, F. and Misiti, D., *J. High Res. Chrom.*, 17:43–45, 1994.
184. Song, Y., Wang, D., Hu, Y., Chen, X., Jiao, Y. and Hou, D., *J. Pharm. Biomed. Anal.*, 31:311–319, 2003.
185. Pirkle, W. H., Welch, C. J. and Lamm, B., *J. Org. Chem.*, 57:3854–3860, 1992.
186. Pirkle, W. H. and Welch, C. J., *J. Liq. Chromatogr.*, 15:1947–1955, 1992.

187. Welch, C. J., Szczerba, T. and Perrin, S. R., *J. Chromatogr. A,* 758:93–98, 1997.
188. Henne, K. R., Gaedigk, A., Gupta, G., Leeder, J. S. and Rettie, A. E., *J. Chromatogr. B,* 710:143–148, 1998.
189. Pirkle, W. H. and Spence, P. L., *Chirality,* 10:430–433, 1998.
190. Kennedy, J. H., *J. Chromatogr. A,* 725:219–224, 1996.
191. Magora, A., Abu-Lafi, S. and Levin, S., *J. Chromatogr. A,* 866:183–194, 2000.
192. Website: Regis Technologies, http://www.chromtech.com/2001catalog/ SeparatePgs/209.pdf or http://www.registech.com/chiral/chiralprod.html.
193. Dyas, A. M., *J. Pharm. Biomed. Anal.,* 10:383–404, 1992.
194. Lämmerhofer, M. and Lindner, W., *J. Chromatogr. A,* 741:33–48, 1996.
195. Vinkovic, V., Kontrec, D., Sunjic, V., Navarini, L., Zanetti, F. and Azzolina, O., *Chirality,* 13:581–587, 2001.
196. Kontrec, D., Vinkovic, V. and Sunjic, V., *J. Liq. Chromatogr. Rel. Technol.* 23:1203–1216, 2000.
197. Zhang, X., Ouyjang, J., Baeyens, W. R. G., Zhai, S., Yang, Y. and Huang, G., *J. Pharm. Biomed. Anal.,* 31:1047–1057, 2003.
198. Clohs, L. and McErlane, K. M., *J. Pharm. Biomed. Anal.,* 31:407–412, 2003.
199. Williams, M. G. and Zhong, W.-Z., *J. Chromatogr. B,* 694:169–177, 1997.
200. Rentsch, K. M., Gutteck-Amsler, U., Bührer, R., Fattinger, K. E. and Vonderschmitt, D. J., *J. Chromatogr. B,* 742:131–142, 2000.
201. Kleidernigg, O. P. and Kappe, C. O., *Tetrahedron: Asymmetry* , 8:2057–2068, 1997.
202. Sellergren, B. and Shea, K. J., *J. Chromatogr. A,* 654:17–28, 1993.
203. Lämmerhofer, M., Di Eugenio, P., Molnar, I. and Lindner, W., *J. Chromatogr. B,* 689:123–135, 1997.
204. Maier, N. M., Nicoletti, L., Lämmerhofer, M. and Lindner, W., *Chirality,* 11:522–528, 1999.
205. Sinibaldi, M., Flieger, M., Cvak, L., Messina, A. and Pichini, A., *J. Chromatogr. A,* 666:471–478, 1994.
206. Franco, P., Lämmerhofer, M., Klaus, P. M. and Lindner, W., *J. Chromatogr. A,* 869:111–127, 2000.
207. Veigl, E., Böhs, B., Mandl, A., Krametter, D. and Lindner, W., *J. Chromatogr. A,* 694:135–150, 1995.
208. Veigl, E., Böhs, B., Mandl, A., Krametter, D. and Lindner, W., *J. Chromatogr. A,* 694:151–161, 1995.
209. Davankov, V. A. and Rogozhin, S. V., *J. Chromatogr. A,* 60:284–312, 1971.
210. Kurganov, A., *J. Chromatogr. A,* 906:51–71, 2001.
211. Davankov, V. A., *J. Chromatogr. A,* 666:55–76, 1994.
212. Galushko, S. V., Shishkina, I. P., Solshonok, V. A. and Kukhar, V. P., *J. Chromatogr.,* 511:115–121, 1990.
213. Ôi, N., Kitahara, H. and Aoki, F., *J. Chromatogr.* 631:177–182, 1993.
214. Vidyasankar, S., Ru, M. and Arnold, F. H., *J. Chromatogr. A,* 775:51–63, 1997.
215. Gübitz, G., Jellenz, W. and Santi, W., *J. Chromatogr.,* 203:377–384, 1981.
216. Zheng, Z.-X., Lin, J.-M. and Qu, F., *J. Chromatogr. A,* 1007:189–196, 2003.
217. Lefebvre, B., Audebert, R. and Quivoron, C., *J. Liq. Chromatogr.,* 1:761–774, 1978.
218. Francotte, E. and Junker-Buchheit, A., *J. Chromatogr. B,* 576:1–45, 1992.
219. Davankov, V. A., *J. Chromatogr. A,* 1000:891–915, 2003.
220. Galushko, S. V., Shishkina, I. P. and Soloshonok, V. A., *J. Chromatogr.,* 592:345–348, 1992.
221. Grierson, J. R. and Adam, M. J., *J. Chromatogr.,* 325:103–109, 1985.
222. Galushko, S. V., Shishkina, I. P., Gerus, I. I. and Kolycheva, M. T., *J. Chromatogr.,* 600:83–85, 1992.
223. Tanaka, Y., Shimomura, Y., Hirota, T., Nozaki, A., Ebata, M., Takasaki, W., Shigehara, E., Hayashi, R. and Caldwell, J., *Chirality,* 4:342–348, 1992.
224. Thomas, S. B. and Surber, B. W., *J. Chromatogr.*, 586:265–270, 1991.

225. Itazaki, H., Nagashima, K., Sugita, K., Yoshida, H., Kawamura, Y., Yasuda, Y., Matsumoto, K., Ishii, K., Uotani, N., Nakai, H., Terui, A., Yoshimatsu, S., Ikenishi, Y. and Nakagawa, Y., *J. Antibiot.*, 43:1524–1532, 1990.
226. Arai, T., Matsuda, H. and Ōizumi, H., *J. Chromatogr.*, 474:405–410, 1989.
227. Nagasawa, T., Ling, J. R. and Onodera, R., *J. Chromatogr. A*, 653:336–340, 1993.
228. Remelli, M., Fornasari, P. and Pulidori, F., *J. Chromatogr. A*, 761:79–89, 1997.
229. Wan, Q.-H., Shaw, P. N., Davies, M. C. and Barrett, D. A., *J. Chromatogr. A*, 765:187–200, 1997.
230. Wan, Q.-H., Shaw, P. N., Davies, M. C. and Barrett, D. A., *J. Chromatogr. A*, 786:249–257, 1997.
231. Ekborg-Ott, K. H., Liu, Y. and Armstrong, D. W., *Chirality*, 10:434–483, 1998.
232. Naidong, W. and Lee, J. W., *J. Pharm. Biomed. Anal.*, 12:551–556, 1994.
233. Fried, K. M., Young, A. E., Usdin Yasuda, S. and Wainer, I. W., *J. Pharm. Biomed. Anal.*, 27:479–488, 2002.
234. Vandenbosch, C., Massart, D. L. and Lindner, W., *J. Pharm. Biomed. Anal.*, 10:895–908, 1992.
235. Berthod, A., Liang Jin, H., Beesley, T. E., Duncan, J. D. and Armstrong, D. W., *J. Pharm. Biomed. Anal.*, 8:123–130, 1990.
236. Ceccato, A., Hubert, Ph., de Tullio, P., Liégeois, J.-F., Felikidis, A., Géczy, J. and Crommen, J., *J. Pharm. Biomed. Anal.*, 18:605–614, 1998.
237. Naidong, W. and Lee, J. W., *J. Pharm. Biomed. Anal.* 11:785–792, 1993.
238. Ekelund, J., van Arkens, A., Brønnum-Hansen, K., Fich, K., Olsen, L. and Vibholm Petersen, P., *J. Chromatogr. A*, 708:253–261, 1995.
239. Armstrong, D. W., Chen, S., Chang, C. and Chang, S., *J. Liq. Chromatogr.*, 15:545–556, 1992.
240. Chang, S. C., Reid III, G. L., Chen, S., Chang, C. D. and Armstrong, D. W., *TrAC Trends Anal. Chem.*, 12:144–152, 1993.
241. Hargitai, T. and Okamoto, Y., *J. Liq. Chrom.* 16(4):843–858, 1993.
242. Bressolle, F., Audran, M., Pham, T.-N., and Vallon, J.-J., *J. Chromatogr B*, 687:303–336, 1996.
243. Wainer, I. W., HPLC Chiral Stationary Phases for the Stereochemical Resolution of Enantiomeric Compounds, In *Drug Stereochemistry: Analytical Methods and Pharmacology*, 2nd ed., Wainer, I. W., Ed., Dekker, New York, pp.139–181, 1993.
244. Armstrong, D. W., Ward, T. J., Armstrong, R. D. and Beesley, T. E., *Science*, 232:1132–1135, 1986.
245. Armstrong, D. W. and DeMond, W., *J. Chromatogr. Sci.*, 22:411–415, 1984.
246. Poole, C. F., *The Essence of Chromatography*, Elsevier, Amsterdam, 2003.
247. Beeson, M. D. and Vigh, G., *J. Chromatogr.*, 634:197–204, 1993.
248. Stalcup, A. M. and Gahm, K. H., *Anal. Chem.*, 68:1369–1374, 1996.
249. Stalcup, A. M., Chang, S.-C., Armstrong, D. W. and Pitha, J., *J. Chromatogr.* 513:181–194, 1990.
250. Duval, R., Lévêque, H., Prigent, Y. and Aboul-Enein, H. Y., *Biomed. Chromatogr.*, 15:202–206, 2001.
251. Wang, X. and Ching, C. B., *Chirality*, 14:798–805, 2002.
252. Pham-Huy, C., Villin-Pautet, G., Hua, H., Chikhi-Chorfi, N., Galons, H., Thevenin, M., Claude, J.-R. and Warnet, J.-M., *J. Biochem. Biophys. Methods*, 54:287–299, 2002.
253. Rudaz, S. and Veuthey, J.-L., *Chirality*, 11:319–325, 1999.
254. Vaccher, C., Berthelot, P. and Debaert, M., *J. Chromatogr.*, 645:95–99, 1993.
255. Vaccher, C., Berthelot, P. and Debaert, M., *J. Chromatogr. A*, 657:213–218, 1993.
256. Makino, Y., Ohta, S. and Hirobe, M., *Forensic Sci. Int.*, 78:65–70, 1996.
257. Brodfuehrer, J. I., Priebe, S. and Guttendorf, R., *J. Chromatogr. B*, 709:265–272, 1998.
258. Vaccher, C., Berthelot, P. and Debaert, M., *J. Chromatogr. A*, 715:361–365, 1995.
259. Francotte, E., Davatz, A. and Richert, P., *J. Chromatogr. B*, 686:77–83, 1996.

260. Hyun, M. H., Jin, J. S. and Lee, W., *J. Chromatogr. A,* 822:155–161, 1998.
261. Aboul-Enein, H.Y., Ali, I., Hyun, M. H., Cho, Y. J. and Jin, J. S., *J. Biochem. Biophys. Methods,* 54:407–413, 2002.
262. Jin, J. S., Stalcup, A. M. and Hyun, M. H., *J. Chromatogr. A,* 933:83–90, 2001.
263. Hyun, M. H., Jin, J. S., Han, S. C. and Cho, Y. J., *Microchem. J.,* 70:205–210, 2002.
264. Hyun, M. H., Jin, J. S., Koo, H. J. and Lee, W., *J. Chromatogr. A,* 837:75–82, 1999.
265. Lee, W., Baek, C.-S. and Lee, K., *Bull. Korean Chem. Soc,.* 23(11):1677–1679, 2002.
266. http://www.astecusa.com.
267. Ekborg-Ott, K. H., Kullman, J. P., Wang, X., Gahm, K, He, L. and Armstrong, D. W., *Chirality,* 10:627–660, 1998.
267a. Péter, A., Vékes, E. and Armstrong, D. W., *J. Chromatogr. A,* 958:89–107, 2002.
268. Tesařová, E., Bosáková, Z. and Pacáková, V., *J. Chromatogr. A,* 838:121–129, 1999.
269. Török, G., Péter, A., Armstrong, D. W., Tourwé, D., Tóth, G. and Sápi, J., *Chirality,* 13:648–656, 2001.
270. Visegrády, B., Konecsni, T., Grobuschek, N., Schmid, M. G., Kilár, F., Aboul-Enein, H. Y. and Gübitz, G., *J. Biochem. Biophys. Methods,* 53:15–24, 2002.
271. Péhourcq, F., Jarry, C and Bannwarth, B., *Biomed. Chromatogr.,* 15:217–222, 2001.
272. Kosel, M., Eap, C. B., Amey, M. and Baumann, P., *J. Chromatogr. B,* 719:234–238, 1998.
273. Lehotay, J., Hrobonová, K., Cizmárik, J., Renèová, M. and Armstrong, D. W., *J. Liq. Chromatogr. Rel. Technol.,* 24(5):609–624, 2001.
274. Gübitz, G., *Chromatographia,* 30(9/10):555–564, 1990.
275. Levin, S., Abu-Lafi, S., The Role of Enantioselective Liquid Chromatography Separations Using Chiral Stationary Phases in Pharmaceutical Analysis, In *Advances in Chromatography,* Grushka, E. and Brown , P. R., Eds., Vol. 33, Marcel Dekker, New York, pp. 233–266, 1993.
276. Choi, S. O., Lee, S. H., Kong, H. S., Kim, E. J. and Park Choo, H.-Y., *J. Chromatogr. B,* 744:201–206, 2000.
277. Watt, A. P., Hitzel, L., Morrison D. and Locker, K. L., *J. Chromatogr. A,* 896:229–238, 2000.
278. Olsen, B. A., Wirth, D. D. and Larew, J. S., *J. Pharm. Biomed. Anal.,* 17:623–630, 1998.
279. Vermeulen, B. and Remon, J. P., *J. Chromatogr. B,* 749:243–251, 2000.
280. Kommuru, T. R., Khan, M. A. and Reddy, I. K., *J. Pharm. Sci.,* 87:833–840, 1998.
281. Baeyens, W. R. G., Van der Weken, G., Haustraete, J., Aboul-Enein, H. Y., Corveleyn, S., Remon, J. P., Garcia-Campaña, A. M. and Deprez, P. J., *J. Chromatogr. A,* 871:153–161, 2000.
282. Ortelli, D., Rudaz, S., Chevalley, A.-F., Mino, A., Deglon, J.-J., Balant, L. and Veuthey, J.-L., *J. Chromatogr. A,* 871:163–172, 2000.
283. Wanner-Olsen, H., Gaarskaer, F., Mikkelsen, E.O., Jakobsen, P. and Voldby, B., *Chirality,* 12:660–664, 2000.
284. Udata, Ch., Tirucherai, G. and Mitra, A. K., *J. Pharm. Sci.,* 88:544–550, 1999.
285. Mišlanowá, C., Štefancová, A., Oravcová, J., Horeck, J., Trnovec, T. and Lindner, W., *J. Chromatogr. B,* 739:151–161, 2000.
286. Blaschke, G., *J. Liq. Chromatogr.,* 9:341–368, 1986.
287. Blaschke, G. and Maibaum, J., *J. Chromatogr.,* 366:329–334, 1986.
288. Okamoto, Y. and Hatada, K., *J. Liq. Chromatogr.,* 9:369–384, 1986.
289. Okamoto, Y., Honda, S., Hatada, K. and Yuki, H., *J. Chromatogr.,* 350:127–134, 1985.
290. Bartsch, R. A. and Maeda, M., Eds., *Molecular and Ionic Recognition With Imprinted Polymers,* ACS Symposium Series N. 703, Oxford University Press, New York, 1998.
291. Mosbach, K., *Trends Biochem. Sci.,* 19:9–14, 1994.

292. Sellergren, B., *J. Chromatogr. A,* 906:227–252, 2001.
293. Kempe, M. and Mosbach, K., *J. Chromatogr. A,* 694:3–13, 1995.
294. Sellergren, B., *TrAC Trends Anal. Chem.,* 16:310–320, 1997.
295. Ramström, O., Yu, C. and Mosbach, K., *J. Mol. Recognition,* 9:691–696, 1996.
296. Kempe, M. and Mosbach, K., *J. Chromatogr. A,* 664:276–279, 1994.
297. Owens, P. K., Karlsson, L., Lutz, E. S. M. and Andersson, L. I., *TrAC Trends Anal. Chem.,* 18:146–153, 1999.
298. Fischer, L., Muller, R., Ekberg, B. and Mosbach, K., *J. Am. Chem. Soc.,* 113:9358–9360, 1991.
299. Blackwell, J. A., Stringham, R. W. and Weckwerth, J. D., *Anal. Chem.,* 69:409–415, 1997.
300. Sharma, S. C., Evans, M. B. and Evans, S. J., *J. Pharm. Biomed. Anal.,* 13:129–137, 1995.
301. Okamoto, Y., Aburatani, R., Kaida, Y. and Hatada, K., *Chem. Lett.* 7:1125–1128, 1988.
302. Tang, Y., *Chirality,* 8:136–142, 1996.
303. Okamoto, Y. and Yashima, E., *Angew. Chem. Int. Ed.,* 37:1020–1043, 1998.
304. Aboul-Enein, H. Y. and Wainer, I. W., Eds., *The Impact of the Stereochemistry on Drug Development and Use,* Wiley, New York, 1994.
305. Yashima, E., Yamamoto, C. and Okamoto, Y., *Synlett.* Accounts and Rapid Communication in Synthetic Organic Chemistry, 4:344–360, 1998.
306. Okamoto, Y. and Kaida, Y., *J. Chromatogr. A,* 666:403–419, 1994.
307. Yashima, E. and Okamoto, Y., *Bull. Chem. Soc. Jpn,* 68:3289–3307, 1995.
308. Okamoto, Y., Aburatani, R., Hatano, K. and Hatada, K., *J. Liq. Chrom.,* 11:2147–2163, 1988.
309. Okamoto, Y., Kawashima, M., Yamamoto, K. and Hatada, K., *Chem. Lett.* 5:739–742, 1984.
310. Okamoto, Y., Aburatani, R. and Hatada, K., *J. Chromatogr.,* 389:95–102, 1987.
311. Mori, M., Nukui, S. and Shibasaki, M., *Chem. Lett.,* 10:1797–1800, 1991.
312. Corey, E. J. and Lee, D. H., *J. Am. Chem. Soc.,* 113:4026–4028, 1991.
313. Technical Report, Daicel Chemical Industry, Tokyo, (1993).
314. Okamoto, Y., Aburatani, R. and Hatada, K., *Bull. Chem. Soc. Jpn.,* 63:955–957, 1990.
315. Okamoto, K, Kawashima, M., Aburatani, R., Hatada, K., Nishiyama, T. and Matsuda, M., *Chem. Lett.,* 7, 1237–1240, 1986.
316. Blackwell, J. A., *Chirality,* 10:338–342, 1998.
317. Ye, Y. K. and Stringham, R. W., *J. Chromatogr. A,* 927:47–52, 2001.
318. Ye, Y. K. and Stringham, R., *J. Chromatogr. A,* 927:53–60, 2001.
319. Wang, T., Chen, Y. W. and Vailaya, A., *J. Chromatogr. A,* 902:345–355, 2000.
320. Yashima, E., *J. Chromatogr. A.,* 906:105–125, 2001.
321. Taschibana, K. and Ohnishi, A., *J. Chromatogr. A,* 906:127–154, 2001.
322. Ishikawa, A. and Shibata, T., *J. Liq. Chromatogr. Rel. Technol.,* 16:859–878, 1993.
323. Chankvetadze, B., Kartozia, I., Yamamoto, C. and Okamoto, Y., *J. Pharm. Biomed. Anal.,* 27:467–478, 2000.
324. Miller, L., Orihuela, C., Fronek, R. and Murphy, J., *J. Chromatogr. A,* 865:211–226, 1999.
325. De La Puente, M. L., White, C. T., Rivera-Sagredo, A., Reilly, J., Burton, K. and Harvey, G., *J. Chromatogr. A,* 983:101–114, 2003.
326. Perrin, C., Vu, V. A., Matthijs, N., Maftouh, M., Massart, D. L. and Vander Heyden, Y., *J. Chromatogr. A,* 947:69–83, 2002.
327. Perrin, C., Matthijs, N., Mangelings, D., Granier-Loyaux, C., Maftouh, M., Massart, D. L. and Vander Heyden, Y., *J. Chromatogr. A,* 966:119–134, 2002.
328. Andersson, M. E., Aslan, D., Clarke, A., Roeraade, J. and Hagman, G., *J. Chromatogr. A,* 1005:83–101, 2003.
329. Wang, A. X., Lee, J. T. and Beesley, T. E., *LC-GC,* 18:626–639, 2000.

330. Vander Heyden, Y., Perrin, C. and Massart, D. L., Optimization Strategies for HPLC and CZE, In *Handbook of Analytical Separations:* Vol. 1, Separation Methods in Drug Synthesis and Purification, Valko, K., Ed., Elsevier, Amsterdam, pp. 163–212, 2000.
331. Smet, E., Staelens, L., Vander Heyden, Y., Aboul-Enein, H. Y., Van der Weken, G., García-Campaña, A. M. and Baeyens, W. R. G., *Chirality,* 13(9):556–567, 2001.
332. Baeyens, W. R. G., Van der Weken, G., Vander Heyden, Y., Van Bossche, V., Aboul-Enein, H. Y., García-Campaña, A. M. and Smet, E., *Anal. Chim. Acta,* 498:9–24, 2003.
333. Keller, H. R., Massart, D. L. and Brans, J. P., *Chemom. Intell. Lab. Systems,* 11:175–189, 1991.
334. Smilde, A. K., Knevelman, A. and Coenegracht, P. M. J., *J. Chromatogr.,* 369:1–10, 1986.
335. Bourguignon, B. and Massart, D. L., *J. Chromatogr.,* 586:11–20, 1991.
336. Harrington, E. C., *Ind. Qual. Control,* 21:494, 1965.
337. Derringer, G. and Suich, R., *J. Qual. Technol.,* 12:214–219, 1980.
338. Derringer, G. C., *Qual. Prog.* 27:51–58, 1994.

19

APPLICATIONS OF LC/MS IN PHARMACEUTICAL ANALYSIS

LEON Z. ZHOU

Analytical Sciences, Purdue Pharma, NY

Handbook of Pharmaceutical Analysis by HPLC,
S. Ahuja and M.W. Dong, editors.

ABSTRACT

The power and versatility of liquid chromatography–mass spectrometry (LC/MS) and LC/MS/MS is discussed in terms of "successful practices," and how individual investigators can follow these practices to obtain useful information during different stages of the drug development process. This chapter focuses on various applications of LC and electrospray mass spectrometry for pharmaceutical analysis. The objective is to provide ideas and experimental designs for the solution of qualitative problems including drug impurity profiling, HPLC method development, deficit in drug mass balance, drug discoloration, and drug counterfeiting. Many examples are presented to illustrate how LC/MS and LC/MS/MS can be used to provide answers to these real-world problems.

I. INTRODUCTION

The year 2002 was an extraordinary year for liquid chromatography–mass spectrometry (LC/MS) practitioners. On October 9, 2002, the Royal Swedish Academy of Sciences announced their decision to award the Nobel Prize in Chemistry to John B. Fenn, Koichi Tanaka, and Kurt Wüthrich for their development of analytical methods for the identification and structural analysis of biological macromolecules. Fenn and Tanaka shared the prize for developing electrospray and soft-laser desorption, respectively. These soft-ionization techniques allow macromolecules to be ionized without fragmentation.

The application of electrospray ionization (ESI) to macromolecules was first described by Yamashita and Fenn[1] in 1984, and he later applied this technique to the study of proteins in 1988. In ESI, charged droplets of protein were produced and the solvent was stripped away, leaving only the free protein ion. In 1987, Tanaka demonstrated that laser pulses from a low-energy nitrogen laser could be used to ionize proteins from a surface. As described in a paper presented during the Second Japan–China Joint Symposium on Mass Spectrometry, proteins are desorbed from a glycerol matrix containing cobalt powder. The cobalt powder was necessary to increase absorption of the laser photons.[2]

The LC/MS combination has been practiced for many years. Various types of interfaces have been developed for LC/MS, including the fast atom bombardment (FAB) source by Barber et al.[3] in 1981 and thermospray by Blakely and Vestal[4] in 1983. However, a broad acceptance of this hyphenated technique in the pharmaceutical industry did not occur until the advent of new ionization methods such as electrospray (ES), which was developed by Yamashita and Fenn[1] in the 1980s. Since then, LC/MS has become indispensable to the pharmaceutical industry. In addition to new ionization methods, the driving force behind the growth and success of LC/MS was its versatility in experimental designs, its extraordinary capability to generate both qualitative and quantitative information, as well as the reliability and affordability of commercially available systems. Although many applications became routine, even to the point of "walk-up" automated experiments (open access system and open access lab), other applications continued to demand the attention of highly skilled scientists for the purpose of designing experiments and interpreting data. Table 1 summarizes currently available mass spectrometers from major vendors that can be interfaced to liquid chromatographs.

LC/MS is notable for its diversity in instrumentation and applications. Although this chapter provides an overview of basic theory and instrumentation, the focus is on applications of LC combined with electrospray mass spectrometry in pharmaceutical analysis. Most types of new molecular (medicinal) entities (NMEs) evaluated for drug purposes can be analyzed using electrospray mass spectrometry, including medium polar to polar compounds, and ranging from low-molecular-weight amines or acids to much larger peptides and proteins. Figure 1 illustrates the compatibility of different compounds with various interfaces/ionization techniques.

Different mass analyzers may impose unique technical requirements when interfaced to LC. Understanding the operating principles and technical properties of both LC/MS interfaces and mass analyzers is deemed beneficial. A brief overview of the history of the development of LC/MS interfaces is given in Section II, which is followed in Section III by a summary of working principles and characteristics of commonly used mass analyzers.

Bringing a promising pharmaceutical candidate to the market can take as long as 10 years and cost $300–700 millions. Many different analytical techniques are employed to obtain necessary information during different developmental phases, from drug discovery to chemical/formulation development, clinical studies, stability evaluation, and quality control. LC/MS has a particularly wide range of applications in the pharmaceutical industry, where the quantitation and structural characterization/identification of thermally labile and hydrophilic compounds are routinely required—for example, structural proof and characterization of drug lead compounds, kinetic studies and impurity profiling at the various steps and scales of chemical synthesis, stability and impurity testing of pharmaceutical formulations and packaging components, toxicological testing of drug candidates in different dosing vehicles, and pharmacokinetic/metabolic investigations for pre-clinical and clinical

TABLE I Manufacturers of Mass Spectrometers Interfaceable to LC

Vendor	System name	Mass analyzer	Interface
Agilent	LC/MSD Trap	IT	API
http://www.chem agilent.com	LC/MSD TOF	TOF	API
Bruker Daltonics	Esquire 2000	IT	API
http://www.bdal.com	Esquire 3000 plus	IT	API
	EsquireHCT	IT	API
	BioTOF II	TOF	ESI
	BioTOF Q	Q-TOF	ESI
	Apex IV and Apex-Q	FTMS	Multiple inlets
IonSpec	HiResESI FTMS	FTMS	ESI
http://www.ionspec.com	Explorer FTMS		Multiple
JEOL	AccuTOF	TOF	API
http://www.jeol.com	LCmate	B	Multiple inlets
Waters/Micromass	Q-Tof API-US	Q-TOF	API
http://www.waters.com/	Q-Tof Ultima API	Q-TOF	API
WatersDivision	Q-Tof micro	Q-TOF	API
	Quattro micro API	QQQ	API
	Quattro Ultima Pt	QQQ	API
	Quattro Premier	QQQ	API
	LCT	TOF	API
Applied Biosystem	API 150EX LC/MS	Q	API
and Sciex	API 2000/3000/4000, LC/MS/MS	QQQ	API
http://www.appliedbiosy stems.com	API QSTAR™ Pulsar i Hybrid LC/MS/MS	Q-TOF	Multiple
	System 4000 Q TRAP™ LC/MS/MS System	Q-IT	API
Thermo Finnigan	LCQ Deca XP MAX	Ion-trap	API
http://www.thermo.com	LCQ Advantage MAX	IT	API
	MAT95XP-Trap	IT	API
	LTQ FT MS	FTMS	API
	TSQ QUANTUM	QQQ	API
	Surveyor MSQ	Q	API

studies. It is worth noting that the chosen analytical techniques continue to play significant roles even after the launch of the drug product in the market. LC/MS is one of the few tools which continues to play a decisive role throughout all of the product's life cycle stages.

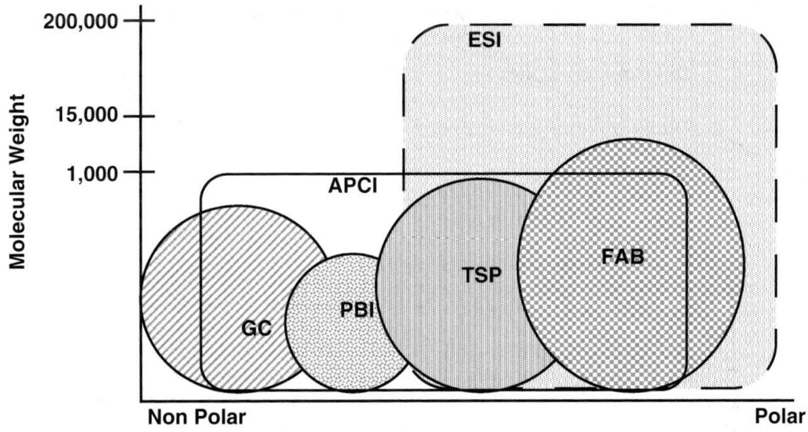

FIGURE 1 **Analyte compatibility with different interfaces/ionization techniques (courtesy of Thermo Finnigan).**

Like gas chromatography–mass spectrometry (GC/MS), integrating LC with MS brings together the potential of utilizing the intrinsic power of both LC and MS, thus expanding the analytical capabilities of both techniques. The importance of LC/MS in the pharmaceutical industry compared to GC/MS is the expansion of applications to new classes of compounds that are not amendable to GC/MS. For compounds of low molecular weights, with good thermal stability and adequate volatility, GC/MS is still a technique of choice. However, for compounds with critical thermal and chemical stability and low volatility the emergence of LC/MS has allowed both qualitative and quantitative analysis that no one had dreamed possible only a decade ago. This chapter is not intended to be a comprehensive compilation of reported applications in the pharmaceutical industry. Rather, it concentrates on several selected examples to provide a useful description of the capabilities and limitations of the technique. Applications to problems such as complex mixture analysis and general structural identification and characterization are covered. A listing of key LC/MS applications can be found in Table 2.

II. LC/MS INTERFACES

A. Overview

1. Why an Interface Is Needed

LC/MS is an integrated hybrid analytical system rather than a hyphenated device. The LC plays a role in preparation, separation, and introduction of sample components to the mass spectrometer; the mass spectrometer is a detector for the LC, garnering both qualitative and

TABLE 2 Applications of LC/MS in the Pharmaceutical Industry

Developmental stages	Information needed
Discovery of drug compound	(a) fast LC/MS for monitoring of target compounds (b) high-resolution mass spectrometry for structural proof and characterization of lead compounds and drug candidates (c) *in vitro* pharmacokinetic and metabolic investigations
Chemical development and scale-up synthesis of drug compound	(d) structural verification of starting materials, intermediates and final product (e) monitoring of impurity profiles from batch to batch (f) characterization and qualification of drug compound/synthetic impurities as reference standards
Stability testing of drug compound; Stability testing of pharmaceutical formulations and packagings	(g) monitoring of changes in purity and impurity profiles over the stability shelf life formulations and packagings under recommended storage conditions (h) investigation of compatibility between drug compound, formulation excipients, and packaging materials (i) understanding of intrinsic stability and degradation pathways of drug compounds and drug products
Development of stability-indicating analytical methods	(j) testing of chromatographic peak purity (k) establishing of impurity profiles under stressed/accelerated conditions
Drug metabolism studies	(l) identification of drug metabolites under physiological conditions
Pharmacokinetic studies	(m) quantitation of drug compounds and metabolites in biological matrix
Toxicological studies	(n) monitoring of impurity profiles throughout the course of study
Marketed product support	(o) continuous monitoring of impurity profiles on drug products and drug substances under long term storage conditions (p) investigation of consumer complaint samples such as drug discoloration issue (q) investigation of counterfeit drug products

quantitative information. The use of a mass detector for peak detection in a chromatographic process offers some special advantages. The most important one is that it can provide a wealth of structural information on the analyte. This information is much more specific than that which can be obtained using a photodiode array detector. For LC, there are a number of widely used detectors such as the ultraviolet-visible spectrophotometer, fluorimeter, refractive index, electrochemical and radioactivity detectors. Each detector has its own niche and offers a range of applications based on its sensitivity, selectivity, and specificity. The successful development of various interface technologies for coupling LC with MS has made this method of detection indispensable in many industries.

Figure 2 shows a chronological display of significant events in the history of MS. However, to define the exact origin of LC/MS is rather difficult. The recognition of the potential of on-line coupling of LC with MS dates back to the 1960s.[5] Interfacing LC with MS is inherently much more difficult than the coupling of GC with MS. The most challenging technical problems arise from the introduction of liquids into a high vacuum system. The mass flow in an LC is tens of thousands times higher than that in a GC. For appropriate performance, typical operating pressures in the mass analyzer region of a mass spectrometer should be in the range of 10^{-8} to 10^{-10} atm (10^{-3} to 10^{-5} Pa). Otherwise, ions will collide with neutral molecules or atoms and be pumped away before reaching the detector. The precipitous drop in pressure at the end of an LC column to

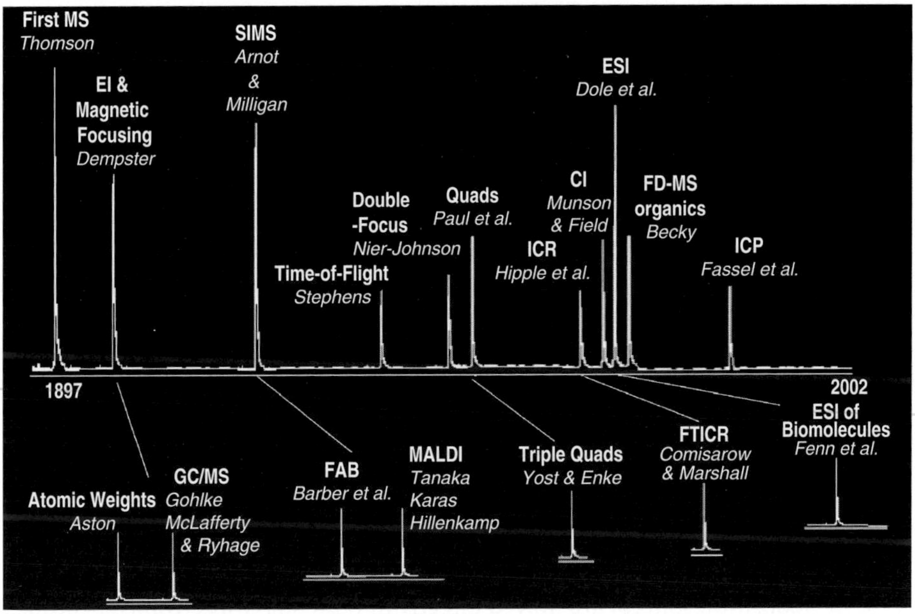

FIGURE 2 MS history (courtesy of http://masspec.scripps.edu/information/history/index.html).

the mass analyzer inside a mass spectrometer would cause enormous problems for the vacuum system if there were no interface.

For example, if the mobile phase is water and a narrow-bore column is used with a flow rate of 0.2 mL/min, it would correspond to 11.2 mmol/min of water, or an equivalent flow rate of 0.272 L/min of gas at ambient condition. This volume of water would expand to 2.72×10^7 L/min of gas at 10^{-8} atm. Such a flow is exceedingly high and cannot be accommodated by even the most sophisticated pumping systems (the theoretical pumping capacity of a cryogenic pumping system is 10^7 L/min). Furthermore, in this example, the tremendous expansion of the HPLC solvent would dilute the analytes to such an extent that even if the pumping system could handle the expanding solvent vapor, the sensitivity would be reduced catastrophically and render the technique useless for many practical applications. An LC/MS interface needed to be invented to remove most of the liquid eluting from the column, and at the same time to prevent dilution and loss of analytes.

2. Characteristics of LC/MS Interfaces

Table 3 shows the objectives and requirements for an ideal interface from the perspectives of both the mass spectrometer and the liquid chromatograph. An ideal LC/MS interface should have the following characteristics:

(1) Allow efficient and precise sample transfer from the LC into the MS with little destruction or loss of analytes.
(2) Permit a wide choice of LC methods and MS operating conditions.
(3) Retain chromatographic peak integrity with minimal peak broadening.
(4) Provide speed, convenience and reliability.
(5) Require minimal instrument maintenance and operator training.

Many interfaces have been developed to meet these demanding challenges. Some of these coupling methods, such as the moving belt or the particle beam interface, are based on the concomitant elimination of the solvent before it enters the mass spectrometer. Other methods such as direct liquid introduction (DLI) or continuous flow FAB rely on splitting the flow of the liquid that is introduced into the interface in order to obtain a flow that can be directly infused into the ionization source. However, these types of interfaces can only handle a fraction of the liquid flow from the LC.

A number of interfaces such as thermospray (TSP), ionspray (IS), atmospheric chemical ionization (APCI) and electrospray (ES) can tolerate much higher flow rates without requiring that the flow be split at the end of the LC column. Ions that are produced in atmospheric pressure ionization sources are moved directly into the mass spectrometer through small apertures.

TABLE 3 Objectives and Requirements for the LC/MS Interfaces

Objectives	Requirements
MS as a detector	
(a) Tolerant of high pressures	(a) Minimal restriction to LC conditions
(b) Selectivity	(b) Sufficient resolution for unambiguous peak identification
(c) Sensitivity	(c) High efficiency in the ionization step and the subsequent ion transmission
(d) Structural identification and characterization	(d) Capable of handling chemically and thermally labile compounds; capable of handling low volatility compounds; versatile in performing various MS/MS experiments
(e) Reliability and reproducibility	(e) Consistent in ionization, ion transmission and detection; user friendly software; and good dynamic range
LC as a separation and sample introduction technique	
(f) Selectivity of separation: sufficient separation so that isobaric compounds can be resolved for unambiguous peak identification by mass spectrometer	(f) Minimal restriction to MS conditions: desirable volatile buffer and compatible flow rate
(g) Sample clean-up and concentration device, eliminate solvent front and salts which cause ion suppression	(g) Retain chromatographic peak integrity; no band-broadening on the chromato graphic peaks; the response should be proportional to the concentration or amount of analyte

B. Development of LC/MS Interfaces

In this section, a brief overview is presented of the development of various LC/MS interfaces. However, it is not intended to include an exhaustive presentation of the working principles of the interfaces. Some excellent reviews of these techniques can be found in a number of articles published by Niessen and co-workers.[6–8]

1. Moving Belt Coupling

The moving wire interface was developed by Scott et al.,[9] and the moving belt interface by McFadden et al.[10] This was the first commercial interface for LC/MS, introduced in 1977. In both of the techniques, the eluent is deposited onto a stainless-steel wire, or a plate usually made of polyimide (known as Kapton), followed by the removal of the solvent in vacuum. The residual solid analyte is vaporized into an ionization source. The heating for solvent removal and volatilization of solid analyte is usually achieved by passing an electric current through the wire.

The moving wire device has a number of major shortcomings. Due to the small surface area of the stainless-steel wire, such as available from a 0.1 mm diameter wire, the device can only accommodate about 10 μL/min eluent which results in poor sensitivity. The system is difficult to operate in a continuous mode. Modification of the moving wire approach has led to the invention of a continuous moving belt, which offers improved transfer efficiency and therefore higher sensitivity. The moving belt interface is capable of handling up to 1 mL/min of mobile phase.

One of the advantages of the continuous moving belt interface is its ability to ionize the solid analyte using either electron ionization (EI) or chemical ionization (CI). In addition to EI and CI, direct ionization from the transport surface has been reported using FAB,[3] which normally reveals both relative molecular mass and structurally fragmented ions. Applications using surface ionization techniques such as laser desorption (LD) together with the moving belt interface have also been successfully reported.[11,12]

2. Direct Liquid Introduction (DLI)

The DLI interface became the second commercially available LC/MS interface in 1981. The liquid eluent is introduced into the ion source through a capillary or a pinhole diaphragm.[13,14] As the name implies, the analyte in DLI LC/MS is introduced from solution into the MS ion source.[15] When sufficient energy is given to the solution, the preformed ions in solution such as protonated molecules, deprotonated molecules, cationized molecules and solvated ions can be desorbed into a mass spectrometer while the bulk solvent is vaporized and eliminated by the vacuum system.

Maximum flow-rates compatible with DLI interfaces are in the range of 50 to 100 μL/min. Microbore columns (<1.0 mm i.d. column) operating at 5 to 100 μL/min are ideally suited for DLI LC/MS. A flow splitter is required to couple conventional LC with a DLI interface so that only a fraction of the total eluent is introduced into the mass spectrometer. Splitting the flow outside the mass spectrometer results in loss of sensitivity, an undesirable consequence. Henion and co-workers[16] have reported an approach in which the splitter is incorporated into the desolvation chamber of the mass spectrometer. The removal of solvent is achieved by diverting the vapor generated by the solvent without loss of sample and, therefore, sensitivity. Excess pressure inside the mass spectrometer is therefore avoided while higher flow-rates can be accommodated.

3. Thermospray

TSP is a breakthrough LC/MS interface capable of effectively removing solvent from the sample matrix (mobile phase) through a heated capillary vaporizer. It readily accommodates reversed-phase LC eluents at conventional flow-rates (0.5–2.0 mL/min). Sensitivity of on-column mass at picogram levels has been reported.[4] The basic TSP interface is comprised of a heated vaporizer, a desolvation chamber, and an ion extraction skimmer. When a sample solution is pumped into a heated stainless-steel capillary, rapid evaporation of solvent from the liquid surface occurs, resulting in an ultrasonic spray of vapor and charged droplets. Disintegration of the charged droplets occurs repetitively due to continuous evaporation of solvent and the Coulombic repulsion between like charges. The process eventually causes ions, as well as neutral molecules, to be released from the surface of the microdroplets. The ions are extracted and accelerated toward the analyzer by an electrostatic system voltage. Therefore, TSP is both an interface and an ion source, which makes a separate ionization source unnecessary.

Ions in the TSP process can also be produced in a two-step manner similar to conventional CI.[17] The reagent ions, e.g., NH_4^+, formed from electrolytes such as ammonium acetate, react with analyte molecules in the gas phase to generate positive analyte ions. This process can also be used with equal facility to generate negative analyte ions.

The TSP interface was very popular and attractive to chromatographers in the 1980s, as a result of its ease of operation and dependable performance. Commercial TSP LC/MS systems are equipped with an electron emitter filament to enhance the CI process.

A modification/enhancement of the TSP is the plasmaspray interface for discharge ionization which has also been commercialized. A TSP system can be operated in three different modes: (a) filament-off mode, i.e., TSP ionization mode, (b) the filament-on mode, and (c) the discharge ionization mode.

4. Continuous-Flow FAB

The principle of FAB, less frequently referred to as liquid secondary ionization mass spectrometry (LSIMS), is very similar to secondary-ion mass spectrometry (SIMS). However, FAB utilizes a liquid matrix, such as glycerol, in which a sample is dissolved. The matrix is used to enhance sensitivity and ion current stability.

Barber et al.[3] introduced FAB in 1981. In this technique, bombardment of a liquid target surface by a beam of fast atoms such as xenon or argon, causes the continuous desorption of ions that are characteristic of the liquid. In a typical FAB spectrum, the analyte ion is usually formed as protonated or cationized ions in positive FAB, and deprotonated ions in negative FAB mode. A few fragmented ions may also be formed. The spectrum usually contains peaks from the matrix, such as protonated matrix clusters of glycerol if it is used as the matrix solvent. FAB utilizes a liquid matrix such as glycerol. The matrix is used to enhance sensitivity and ion current stability.

The most commonly used FAB interface in LC/MS is known as continuous-flow fast-atom bombardment (CF-FAB) ionization, in which the fast atoms or ions are directed at a target along which the LC eluent flows.[18] In a CF-FAB, the LC eluent, mixed with a FAB matrix such as 5% aqueous glycerol, is continuously introduced and deposited on the tip of a FAB probe. The maximum flow rate is in the range of 5 to 15 μL/min. A comprehensive review of the principles and application of CF-FAB for LC/MS has been written by Caprioli.[19]

5. Particle Beam Interface

Particle beam interface is a device capable of separating solvent from solute without losing much of the solute. The interface has been developed based on the original work of Browner and co-workers.[20,21] The eluent from LC is passed through a nebulizer and converted into a spray of fine liquid droplets with high velocity. The nebulization is either assisted pneumatically, or thermally. Solvents begin to evaporate and the solute starts to concentrate. When the liquid droplets exit the heated chamber, they leave as a fast-moving particle beam. The beam entering the ion chamber is subjected to EI or CI ionization.

6. Electrospray Ionization

ESI has become the most commonly used interface for LC/MS. It was recognized by John Fenn and co-workers[22] as an important interface for LC/MS immediately after they developed it as an ionization technique for MS. ESI transforms ions in solution to ions in the gas phase and may be used to analyze any polar molecule that makes a preformed ion in solution. The technique has facilitated the ionization of heat-labile compounds and high-molecular-weight molecules such as proteins and peptides. ESI is a continuous ionization method that is particularly suitable for use as an interface with HPLC. It is the most widely accepted soft-ionization technique for the determination of molecular weights of a wide variety of analytes and, has made a significant impact on drug discovery and development since the late 1980s.

A couple of excellent reviews on the ionization mechanism of ESI can be found in the literature.[23,24] The following major events occur in an ESI process:

(a) Nebulization of sample solution into electrically charged droplets.
(b) Drastic reduction in the volume of the charged droplet by solvent evaporation.
(c) Disintegration of the droplets which results in highly charged micro-droplets capable of ion formation.
(d) Production of gas phase ions from these charged droplets.
(e) The ions formed are extracted, focused, and transported into the mass spectrometer for mass analysis.

A solution of the sample is sprayed through a capillary needle maintained at a high electric potential of approximately 5 kV. The voltage on the needle causes the liquid spray to be charged as it is nebulized. The droplets evaporate in a stream of dry gas and applied heat, in a region maintained at a vacuum of several Torr. As the droplet decreases in size, the charge density on the droplet surface increases. When the Coulombic repulsion between like charges on the surface overcomes the forces of surface tension (the Rayleigh limit[25]), the droplet disintegrates explosively to form second-generation liquid droplets. This process occurs repeatedly so that ions leave the droplet and are directed into the mass analyzer. A schematic presentation of the ESI process is illustrated in Figure 3. The ESI process bears some similarity to other LC/MS interfaces such as TSP[26,27] and ion evaporation.[28]

The most important feature of an ESI spectrum is that multiply charged ions can be formed. As a result, mass analysis of large molecules becomes feasible. For example, under electrospray positive ion monitoring, apo-myoglobin, which has a molecular weight of 16 951.5 atomic mass unit (amu), usually produces a series of ions with charge states from +8 to +27, and mass peaks from about 600 to 2000 amu. Figure 4 shows the ESI positive ion mass spectra for angiotensin I corresponding to charge states from 1, 2, 3, and 4. The charge state can be recognized by the difference in the *m/z* values between two adjacent isotopic peaks. A difference in the *m/z* values of 1 indicates a charge state of 1; a difference of 0.5 corresponds to a charge state of 2; and so on.

Typical mobile phases compatible with electrospray are methanol, acetonitrile, isopropanol, and volatile buffers. Many other solvents can also be used in ESI, including $CHCl_3$ and THF. Although 100% water can be used, better sensitivity is obtained with some organic modifier

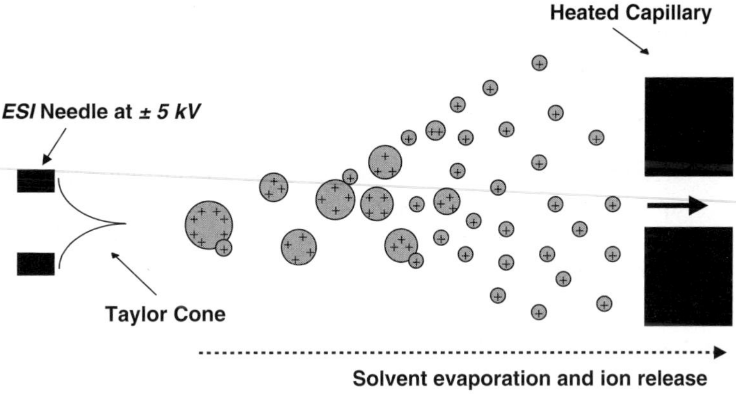

FIGURE 3 Schematic presentation of ESI process.

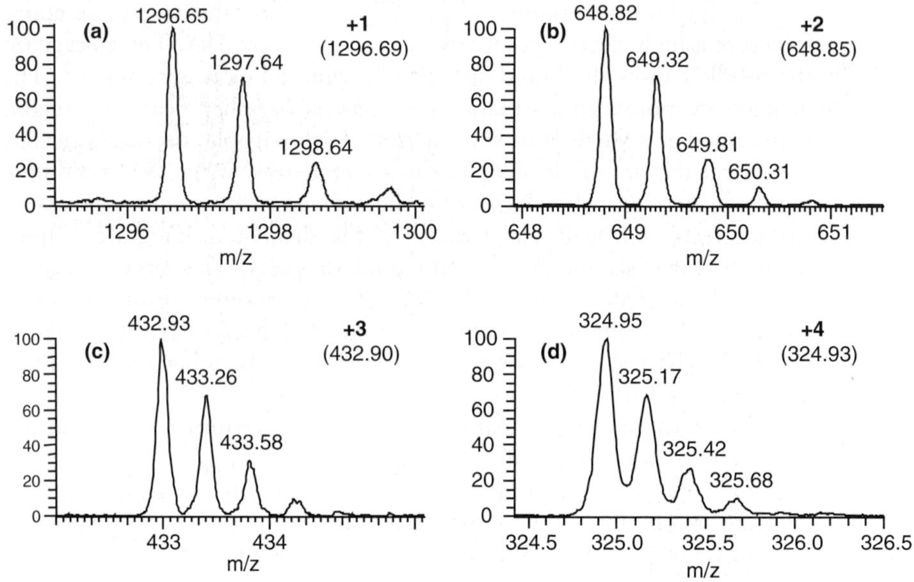

FIGURE 4 ESI positive ion mass spectra of angiotensin I with charge states of +1, +2, +3, and +4 (courtesy of Thermo Finnigan).

being present. Even 5–10% of MeOH or ACN significantly increases the stability of the nebulization process.

7. Atmospheric Pressure Chemical Ionization

In contrast to the ESI process where ions are primarily formed by desorbing preformed ions from a solution, APCI is an ionization technique in which ions are produced at atmospheric pressure, by gas-phase ion-molecule reactions between analyte molecule and solvent-based reagent gas. A graphical illustration of the APCI process is displayed in Figure 5. Ionization of solvent molecules is initiated by a corona discharge at the tip of the corona needle. The LC eluent is introduced into a heated pneumatic nebulizer,[29,30] where the liquid is nebulized pneumatically into a heated tube, allowing the droplets to collide with the hot walls. The interface takes advantage of the large gas and solvent vapor throughput tolerated by the API source and provides routine operation using LC flow rate up to 2 mL/min with reversed-phase eluents. The analytes arriving in the ionization source are chemically ionized through proton transfer in the positive mode and through proton loss in the negative mode.

C. Ionizations

As discussed above, ionization may occur in the LC/MS interface or in a separate ion source. During the development of commercial LC/MS

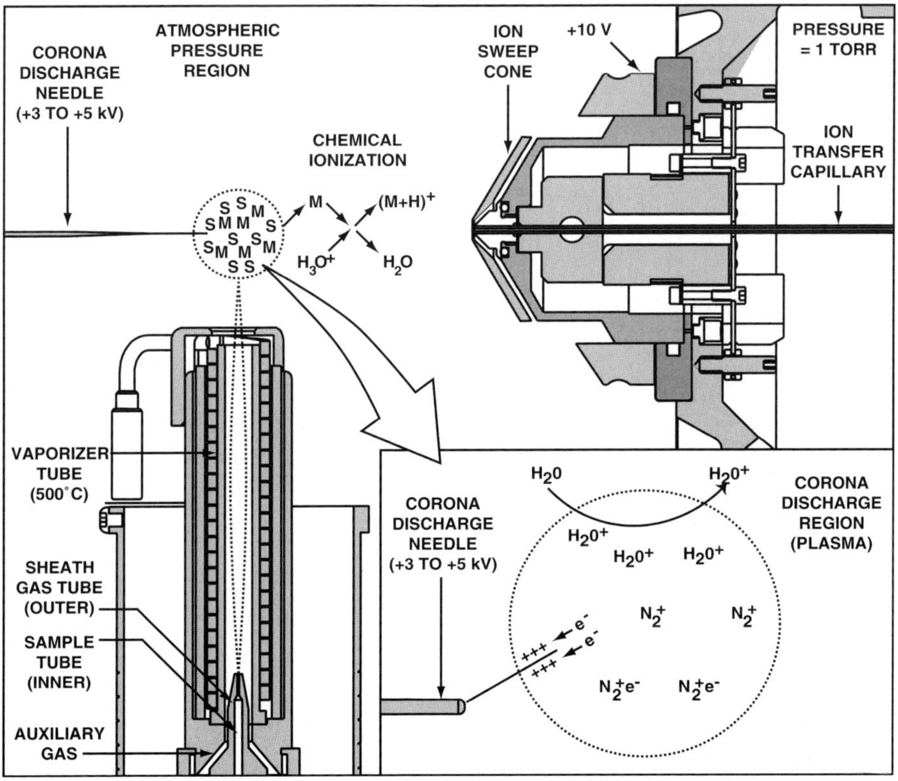

FIGURE 5 **Schematic presentation of APCI process (courtesy of Thermo Finnigan).**

interfaces, much emphasis was placed on the designs of interface between LC and MS. For example, the moving belt and particle beam systems are interfaces between the LC and EI and/or CI sources. DLI is an interface in which the ionization is achieved by solvent-mediated CI. The thermospray nebulizer is the interface for electron beam- or electric discharge-induced CI. The heated pneumatic nebulizer is the interface of choice for LC/MS with APCI.

EI produces predominantly +1 charge ions by ejecting an electron off the gaseous, neutral molecules. For compounds with high electron affinity, capture of an electron by the neutral analyte molecules to produce −1 charge ionic species may also be observed in the EI process. The analyte molecules must be transferred into the gas phase from condensed phases (solid or solution) before interacting with the electron beam in the ionization region. Heat used for this phase transformation can lead to thermal decomposition of analyte molecules prior to vaporization. Although EI was the primary ionization source for mass spectrometric analysis until the 1980s, its limitation for thermally labile molecules and large biomolecules left much to be desired and inspired many MS pioneers to develop alternative ionization techniques known today as electrospray and matrix-assisted laser desorption and ionization.

ESI desorbs the preformed ions from the solution phase into the gaseous phase. Many compounds in solution establish an equilibrium with their ionic species. Common ionic forms may include protonated, cationized, or deprotonated molecules. When ES is operated in the positive ion mode, a partial separation of positive and negative ions of the analyte and other electrolytes present in the solution occurs near the capillary tip. The electrosprayed droplets are found to contain excess positive ions. Under the effect of drying gas and heating, the continuous evaporation of solvent molecules occurs, which results in increasing charge density that makes the droplets electrohydrodynamically unstable. The droplets disintegrate when electronic repulsion reaches the Rayleigh stability limit. Further evaporation of solvent from the resulting smaller droplets leads to repeated droplet disintegration, which eventually results in the generation of ions in the gas phase. The ions are often solvated. The solvated ions can be preserved under soft ionization conditions such as electrospray ionization. Therefore in an ESI mass spectrum solvent-adduct ions are frequently detected.

III. MASS ANALYZERS

A. Characteristics of a Mass Analyzer

A mass analyzer is the device that resolves different gas-phase ions according to their mass-to-charge ratios. There are many different mass analyzers, and most of them can be easily coupled to an HPLC system and function as HPLC detectors. The most commonly used mass analyzers include Sectors (magnetic and electrostatic), Quadrupoles (single or triple quadrupole), Quadrupole Ion Trap, Fourier Transform Ion Cyclotron Resonance and Time-of-Flight. The separation of ions can be achieved either in a temporal or spatial dimension. Some mass analyzers allow the simultaneous detection of all ions, such as the Fourier transform ion cyclotron resonance mass analyzer. Other mass spectrometers, such as the quadrupole mass analyzers, act as mass filters so that only ions with particular mass-to-charge ratio can register a signal at any particular time. The time-of-flight (TOF) mass analyzer requires the ions to be produced simultaneously and all ions to be accelerated to the same kinetic energy before they can be separated in the flight tube. However since ionization in a typical LC/MS operation generates a continuous stream of ions, the ions are generally introduced into the TOF mass analyzer orthogonally before they are pulsed out of the source region.

The type of mass analyzer largely determines the information obtained from a given experiment, and to what extent that information can be translated into knowledge. The performance of a mass analyzer is evaluated by its capability to resolve neighboring ions, mass accuracy, upper mass limit, and transmission efficiency. A summary comparing the advantages and disadvantages of different mass analyzers is given in Table 4.

Instruments combining several analyzers in sequential order are very common. This combination allows mass spectrometry and mass spectrometry experiments (MS/MS) to be carried out. Modern MS/MS includes many different experiments designed to generate substructural information or to quantitate compounds at trace levels. A triple quadrupole mass spectrometer allows one to obtain a daughter ion mass spectrum resulting from the decomposition of a parent ion selected in the first quadrupole. The MS/MS experiments using an FTICR or ion trap, however, are carried out in a time-resolved manner rather than by spatial resolution.

B. Working Principles of Mass Analyzers

A brief description of the working principles of commonly used mass analyzers is given below. For a more comprehensive discussion of the principles of these mass analyzers, excellent reviews can be found in the literature.[31,32]

1. Sector Mass Analyzers

In sector instruments, magnetic sectors are generally coupled with electric sectors to correct the difference in the initial energetic and spatial distribution for the incoming ions. Magnetic sector mass analyzers are fundamentally momentum analyzers. Ions in a magnetic field traverse in a circular path in a plane perpendicular to the direction of the magnetic field. Ions are dispersed spatially according to their momentum-to-charge ratio. Ions of different mass but of the same kinetic energy follow different trajectories in a fixed magnetic field. The strength of the magnetic field, if varied, will allow ions with different masses to follow the trajectory that will lead them eventually to the detector. Therefore only ions with a given mass-to-charge ratio will be able to reach the detector in any given magnetic field.

2. Quadrupole Mass Analyzers

The quadrupole is a device which takes advantage of the stability of the ion trajectories to separate ions in accordance with their mass-to-charge ratio. Quadrupole analyzers consist of four rods with circular or, ideally, hyperbolic surface sections. The four rods are arranged

TABLE 4 Pros and Cons of Mass Analyzers

Mass analyzers	Advantages	Disadvantages
Sector	• High resolution and accurate mass measurement, generation of elemental compositions • High-energy CID MS/MS for generation of sub-structure information • Mass range up to 20 000	• The system is expensive and often complex • Low sensitivity • Coupling to LC is technically demanding • Slow scan speed
Quadrupole	• Energy and spatial distribution of ions produced in the ion source is not critical • Low cost and easy to couple to LC • Tandem MS experiments available in triple quadrupole or Q-TOF systems for sub-structure information and/or quantitative analysis • Vacuum system demands are minimum	• Low resolution and low accuracy in mass measurement except in Q-TOF systems • Mass range limited to approximately 4000
Ion-trap	• Energy and spatial distribution of ions produced in the ion source is not critical • Inherent tandem MS capabilities for generation of sub-structure information • Low cost and easy to couple to LC • Vacuum demand is minimum	• Low resolution and low accuracy in mass measurement
FTICR	• Ultra high resolution and high mass accuracy, elemental composition can be obtained • Inherent tandem MS capabilities for generation of sub-structure information • Mass analysis is nondestructive • High sensitivity • Mass range > 20 000 • Fast scan speed	• High purchase price and high cost in maintenance • Vacuum requirements are demanding and coupling to LC is difficult • Fast computing required
TOF	• High resolution and mass accuracy when operated in reflectron mode or in Q-TOF systems, elemental composition can be obtained • Tandem MS available for generation of sub-structure information or quantitative analysis • High sensitivity • Very fast scan speed • Unlimited mass range	• Initial energy and spatial distribution must be corrected for ions • High-performance electronics needed

precisely parallel, and equally spaced, around a central axis, i.e., in a square array. The ion beam is focused down the axis of the array and an electrical potential of fixed (DC) and radio frequency (RF) components is applied to the diagonally opposed rods. For a given combination of DC and RF components, ions of one specific m/z ratio have a stable path down the axis. All others are deflected to the sides and lost before reaching the mass detector—ions with too low a value of m/z spiral out of control and crash into the positively charged rods, while ions with too high a value of m/z spiral out of control and crash into the negatively charged rods. Mass scanning spectrum is generated by changing the DC and RF components of the voltages while maintaining a constant ratio.

3. TOF Analyzers

Ion separation in a TOF mass analyzer is based on Newton's third law. The underlying principle is that ions of different masses with equal kinetic energy have different velocities. If there is a fixed distance for the ions to travel, the time of travel is proportional to the square root of the mass-to-charge ratio of the ions. To measure the time of flight, ions are introduced into the mass spectrometer in discrete packets so that a starting point for the timing process can be established. Ion packets are generated either through a pulsed ionization process or through a gating system in which ions are produced continuously, but are introduced only at given times into the flight tube.

4. Ion-Trap Analyzers

The analyzers discussed above are all mass filter analyzers, with spatially separated input and output. There is another class of mass spectrometers in which mass analyzers store the ions for subsequent mass analysis. The most common of these is the quadrupole ion-trap, which is effectively a quadrupole mass filter bent around on itself. This quadrupole-type device is composed of a ring electrode placed between two end-cap electrodes. The end caps are either held at ground potential or have an RF voltage applied to them, while an RF voltage is placed on the ring electrode. As a result of this specific geometric arrangement, the hyperbolic surfaces of the three elements form a three-dimensional quadrupole analyzer. Holes in the end caps allow ion injection into the device, and ion ejection out of the device to the detector. In the ionization step, the RF voltage on the ring electrode is set low enough so that the ions within the mass range of interest are trapped within the device. The RF voltage on the ring electrode is raised following ionization, and the ions of successively higher masses are ejected from the ion trap into an electron multiplier detector. A particular advantage of ion-trap MS is the so-called MS^n technique; in such a process, a fragmented ion from its precursor is selected and subjected to collision induced dissociation (CID) so that structural information on the precursor ion can be obtained. Low mass cut off is a problem in MS ^{n}, but may be overcome by using additional isolation and excitation steps.

5. Ion Cyclotron Mass Analyzers

Fourier transform ion cyclotron mass analyzers (FTMS) are based on the ion cyclotron resonance (ICR) principle. Ions are stored and analyzed inside an ICR analyzer cell that is located in a homogeneous region of a superconducting magnet. Under the influence of a magnetic field, ions experience a force perpendicular to their motion that makes them move in a helical direction. The time required for ions to complete one helical movement is proportional to the mass-to-charge ratio of the ions. Faster ions move in larger helices (or orbits), and slower ions in smaller helices. Therefore, the cyclotron frequency for ions with the same mass-to-charge ratio is always the same. A frequency analysis coupled with a Fourier transform of the signal generates information for the mass-to-charge ratio of the ions. The advantages of an FTMS instrument are its remarkably high resolution and the ability to measure molecular mass accurately, from which the elemental composition of an unknown compound can be derived.

IV. PRACTICAL CONSIDERATIONS IN SELECTION OF LC/MS PARAMETERS

The most commonly used LC/MS interfaces in pharmaceutical analysis are ESI and APCI. An ESI interface on the majority of commercial mass spectrometers utilizes both heat and nebulization to achieve conditions in favor of solvent evaporation over analyte decomposition. While ionization in APCI occurs in the gas phase, ionization using ESI occurs in solution. Attributes of a mobile phase such as surface tension, conductivity, viscosity, dielectric constant, flow rate and pH, all determine the ionization efficiency.[33,34] They therefore need to be taken into consideration and controlled.

A. Flow Rate

It is well known that UV detectors used in liquid chromatographs are concentration-sensitive devices. Injection of the same mass of a particular compound onto two columns with identical plate number and length but different inner diameters, will result in a higher response from the column with the smaller inner diameter. The gain in the signal is inversely proportional to the square of the ratio of the inner diameters of the two columns. The situation is different for a mass spectrometer, which is a mass-flow sensitive detector. Under constant flow conditions, the mass spectrometric response is proportional to the concentration of analyte ions reaching the mass detector. If the analyte ions are continuously introduced into a mass spectrometer, the signal will decrease as the flow rate decreases.

Niessen[31] recommended that efforts to improve concentration detection limits should be directed at improving the mass-flow to the mass spectrometer. Hopfgartner et al.[35] also confirmed that there was no gain in concentration detection limit by using microbore columns vs. a conventional column with a flow split as the gain in the analyte concentration obtained by using a smaller diameter column comes at the expense of a reduced flow-rate entering the electrospray interface. Another factor arises from the requirement that similar reduction of the injection volume must be adjusted with the reduced column diameter. The two factors work concurrently but antagonistically, leading to reduced mass reaching the mass detector.

ESI, however, has shown that ionization and sampling efficiency is inversely proportional to the flow rate.[36] Observations of reverse mass-flow dependence behavior for the ESI interface have been reported.[37,38] The explanation of these disparate observations can be attributed to an improvement in the ionization and sampling efficiency achieved in the ESI process at lower flow rates. When the gain due to higher ionization and ion sampling efficiencies exceeds the loss of sensitivity due to mass-dependence from mass detector, an improvement in the overall response results. An important event in electrospray mass spectrometry is the sampling of analyte ions formed in the atmospheric-pressure ionization chamber. Although modern ESI interfaces can handle flow rates up to 1 mL/min without flow splitting prior to the interface, the sampling of analyte ions through either a heated capillary or an orifice is actually a split process, which allows only a fraction of the formed ions to eventually reach the mass detector. According to literature research for the period of 1992–1997 reported by Abian et al.,[38] narrow-bore columns were more often used for LC/MS applications. Narrow-bore columns are those with 1 to 2 mm in internal diameters. The popularity of narrow-bore columns can be attributed to their compatibility with common HPLC pumping systems and pneumatically-assisted ESI source requirements. Compared to conventional columns, typical flow rates used in narrow-bore columns are reduced by as much as 90%. Flow rates in the range of 0.2 to 0.4 mL/min are commonly used for ESI LC/MS applications.

Utilization of capillary columns in conjunction with micro ESI devices is becoming a new trend in the field of LC/MS. Capillary HPLC has become a particularly important technique in situations where the supply of analyte is limited, such as in proteomic analysis. According to studies conducted by Smith et al.,[39] only one in a hundred thousand of analyte molecules present in solution eventually reach mass detection in a conventional ESI interface. Smith et al. attributed this poor electrospray performance to less efficient sampling process. The efficiency can be improved to one of every 390 analyte molecules being detected using nano-ESI tip which is placed 1–2 μm from the MS entrance. However, the use of capillary columns in combination with nano-ESI in pharmaceutical analysis is not required because abundant sample is usually available. Since capillary column/nanoESI requires special configuration of the instrument, it has not yet been accepted as a routine analytical technique for pharmaceutical analysis.

B. Organic Solvent in the Mobile Phase

Mobile phases used in LC/MS for pharmaceutical analysis mostly consist of methanol and/or acetonitrile, and a volatile buffer such as ammonium formate. The viscosity of solvent mixtures varies when the composition ratio changes. It is well known that less viscous solutions lead to more efficient evaporation and ion desorption. Since organic solvents such as methanol and acetonitrile have lower viscosity and higher vapor pressure than water, these solvents are more easily removed in the ESI interface and thus enhance ESI performance. Methanol has also been reported to generate stronger signals than acetonitrile[40,41] in positive ion mode MS. The need to maintain low concentrations of buffer can be understood from a number of publications by Kebarle and Tang.[42–44] They monitored the response of analyte ions as a function of the background electrolyte content in the solution. They reported that the formation of a particular analyte ion can be significantly suppressed by the presence of a second analyte. For example, the intensity of analyte ions such as Bu_4N^+ and $CodeineH^+$ decreased by more than a factor of 10 when the concentration of ammonium acetate was increased from 0.05 to 1 mM. Other references found in the literature[45] discuss the relationship of mobile phase additives such as formic acid, ammonium formate content and the response of an analyte. The optimal response was reported when the additive concentration was between 0.1 and 1 mM. The presence of formic acid enhances the formation of protonated analyte ion, while the introduction of a metal alkali such as sodium acetate can lead to an abundance of sodiated analyte ions. Nevertheless, the concentration of volatile buffers such as formate in the mobile phase is recommended at 10 to 30 mM for practical reasons of having sufficient buffering capacity.

Straub and Voyksner[46] developed a schematic approach for ionization optimization at a variety of pH values and organic modifiers. As a general rule, for every 5–10% increase of organic modifiers in the mobile phase, the mass response will double. A compromise approach must be adopted in choosing the type and ratio of organic modifier so that desirable separation and ionization can be achieved simultaneously. As a general rule, in either methanol- or acetonitrile-based mobile phases, a concentration range of 20% to 80% of either methanol or acetonitrile is optimal.

C. Additives and Buffers

It is well known that nonvolatile HPLC additives should be avoided. These include alkali metal phosphates, borates, citrates, hydrochloric acid, sulfuric acid, alkali metal bases, etc. The influence of pH on the ionization is obvious in the ESI process. Analyte molecules must enter the ESI interface under conditions favorable to the formation of the ions in the solution phase. The general rule is that basic compounds are chromatographed in acidic mobile phases to produce protonated analyte ions, and that acidic compounds are chromatographed in basic mobile phase to generate deprotonated analyte ions. However, the pK_a value of a compound when in the gas phase may be drastically different from its pK_a in solution. The phenomenon of "wrong-way-round" in electrospray ionization[47] may be partially due to this shift in pK_a values during phase transformation. The fact that amino acids can be protonated as well as de-protonated during ESI process in the pH range between 3 and 11 gives an analyst a broad choice of pH values of mobile phase should separation become a more critical issue than sensitivity. Under such circumstances the pH value can be optimized to achieve optimal separation without severely compromising MS sensitivity.

Post-column introduction of a sheath liquid gives more flexibility to the selection of pH values for a mobile phase. Chiron et al.[48] applied post-column addition of tripropylamine to reduce the acidicity of the mobile phase in their study of bentazone and chlorophenoxyacetic acids to enhance ESI negative ion signals. Apffel et al.[49] on the other hand, introduced propionic acid/isopropanol to displace TFA in the mobile phase, and improved the ESI positive ion signals.

Using seven nucleoside antiviral compounds, Kamel et al.[50,51] have reported that ES sensitivity for all seven compounds was increased by a factor of at least 2 when the mobile phase was modified with 1% acetic acid (pH 3.1) rather than 0.1% trifluoroacetic acid (pH 2.3). At constant pH, intensity of the protonated ion increased with increasing pK_a of the analyte. In the positive ionization mode, the sensitivity of these nucleoside compounds as $[M+H]^+$, did not depend on the pH of the solution. In the negative ion mode however, increased sensitivity in $[M-H]^-$ ion was indeed observed with increasing pH values. Temesi and Law[52] also reported a study of the effect of LC eluent composition on mass spectrometric response using electrospray ionization. The study was carried out using 35 compounds with diverse structures. They included strong and weak acids and bases, amphoterics, and polar neutral compounds. The authors concluded that formic acid provided the best all-round responses for all 35 compounds tested, regardless of what organic modifiers were used. Trifluoroacetic acid resulted in reduced responses in the positive ionization mode, and a total absence of signals was the result in the negative ionization mode with trifluoroacetic acid in the mobile phase.

Figure 6 shows the signal response in ESI of Leucine Enkephalin, a pentapeptide, as a function of solvent and additive variation. Although the peptide can be ionized in basic media, acidic pH is much more favorable. This result also confirms that formic acid and acetic acid promote ESI signals far more efficiently than TFA.

D. MS Operating Conditions

Although modern mass spectrometer has evolved to a point that a fully automated system can be set up in an open-access laboratory, some general aspects in their operation should be mentioned. Since large amounts of nitrogen gas are consumed as nebulizing and drying gas, most pharmaceutical companies provide it through internal plumbing systems. Typically, these systems deliver nitrogen gas from a remote nitrogen generator. It is worthwhile to install an in-line filtration device so that moisture and grease can be removed before they reach the mass spectrometer ion source. The highest research-grade of argon gas should be used as the CID gas to avoid possible contamination. For systems in which the spray needle is adjustable, one must examine the influence of needle position on the ionization response. The API probes used in Thermo Finnigan LCQ systems are shown in Figure 7. The *xy* position of the spray needle can be adjusted by using the screw on the ESI flange.

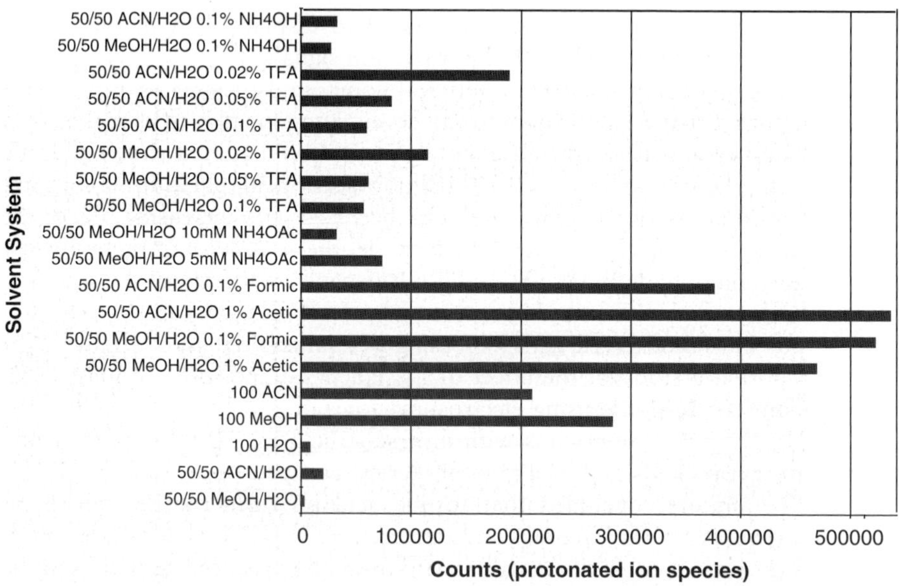

FIGURE 6 Effects of solvents and additives on ESI response of Tyr–Gly–Gly–Phe–Leu (Leucine Enkephalin) (courtesy of Thermo Finnigan).

The scan range can have a great impact on the quality of the mass spectrum. Ionization techniques for LC/MS such as particle beam, thermospray, and electrospray inherently generate high levels of chemical noise in their total ion chromatograms and in their mass spectra. Although background subtraction is effective in eliminating most of the chemical noise, it is wise to limit the scan range thereby enhancing the quality of the spectrum. If the molecular mass of the drug is designated as M_r, then the upper *m/z* limit is recommended as $2M_r + 50$ since dimerization of drug molecules may be one of the degradation pathways. The low end of the *m/z* limit should be set at a point so that all of the synthetic intermediates and starting materials can be detected. If the method is intended to screen for leachables from packaging materials, the *m/z* range should also be set accordingly. A number of trial or scouting LC/MS runs can be made to obtain preliminary information so that more optimal parameters can be set up for subsequent experiments.

E. Sample Solution Effects

In principle, enhanced sensitivity can be achieved when the on-column focusing of the analyte is possible. The sample is preferably dissolved in a solvent with lower eluotropic strength compared to that of the mobile phase at the start of a chromatographic run. A solvent with

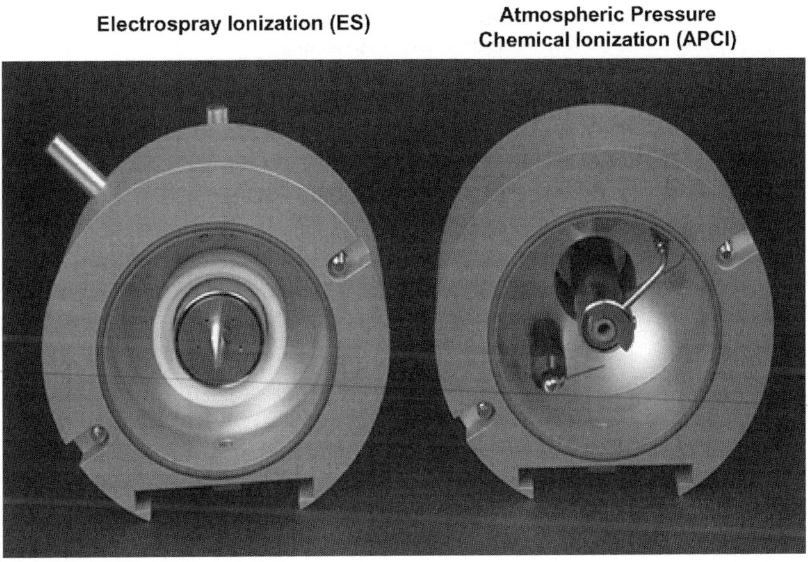

FIGURE 7 API probes used in Thermo Finnigan LCQ systems (courtesy of Thermo Finnigan).

much higher eluotropic strength, will likely cause chromatographic anomalies such as peak distortion, splitting or broadening. Another factor that may cause peak distortion or splitting is a mismatch of the pH of the sample solution and that of the mobile phase.

The majority of components in a pharmaceutical formulation are excipients. Some of them, when extracted, may have a deleterious effect on the mass analysis process. For example, polyethylene glycol, a water-soluble polymer, can span several minutes of the total ion chromatogram. The presence of PEG ions makes identification of trace level impurities extremely difficult. Figures 8(a) and (b) compare the UV and TIC traces. Three unknown peaks in between retention times of 8 and 12 min were observed in the LC/UV chromatogram. Unfortunately, in the same time region, the total ion chromatogram is dominated by the signals from polyethylene glycol, making detection and the assignment of signals to the unknown peaks an arduous task. Figures 9(b)–(d) represent the extracted mass spectra from the retention time regions corresponding to these three unknown peaks. Five strong signals in Figure 9(b) have been observed. Their *m/z* values are 236, 459, 476, 503, and 520. The ion at *m/z* of 236 is "chemical background noise"; the ions at *m/z* of 459 and 476, are the protonated and ammoniated ions of an oligomer of polyethylene glycol; the ions at *m/z* of 503 and 520 are another pair of

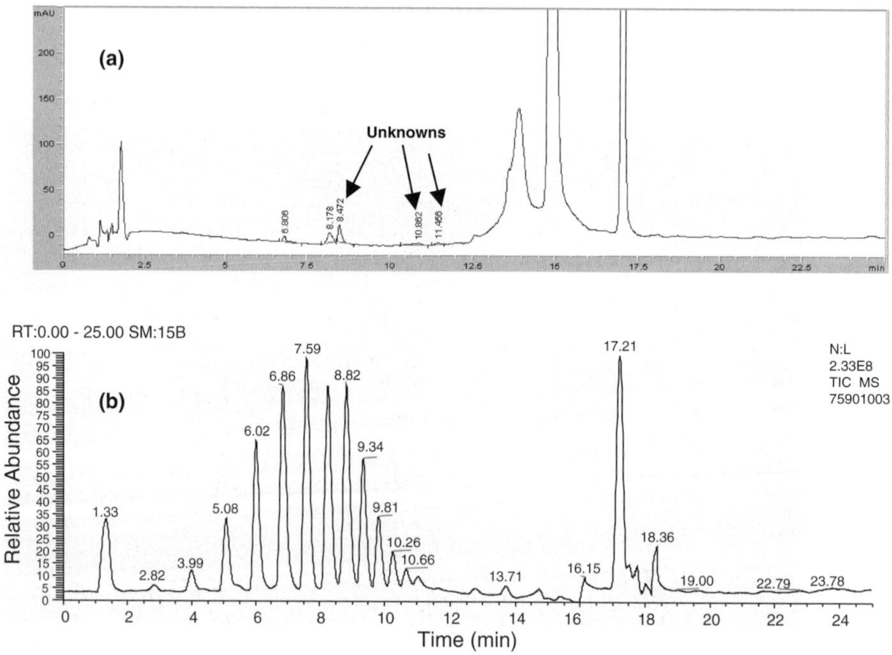

FIGURE 8 **PEG effect: (a) LC/UV chromatogram and (b) total ion chromatogram.**

signals corresponding to another oligomer of polyethylene glycol. The oligomers of two adjacent polyethylene glycol peaks are characterized by an increment of 44 amu. In addition, each oligomer shows a pair of ions, one is the protonated molecular ion, while the other is the ammoniate molecular ion. The signals for all of the polyethylene glycol oligomers have been assigned and labeled in Figure 9(a). There is a weak signal in Figure 9(b) with an *m/z* value of 302. This signal does not fit the pattern for polyethylene glycol oligomers. The *m/z* value of this ion is 40 amu lower than that of the API, suggesting that it is a degradation product formed through a dealkylation reaction leading to the loss of an allyl group from the drug compound.

Other drug excipients, such as plasticizers, may also have an adverse impact on LC/MS performance and interpretation. Plasticizers such as triethyl citrate, *tert*-butyl phthalate, and some others, can act as bases to compete for protons during the ion generation process. Selecting the optimal solvent or designing sample work-up strategies to discriminatorily dissolve components of interest is a very important step towards the generation of high quality mass spectrometric data. A more detailed discussion regarding sample preparation for pharmaceutical dosage forms is presented in chapter 5.

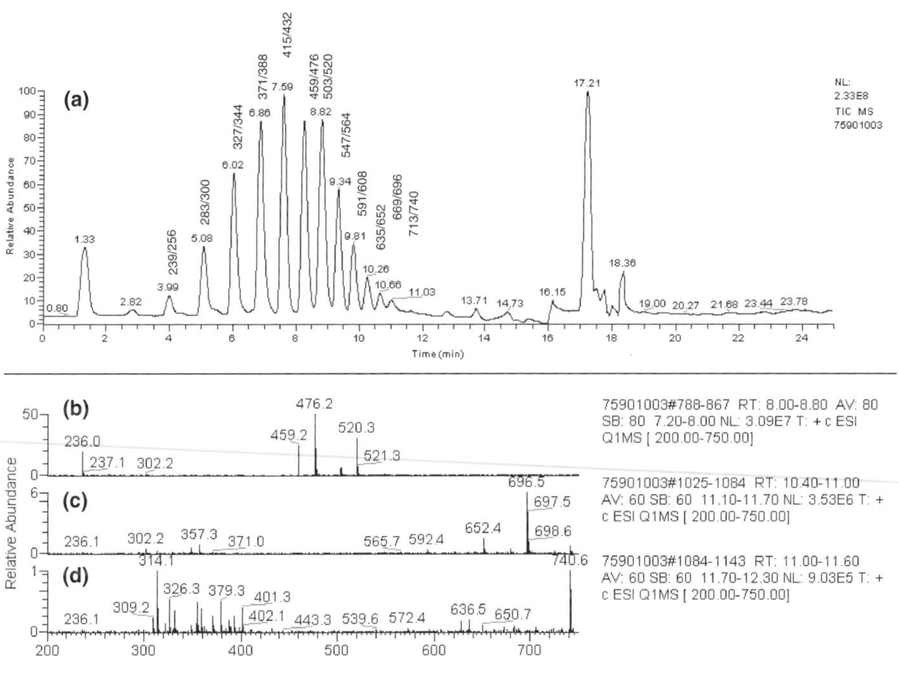

FIGURE 9 Extracted mass spectra of peaks of interest in the presence of PEG signals.

V. PEAK TRACKING BETWEEN LC/UV AND LC/MS METHODS

Preferably, identification of impurities in drug substances and drug formulations by LC/MS is carried out using the same chromatographic method used to detect the unknown initially. However, nonvolatile buffers and ion-pairing reagents are commonly used to mediate chromatographic separation and improve its performance such as high sensitivity in the HPLC methods designed for assay and the determination of impurities. Since nonvolatile buffers are not compatible with LC/MS interfaces, peak tracking between different chromatographic methods has become one of many challenges one must deal with.

A. Developing an Equivalent HPLC Method

A practical approach to peak tracking is to develop an LC method for LC/MS application which has the same or similar peak elution order as the original method. Phosphate buffers can be replaced by ammonium formate (pH 2.8–4.8), ammonium acetate (pH 3.8–5.8), or ammonium bicarbonate (pH 8.2–10.2). Figure 10(a) shows a chromatogram for a neutraceutical mixture using a phosphate-buffered mobile phase. When phosphate is replaced with formate at the same pH, chromatographic selectivity is often preserved. Identical elution order, and even the same retention times, can be seen when comparing Figure 10(b) with (a). Ion-pairing agents such as sodium dodecyl sulfate (SDS) can be substituted with one that is MS friendly. There are a number of ion-pairing reagents, which at appropriate concentration levels, can maximize HPLC separation without compromising ESI/MS sensitivity. Trifluoroacetic acid has been known since the early 1980s for its characteristics in achieving better separation for peptides in reversed-phase HPLC process. In the last few years heptafluorobutyric acid and pentafluorobutyric acid have been used to improve HPLC separation at concentration levels such as 0.05% for small molecules and 0.005% for peptides.

Once an appropriate, MS-friendly, LC method is developed, information such as the UV spectrum and percent peak area can be used to establish peak correlation between LC/UV and LC/MS methods.

B. "On-line" Peak Trapping and Elution

A well-known method for peak tracking is based on the "phase-system switching" idea,[53–55] which was developed to solve problems of mobile phase incompatibility in LC/MS target compound analysis. An analytical column is usually connected to a trapping column in tandem mode. A switching valve is placed after the UV detector, and the flow of nonvolatile eluents is directed through the trapping column to waste. When the peak of interest elutes from the analytical column it is

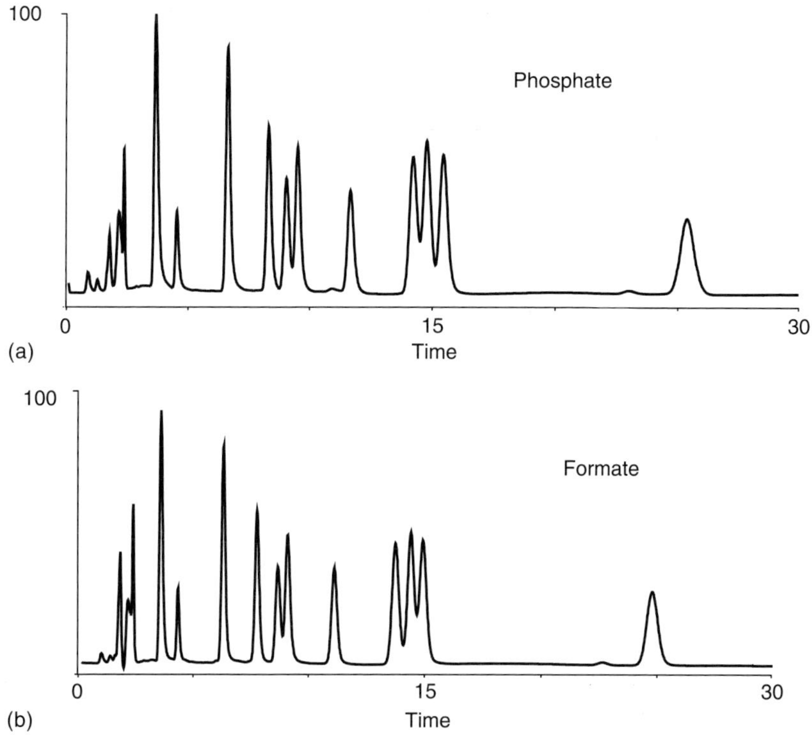

FIGURE 10 **Separation of a neutraceutical mixture in phosphate- and formate-buffered mobile phases (courtesy of Waters Corp.).**

switched into the trapping column where desalting takes place by pumping aqueous eluents through the column. The trapped peak can then be re-chromatographed using a mass spectrometer-friendly mobile phase for optimal mass detection. A typical setup of "on-line" peak trapping and elution is illustrated in Figure 11.

The advantage of this technique compared to the off-line fractionation approach is that it directly transfers the peak of interest for MS and MS/MS analysis to the mass spectrometer. The trapping column can be used as a concentration device if necessary. Multiple injections can be carried out on the first column to accumulate a peak of interest onto the trapping column. The trapping, however, can be a little tricky. Ermer[56] reported a successful on-line trapping and identification of an impurity at 0.1 UV-area% from a phosphate HPLC mobile phase. He suggested that the content of organic modifier used for elution of the trapped peak ought not be significantly lower than its content in the mobile phase from the nonvolatile separation. Asakawa et al.[57] reported a three-column switching configuration used in frit-FAB-MS determination of tocopherol and riboflavin. The three columns were made of the same stationary phase

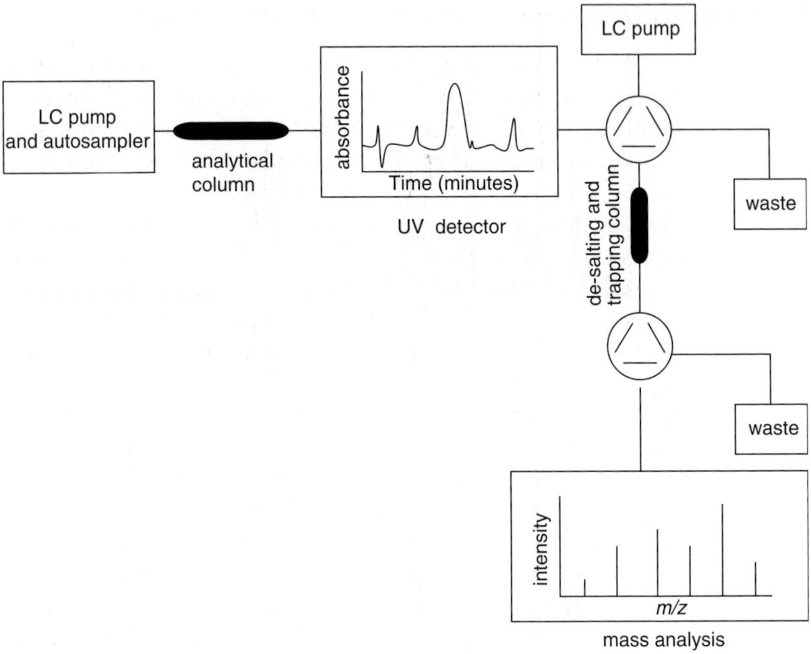

FIGURE 11 **"On-line" peak trapping and elution.**

and were run independent to each other. LC-1 with a conventional analytical column served to separate the compounds of interest; LC-2 and LC-3 trapped and re-chromatographed the compounds for MS analysis.

VI. DEVELOPMENT OF GENERIC LC/MS METHOD AND LC/MS DATABASE

Development and validation of HPLC methods for the assay of a drug compound and its related substances is central in pharmaceutical analysis. From pre-clinical to clinical development to final NDA submission, the methods usually go through many changes, from one set of chromatographic conditions to another set. The changes in the methods can be necessitated by changes in chemical synthesis, scale-up, modifications of drug formulations during the drug product development process, or a number of other reasons. Therefore the impurity profile of a drug substance and drug product formulation can be changing until the manufacturing processes of the drug substance and the drug product formulation are finalized. A strategic approach to dealing with this constantly changing situation was introduced by Ayrton et al.[58,59] The idea is known today as early locking-in of a "generic method." This strategy allows chemists to develop "generic HPLC methods" to use on diverse samples throughout the development lifecycle of a drug product.

A. "Generic LC/MS Method" with Basic pH Mobile Phase

A "generic HPLC method" must possess characteristics such that compounds with a wide polarity range can be chromatographed and monitored by LC/MS and LC/MS/MS. Drug degradants, such as hydroxylated derivatives of drug compounds, could be much more polar than their precursors. Therefore, a wide gradient program is usually used as the "generic HPLC method". Lee,[60] for example, suggests using a fast gradient program from 95% aqueous/5% organic to 5% aqueous/95% organic at neutral pH values (i.e., 6–7). The conditions can be refined to best suit one's need so that the desired performance is achieved. A generic gradient method generates maximum information in a short time-period rather than trying to obtain optimal resolution among all the impurity peaks. Figure 12 shows an example of how this approach was successfully used in support of formulation development of a water-soluble basic drug substance. Using a gradient more than a dozen impurity/degradation peaks were detected in a 30-min run. A high pH mobile phase was chosen to enhance the retention of the drug substance, and selectivity of the drug substance from its related substances. Several key process parameters need to be optimized during formulation development. Degradation pathways of API and the levels of degradants formed are often dependent on these parameters. The degradation peak at 3.49 min (Figure 12) was the result of *N*-dealkylation from the drug substance. The degradant is much less hydrophobic and therefore less retentive than the drug compound which elutes at a retention time of 12.25 min. On the other hand, the peaks at RT of 14.2, 15.8, 17.2 and 18.0 min were API dimerization products through either free-radical C–C coupling, the Aldol reaction, or Michael addition. There is a significant difference in the polarity of degradation products from *N*-dealkylation

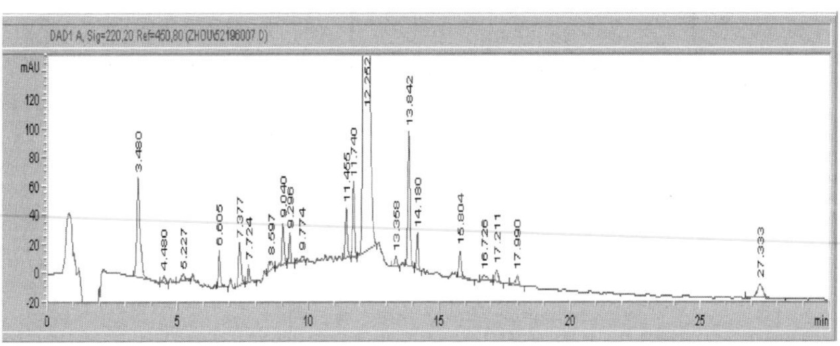

FIGURE 12 **"Generic" LC/MS method with basic pH mobile phase. Column: Waters XTerra 3.5 μm MS C18, 150 x 2.1 mm at 50°C. Mobile phase: A = 0.16% of ammonium carbonate at pH 8.9 and B = acetonitrile. Linear gradient from 6% of B to 60% of B in 15 min and held for 15 min. Detection: 220 nm.**

or dimerization. Figure 13 shows a total ion mass chromatogram using this "generic HPLC method". The retention time of peaks in the TIC is usually a few tenths of a minute later than their corresponding peaks in the LC/UV trace because there is a time delay for a peak to elute from the UV detector and travel to the mass detector.

B. "Generic LC/MS Method" with Acidic pH Mobile Phase

Figure 14 shows another "generic LC/MS method" used to establish the component profile for a natural product, which contains six major active components (denoted as A1, A2, A3, A4, A5, and A6). Our first trial was the use of a fast gradient with an apparent mobile phase pH of about 9, since all of these known components are acidic compounds. However, the method failed to separate A2, A3, and A4 from each other, and A6 from A5. In addition, the degradant peak D1 was not baseline resolved from the excipient peak B. When the pH of the mobile phase was lowered from 9 to 3.5, more than 35 peaks were baseline resolved in a 30-min run. The method was successfully applied to monitoring the impurity profile in different formulations. Although ESI of acidic compounds is typically performed using neutral or basic pH values, we did have adequate sensitivity at acidic pH in detecting signals for both the active components and their respective degradation products. Figure 15 illustrates a daughter ion mass spectrum for a naturally occurring impurity at a retention time of about 3.8 min. The spectrum clearly shows two

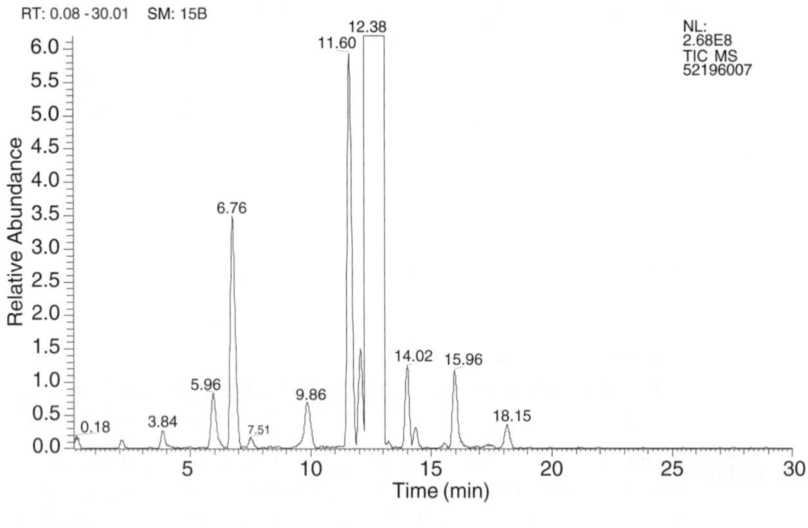

FIGURE 13 TIC of the "generic" LC/MS method with basic pH mobile phase.

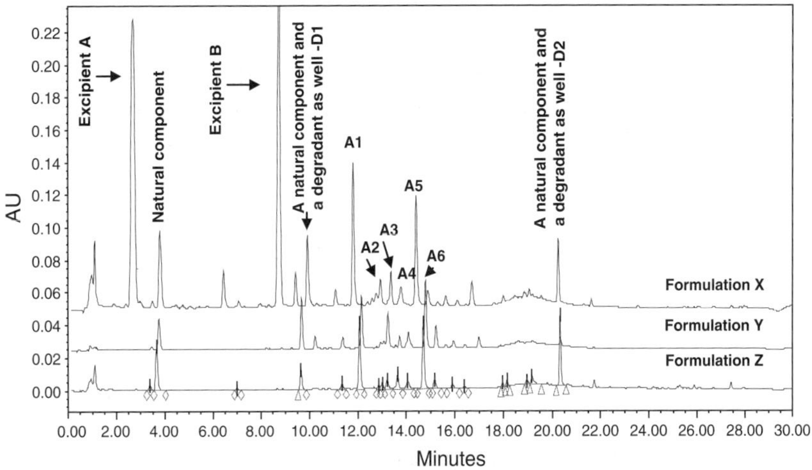

FIGURE 14 "Generic LC/MS method" using acidic pH value to monitoring of component profiles for different drug formulations. Column: Prodigy ODS(3) 3 µm, 150 x 2.0 mm. Mobile phase: A = pH 3.5 20 mM HCO_2NH_4 and B = acetonitrile. 95%A/5%B to 5%A/95%B in 15 min and hold for 15 min. Flow rate: 0.25 mL/min. Detection: 280 nm.

functional groups present in this compound. The neutral loss of 162 amu reveals that the compound bears a glucosyl group; another neutral loss of 44 amu is clearly due to the cleavage of CO_2, which resulted from the decarboxylation of the carboxylic acid group.

In summary, the strategy of using a "generic LC/MS method" in drug development provides multiple advantages. It provides a standardized starting point so that impurity profiles of drug substances and drug products can be monitored and tracked throughout the life cycle of the drug product development process. As recognized by Lee,[60] it eliminates iterative cycles of method development and refinement and provides faster project start-up.

C. Constructing a Database for Impurity Profiles

Webster's definition for "database" is "a large collection of data in a computer, organized so that it can be expanded, updated, and retrieved rapidly for various uses." An LC/MS database established for drug impurities contains multi-dimensional information such as relative retention times, UV spectra, molecular mass and substructural information. In order for the information to be updated and expanded, the methods used for information collection need to be unified. A "generic LC/MS method" allows relevant information to be collected in a consistent

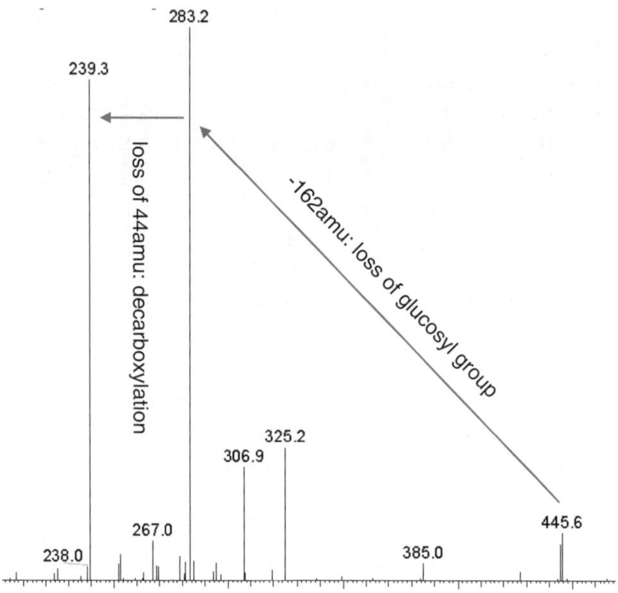

FIGURE 15 ESI/MS/MS daughter ion mass spectrum of the impurity peak at RT = 3.8 min. The impurity is a weak acidic compound, and LC/ESI MS was operated at acidic condition.

manner so that the impurity profile of one batch of drug substance or drug product can be compared to that of another batch of drug substance or drug product. This is extremely important when a drug enters the stages of clinical studies. The safety of a drug product is dependent not only on the toxicological properties of the active drug substance itself but also on the impurities therein. Qualification status established through the safety-batches ought to be maintained before further clinical studies are carried out. Analytical monitoring of impurity profiles in the drug substance and drug product plays an important role in tracking the qualification status of the API and drug product. According to the latest survey presented by the Product Quality Research Institute in its August 4–6, 2003 meeting of "Good Regulation Through Good Science" in Arlington, VA, 44% and 40% of the survey respondents said that LC/MS was the technique used for impurity characterization during phase 1 and 2 of drug development, respectively. When the development enters phase 3, 48% of survey respondents preferred isolation of the unknown impurities and complete characterization of the impurity by spectroscopic methods; 40% used LC/MS routinely to determine the content of impurities in the API and drug product.

Table 5 illustrates typical information available in an LC/MS database. More structural information such as UV spectra, parent ion and daughter ion mass spectra can be built into the database as hotlinks tied to the identity column. For example, double-clicking on the identity column for the impurity peak at RT of 11.5 min will bring up its UV spectrum, parent ion and daughter ion mass spectra (Figure 16). Rourick et al.[61] developed an elegant predictive strategy for rapid structural identification of drug degradants using LC/MS and LC/MS/MS databases. In order to generate a collection of possible drug degradation products, an expired lot of drug substance (cefadroxil) was stressed at a variety of conditions to induce degradation. The stressed samples were chromatographed using a gradient reversed-phase HPLC method, followed by on-line LC/MS and LC/MS/MS analysis to obtain structural information for the impurities and degradants. Structural data and chromatographic behavior of 18 impurities and degradants formed under these stressed conditions were built into the database. The database contained proposed structures, relative retention time, molecular weight, and diagnostic substructures. A structural library such as the one proposed by Rourick, can provide a foundation for predicting drug stability monitoring, chemical synthesis optimization, and formulation development, etc. during the pre-clinical and developmental phases of the drug development project. The library can also provide information and predict expected degradants under typical

TABLE 5 Typical Information in an LC/MS Database

Entry	*m/z*	RT (min)	UV shift?	Identity
1	288	3.47	No	Dealkylation degradation product
2	360	3.47	No	Hydroxy-substituted product
3	344	5.23	No	*N*-oxide
4	328	6.59	No	Geoisomer, process impurity
	328	12.25		API
5	669	7.36	Yes	Aldol adducts between API and 10-oxo-API
6	669	11.76	Yes	
7	671	9.04	No	Aldol adducts between API and 10-hydroxy-API
8	671	13.85	No	
9	326	11.47	Yes	Quinone like API
10	342	Shoulder peak	N/A	Di-keto
11	653	14.17	Yes	Bis-coupling API (2,2' C–C coupling, or ether linkage)
12	653	18.00	Yes	
13	655	15.82	No	Aldol adducts (5,6' , or 7,6' C–C coupling)
14	655	17.23	No	

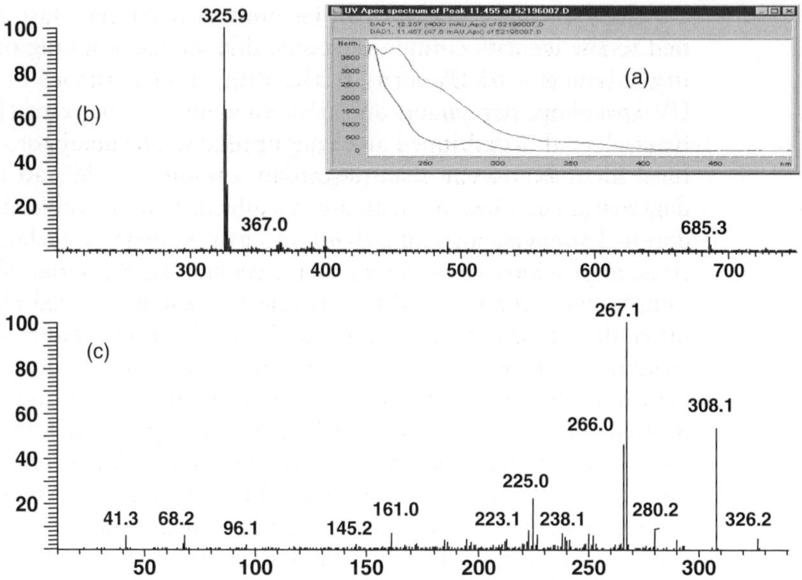

FIGURE 16 (a) UV spectrum, (b) parent ion and (c) daughter ion mass spectra of an impurity peak can be built into an LC/MS database.

drug processing and storage conditions, and sometimes drug metabolites under physiological conditions.

Xu et al.[62] also generated a degradant database for sumatriptan succinate by subjecting the drug substance to different stressed conditions followed by subsequent structure elucidation studies using LC/MS and LC/MS/MS. Sumatriptan succinate is a serotonin agonist effective in the acute treatment of migraine headaches. The study concluded that the drug was stable on exposure of acid, base, oxidation, and UV radiation at ambient conditions.

One of the key elements in such structural databases is the MS/MS daughter ion mass spectra of impurities. When impurity standards are not available, the spectra can serve as reference in prediction and identification of unknown degradants. Nicolas and Scholz[63] developed a method to assess and test the feasibility of using LC/MS/MS daughter ion mass spectra as "fingerprints" for structure identification. The fingerprints included a precursor ion mass, and at least three daughter ion masses. The authors generated MS/MS fingerprints for a drug substance and several of its impurities using either available standards or on-line LC/MS/MS analysis of a drug substance lot used in the earliest safety study. Several subsequent lots of drug substance were examined and the fingerprints of the impurities were obtained. When the same collision energy was used in the analysis of the same impurity peak, its MS/MS fingerprint matched very well with that obtained for the impurity from the safety lot. Nicolas and Scholz were successful even when they studied impurities at very low levels, such as 0.01 UV area percent of the drug substance.

Drug impurity databases, as discussed above, can be generated based on stressed drug substances or drug products. The databases can also be built using long-term storage stability samples. While various forced degradation studies, in which a drug substance or a drug dosage form is exposed to environmental factors such as acid, base, heat, light, oxidant, etc. can provide a diversity of degradant structures, overstressing of the sample can occur and lead to degradation profiles that are not representative of real storage conditions and perhaps not relevant for stability monitoring. Volk et al.[64] reported their LC/MS and LC/MS/MS investigation results on an aged sample of butorphanol tartrate that was stored in the dark for 170 weeks at 30°C. The investigation used the same analytical approach, i.e., acquiring molecular weights for the impurities during LC/MS profiling followed by LC/MS/MS substructural analysis.

Once a database is established, it is made available to other laboratories through the company's secured intranet, so that the information therein can be updated, retrieved and reviewed. The resulting structural library can be referenced throughout the lifetime of the drug for rapid identification of impurities, degradants, and metabolites.

VII. MS-ASSISTED HPLC METHOD DEVELOPMENT AND VALIDATION

A. Peak Purity Assessment

HPLC with UV detection is the analytical technique most often used for the assay of the active pharmaceutical ingredient (API) and determination of impurities in API and drug products. HPLC methods designed for assay of API, and the determination of impurities must be stability-indicating. A minimum requirement in chromatographic method development is to ensure specificity of the method, i.e., to achieve a set of chromatographic conditions in which the analyte peak is free from interference or contamination by other components of the formulation. The suitability of an HPLC method must be demonstrated in the validation step. For example, specificity of the method must be demonstrated to show that the API peak is separated from impurities; or that a degradant peak is well resolved from other degradants or synthetic impurities.

Diode-array UV detectors (DAD) are generally used to conduct peak-purity examination. The purity check is based on the comparison and matching of UV spectra at various points across the HPLC peak to that at the apex of the peak. The inherent disadvantage of using DAD for peak purity investigations is that it can only reveal spectral homogeneity, but not chemical homogeneity, across a chromatographic peak. This approach assumes that the UV spectrum obtained at the apex is free of any co-eluting impurities, and that the UV spectra of impurities are at least slightly different from those of the API. However, many impurities such as degradants and synthetic impurities are structurally related to the API, and contain in their structure very similar chromophores, making purity assessment based solely on DAD data not reliable. Therefore, an LC peak with homogeneous UV-spectral characteristics does not necessarily prove chemical purity of the peak. Other concerns in using DAD alone may include concerns that analytes many not contain chromophores, analytes may be present at very low concentration levels, etc.

Coupling a mass spectrometer to a liquid chromatograph brings a new dimension to specificity studies. Since a mass spectrometer separates compounds by their respective mass-to-charge ratios, any difference in the *m/z* values between the impurities and the drug substance will allow an unambiguous detection regardless of similarities in their UV spectra. Therefore chromatographic co-eluting components will be separated in MS as long as their *m/z* values are different.

Figure 17 shows an example where LC/DAD-UV versus LC/MS is contrasted in the establishment of chromatographic peak purity. Comparison of the UV spectra taken on the upslope, apex, and downslope of the peak reveals that the peak is spectrally nonhomogeneous. However, the method cannot address questions as to how many components are co-eluting, and structural information for these components is unavailable. LC/MS, on the other hand, gives unequivocal peak identification by comparing mass spectra extracted from the same three locations of the chromatographic peak. The mass spectrum extracted from the onset of the chromatographic peak shows a single chemical entity with mass-to-charge ratio of 309. A second ion with *m/z* of 287 is seen in the spectrum extracted from the apex of the peak; the relative intensity of this second ion vs. the ion at *m/z* of 309 increases in the spectrum on the offset of the peak.

Bryant et al.[65] have reported the suitability of using ESI-MS for assessment of peak homogeneity. The HPLC system was designed to intentionally co-elute a number of impurities within the peak of the drug substance. Mass spectra across an HPLC peak were summed followed by background subtraction to examine the co-eluting impurities. The co-eluting impurities can be rapidly detected to a level of 0.02% of the peak of the drug substance. The method is shown to be superior to the UV method in that compounds with identical and similar UV spectra can be distinguished.

B. Peak Tracking in Method Development

LC/UV detection, because of its sensitivity, robustness and precision, is the most commonly used technique for the detection and quantitation

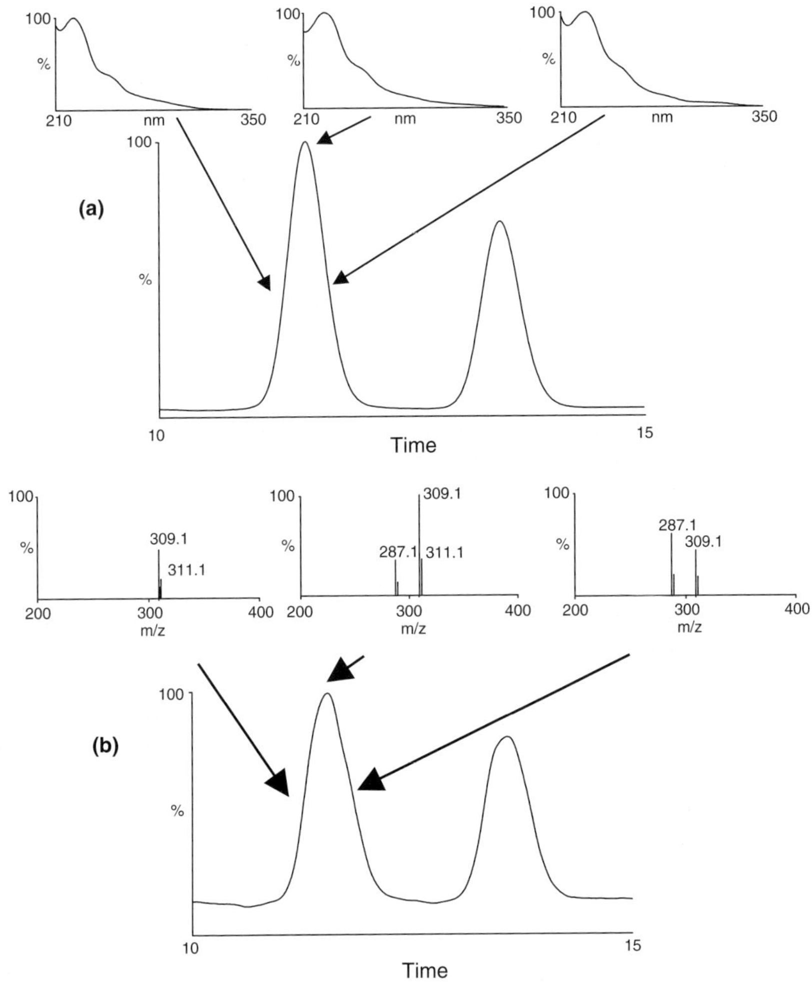

FIGURE 17 Peak purity detection using (a) LC/DAD-UV vs. (b) LC/MS (courtesy of Waters Corp.).

of impurities in drug substances and drug products. As discussed in Section VI, LC/UV methods designed for the determination of impurities usually undergo many changes from one set of chromatographic conditions to another during the drug product development process. Each such change may result in different elution order of impurity peaks. Figure 18 shows two chromatograms with (a) corresponding to an acetonitrile-based mobile phase and (b) corresponding to a methanol-based mobile phase. During early drug development, reference standards for many of the drug impurities may not be available. Correlation for the group of peaks at RT from 12 to 15 min could not have been established if the only information available was the percent of peak area and the UV spectra of the peaks. Conversely, full scan LC/MS allowed simultaneous detection and molecular mass determination of these three peaks in both mobile phase systems. Figure 19 presents full scan LC/MS results for impurities separated using both chromatographic methods. The molecular ions, inarguably, provide adequate information to allow peak correlation.

C. Development of an HPLC Method to Account for Mass Balance

Execution of stability programs for drug substances and dosage forms is a central focus during the development of drug products. One of the objectives that needs to be achieved from the stability data is to establish a mass balance equation for the API. A viable pharmaceutical formulation is expected not only to deliver desired pharmacokinetics but also to demonstrate good stability behavior for the API under recommended storage conditions. However it is inevitable for the API to experience some degree of degradation when the formulation is exposed to environmental factors such as heat, humidity and UV light. The degradation of API in formulations can be very complex and sometimes unpredictable. It is not unusual for the decrease in the assay value not to be accompanied by a proportional increase of degradation products. The

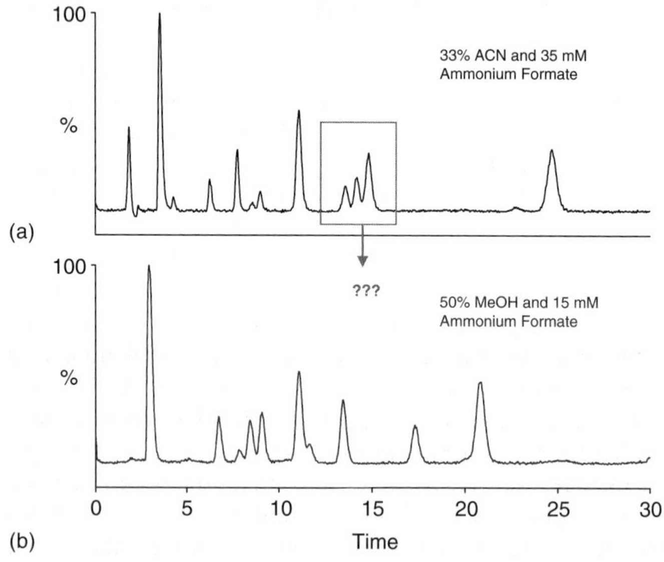

FIGURE 18 **A challenging case in which peak correlation between two LC methods cannot be achieved based on the percent of peak area and the UV spectra of the peaks (courtesy of Waters Corp.).**

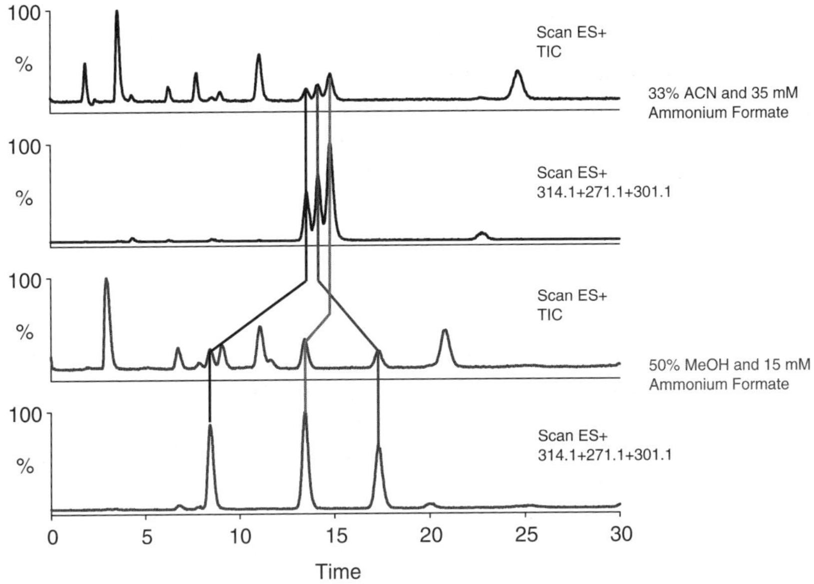

FIGURE 19 Detection of molecular ions from full scan LC/MS snap the peak correlation between two chromatographic methods (courtesy of Waters Corp.).

detection limit of LC and the change in the UV chromophores of degradation products make the justification of mass balance a very challenging task during drug development.

The degradation of the API can be of unimolecular reactions such as dehydration, epimerization of API. More often encountered are bimolecular reactions such as oxidation, hydrolysis, and dimerization of the API. Reactions of API with excipients and packaging components are also possible. In some cases, a degradant derived from multiple-stage degradation processes can still react with the API. Ginsburg et al.[66] reported a degradation product in a parenteral formulation as a result of a reaction between the API and an oxidation product of an excipient. Butamben (butyl *p*-amino-benzoate) was formulated to provide local anesthesia for the treatment of chronic pain. The formulation was found to undergo discoloration when the formulation was stored at ambient conditions. The compound responsible for the discoloration was isolated and identified by positive FAB-MS and $^{1}H/^{13}C$ NMR experiments as an oxalamidine. The oxalamidine was the condensation product of oxalic acid and four equivalences of butamben. Oxalic acid, according to Ginsburg et al.,[66] was the oxidation product of poly(ethylene) or/and polysorbate 80 – both were excipients in the formulation.

Direct infusion mass spectrometry is a popular technique used by medicinal chemists to check the molecular mass of target compounds. In the author's laboratory, it is often used to build an "impurities inventory" for a stressed drug substance or drug formulation. In order to establish mass balance for the API in a drug formulation, all the degradation peaks must elute from the HPLC column and be detected at the chosen wavelength. The polarity and solubility of some degradants can be drastically different from that of the API. Poor recovery of impurities and degradation products during sample preparation, combined with inappropriate choices of organic strength and pH values in the mobile phase can make impurities and degradation products undetectable. If one or more of the drug degradants were not detected in an HPLC chromatogram, one would likely experience an unexplainable mass loss.

Forced degradation of a drug substance or a drug product is usually conducted to demonstrate method specificity when developing a stability-indicating chromatographic method. Forced degradation can be carried out using different combinations of heat, humidity, light, acid, base, or oxidant. These studies also provide information about what degradation products could form during storage. Impurity profiling studies using MS and MS/MS approaches on the stressed samples can lead to an inventory of impurities. Data built in this "inventory" provides information on not only the degradation pathways but also on chemical markers to ensure the successful development of methods for assay and impurities determination.

Although ionization suppression of minor components can occur during the analysis of concentrated solutions of API, the ultimate resolving power and accurate mass measurement of the FTMS can still be the best choice to generate qualitative impurity profiles. Direct infusion of a sample allows the detection of all ionizable constituents. It also bypasses the LC separation, and avoids selective loss of components at extreme ends of the polarity scale. In one instance, during formulation development, up to a 20% loss of API was observed after storage for 6 months at 40°C/75% RH. A reversed-phase LC method was used for the assay of the API. The mobile phase consisted of 80% of 20 mM HCO_2NH_4 at pH 3.2 and 20% of CH_3OH and CH_3CN in a 1:1 ratio. A total of 5% of three degradation products was observed using these HPLC conditions. However, a significant 15% loss of drug compound could not be accounted for. A stressed sample, heated at 80°C in an oven chamber saturated with moisture, was analyzed by the direct infusion into it of a Bruker Apex II 4.7 Tesla FTMS. Figure 20 shows the mass spectrum for the stressed sample. Eighteen likely degradation products were observed. Based on their exact masses and chemical reactivity of the API, structures for 14 of these degradants were proposed (Table 6). An improved HPLC assay method was developed based on the information obtained from the direct infusion FTMS results. Figure 21 presents a comparison of the original assay method to the improved method. One dominant degradation product was identified as the result of dealkylation of the drug molecule. The dealkylated degradant was much more hydrophilic, and as a result was eluting in the solvent front in the original HPLC method. A Waters XTerra column was chosen to replace the originally used Symmetry column because polar compounds are more retained with XTerra columns. Other adjustments in order to retain this degradant included lowering the organic strength at the beginning of the chromatographic run, and increasing the pH value of the mobile phase from 3.5 to 9. Many of the degradation products were identified as dimerized compounds; some of them did not elute from the column in the original, isocratic method. A gradient elution program was therefore used to end the chromatographic run with 60% of acetonitrile, and simultaneously the temperature for the column was raised to 50°C from 40°C. With all the modifications, seven more degradant peaks were detected, which brought the mass balance to about 98% when the quantitative levels of impurities were expressed as a percentage of the peak area of the API peak.

VIII. IMPURITY PROFILING FOR DRUG SUBSTANCES AND PHARMACEUTICAL PRODUCTS

Toxicity and pharmacological effects of impurities in a pharmaceutical formulation are monitored in a variety of safety and clinical studies.

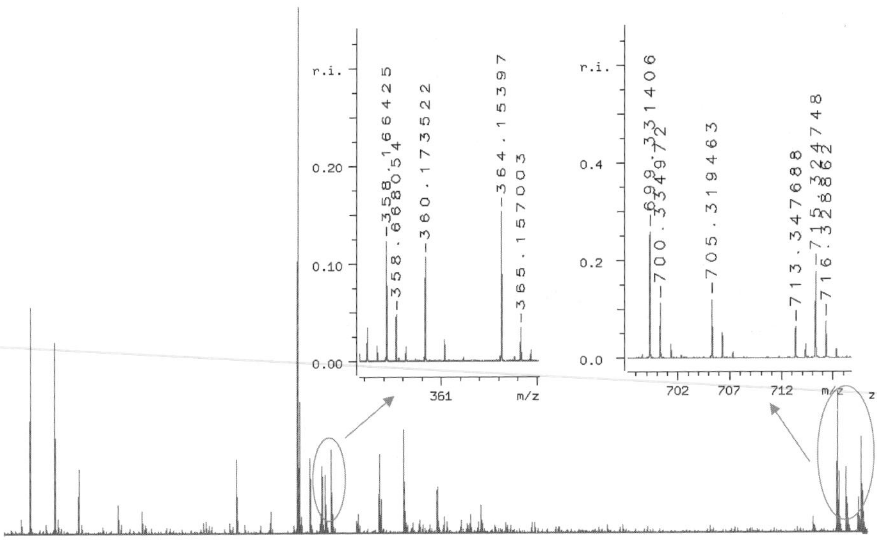

FIGURE 20 FTMS Spectrum of a stressed sample: the API signal appeared at *m/z* of 342, 18 degradants were detected.

TABLE 6 Direct Infusion FTMS Results on a Stressed Solid Dosage Formulation

Peak#	Experimental	Theoretical	Error (ppm)	Structural information
1	149.02334	149.02332	0.16	Environment contaminant—phthalate
2	288.12289	288.12303	0.47	Degradant resulted from *N*-dealkylation
3	324.15956	324.15942	0.43	A likely dehydration degradant
4	340.15488	340.15433	1.62	Quinone-type degradant
5	356.14842	356.14925	2.33	Two steps of oxidation to give rise to keto-degradant
6	358.16468	358.16490	0.61	Hydroxylated degradant
7	360.18082	360.18055	0.75	Addition of water
8	388.17496	N/A	N/A	?
9	390.15434	N/A	N/A	?
10	396.21659	396.21693	0.86	*Trans*-alkylation degradant
11	406.18485	N/A	N/A	?
12	665.32117	665.32213	1.44	Aldol product of API
13	681.31830	681.31704	1.85	Free radical dimerization of API; pesudo product of API
14	683.33283	683.33269	0.20	Aldol adduct of API
15	679.30221	679.30139	1.21	Aldol product of API and *m/z* 356
16	695.29974	695.29631	4.94	Free radical coupling of API and *m/z* 356
17	697.31573	697.31196	5.41	Aldol product of API and m/z 356
18	705.31765	N/A	N/A	?
19	713.30745	713.30687	0.81	Free radical coupling of *m/z* 356 and *m/z* 360

The degradation and impurity profiles are critical to the correct safety and potency assessment of the drug product for clinical trials. LC/MS and LC/MS/MS have given an impetus to both qualitative and quantitative analyses of drug impurities and degradants.[67–74] When fragmentation information in combination with molecular weight is not sufficient to differentiate between isomeric and isobaric structures, other techniques are sometimes used, such as the rapidly advancing, on-line coupling LC-NMR for structural proposals.[75,76] The applications of LC/MS and LC/MS/MS for pharmaceutical development are discussed in a number of review articles.[77–81] There are similarities and differences in the impurity profiles in a drug substance and the drug product. Therefore, the qualification and the strategy for their identification should be treated accordingly. For drug substances, impurities are subdivided into

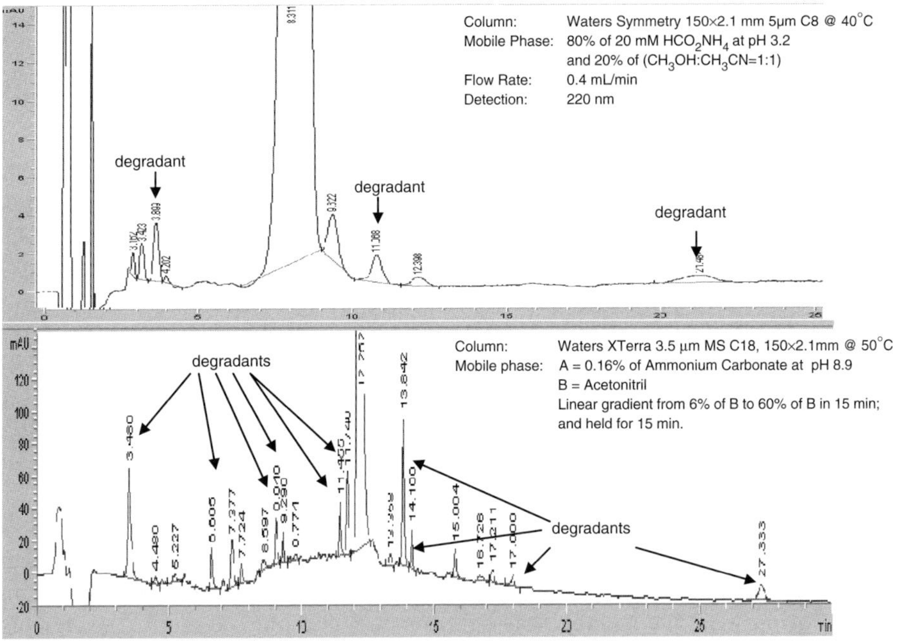

FIGURE 21 (a) An isocratic reversed-phase HPLC method developed for assay, with which 15% of mass balance was not accountable. (b) An improved gradient HPLC assay method based on infusion FTMS results, with which seven more degradation product peaks can be detected.

inorganic impurities, organic impurities, and residual solvents. Organic impurities may arise during the manufacturing process and/or storage of the drug substance. Typical impurities include synthetic impurities such as starting materials, by-products, intermediates, ligands and catalysts. Typical degradation products can be oxidation products, hydrolysis products, epimerization products and dimerization products of the drug molecule. For drug products, only the impurities formed by the degradation of the active pharmaceutical ingredient and by reaction of the API with an excipient and/or some component of the container closure system have to be taken into consideration. However, the complexity of the matrix in pharmaceutical formulations poses a significant challenge for LC/MS analysis. Many excipients used in the formulation can cause detection problems. For example, signals from PEG, as discussed in Section IV. E. and illustrated in Figure 8, can dominate the mass chromatogram.

During the course of chemical development, impurity profiles in drug substances may change due to changes in synthetic route and changes in the size of the batch. ICH guidelines for Impurities in New Drug Substances (ICH Q3A), require that impurity test results for batches designed to scale-up, optimize, and validate the manufacturing process of a new drug substance be summarized in the NDA. Detailed laboratory studies conducted to detect and identify impurities at 0.1% and higher levels must also be included.

Ermer,[56] for example, utilized LCQ to monitor impurity profiles of various batches of ramorelix used in toxicological studies, clinical studies and scale-up. Ramorelix is a synthetic glycosylated decapeptide with monoisotopic M_r of 1530.7. The toxicological batch served as the benchmark against which all other batches were compared. Molecular weights of impurities were determined by ESI mass spectrometry, and used in conjunction with UV peak area % to gauge impurities in batches used in clinical trials. These impurity profiles were compared to those of batches used in the toxicologically qualified batch. Four impurities were detected with the same M_r value. They were believed to be diastereoisomers of ramorelix, i.e., a peptide sequence with one of the amino acids in the opposite enantiomeric form.

Similarly, impurity profiles may change when the formulation is modified or a scale-up of a specific formulation is made. Pharmaceutical formulations are a complex physiochemical system that may result in impurities due to reactions between API and pharmaceutical excipients and/or packaging materials. In some cases, degradants that were generated by multiple-step degradation pathways can still react with the API leading to the formation of degradants that can be difficult to identify.

In a package screening study, a low level degradant of famotidine was detected when stored in child-resistant foil pouches. The identity of the impurity was investigated using HPLC/APCI mass spectrometry.[82] The molecular weight of the impurity was reported as 12 amu higher than that of famotidine, indicating the insertion of one C atom to the drug compound. Detailed substructural analysis by comparison of the fragmentation pattern with that of famotidine suggested that the carbon insertion occurred at the *N*-(aminosulphonyl)-propanimid-amide end of famotidine. When famotidine was exposed to a source of formaldehyde, a degradation product with the same molecular weight was formed. Unfortunately, there was no discussion in the paper as to whether the synthesized compound was the one detected in the child-resistant packaged drug product.

A distinction needs to be made between identification and qualification of degradation products. Identification means that the structure has been elucidated, while qualification is defined as the process of acquiring and evaluating data to establish the biological safety of an individual impurity or a given impurity profile. The analytical method must be sensitive enough to quantitate known impurities at the specified limits and unknown impurities at the levels required by the ICH guidelines. Successful identification of drug impurities using LC/MS and LC/MS/MS always relies on the generation of molecular weight, substructural information, UV spectrum, and the chemical reactivity of the drug compound.

A. Determination of Molecular Mass

Protonated and cationized species are the most commonly detected ions using the ES process if a positive ionization mode is selected. Protonation is a result of the addition of proton(s) to a neutral molecule; for every proton added, a net charge of +1 will result. Similarly, cationization is due to the addition of cation(s) to a neutral molecule. Detection of cationized ions can be useful in the determination of molecular mass of unknown analytes.[83] If ESI is operated under negative ion mode, deprotonated ions will usually be the most dominant ions. For either operation mode, solvent adducts of the protonated/deprotonated ions are frequently detected in ESI/MS mass spectra.

Assignment of various ions in an ESI/MS spectrum is usually straightforward. Although high levels of chemical noise in the total ion chromatogram are inherent in the ES process, many algorithms are available in commercial instruments which allow background subtraction to correct for part of the chemical noise. The chemical noise results from a number of sources, including the LC mobile phase and buffers that give rise to high contributions in the signal. Figures 22(a) and (b) show LC/UV and ESI–TIC traces, respectively. The mass spectrometric analysis was conducted on a Finnigan MAT TSQ 7000 with ESI positive ion monitoring. Figure 22(c) is the extracted mass spectrum for the peak at retention time of 25.35 min without background subtraction, while Figure 22(d) shows the mass spectrum for the same peak with background subtraction. Despite the noisy appearance of the TIC trace, individual mass spectra, as evident in Figure 22(c) and (d), obtained from ESI LC/MS are generally of good quality.

The first step in establishing an impurity profile is to extract mass spectra at the positions of impurity peaks in the corresponding UV chromatogram. A profile of selected ion mass chromatograms can be obtained by extracting ions with relevant mass-to-charge ratios and compared to the UV trace. Assignment of various ions will lead to the determination of molecular masses of impurities. The ions observed in Figure 22(c) and (d) can be designated in Table 7. One of the characteristics in the ESI mass spectrum is that cationized ions as well as solvent adducts of protonated pseudo molecular ions are frequently observed. Since the protonated pseudo molecular ion was detected with mass-to-charge ratio of 344, the molecular mass for the impurity peak at RT=25.35 min is 343 amu. Molecular masses of impurities may reveal rich structural information. For example, an

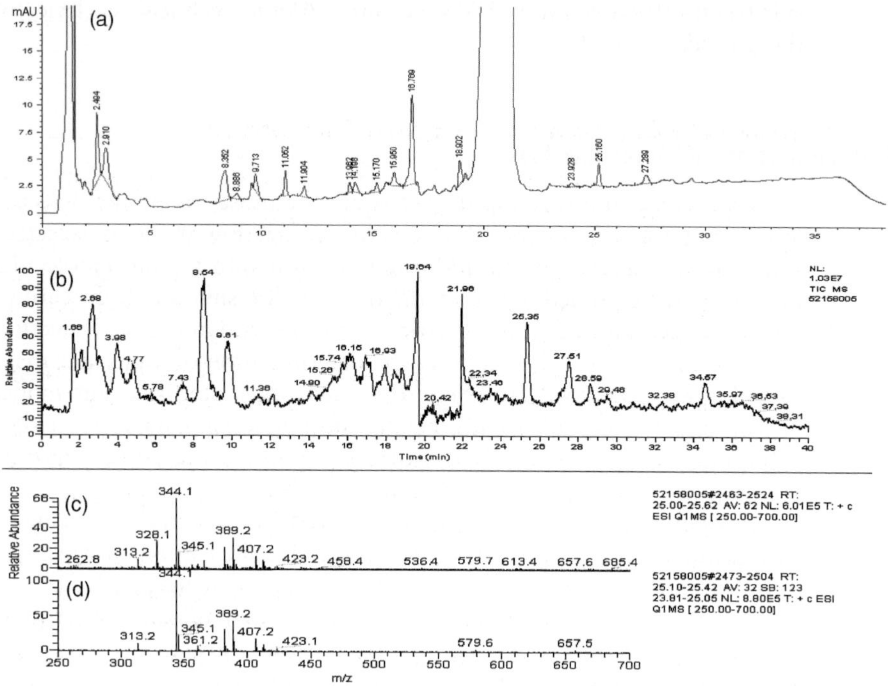

FIGURE 22 **(a) HPLC/UV chromatogram. (b) ESI LC/MS total ion chromatogram. (c) extracted mass spectrum for RT = 25.35 min peak without background subtraction. (d) Extracted mass spectrum for the same peak with background subtraction.**

TABLE 7 **Assignment of Ions Detected in ESI Parent Ion Mass Spectrum**

Ions (*m*/*z*)	Assignment of ions
313	Chemical noise or thermally degradated ion
328	Chemical noise
344	$M + H^+$, protonated pseudo molecular ion
345	"A+1" isotope of $M + H^+$
361	$M + NH_4^+$, ammoniated pseudo molecular ion
382	$M + K^+$, potassiated pseudo molecular ion
389	$M + C_2H_5NH_2 + H^+$, ethyl amine adduct of the protonated pseudo molecular ion
407	$M + HCOONH_4 + H^+$, ammonium formate adduct of the protonated pseudo molecular ion

increment of 16 amu in the molecular mass of a degradant relative to the drug compound usually suggests oxidation of the drug compound to form hydroxy or *N*-oxide compounds; an increment of 14 amu suggests the formation of ketones via dual-stage oxidation of the drug compound.

B. Determination of Exact Mass and Elemental Composition

Determination of elemental compositions is an important step towards the structural elucidation of drug impurities. Elemental composition of an unknown compound can only be derived from the measurement of its exact mass. In combination with the knowledge of the physiological or chemical process, elemental compositions alone may be enough to propose a correct structure.

There are three kinds of mass spectrometers capable of generating high-resolution mass spectra and determining accurate mass: the TOF mass spectrometers or the hybrid quadrupole time-of-flight (Q-TOF) mass spectrometers, Fourier transform ion cyclotron resonance (FTICR) mass spectrometers and the double-focusing sector mass spectrometers. In order to achieve high accuracy in mass measurement, mass spectrometers must be calibrated with calibrants of known exact masses. With internal or external calibration, good mass measurement can be achieved. For example, FTMS can generate mass measurement with accuracy of less than 1 ppm in mass error between the experimentally determined value and the theoretical exact mass.

All elements have uniquely assigned masses using a reference standard of ^{12}C as 12.00000. Chemical compounds with unique elemental compositions will, therefore, have a distinguishing exact mass. Computer programs are available to calculate possible elemental compositions for the measured accurate mass. Many different elemental combinations may exist for a molecule with a specific accurate mass. The number of possible compositions increases with increasing mass and the increasing number of elements expected to be present in a molecule. The more accurate the measured mass, the smaller the number of possible elemental compositions that will be derived. As important as mass accuracy, the isotopic pattern in a full scan high-resolution mass spectrum can be used as a fingerprint to help derive the elemental composition of the unknown. The abundance of isotope peaks is governed by the natural occurrence of the elements in the ions. The presence of "A+2" elements such as Cl, Br, Si and S can usually be recognized from the higher-than-normal intensity of the "A+2" peak. The maximum number of carbon atoms can be determined by dividing the peak ratio of "A+1"/"A" by the natural abundance of ^{13}C - 0.011.

Since FTMS provides high resolution and high mass accuracy, it enables detailed structural investigation for small molecules and proteins as well. There has been a publication[84] examining the effectiveness of accurate mass measurements in minimizing false protein matches by varying the mass error allowed in the search over a range from 2 to 500 ppm. The report evaluated the utility of accurate mass tags by calculating the number of possible enzymatically digested polypeptide fragments for all predictive proteins from *C. elegans*. Out of 19 100 predicted proteins, or 918 655 possible tryptic fragments, approximately 60% of the peptides with molecular mass of 2000 Da or less have unique masses at a mass measurement accuracy of 0.1 ppm; at an accuracy of 1 ppm, this value decreases to 40%. For the peptides with mass of 2500 Da and higher, more than 80% have unique masses at the sub-ppm error level.

Eckers et al.[85] used on-line coupled reversed-phase LC to a Q-TOF mass spectrometer, and have successfully identified trace impurities in cimetidine. In an earlier paper published by Haskins et al.,[86] they successfully identified four chromatographically unresolved reaction by-products in cimetidine, using LC-FTMS. Another example using fast LC coupled with a Q-TOF mass spectrometer was the identification of cimetidine-related drug substance impurities reported by Lee et al.[87] The exact masses for six impurities were determined with an experimental error of less than 3.1 ppm.

A case study of the identification of a counterfeit drug molecule is discussed in Section IX. C. This is a step-by-step discussion of the experimental procedure using FTMS to address this important issue. After the exact mass of the unknown compound has been determined, the next step is to derive its elemental composition. The minimum and maximum number of expected atoms present in the compound must be specified in the search criteria to allow the computer program to calculate possible elemental compositions. A unique fit of only one elemental composition is rarely obtained. The number of possible compositions increases with the increasing number of elements present and with increasing mass. However, other information, such as the number of double bond equivalency and the isotopic distribution of the parent ion mass spectrum, can be used to reduce the possible elemental compositions to a reasonable number. Further discussion can be found in Section IX. C.

C. Structural Identification and Elucidation

Traditional methodologies for structural identification of trace level impurities in drug substances/products usually involve fractionation of each impurities using a scaled-up analytical chromatographic method, followed by off-line spectroscopic analysis. Coupling of HPLC separation and electrospray mass spectrometry allows on-line acquisition of full scan mass spectra and generation of tandem mass spectrometric data. LC/ESI MS has revolutionized trace analysis for qualitative and quantitative studies in pharmaceutical analysis.

1. Use of the Drug Molecule as a Structural Template

The interpretation of the daughter ion mass spectra is often hampered by a lack of knowledge of the fragmentation rules. ESI MS/MS spectra are often characterized by frequent and complicated hydrogen-shifts. Hence, an alternative method for the elucidation of mass spectra is vital. Fortuitously, when a drug compound degrades, most of its degradants would be expected to retain a major portion of the substructure of the parent drug compound. A substructural analysis based on specific product ions and neutral losses from the drug compound can serve as "templates" for the interpretation of the structures of unknown degradants. This is a widely used method[88,89] to help identify the structures of unknown degradation products. Precursor ions, usually protonated pseudo molecular ions, are selected and subjected to low-energy CID in the presence of collision gas. The dissociations produce fragmented ions that are recorded in the MS/MS spectrum. The comparison of MS/MS patterns of unknown degradants to those of the drug compounds is likely to reveal similarities and the differences. Common product ions and neutral losses observed in the drug compound and the degradants suggest common substructures, and the differences indicative of the modifications of substructures and locations of such modifications have occurred.

Figures 23(a) and (b) show daughter ion mass spectra of a drug molecule and one of its impurities. The spectra were acquired using a Finnigan MAT TSQ 7000 interfaced to an Agilent 1090 LC, operated in the ES negative ion mode. The *m/z* of 861 was the de-protonated pseudo molecular ion. The base peak, with m/z of 699, was 162 amu lower than that of the deprotonated pseudo molecular ion. The loss of 162 amu was due to deglucosylated fragmentation, a signature loss that is consistent with the structure of the drug molecule, which bears the glucosyl functionality. Another abundant daughter ion was observed at *m/z* of 817, corresponding to a loss of 44 amu from the parent molecular ion. This loss was indicative of decarboxylation from the drug molecule. This type of fragmentation pattern was expected because the drug molecule contains a carboxylic acid moiety. The impurity, however, did not show the loss of 162 amu. Instead, a loss of 324 amu was detected as the predominant fragmentation ion. The loss of 324 amu is consistent with the molecular weight of lactose, which suggests that this impurity contains a lactosyl functionality.

2. Hydrogen/Deuterium Exchange Experiment

Hydrogen/deuterium (H/D) exchange experiments can provide critical information for structure elucidation. The use of H/D exchange is a well-accepted protocol in the field of proteomics for identification of proteins.[90,91] The generation of proteolytic peptide mapping for protein identification has been reported. This involves labeling selected amino acids with ^{13}C, ^{15}N, or ^{2}H and incorporating them into proteins during cell culture.[92] Each of the labeled amino acids carries a defined change in

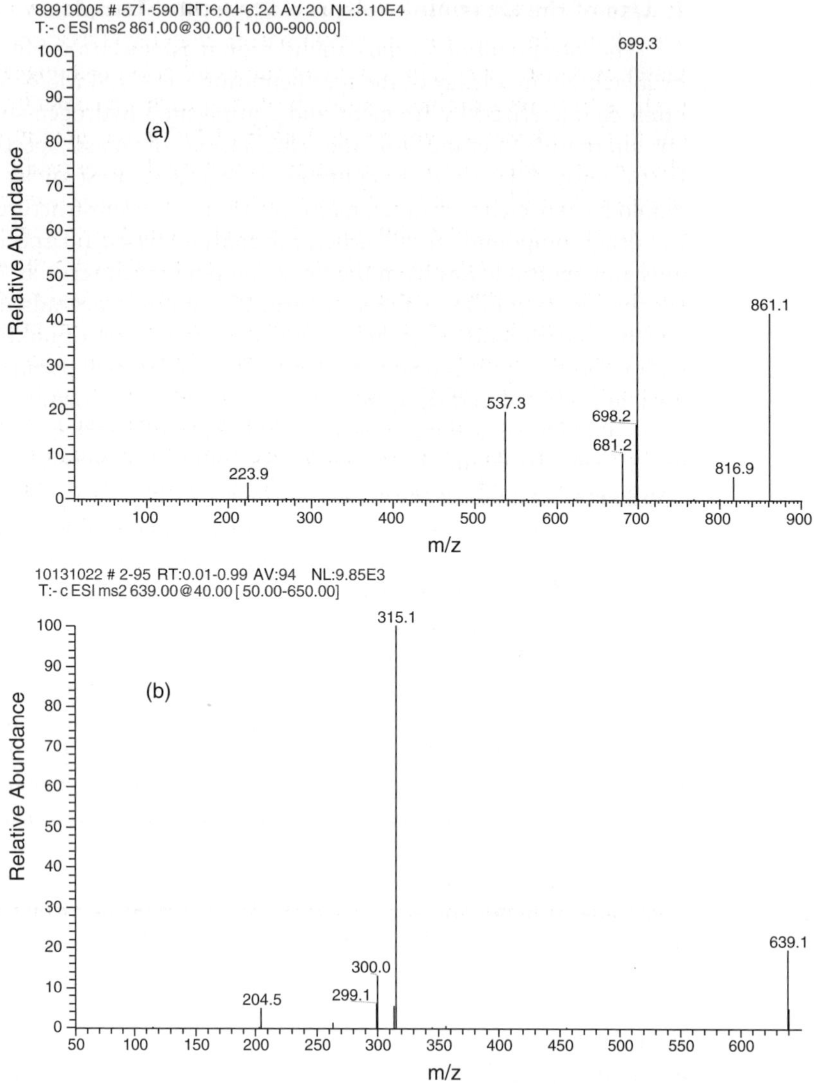

FIGURE 23 Comparison of daughter ion mass spectra of drug molecule (a) and its impurity (b).

mass, and a change in the appearance of the monoisotopic distribution pattern in the mass spectrum. These changes in the mass spectrum permit the isotopically labeled peptides to be distinguished from other peptides after proteolysis of the protein.

The application of H/D exchange experiments in the identification of small molecules is of great importance as well, and has also been reported.[93,94] The experiment is usually carried out using deuterium oxide as a source of deuterium for incorporation into the unknown compound. The change in mass for the unknown in deuterium oxide (compared to water as the solvent) is indicative of the number of exchangeable protons such as hydroxyl, amine, amide, thiol, and carboxylic acid groups present in the unknown compound. A shift of "n" amu units in the mass to charge value when mass spectrometer is operated under positive ion mode suggests "$n-1$" exchangeable protons because the proton carrying the charge of the ion is also shifted. On the other hand, a shift of "n" in negative ionization mode reveals that "$n+1$" exchangeable protons are present in the unknown compound. In our laboratory, the technique is frequently employed to aid in the identification of drug impurities. For example, an unknown impurity peak was observed in a natural product during the development of an HPLC assay method. LC-ESI negative ion mass spectrometric analysis confirmed the nominal mass of this impurity as 480 Da. The daughter ion mass spectrum of this unknown compound was acquired on a Bruker Apex II 4.7 Tesla FTMS. As seen in Figure 24(a),

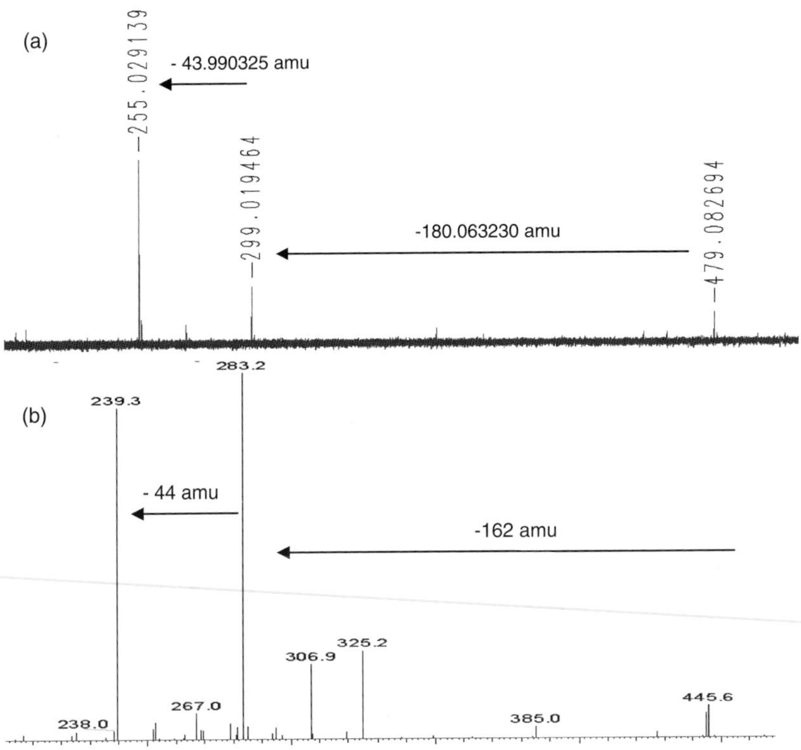

FIGURE 24 **(a) Daughter ion mass spectrum of the unknown with nominal mass of 480 Da, acquired using Bruker Apex II 4.7 T FTMS. (b) Daughter ion mass spectrum of a typical impurity, acquired using Finnigan MAT TSQ 7000.**

an intense daughter ion was detected at *m/z* of 299.019424. The neutral loss from *m/z* of 479.082594 to 299.019424 corresponded to 180.06323 Da, which agreed very well with the theoretical exact mass of 180.06339 Da for the glucose molecule. The daughter ion mass spectrum acquired using a Finnigan MAT TSQ 7000 mass spectrometer for a typical impurity is given in Figure 24(b). The impurity was used as a control in the H/D exchange experiment. A neutral loss of 162 Da was observed (Figure 24(b)). This is the same neutral loss as for the drug compound and most of its impurities. The loss of 44 Da was observed in both the unknown and the control. Therefore, two possible structures were proposed and are shown in Figure 25.

Structure A contains nine exchangeable protons, while structure B contains eight such protons. A H/D exchange experiment was performed in order to provide the answer. The validity of H/D results was examined by carrying out the H/D exchange for the control compound in a parallel fashion. The control compound contained six exchangeable protons; four of them were the hydroxyl groups in the glucosyl substituant, one was the phenol proton, and the last one was the carboxylic acid proton. A shift in molecular mass of 5 amu was observed when the spectrum was acquired using ESI in the negative ion mode. This demonstrated that the exchange experiment was successful. For the unknown compound a shift of 7 amu was observed, which strongly suggested the presence of eight exchangeable protons rather than nine. It was therefore concluded that

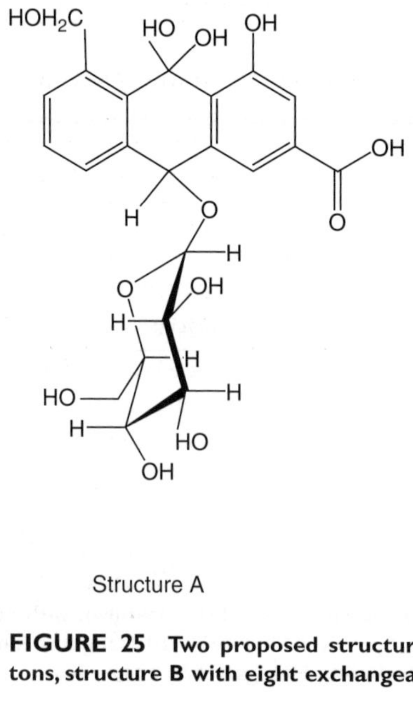

Structure A

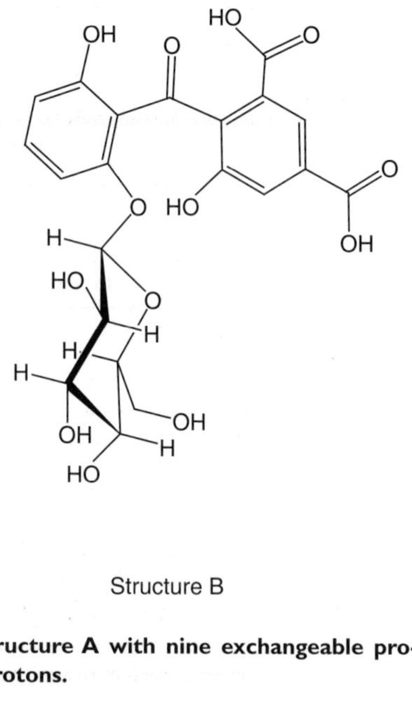

Structure B

FIGURE 25 **Two proposed structures: structure A with nine exchangeable protons, structure B with eight exchangeable protons.**

structure B was the most likely structure for this unknown impurity. Tables 8 and 9 summarize MS/MS results comparing fragmentation patterns of the D-labeled entities with their H-labeled counterparts. The structure was supported by various 1D and 2D NMR experiments (data not shown). A literature search found that the unknown impurity was recently reported by Terreaux et al.[95] who isolated it from a Tinnevelli Senna pod extract.

IX. SUPPORTING MARKETED DRUG PRODUCTS

A. Continuous Monitoring of Impurity Profile in Drug Product

Once an NDA is approved and the drug product is successfully launched, corporate resources shift to production, storage, distribution and marketing of the drug product. The quality and integrity of the drug

TABLE 8 MS Results of H/D Exchange Experiment: Mass Spectrometer Was Operated at Negative Ion Mode

Controlled compound $[M-H]^-$		Unknown $[M-H]^-$	
In H_2O	In D_2O	In H_2O	In D_2O
445	450	479	486
6 exchangeable protons		8 exchangeable protons	

TABLE 9 MS/MS Results of H/D Exchange Experiment: Mass Spectrometer Was Operated at Negative Ion Mode

Controlled compound		Unknown	
In H_2O	In D_2O	In H_2O	In D_2O
162(445/283)	165(450/285)	180(479/299)	185(486/301)
44(283/239)	44(285/241)	44(299/255)	44(301/257)
The ether bond between glucosyl and phenyl was cleaved; and glucosyl leaves as a tautomer of 2-hydroxyglucal.		The ether bond between glucosyl and phenyl was cleaved; and glucosyl leaves as glucose.	
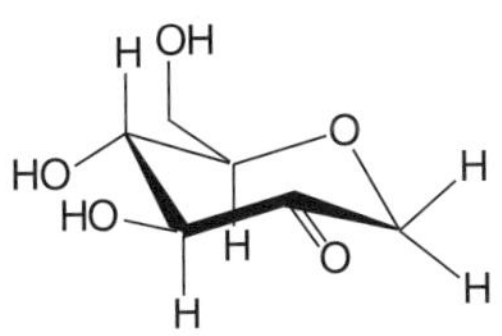		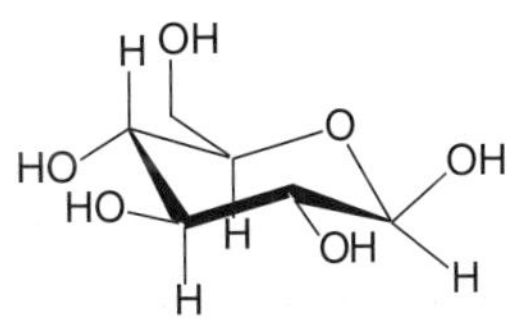	

product is under the increasingly tightened scrutiny of regulatory agencies. Maintaining the standards given in the NDA is the key to protecting the franchise. In case new impurities or degradants are detected during the post-approval stability program, identification and qualification must follow ICH guidelines given in Q3B. A positive identification and quick turn around time help appropriate decision-making. Due to its inherent sensitivity, specificity, and speed, LC/MS is the technology of choice to identify unknown impurities in production batches. New impurities may come from many different sources and may not be related to the drug substance. Cross-contamination due to inadequate cleaning of the manufacturing apparatus, and contaminants due to leachables from packaging components are among the common causes for detection of new impurities. Identification of new impurities in a timely fashion has become a business necessity in defending the drug product franchise. Toxicological studies can only be designed rationally after the structures of impurities are known in order to address issues such as potential toxicity and side effects.

B. Investigation of the Cause for Drug Discoloration

According to the *FDA/Center for Drug Evaluation and Research* the leading causes for drug recall in fiscal year 2001 were:

- Deviations from current good manufacturing practices
- Subpotency
- Failure of stability data to support expiration date
- Failure of drug to dissolve properly
- Correctly labeled product in the wrong carton or package
- Strength of product incorrectly labeled
- Microbial contamination of nonsterile products
- Drug product marketed without an approved new or generic application
- Lack of assurance of sterility in production or testing of sterile drug products
- Discoloration
- Counterfeit dosage form.

Drug discoloration, listed as one of the top 10 causes for drug recall, manifests the stigma that pharmaceutical manufacturers have to tackle with. Discoloration refers to the change or shift in the color on the dosage form from the specified appearance. Discoloration may occur during drug products' transportation, distribution, and/or storage. It can also occur as a result of cross-contamination during the manufacturing process. Discoloration of a drug product is potentially a risk to public health. In cases where discoloration has occurred, the drug product needs to be recalled, and the chemical entity causing discoloration must be identified.

Discoloration, observed as isolated spots on the dosage form, tends to suggest a manufacturing problem such as inadequate cleaning of manufacturing equipment, leading to cross-contamination between different strengths of the same product line, or between different product lines. On the other hand, if discoloration is homogeneous throughout the dosage form, the API and/or excipients must have experienced chemical degradation so that chromophores are altered. Comparison of LC/UV, and/or LC/MS profiles between a normal dosage form with the discolored dosage form may uncover compound(s) responsible for the discoloration. Studies of discoloration, possibly due to the Maillard or browning reaction between lactose and the amine-containing diuretic hydrochlorothiazide, have been reported by Harmon et al.[96] The Maillard reaction may occur, and is always characterized by the formation of brown pigments, when carbonyl compounds are mixed with amines, amino acids, or proteins and are subjected to environmental factors such as high humidity and heat. Diuretic hydrochlorothiazide, after being heated at 60°C for 2 weeks in the presence of lactose and water, was found to form cyclic *N*-substituted glycosylamines as primary condensation products. Blaug and Huang[97] also reported a discoloration problem in a formulation containing dextroamphetamine sulfate and spray-dried lactose. The discoloration was enhanced by the presence of amines and by storage at elevated temperatures. The compound responsible for the discoloration was identified as dextroamphetamine-hydroxymethylfurfural, a reaction product between the active pharmaceutical ingredient and lactose.

Quinidine polygalacturonate (structure illustrated in Figure 26) has been formulated as tablets and is used as an antiarrhythmic agent. The tablet was found to develop a lightly beige color, different from its off-white color, after prolonged storage at ambient conditions. The discoloration was observed in the entire tablet. Chromatographic analysis of

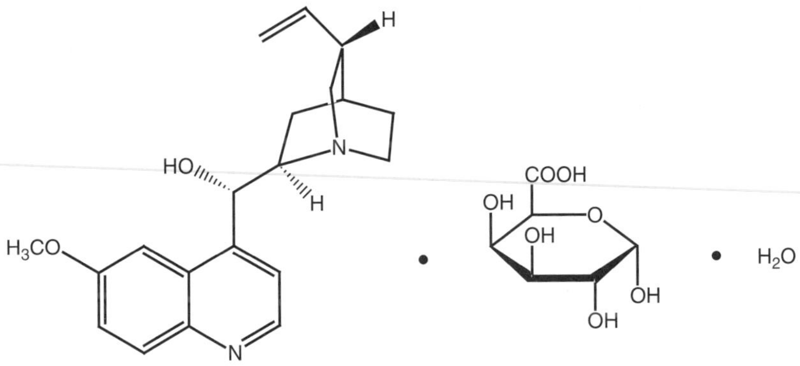

FIGURE 26 **Quinidine polygalacturonate.**

an extract of the discolored tablets uncovered two late eluting peaks, which were either absent or detected at significantly lower levels in the native tablets whose color was in conformance to specification. The late eluting peaks are shown in Figure 27 at RT of 24.2 and 30.4 min. Mass spectrometric analysis of these two peaks was conducted on a Finnigan MAT TSQ 7000 mass spectrometer equipped with an electrospray interface. Molecular mass for both peaks was determined as 324 Da, which is the same as that of quinidine itself. MS/MS results, shown in Figures 28(a) and (b), suggested that they could be hydroquinidinone and hydroquninone, respectively. These compounds are believed to be oxidation products of hydroquinidine and hydroquinine, which are also process impurities present at low levels in the drug substance. Figure 29 shows the structures for all four compounds. UV spectra comparing quinidine and the hydroqunidinone/hydroquninone are given in Figure 30. The formation of the keto-functionality at a carbon α to an aromatic system resulted in a bathochromic shift in UV absorbance, which is believed to have caused discoloration of the tablets.

C. Investigation of Counterfeit Drugs

Counterfeit drugs pose a serious threat to public health. According to the World Health Organization,[98] a counterfeit medicine is one

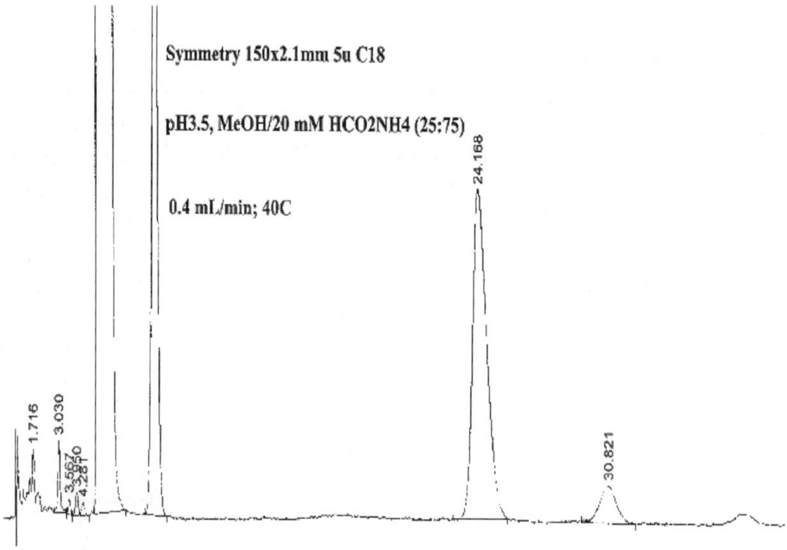

FIGURE 27 LC/UV chromatogram shows two late eluting peaks for the discolored tablets at retention time of 24.2 and 30.4 min.

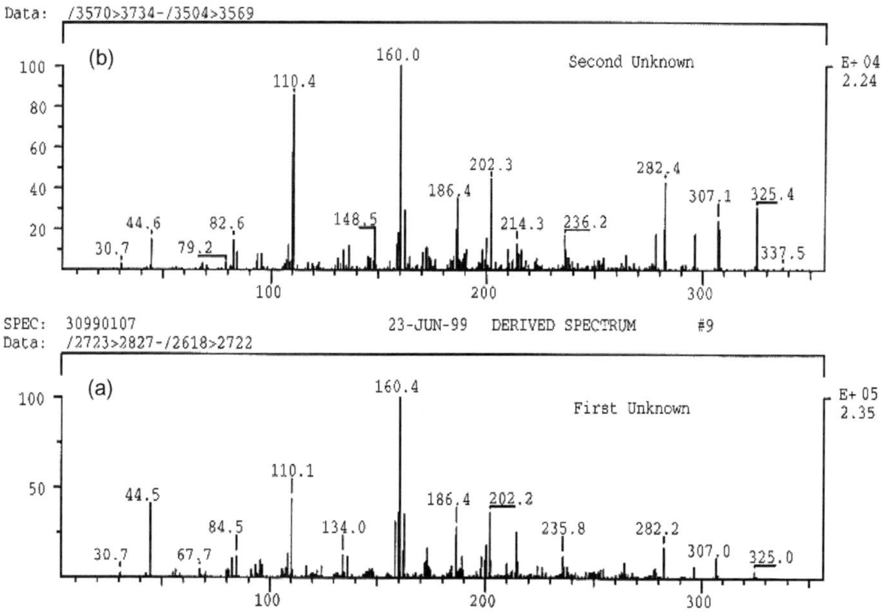

FIGURE 28 **MS/MS spectra of the late eluting peaks. Similar fragmentation patterns have been observed for both peaks, indicating that they are likely diastereomers.**

"which is deliberately and fraudulently mislabeled with respect to identity and/or source." Counterfeiting can apply to both branded and generic products and counterfeit products may include products with the correct ingredients or with the wrong ingredients, without any active ingredients, with insufficient active ingredients (sub-potent) or with fake packaging." The United States Federal Food, Drug and Cosmetic Act defines a counterfeit drug as "...a drug which, or the container or labeling of which, without authorization, bears the trademark, trade name, or other identifying mark, imprint, or device or any likeness thereof, of a drug manufacturer, processor, packer, or distributor other than the person or persons who in fact manufactured, processed, packed or distributed such drug, and which thereby falsely purports or is represented to be the product of, or to have been packed or distributed by, such other drug manufacturer, processor, packer, or, distributor."[99]

Although there is no universal definition of counterfeit drugs, they are however, deliberately and fraudulently mislabeled with intent to deceive and profit. Counterfeit drugs have rarely been efficacious but are often dangerous and detrimental to public health. Low potency or absence of API altogether will fail to provide the expected and desired therapeutic effect. Others may even have contaminants or adulterants that may be toxic. As the profits from selling a counterfeit drug can be

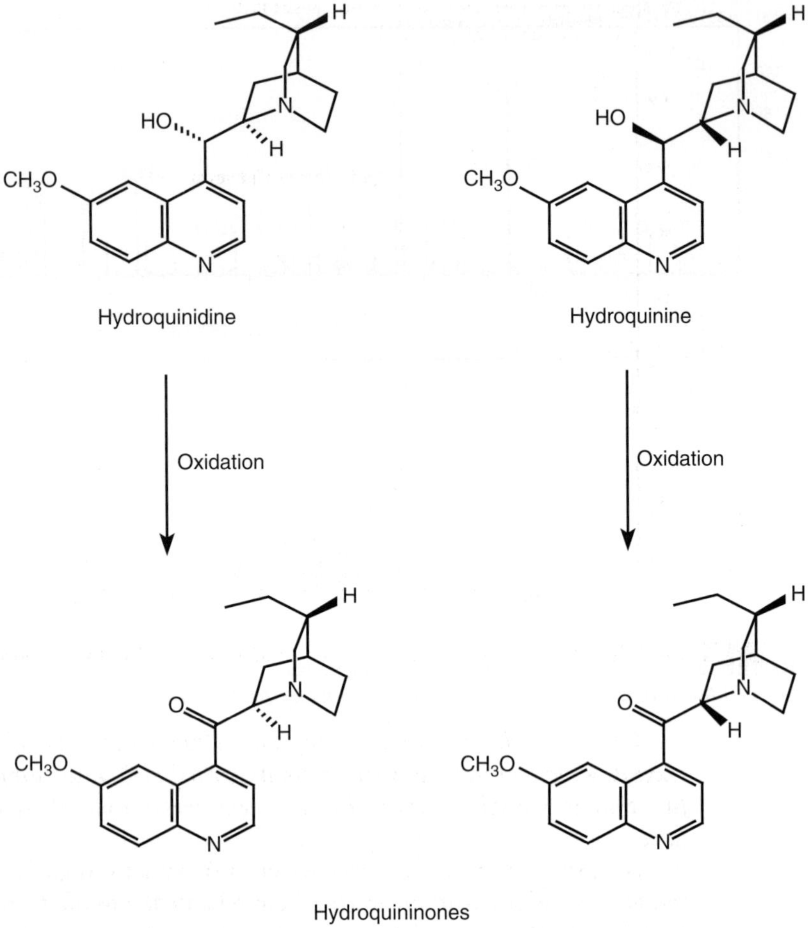

FIGURE 29 **Formation of hydroquinidinone/hydroquininone.**

as high as the sale of illegal narcotics, organized crime has become involved in the "counterfeit drug product industry". Harvey Bale, Director General of the International Federation of Pharmaceutical Manufacturers Associates (IFPMA), said that organized crime faces "less risk in moving into counterfeit medicine than illegal drugs." The World Health Organization and the Pharmaceutical Research & Manufacturers of America estimate that 8–10% of prescription drugs on the world market are counterfeit.[100] Some of the most recent drug recalls due to counterfeiting problem include the following:

Lipitor® recall in Spring 2003—millions of fake pills were smuggled into the United States from abroad. The Lipitor® case highlights the problem of increasing and increasingly sophisticated counterfeit drugs.

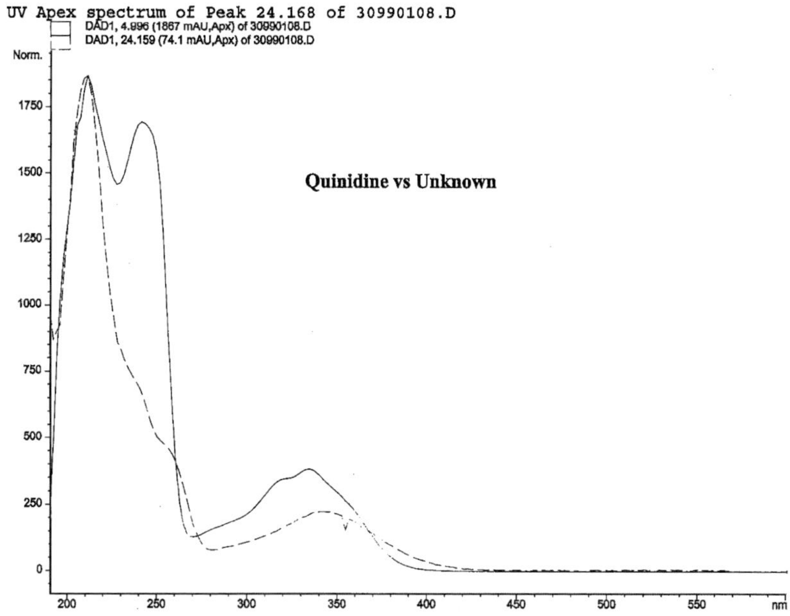

FIGURE 30 **Comparison of UV spectum of quinidine with that of hydroquininoness.**

Glaxo-SmithKline recalled in May 2002 three lots of Combivir® (Lamivudine/zidovudine, used in treatment of HIV infection) due to the fact that it actually contained a drug known as Ziagen® (abacavir sulfate).

Amgen, in May 2001, was forced to recall its Epogen® when counterfeit vials of the product were discovered in the market. Investigation revealed that the counterfeit drug contained only one-twentieth of the labeled strength.

Patients are not the only victims of counterfeit drugs. Pharmaceutical companies can lose income from counterfeit drug competition. Public confidence in the company can be undermined, leading it to stop buying the product and instead purchase a competitor's product even after the counterfeit has been destroyed. A fast identification of the counterfeit drug is in the interest of pharmaceutical companies and the protection of patients' rights.

In July 2002, a DEA regional office confiscated tablets that were counterfeits of a leading brand name prescription analgesic drug. The author's laboratory participated in the investigation. Unlike the identification of drug impurities and degradants, for which a body of knowledge such as the synthetic route, drug product formulation, reactivity of drug compound, etc. is available, the identification of a counterfeit drug molecule begins with no prior information. Identification of counterfeit drug molecules is akin to "looking for a needle in a haystack." There are hundreds of thousands of organic compounds in the ACS registries. Even drug-related entries in the Merck Index still stand at about 10 000. Unarguably, the generation of data using high-resolution mass spectrometry is the best choice in tackling such a challenging task.

Figure 31 shows the ESI/FTMS parent ion mass spectrum of the API in the counterfeit tablet. Inspection of the mass spectrum suggested that this compound contains less than nine carbon atoms because the "A+1" ion intensity is less than 10% of that of monoisotopic peak. The intensity of the "A+2" ion at *m/z* of 340.04810 was estimated at ~13% of the monoisotopic signal. This observation suggested that the counterfeit compound contained some "A+2" element(s) other than Cl and Br. The doublet peaks of the "A+1" isotopic signal at *m/z* of 399.04828 and *m/z* of 399.05638 were observed, with relative intensity at about 1 to 2. This important information led to the supposition that the unknown compound may contain a significant number of atoms from the "A+1" elements such as nitrogen, and/or sulfur. Other common elements in an organic compound may include fluorine and phosphorous. When all the elemental information was built into the search criteria, a list of the top 30 candidates was obtained and is displayed in Table 10. Our procedure for the rapid identification of the best-fit elemental composition can be summarized as follows:

- Build search criteria that include information such as: possible elements and the numbers of each element; maximum and minimum number of double bond equivalency (DBE); maximum tolerance of experimental error in the determined exact mass.
- Examine the DBE value for each candidate from the elemental search report. If a compound shows UV absorbance at a wavelength

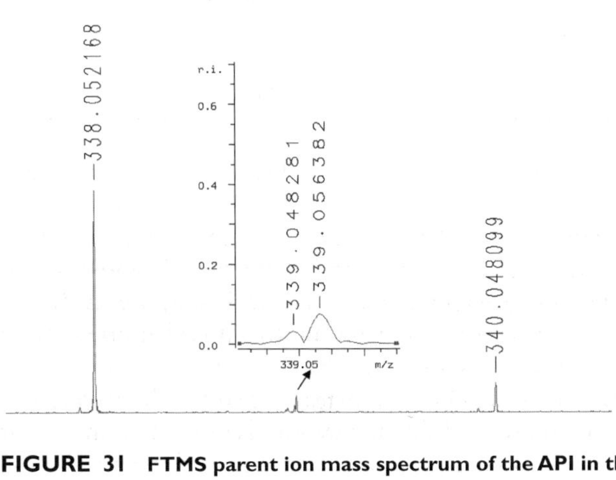

FIGURE 31 **FTMS parent ion mass spectrum of the API in the counterfeit tablet.**

TABLE 10 Possible Elemental Compositions: Mass Analysis for mass 338.0521680

#	C	H	O	N	S	P	F	Mass	DBE	error
1	7	22	4	3	2	2	0	338.0521470	−0.5	6.203e-08
2	3	12	5	6	0	1	5	338.0521465	−1.0	6.366e-08
3	2	6	0	13	0	1	5	338.0521412	4.5	7.924e-08
4	0	15	2	10	1	2	3	338.0521995	−2.0	9.325e-08
5	10	25	1	0	3	2	1	338.0521301	−1.0	1.120e-07
6	6	15	2	3	1	1	6	338.0521296	−1.5	1.136e-07
7	8	16	2	7	3	0	0	338.0522113	4.5	1.280e-07
8	9	22	7	0	3	0	0	338.0522165	−1.0	1.436e-07
9	7	16	10	2	0	1	1	338.0521109	1.0	1.690e-07
10	6	10	5	9	0	1	1	338.0521056	6.5	1.846e-07
11	9	13	2	6	1	1	2	338.0520887	6.0	2.345e-07
12	5	22	1	6	1	4	0	338.0520775	0.0	2.677e-07
13	1	9	0	14	2	0	3	338.0520638	3.0	2.833e-07
14	8	6	0	6	0	0	8	338.0520713	5.0	2.861e-07
15	2	15	5	7	2	0	3	338.0522690	−2.5	2.989e-07
16	0	12	5	13	2	0	0	338.0520284	1.5	4.131e-07
17	8	19	4	2	0	3	2	338.0520245	1.0	4.246e-07
18	3	15	2	10	3	0	1	338.0520115	1.0	4.630e-07
19	7	7	0	10	0	1	4	338.0523410	8.0	5.118e-07
20	8	13	5	3	0	1	4	338.0523463	2.5	5.274e-07
21	4	9	0	7	0	1	8	338.0523819	0.5	6.327e-07
22	5	16	2	7	1	2	2	338.0523993	1.5	6.843e-07
23	1	7	0	17	0	2	1	338.0524110	7.5	7.187e-07
24	2	13	5	10	0	2	1	338.0524162	2.0	7.343e-07
25	2	15	10	5	0	1	2	338.0519111	−2.5	7.600e-07
26	1	9	5	12	0	1	2	338.0519058	3.0	7.756e-07
27	0	3	0	19	0	1	2	338.0519005	8.5	7.912e-07
28	4	12	2	9	1	1	3	338.0518889	2.5	8.256e-07
29	10	19	2	1	3	0	3	338.0524520	0.5	8.400e-07
30	6	10	0	11	2	0	2	338.0524636	6.5	8.744e-07

higher than 230 nm, then it must bear a reasonable number of DBE. In this case, the compound showed a UV absorbance maximum at 260 nm. Therefore, out of the top 10 candidates, 1, 2, 4, 5, 6, 8, and 9 can be eliminated from consideration because these candidates do not meet the DBE requirement.

- Of the remaining candidates, from top to bottom, generate their theoretical isotopic distributions and compare them to the experimental one. Theoretical isotopic distribution of candidate 3 was an obvious mismatch because much lower intensities for both the "A+1" and the "A+2" signals were observed. On the other hand, the isotopic fingerprinting of candidate 7 (Figure 32) agreed very well with the pattern seen in Figure 31. Therefore candidate 7 was chosen as the candidate with the best-fit elemental composition.

- The rest of the candidates in the table can be eliminated when the above-mentioned procedure is repeated.

Since the monoisotopic signal is the protonated ion, the elemental composition for the API in the counterfeit tablets was therefore determined as $C_8H_{15}O_2N_7S_3$. A computerized search of The Merck Index using the elemental formula suggested the unknown compound to be Famotidine, an active pharmaceutical ingredient widely used in over-the-counter antiulcerative medicine.

Daughter ion mass spectrum of the compound is shown in Figure 33. A dominant daughter ion was observed at *m/z* of 259.07757. Observation of this ion conformed to the likely fragmentation pattern expected for the Famotidine molecule. Scheme 1 illustrates the fragmentation pathway leading to the formation of this dominant daughter ion.

Confirmative data were obtained by acquiring FTMS and FTMS/MS spectra for a solution made from an authentic commercial product containing Famotidine. Both spectra were identical to those acquired for the counterfeit tablets.

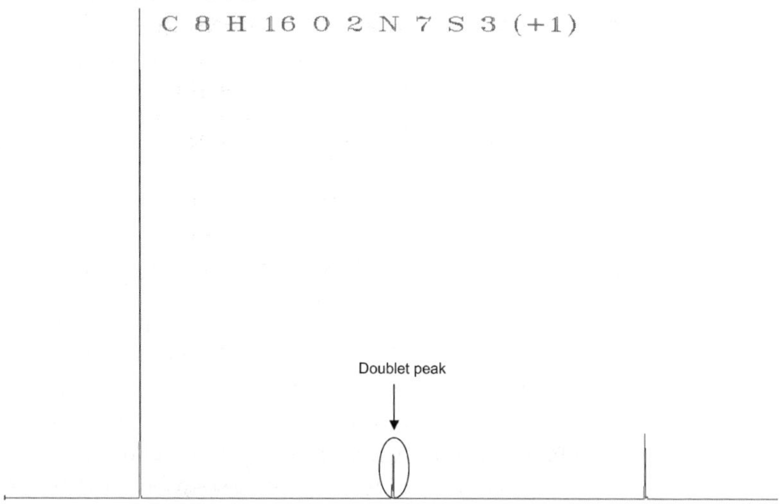

FIGURE 32 **Theoretical isotopic distribution of elemental composition candidate #7.**

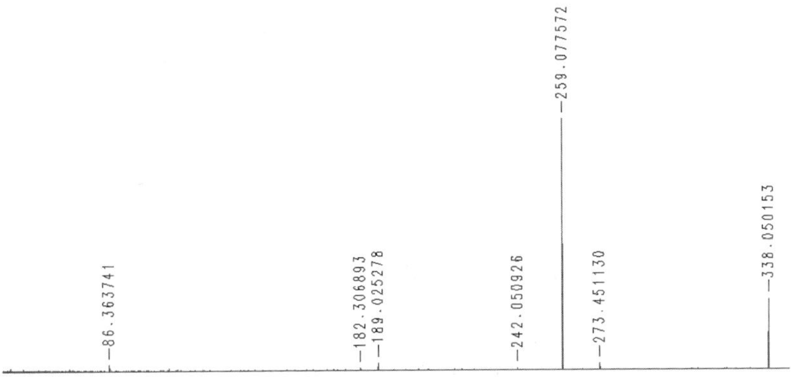

FIGURE 33 FTMS/MS daughter ion mass spectrum of the API in the counterfeit tablets.

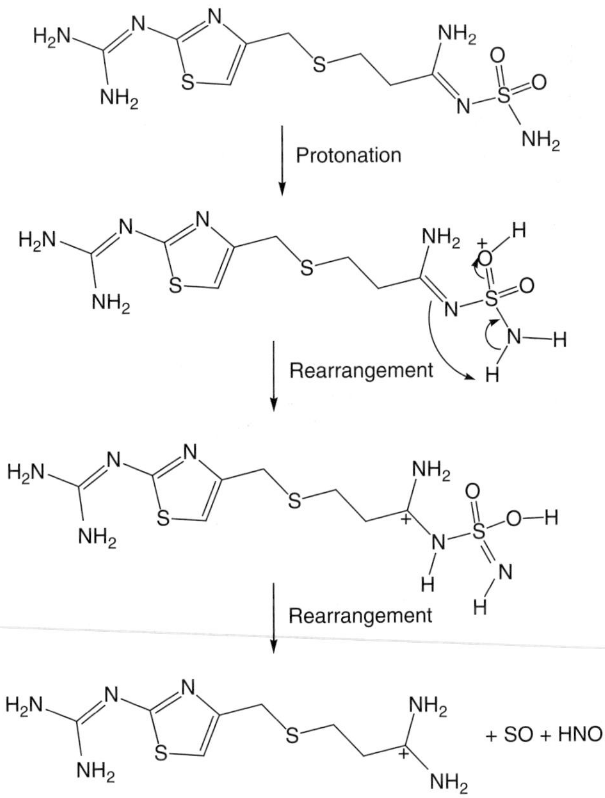

SCHEME 1 Fragmentation mechanism leading to the formation of *m/z* of 259.07757.

D. Patent Protection

Pharmaceutical research and development is a very time-consuming and costly endeavor. The average time required to bring a new molecular entity to market is about 10 years and may cost up to 700 million dollars. Protection of intellectual properties through patent protection is taking high priority in the pharmaceutical industry. Due to its high sensitivity and specificity, LC/MS is a useful tool in the generation of structural information for patent protection of proprietary chemical processes. Drug impurities, especially those specific to the synthetic route can be used as chemical markers to determine whether the synthesis process is being infringed. Almudaris et al.[101] reported an LC-thermospray-MS method used for the identification of three synthetic-route indicative impurities at levels of 50 to 100 ppb (w/w) for the purpose of patent protection. The impurities were extracted and pre-concentrated by normal-phase LC procedures, followed by an on-line structural study by reversed-phase HPLC-TS-MS.

X. CONCLUSION

LC coupling with MS has emerged as one of the most important technologies in pharmaceutical analysis. It is used to ensure the quality and safety of pharmaceutical products. MS can detect molecules in the positive and negative ionization modes, in which molecules are transformed, primarily, into protonated or deprotonated pseudo molecular ions. Due to its inherent analytical capabilities, the technique is widely used in applications ranging from monitoring, identifying, and characterizing drug impurities to investigating deficit mass balance, drug product discoloration, and drug product counterfeiting. The obvious benefit of using LC/MS is the opportunity it affords for rapid identification of unknown compounds by generating compound specific information such as molecular weight and elemental composition. Substructural information can also be generated from CID experiments. Many other benefits can also be realized, such as increased gain in confidence and decreased time required for HPLC method development. Using a "generic HPLC/MS method," data can be collected in a consistent manner throughout the development life cycle of a pharmaceutical product.

However, LC/MS is not a panacea. In mass spectrometry, molecules must exist as ionic species in order to be detected. The ionization efficiency can vary over many orders of magnitude depending on the structure of the compounds studied. For nonpolar and neutral drug compounds, an ESI or APCI interface may not be sensitive enough to allow meaningful structure determination. Although the assignment of various signals in an ESI/MS mass spectrum is usually straightforward, artifacts can arise from electrochemical reactions during the electrospray ionization process. Thermally induced degradation is also a common phenomenon with both ESI and APCI interfaces due to the heat input to aid in the evaporation of solvents. The presence of artifacts will obviously complicate accurate determination of the molecular weight for trace level drug impurities. Knowledge of the sample in terms of its formulation, packaging materials, synthetic route, chemical activity of drug compound, etc. is always beneficial in the identification of unknown impurities.

ACKNOWLEDGMENTS

The author is grateful to Raphael Ornaf and Michael Dong of Purdue Pharma, and Connie Ye of Novartis Pharmaceuticals Corp. for their critical review and insightful comments. The author would like to acknowledge David Wu of Purdue Pharma for the data used in the discussion of drug discoloration. The author also thanks Helen Yun of Purdue Pharma for her help in literature research, and Vladimir Binshtock of Waters Corp. for providing much needed technical information.

REFERENCES

1. Yamashita, M. and Fenn, J. B., *J. Phys. Chem.*, 88:4451, 1984.
2. Tanaka, K., Ido, Y., Akita, S., Yoshida, Y. and Yoshida, T., *Second Japan–China Symposium on Mass Spectrometry*, Abstract, pp. 185–188, 1987.
3. Barber, M., Bardoli, R. S., Sedgwick, R. D. and Tyler, A. H., *J. Chem. Soc., Chem. Commun.*, 7:325–327, 1981.
4. Blakely, C. R. and Vestal, M. L., *Anal. Chem.*, 55:750, 1983.
5. Tal'roze, V. L., Skurat, V. E. and Karpov, G. V., *J. Phys. Chem. (Moscow)*, 43:241, 1969.
6. Niessen, W. M. A., Tjaden, U. R. and Greef, J. Van der, *J. Chromatogr.*, 554:3, 1991.
7. Niessen, W. M. A. and Tinke, A. P., *J. Chromatogr. A*, 703:37, 1995.
8. Niessen, W. M. A., *J. Chromatogr. A*, 794:407, 1998.
9. Scott, R. P. W., Scott, C. G., Munroe, M. and Hess, J., *J. Chromatogr.*, 99:395, 1974.
10. McFadden, W. H., Schwartz, H. L. and Evans, S., *J. Chromatogr.*, 122:389, 1976.
11. Noeller, H. G., Olaschegg, H. D. and Wechung, R., *German Pat.*, 2,837,799 1980.
12. Hardin, E. D., Fan, T. P. and Vestal, M. L., *Anal. Chem.*, 56:2, 1984.
13. Krien, P., Devant, G. and Hardy, M., *J. Chromatogr.*, 251:129, 1982.
14. Tijssen, R., Bleumer, J. P. A., Smit, A. L. C. and van Kreveld, M. E., *J. Chromatogr.*, 218:137, 1981.
15. Arpino, P. J., Baldwin, M. A. and McLafferty, F. W., *Biomed Mass Spetrom.*, 1:80, 1974.
16. Sugnaux, F. R., Skrabalak, D. S. and Henion, J. D., *J. Chromatogr.*, 264:357, 1983.
17. Iribarne, J. V. and Thomson, B. A., *J. Chem. Phys.*, 64:2287, 1976.
18. Caprioli, R. M., Fan, T. and Cotrell, J. S., *Anal Chem.*, 58:2949, 1986.
19. Caprioli, R. M., *Anal. Chem.*, 62:477A, 1990.
20. Willoughby, R. C., Browner, R. F., *Anal. Chem.*, 56:2626, 1984.

21. Winkler, P. C., Perkins, D. D., Williams, D. K. and Browner, R. F., *Anal. Chem.*, 60: 489, 1988.
22. Whitehouse, C. M., Dreyer, R. N., Yamashita, M. and Fenn, J. B., *Anal. Chem.*, 57:675, 1985.
23. Bruins, A. P., *J. Chromatogr. A*, 794:345, 1998.
24. Niessen, W. M. A., *J. Chromatogr. A*, 856:179, 1999.
25. Lord Rayleigh, *Phil. Mag.*, 14:184–186, 1882.
26. Liberato d. J., Fenselau, C. C., Vestal, M. L. and Yergey, A. L., *Anal. Chem.*, 55:1741, 1983.
27. Blakney C. R., Carmody, J. J. and Vestal, M. L., *Anal. Chem.*, 52:1636, 1980.
28. Thomson, B. A., Iribarne, J. V., and Dziedic, P. J., *Anal. Chem.*, 54:2219, 1982.
29. Thomson, B. A., Danylewych-May L. and Ngo, A., *Proceeding of the 31st Annual Conference on Mass Spectrometey Allied Topics*, Washington, DC, pp. 65–66, 1983.
30. Thomson, B. A. and Danylewych-May L., *Proceeding of the 31st Annual Conference on Mass Spectrometey Allied Topics*, Washington, DC, pp. 852–853, 1983.
31. Niessen, W. M. A., *Liquid Chromatography–Mass Spectrometry*, 2nd ed., Marcel Dekker, Inc., New York, 1999.
32. Brunnee, C., *Int J. Mass Spectrom. Ion Proc.*, 76:125, 1987.
33. Hayati, I., Bailey, A. I. and Tadros T. F., *Nature,* 319:41, 1986.
34. Tyczkowska, K. L., Voyksner, R. D. and Aronson, A. L., *J. Chromatogr.,* 594:195, 1992.
35. Hopfgartner, G., Bean, K., Henion, J. D. and Henry, R. A., *J. Chromatogr.,* 647:51, 1993.
36. Bruins, A. P., *J. Chromatogr. A,* 794:345, 1998.
37. Banks, J. F., *J. Chromatogr. A*, 743:99, 1996.
38. Abian, J, Oosterkamp, A. J. and Gelpí, E., *J. Mass Spectrom.*, 34:244, 1999.
39. Smith, R. D., Loo, J. A., Edmonds, C. G., Barinaga, C. J. and Udseth, H., *Anal. Chem.*, 62:882, 1990.
40. Jemal, M. and Hawthorne, D. J., *Rapid Commun. Mass Spectrom.*, 13:61, 1999.
41. Law, B. and Chan, P. F., *J. Pharm. Biomed. Anal.*, 9:271, 1991.
42. Tang, L. and Kebarle, P., *Anal Chem.*, 63:2709, 1991.
43. Tang, L. and Kebarle, P., *Anal Chem.*, 65:3654, 1993.
44. Kebarle, P. and Tang L., *Anal Chem.*, 65: 972A, 1993.
45. Zhou, S. and Hamburger, M., *Rapid Commun. Mass Spectrom.*, 9:1516, 1995.
46. Straub, R. F. and Voyksner, R. D., *J. Am. Soc. Mass Spectrom.,* 4:289, 1993.
47. Mansoori, B. A., Volmer, D. A. and Boyd, R. K., *Rapid Commun. Mass Spectrom.*, 11:1120, 1997.
48. Chiron, S., Papilloud, S., Haerdi, W. and Barceló, D., *Anal. Chem.,* 67:1637, 1995.
49. Apffel, A., Fischer, S., Goldberg, G., Goodley, P. C. and Kuhlmann, F. E., *J. Chromatogr. A*, 712:177, 1995.
50. Kamel, A. M., Brown, P. R. and Munson, B., *Anal. Chem.,* 71:5481, 1999.
51. Kamel, A. M., Brown, P. R. and Munson, B., *Anal. Chem.,* 71:968, 1999.
52. Temesi, D. and Law, B., *LC-GC*, 17(7):626, 1999.
53. Verheij, E. R., Reeuwijk, H. J. E. M., Niessen, W. M. A., Tjaden, U. R., van der Greef J. and LaVos, G. F., *Biomed. Environ. Mass Spectrom.,* 16:393, 1989.
54. Ermer, J. and Kibat, P. G., *Pharm Sci. Technol. Today*, 1:76, 1998.
55. Greef, J. Van der, Niessen, W. M. A. and Tjaden, U. R., *J. Pharm. Biomed. Anal.*, 6:565, 1988.
56. Ermer, J., *J. Pharm. and Biomed. Anal.,* 18:707, 1998.
57. Asakawa, N., Ohe, H., Tsuno, M., Nezu, Y., Yoshida, Y. and Sato, T., *J. Chromatogr.*, 541:231, 1991.
58. Ayrton, J., Plumb, R. S., Leavens, W. J., Mallett, D. N., Dickins, M. and Dear, G. J., *Rapid Commun. Mass Spectrom.*, 12:217, 1998.
59. Ayrton, J., Dear, G. J., Leavens, W. J., Mallet, D. N. and Plumb, R. S., *J. Chromatogr. B, Biomed. Sci. Appl.*, 709:243, 1998.

60. Lee, M. S., *LC/MS Applications in Drug Development*, John Wiley and Sons, Inc., New York, pp. 43–46, 2002.
61. Rourick, R. A., Volk, K. J., Klohr, S. E., Spears, T., Kerns, E. H., Lee, M. S., *J. Pharma. Biomed Anal.*, 14:1743, 1996.
62. Xu, X., Bartlett, M. G. and Stewart, J. T., *J. Pharma. Biomed. Anal.*, 26:367, 2001.
63. Nicolas, E. C., Scholz, T. H., *J. Pharma. Biomed. Anal.*, 16: 825, 1998.
64. Volk, K. J., Klohr, S. E., Rourick, R. A., Kerns, E. H. and Lee, M. S., *J. Pharm. Biomed. Anal.*, 14:1663, 1996.
65. Bryant, D. K., Kingswood, M. D. and Belenguer, A., *J. Chromatogr. A*, 721:41, 1996.
66. Ginsburg, E. J., Stephens, D. A., West, P. R., Buko, A. M., Robinson, D. H., Li, L. C. and Bommireddi, A. R., *J. Pharm. Sci.*, 89(6):766, 2000.
67. Olsen, B. A., Baertschi, S. W. and Riggin, R. M., *J. Chromatogr.*, 648:165, 1993.
68. Denk, O. M., Skellern, G. G. and Watson, D. G., *J. Pharm. Pharmacol.*, 54:87, 2002.
69. Sheldon, E. M., *J. Pharma. Biomed. Anal.*, 31:1153, 2003.
70. Görög, S., Babják, M., Balogh, G., Brlik, J., Dravecz, F., Gazdag, M., Horváth, P., Laukó, A. and Varga, K., *J. Pharma. Biomed. Anal.*, 18:511, 1998.
71. Nuijen, B., Rodrigues, I. M., Noain, C. P., Floriano, P., Manada, C., Bouma, M., den Bosch, J. J. K.-v, Bult, A. and Beijnen, J. H., *J. Liq. Chrom. & Rel. Technol.*, 24(20):3119, 2001.
72. Arzamastsev, A. P., Luttseva, T. Y., Klyuev, N. A. and Sadchikova, N. P., *Pharma. Chem. J.*, 33(10):568, 1999.
73. Matchett, M. W., Berberich, D. W. and Johnson, J. E., *J. Chromatogr. A*, 927:97, 2001.
74. Fuzzati, N., Gabetta, B., Strepponi, I. and Villa, F., *J. Chromatogr. A*, 926:187, 2001.
75. Wolfender, J.-L., Terreaux, C. and Hostettmann, K., *Pharma. Biol.*, 38:Suppl. 41, 2001.
76. Strohschein, S., Rentel, C., Lacker, T., Bayer, E. and Albert, K., *Anal. Chem.*, 71:1780, 1999.
77. Hamdan, M., *Process Control Qual.*, 10:113, 1997.
78. Ermer, J. and Vogel, M., *Biomed. Chromatogr.*, 14:373, 2000.
79. Niessen, W. M. A., *Chimia*, 53(10):478, 1999.
80. Wu, Y., *Biomed. Chromatogr.*, 14:384, 2000.
81. Yang, L., Wu, N. and Rudewicz, P. J., *J. Chromatogr. A*, 926:43, 2001.
82. Qin, X., Ip, D., Chang, K. H.-C., Dradransky, P. M., Brooks, M. A. and Sakuma, T., *J. Pharm. Biomed Anal.*, 12(2):221, 1994.
83. Neubauer, G. and Anderegg, R. J., *Anal. Chem.*, 66(7):1056, 1994.
84. Smith, R. D., Pasa-Tolic, L., Lipton, M. S., Jensen, P. K., Anderson, G. A., Shen, Y., Conrads, T. P., Udesth, H. R., Harkewicz, R., Belov, M. E., Masselon, C. and Veenstra, T. D., *Electrophoresis,* 22:1652, 2001.
85. Eckers, C., Haskins, N. and Langridge, J., *Rapid Commun. Mass Spectrom.*, 11 1916, 1997.
86. Haskins, N., Eckers, C, Organ, A. J., Dunk, M. F. and Winder, B. E., *Rapid Commun. Mass Spectrom.*, 9:1027, 1995.
87. Lee, M. J., Monté, S., Sanderson, J. and Haskins, N. J., *Rapid Commun. Mass Spectrom.*, 13: 216, 1999.
88. Zhao, Z., Wang, Q., Tsai, E. W., Qin, X.-Z and Ip, D., *J. Pharma. Biomed. Anal.*, 20:129, 1999.
89. Lee, M. S., Kerns, E. H., Hail, M. E., Liu, J. and Volk, K. J., *LC-GC*, 15(6):542, 1997.
90. Karlsson, K. E., *J. Chromatogr.*, 647:31, 1993.
91. Gard, E., Green, M. K., Bregar, J. and Lebrilla, C. B., *J. Am. Soc. Mass Spectrom.* 5:623, 1994.
92. Chen, X., Smith, L. M. and Bradbury, E. M., *Anal. Chem.* 72:1134, 2000.
93. Siegel, M. M., *Anal. Chem.* 60:2090, 1998.
94. Ohashi, N., Furuuchi, S. and Yoshikawa, M., *J.Pharm. Biomed. Anal.*, 18:325, 1998.

95. Terreaux, C., Wang, Q., Ioset, J., Ndjoko, K., Grimminger, W. and Hostettmann, K., *Plata Med* , 68:349, 2002.
96. Harmon, P. A., Yin, W., Bowen, W. E., Tyrrell, R. J. and Reed, R. A., *J. Pharma. Sci.*, 89(7):920, 2000.
97. Blaug, S. M. and Huang W., *J. Pharma Sci.*, 61(11):1771, 1972.
98. Report of the WHO Expert Committie on Specifications for Pharmaceutical Preparations, Technical Report Series 863, WHO, Geneva, 1996.
99. The Federal Food, Drug and Cosmetic Act, Subchapter II: reproduced in the United States Pharmacopoeia, USP23, 1995, United States Pharmacopoeial Convention, Inc., Rockville, MD.
100. The Gold Sheet, Vol. 37, No. 3, March 2003.
101. Almudaris A., Ashton D. S., Ray A. and Valko K., *J. Chromatogr. A*, 689:31, 1995.

20 APPLICATION OF LC-NMR IN PHARMACEUTICAL ANALYSIS

MIENG-HUA WANN

Purdue Pharma, Ardsley, NY

ABSTRACT

Hyphenated analytical techniques such as LC-MS, which combines liquid chromatography and mass spectrometry, are well-developed laboratory tools now widely used in the pharmaceutical industry. However, in most cases MS alone is insufficient for complete structural elucidation of unknown compounds. Traditionally nuclear magnetic resonance (NMR) experiments are performed on more or less pure samples, in which the signals of a single component dominate. Therefore, the structural analysis of individual components of complex mixtures is normally time-consuming and less economic. The combination of chromatographic separation techniques with NMR spectroscopy offers advantages for the on-line separation and structural elucidation of unknown compounds. Mixtures such as crude reaction mixtures in drug discovery can be analyzed without prior separation. Experiments in combining an HPLC with NMR on the study of mixtures were introduced to the scientific community in the early 1980s. However, LC-NMR was not widely practical due to its low sensitivity, approximately six orders in magnitude inferior to that of MS. Another challenge comes from the measurement of proton signals in mobile phase. However, recent developments in higher magnetic field strength and electronics that improve the sensitivity of probes, together with advanced solvent suppression techniques, have made LC-NMR measurement practical.

During the last decade, LC-NMR has been fully commercialized and its application in both academia and industry has been growing rapidly. Examples like LC-NMR-MS, a more powerful hyphenated technique combining LC-MS and LC-NMR, provide complementary information about structures simultaneously and offer a method for more accurate and rapid structural analysis. In this chapter, some routine subjects such as the theory and the design of flow probes have been avoided as there is much literature published in the 1990s which has covered these topics comprehensively. The emphasis is therefore on describing the experimental design, the practical applications, and the recent developments in technology. With all the applications to date, LC-NMR spectroscopy is still a relatively insensitive technique due to the poor mass sensitivity of the NMR detection system. To this end, several other hyphenated NMR techniques have been developed to enhance the sensitivity of this technique. LC-SPE-NMR dramatically increases the sensitivity up to a factor of four by utilizing a solid phase extraction device after the LC column. Capillary LC-NMR also significantly lowers the detection limit to a low nanogram range through integration of capillary LC with NMR detection. Other breakthroughs such as cryo-LC-probe technology combine the advantages of sample flow and the enhanced sensitivity from a cryogenically cooled NMR probe. Many of the latest developments in LC-NMR technology will also be covered in this chapter.

I. INTRODUCTION

Traditionally, the characterization of a multi-component sample required the isolation of the pure component followed by off-line spectroscopic analysis. Liquid chromatography–mass spectrometry (LC-MS) significantly simplified this procedure and became a well-established technique for determining the molecular structure of components in mixture.[1,2] However, since the ionization process used for most LC-MS applications is relatively "soft", little structural information can be obtained without further MS-MS analysis.[3] In addition, if there are coeluting peaks present, only the most energetically favorable one is ionized. In the case of isomers, some can be elucidated from their fragmentation pattern, but others may isomerize to a common intermediate upon ionization. It could be challenging to study stereochemistry and further identify the geometric isomers by MS.[4]

On the other hand, nuclear magnetic resonance (NMR) spectroscopy is one of the most powerful tools for the structure elucidation of organic compounds. However, to solve the molecular structure of a novel substance by NMR spectroscopy alone is often time-consuming (when compared to MS). Besides, the identification of components in a complex mixture usually requires the separation and/or isolation of the components of interest prior to NMR analysis. Therefore multiple preparatory chromatographic separations may be needed to provide a sufficient quantity of sample for NMR experiments in a regular NMR tube. If the component is labile, it may decompose or rearrange during the isolation process.

The hyphenated technology, LC-NMR, combines a powerful separation technique with a detector that offers a wealth of structural and stereochemical information. It eliminates the need for off-line isolation, and the NMR data may be acquired on many components within a single chromatographic run. This technique is particularly powerful for the analysis of labile compounds, which cannot be easily isolated. However, LC-NMR is an inherently insensitive technique, with almost six orders of magnitude lower sensitivity than that of mass spectrometry. Recent technical improvements in NMR probe design and higher field magnets, coupled with advanced solvent suppression techniques, have dramatically increased instrumental sensitivity and spectral resolution, making this hyphenated technique a more practical problem-solving tool.[5–10]

II. INSTRUMENTATION AND DESIGN

A complete LC-NMR setup with its various parts is shown in the schematic diagram in Figure 1. In most of the analytical laboratories, the chromatography unit comprises an injection device, an LC pump, a column oven, and an UV detector. The LC system is located at a distance of 1.5 to 2.0 m from an unshielded magnet. The flow of the sample is guided from the detector into the LC-NMR interface, which is normally equipped with additional loops for the intermediary storage of selected LC peaks. The flow from the LC-NMR interface is then directed into the flow cell of NMR probehead. From there, a stopped-flow NMR acquisition of peaks of an LC separation can be performed, triggered by the detection of the UV signal from the detector. Following the completion of acquisition, the flow is routed from the probehead to a fraction collector for recovery or to a waste receptacle.

In the stopped-flow scanning mode, the LC-NMR interface controls the sequential transfer of LC peaks. Following on-line UV detection, the selected LC peak is transferred to the NMR flow cell and the flow is stopped when the peak maximum is exactly in the center of the NMR detection cell. After the NMR acquisition is completed, the LC pump is started again to bring in the next peak for NMR analysis. During every stop, the LC gradient conditions and the detector trace are halted and then resumed with little distortion after the NMR experiment. Bruker's LC-NMR system also features the storage loops for the intermediary storage of LC peaks, which allows uninterrupted chromatography. The peak storage can be performed automatically in the auto-detection mode. Following the LC run, the loop contents can be transferred to the NMR flow cell for NMR experiment automatically.

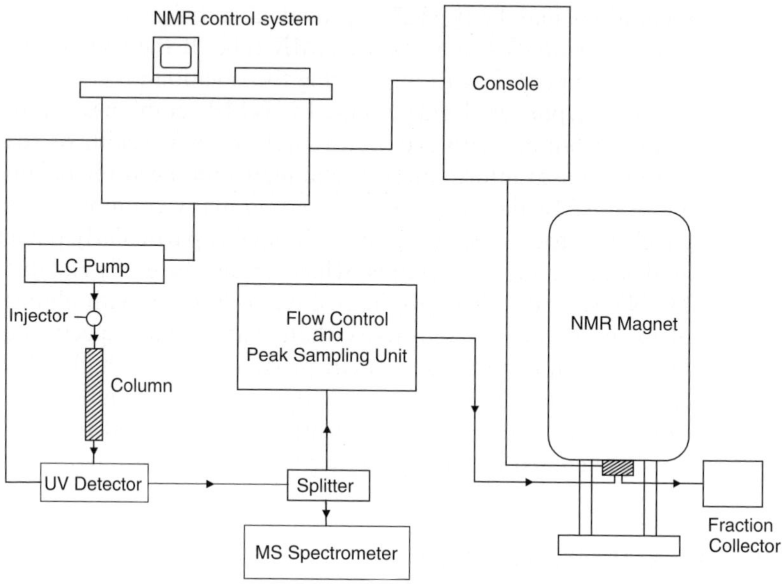

FIGURE 1 A schematic diagram of a complete LC-NMR setup.

LC-NMR can also be run in an "on-flow" mode, where NMR acquisition is started at the chromatography injection and randomly divided into a series of time slices. However on-flow NMR measurement not only suffers from low sensitivity, but LC gradient elution also causes problems on account of the drift of signals relative to the signal lock, where the presaturation frequency of each time slice needs to be defined in a separate run.

A. LC-NMR Probeheads

One of the major prerequisites for the development of LC-NMR is the continuous-flow probe, which contains a dedicated flow cell with a directly fixed double-saddle Helmholtz coil that is centered around the vertically oriented flow cell (Figure 2). The whole setting is centered in the glass dewar of a conventional probe body where a thermocouple is inserted to allow temperature-controlled experiments to be performed. The important parameters are the diameter of the flow cell and its volume. Currently, the inner diameter of the flow cell is available at 2, 3, and 4 mm, which result in active volumes of 60, 120, and 180 μL, respectively. The size of the flow cell should be chosen according to the expected sample quantities and LC peak volumes, the amount of back mixing allowed, and the field of the spectrometer. Generally, the 3 mm probeheads are the better compromise for good sensitivity and low back mixing for higher proton frequency at 400 MHz and above.

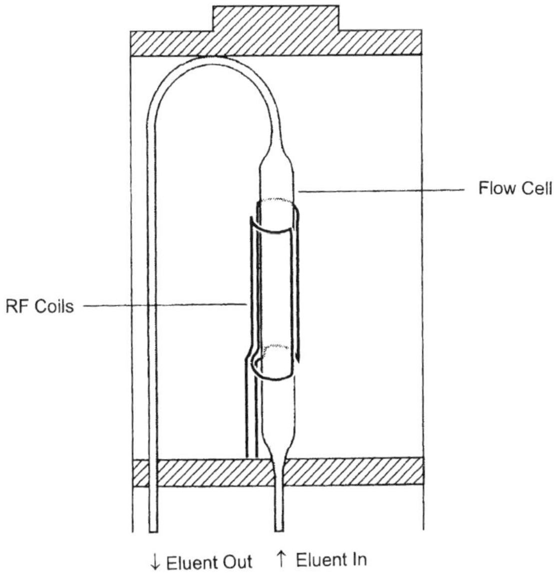

FIGURE 2 **Schematic representation of an LC-NMR probehead.**

The most commonly used probehead is the inverse dual $^{1}H/^{13}C$ probehead, which contains an additional coaxial coil that is tuned to ^{13}C resonance frequency, surrounding the ^{1}H detection coil for $^{1}H/^{13}C$ heteronuclear experiments. A typical NMR resolution with this probe is shown in Figure 3, where the line width at half peak height of chloroform is 8–9 Hz, which allows determination coupling constants up to 1Hz in LC-NMR spectra.

III. PRACTICAL APPLICATIONS

Undoubtedly high-resolution NMR spectroscopy is one of the most powerful analytical tools for the structure elucidation of organic molecules. The coupling of HPLC and NMR spectroscopy recently has further extended its utility in pharmaceutical analysis and other fields. This section will summarize the applications of LC-NMR to studies of pharmaceutical interest.

A. Impurities in Drug Substances

According to the ICH guidelines, any impurity greater than 0.1% of UV intensity by HPLC has to be characterized and identified. Traditionally, the individual impurities have to be isolated by preparative LC, and then subjected to NMR structure elucidation. The disadvantages

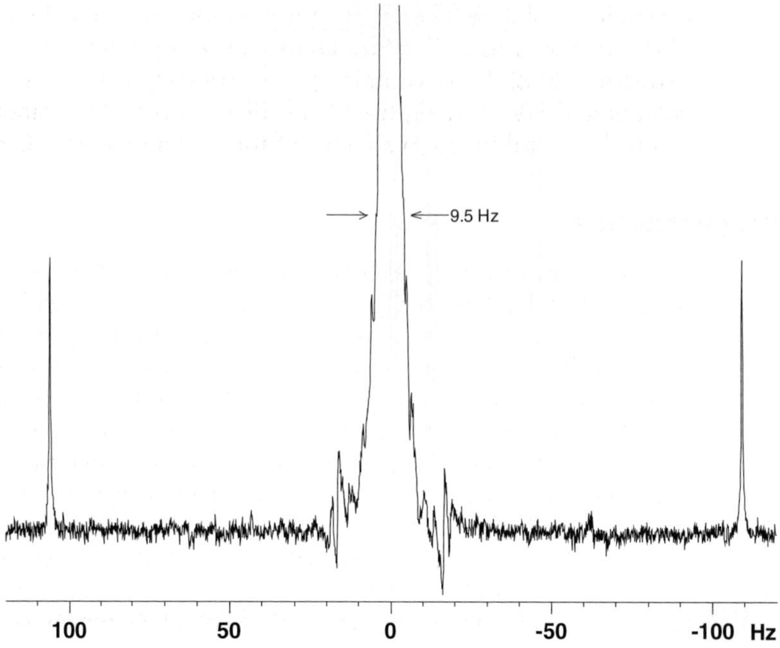

FIGURE 3 **Proton NMR line shape of chloroform in acetone-d6, measured with a 4-mm LC probe on a 600-MHz NMR.**

of this approach include time/labor intensity, and possible degradation during purification of collected impurities. LC-NMR provides a rapid online approach for efficient identification of multiple drug impurities during pharmaceutical development.

LC-NMR is a powerful technique especially for the analysis of compounds which cannot be isolated by preparative HPLC. For the identification of impurities in API from unpublished work from this laboratory, impurities at a 1% level from drug substances can be identified on a 600 MHz NMR spectrometer using the stop-flow mode.

B. Degradants in Dosage Formulations

The identification of degradants from drug products is very important at the formulation development stage. Abnormally high levels of degradants encountered during stability studies may lead to the rejection of a drug candidate or to the modification of the drug's molecular structure and/or the change of formulations.

For LC peaks at 0.1% level or less, a considerably long acquisition time may be required to achieve suitable signal-to-noise ratio. A couple of analytical techniques may be used to concentrate the degradants before the LC-NMR analysis, such as column switching or solid phase extraction. Solid phase extraction has recently been combined with LC-NMR to boost the LC-NMR performance. By inserting a solid phase extraction (SPE) device containing peak trapping cartridges after the LC column and UV detector, the LC-NMR sensitivity for a single injection can be increased by up to a factor of four for individual LC peaks.

C. Drug Metabolites

The study of drug metabolism has been extremely important in drug research and development, particularly those associated with adsorption, distribution, metabolism, excretion (ADME), toxicology, and bioavailability. Typically, the drug metabolites will be isolated from a biological fluid or an in vitro incubation broth using HPLC, and then submitted for structure elucidation by LC-MS or NMR spectroscopy. However, even the isolation of small quantities at the microgram level is often time-consuming and may lead to degradation during the purifying process. The introduction of LC-NMR has made it possible to eliminate the isolation step and provided a new powerful tool for the study of drug metabolism. Researchers can study samples to reveal metabolite identity faster and with greater confidence than with separate analytical instrument offerings.[11–16]

D. Natural Products

LC-NMR can be used to identify natural products in crude plant extracts that usually consist of complex mixtures. The crude natural product extracts normally contain a great number of closely related and difficult-to-separate compounds. The classical separation approach may become very tedious and time-consuming. The directly coupled HPLC-NMR presents an efficient separation technique together with a powerful spectroscopic method to speed up the identification process. LC-NMR has been used extensively for characterization of natural products. More recently, the combination of LC-NMR and LC-MS has been further developed in this field. For example, Wilson et al.[17] have used combined on-flow NMR and electrospray ionization MS to characterize ecdysteroids in extracts of silene otites. After reversed-phase HPLC using D_2O in acetonitrile-d_3 and UV detection, the LC flow was split 95:5 for the simultaneous detection by NMR and MS. The peaks of interest were analyzed by stop-flow NMR to give better quality spectra for structural assignment.

E. Combinatorial Chemistry

The use of combinatorial chemistry to produce libraries of compounds is pivotal in drug discovery. Screen hits need to be analyzed to identify the structure of the individual active component. Large numbers of samples containing only small quantities of complex mixtures require analysis. As a result, the ability to obtain the maximum amount of information from a single analytical run is very important. A recent assessment[18] of this technique was conducted on a mixture of 27 closely related tripeptides with reversed-phase gradient of acetonitrile in D_2O/phosphate buffer, followed by on-flow LC-NMR. The majority of these tripeptides were characterized based on their chemical shifts and peak multiplicities.

IV. RELATED HYPHENATED TECHNIQUES

A. LC-NMR-MS

In practice, the use of either LC-MS or LC-NMR alone can often solve many analytical problems. However, there are numerous occasions when multiple techniques are required to solve the problem. In these cases, using both LC-MS and LC-NMR can provide extra information and structural identity. It seems a natural progression to combine the two systems into one integrated LC-NMR-MS system. The addition of an MS spectrometer to an LC-NMR system provides two primary functions: the acquisition of a mass spectrum of the individual components and the highly selective nature of MS spectrometry are used as discriminating factors for peaks prior to NMR analysis. In addition, MS measurements are possible with the incorporation of a specialized LC-NMR-MS interface. From LC-NMR-MS, one can obtain UV, NMR, and the MS spectral data for components of interest in mixtures as they elute from an HPLC in a single chromatographic run. It can avoid the potential problem of misinterpreting data from chromatographic components, which may change their elution order with small variations in HPLC conditions. This makes LC-NMR-MS an effective alternative for more accurate and rapid structure analysis.

In order to get good LC-NMR-MS results, the LC and the MS spectrometer should be placed as close as possible to the NMR magnet. Unnecessary lengthy tubing should be avoided to minimize band-broadening of the eluted LC peaks. Since a high magnetic field may have an effect on MS performance for lab-built systems using older NMR instruments, the position of the mass spectrometer may need to be adjusted. Some find it helpful to tune and calibrate the MS using an internal reference before and after moving into the magnetic field. It will help to fully investigate the effect of proximity and orientation to the magnet on the ion beam.

In general, MS performance should not be compromised by a static magnetic field. Many factors such as the design of ion optics, selection of interface, and type of MS analyzer should be considered with regard to the construction and configuration of the double hyphenated system. The optimum design should overcome some of the mutual incompatibilities of LC-NMR and LC-MS systems.[19] For LC-MS, ionization propensities vary considerably depending upon solvent, ionization source type, and complex matrix effects. Most NMR analyses, however, are not affected by variations in the matrix. Several research groups have demonstrated that a parallel online mass analyzer could be effective for such application.[20] In other cases, researchers have preferred LC-MS analysis followed by solvent exchange and NMR analysis.[21–25]

B. LC-SPE-NMR

To enhance the sensitivity of LC/NMR, the new hyphenated technology (LC/SPE/NMR) is integrated by inserting an SPE device containing peak trapping cartridges after the LC column and UV detector. It has been reported that the LC-NMR sensitivity for a single injection can be increased by up to a factor of four. The lowest detectable level may reach the low nanogram range.

The system includes automated procedures for washing, drying, conditioning, and transfer of the trapped peaks into the NMR with user-defined deuterated solvents. The NMR structural elucidation of components of any sample with late eluting, broad LC peaks can benefit from this technology. The sensitivity can be further enhanced by making multiple injections of the same sample and storing the resulting peaks onto the same trapping cartridge. Peaks obtained by repeated trapping can be eluted simultaneously into the NMR probe. This hyphenated technology can be further developed into the LC-SPE-NMR-MS system, which is even more powerful in identifying low-level sample components such as metabolites.

C. Capillary LC-NMR

This new hyphenated analytical system integrates capillary LC with NMR detection. The capillary LC-NMR system is comprised of an NMR spectrometer equipped with a capillary flow probe and the capillary LC. The capillary flow probe has a flow-cell design with an active sample volume of only 1 or 1.5 μL. This volume is chosen to match the typical peak volumes of capillary LC separation.

Both the capillary LC and NMR are controlled by the interface software, which enables the operator to use the UV-detector output for peak selection. Only peaks of interest can be subjected to NMR analysis, while minor or unimportant compounds can be directed to waste. NMR acquisition can take place in either on-flow or stop-flow mode. The combination of capillary LC and NMR is suitable for sample-limited applications (e.g., proteomics) and allows for low nanograms detection.

D. Cryo-Flow LC-NMR

This revolutionary technology uses a cryogenically cooled NMR probe for NMR detection, in which the NMR coil assembly and the preamplifier are at cryogenic temperatures leading to a 3–4-fold increase in signal-to-noise ratio (*S/N*). The spectroscopic handling is very similar to that of a conventional LC probe. While the operating sample temperature is stabilized at a user-defined value around room temperature, the NMR coil assembly that is located only a few millimeters from the sample is cooled with cryogenic helium gas.

The cryo-LC probe offers an enhancement of up to four-fold in sensitivity over conventional LC probes. Hyphenated techniques such as LC-NMR and LC-SPE-NMR, as well as high-throughput NMR screening methods, will benefit from this significant increase in sensitivity.

V. CONCLUSION

In general, LC-NMR is an efficient analytical technique for the identification of components in pharmaceutical mixtures. Especially when it is combined with MS, the two on-line detectors are complementary in providing unequivocal structural identification for both expected components and for unknown substances. However the relatively low NMR sensitivity is still the major drawback in the utility of the LC-NMR technique. Hopefully, the newly developed cryogenic LC-NMR probes coupled with recent interface enhancements and the higher magnetic field strengths will further enhance the utility of this technique in solving more challenging problems.

REFERENCES

1. Bowers, L. D., *Clin. Chem.*, 35:1282, 1989.
2. Tomer, K. B. and Parker, C. E., *J. Chromatogr.*, 492:189, 1989.
3. Niessen, W. M. A. and van der Greef, J., *Liquid Chromatography–Mass Spectrometry*, Marcel Dekker, New York, 1992.
4. Strohschein, S., Rentel, C., Lacker, T., Bayer, E. and Albert, K., *Anal. Chem.*, 71:1780, 1999.
5. Bayer, E., Albert, K., Nieder, M., Grom, E., Wolff, G. and Rindlisbacher, M., *Anal. Chem.*, 54:1747, 1982.
6. Laude Jr., D. A. and Wilkins, C. L. *Anal. Chem.*, 56:2471, 1984.
7. Laude Jr., D. A. and Wilkins, C. L., *Anal. Chem.*, 59:546, 1987.
8. Albert, K., Kunst, M., Bayer, E., Spraul, M. and Bermel, W., *J. Chromatogr.*, 463:355, 1989.
9. Morita, T., Suematsu, T. and Takahashi, Y., *Application & Research Center*, *JOEL Ltd.*, 28A:1, 2002.
10. Holt, R. M., Newman, M. J., Pullen, F. S., Richards, D. S. and Swanson, A. G., *J. Mass Spectrom.*, 32:64, 1997.
11. Shockcor, J. P., Unger, S. E., Wilson, I. D., Foxall, P. J., Nicholson, J. K. and Lindon, J. C., *Anal. Chem.*, 68:24, 1996.
12. Gu, T. Z., Deng, Y. and Spooner, N., *Poster*, TP24, CPSA 2001 Digest, 2001.

13. Burton, K. I., Everett, J. R., Newman, M. J., Pullen, F. S., Richards, D. S. and Swanson, A. G., *J. Pharma. Biomed. Anal.,* 15:1903, 1997.
14. Scarfe, G. B., Wilson, I. D., Spraul, M., Hofmann, M., Braumann, U., Lindon, J. C. and Nicholson J. K., *Anal. Commun.* 34:37, 1997.
15. Clayton, E., Taylor, S., Wright, B. and Wilson, I. D., *Chromatographia*, 47:264, 1998.
16. Louden, D., Handley, A., Taylor, S., Lenz, E., Miller, S., Wilson, I. D. and Sage, A., *Analyst*, 125:927, 2000.
17. Wilson, I. D., Morgan, E. D., Lafont, R., Shockcor, J. P., Lindon, J. C., Nicholson, J. K. and Wright, B., *Chromatographia*, 49:374, 1999.
18. Linden, J. C., Farrant, R. D., Sanderson, P. N., Doyle, P. M., Gough, S. L., Spraul, M.and Hofmann, M., *Magn. Reson. Chem.* 33:857, 1995.
19. Shockcor, J. P., Unger, S. E., Wilson, I. D., Foxall, P. J. D., Nicholson, J. K. and Linden, J. C., *Anal. Chem.* 68:4431, 1996.
20. Shockcor, J. P., Unger, S., Savina, P., Nicholson, J. K. and Lindon, J. C., *J. Chromatogr.*, 748:269, 2000.
21. Sandvoss, M., Pham, L. H., Levsen, K., Preiss, A., Mugge, C. and Wunsch G., *Eur. J. Org. Chem.*, 1253, 2000.
22. Bailey, N. J. C., Stanley, P. D., Hadfield, S. T., London, J. C. and Nicholson, J. K., *Rapid Commun. Mass Spectrom.*, 14:679, 2000.
23. Bailey, N. J. C., Cooper, P., Hadfield, S. T., Lenz, E. M., London, J. C., Nicholson, J. K., Stanley, P. D., Wilson, I. D., Wright, B. and Taylor, S. D., *J. Agric. Food Chem.*, 48:42, 2000.
24. Holt, R. M., Newman, M. J., Pullen, F. S., Richards, D. S. and Swanson, A. G., *J. Mass Spectrom.*, 32:64, 1997.
25. Ludlow, M., Louden, D., Handley, A., Taylor, S., Wright B. and Wilson, I. D. *J. Chromatogr. A*, 857:89, 1999.

21 CHROMATOGRAPHY DATA SYSTEMS (CDS) IN THE PHARMACEUTICAL LABORATORY: ITS HISTORY, ADVANCES, AND FUTURE DIRECTION

ROBERT P. MAZZARESE

Waters Corporation, Milford, MA

ABSTRACT

This chapter discusses the history, technological advances of chromatography data systems. Also included here are the particular needs of the pharmaceutical industry and the future directions of CDS.

Handbook of Pharmaceutical Analysis by HPLC,
S. Ahuja and M.W. Dong, editors.

I. INTRODUCTION

Chromatographic techniques represent one of the most significant sources of analytical data found in today's pharmaceutical laboratories. Scientists from all segments of the industry—discovery, development, quality control, and manufacturing—rely on techniques as simple as thin layer chromatography (TLC), to routine analyses using UV-Vis, refractive index (RI), fluorescence, and photodiode array (PDA), to complex, multi-dimensional chromatography using mass spectrometry (MS) as the detector of choice. Although the specific techniques used within each segment may differ, there is one common thread that ties them all together. All of these chromatographic techniques produce data that must be acquired, interpreted, quantitated, compared, reported, and finally archived.

Whether the analysis is qualitative or quantitative in nature, the data must somehow be interpreted and reported so that meaningful decisions can be made. It may be as simple as a qualitative decision that indicates whether or not a reaction has completed successfully, or it may be a series of quantitative analyses that help determine if a batch or lot of product meets its specifications and may now be released.

This chapter will examine the evolution, and perhaps the revolution, that has taken place within the chromatography data system (CDS) marketplace.

II. A HISTORICAL LOOK AT CDS DEVELOPMENT

For pharmaceutical scientists in the 1960s and early 1970s, chromatographic data were often the paper trace that was generated by a strip-chart recorder. Quantitation was difficult, with scientists having to use crude techniques to calculate peak areas, heights, and amounts. Some laboratories used the "cut and weigh" technique where each peak in the chromatogram was cut from the trace and weighed on an analytical balance. This would then be the basis for calculating peak amounts for unknowns when compared to the "weighed peaks" from the standard. Other laboratories would take a pencil and ruler and use the "triangulation" method for determining the area or height of the peak. This technique would require the scientist to draw a "best fit" triangle on the peaks of interest and then calculate the area using the formula: $Area = \frac{1}{2}\, base \times height$. This is a very simplistic calculation that treats each peak as having equal sides. As chromatographers, we recognize the limits of such a calculation when peak tailing comes into play. Using either technique, the results were marginal at best, very labor-intensive, and subject to enormous variations when you take into account that there may be different scientists performing the calculation within the same laboratory.

A. Early Integrators

It was quite apparent from the volume of samples being run within a typical analytical chromatography laboratory that a more precise and automated technology was needed. In the mid-1970s, we saw the development and introduction of the *integrator*. These early devices were capable of automatically calculating the area or height of a peak. They were connected to the analog output of the detector on the HPLC or GC. The analog signal, which is nothing more than a voltage that corresponds to the intensity of the detected signal, was converted to a digital signal that could be processed by the electronics of the integrator. This circuitry is known as an analog to digital converter or A/D converter. Although these early A/Ds were quite crude, the basic concepts are still used in today's high-speed, high-resolution A/D converters. The output from these early integrators was often a numerical printout very similar to that of a cash register. The peaks would be identified by their calculated retention time (RT). The retention time, along with the peak's associated area or height would be printed next to the peak. Although this was a significant improvement over the "Cut & Weigh" or the "Triangulation" methods, it still required a fair amount of work to create calibration curves, check the relative standard deviation (RSDs) of replicate injections, and calculate the final results for the samples that had to be run. In 1974, Hewlett-Packard introduced the HP3380A integrator. It was the first integrator to plot the chromatogram on thermal paper, and it had the math chips from the HP35 calculator that enabled it to process the peak information. As more and more scientists used integrators for their analyses, instrument vendors began to improve the overall quantitation capabilities of these devices.

From a hardware perspective, integrators began to use:

- Battery back-up to maintain integration parameters during a power failure
- Volatile and nonvolatile memory modules that were capable of storing chromatograms for playback and reintegration
- Hard disk or tape storage of chromatograms for playback, reintegration, and more importantly *data archiving*.

From a software perspective, integrators began to offer specific capabilities:

- System suitability calculations like k′, resolution, and plate count (N).
- Specialty calculations for gel permeation chromatography (GPC) where the integrator would calculate the molecular weight distribution, molecular weight average, and polynomial curves for calibration.
- Basic programming for custom calculations and statistics.

Despite these advancements in chromatographic data processing, peak detection and integration algorithms were crude, the user interface was cumbersome, and there was very little flexibility in the types of reports that these systems could generate.

B. Impact of Computers on CDS Development

This is probably an appropriate time to bring up another technology that was evolving at the same time, and would prove to have a significant impact on the development of the CDS, as we know it today. I am, of course, referring to the computer. ENIAC was the first operational electronic digital computer, and was designed for the US Army to calculate ballistic trajectories. Completed in 1946, it used 18 000 vacuum tubes, took up 1800 ft^2 and performed 5 000 additions per second. By the mid-1960s, advancements in transistors and microprocessors, RAM memory, disk drives, and operating systems led to the introduction of the first mainframe and minicomputers. These were powerful, time-sharing systems that could perform sophisticated calculations while supporting a large number of concurrent users. This was the technology that was needed in order for analytical laboratories to improve the accuracy of their analyses, while at the same time providing a common platform for all of the scientists within a particular department.[1]

In the late 1970s, Hewlett-Packard introduced the HP-3300 series data-acquisition system, which was able to connect to 60 chromatographic instruments through an A/D converter. This was the beginning of what would become a revolution in CDS development within the analytical instrument industry. By the mid-1980s, all of the major analytical instrument manufacturers offered network-based data-acquisition systems: Beckman, HP, PE, VG, and Waters. These were multi-user, time-sharing systems that used A/D converters to acquire data from the instruments. Instrument control, both HPLC and GC, was a capability that would soon follow. Several CDS manufacturers offered serial control of the HP 5890 GC while Waters also offered instrument control for their own HPLCs.

Some of the most significant advancements came in the late 1980s. Two pioneering companies who brought chromatography data systems to the desktop were Nelson Analytical and Dynamic Solutions. Both of these companies were delivering desktop PC solutions long before Microsoft introduced its first version of Windows. As their products grew in popularity, so did the awareness by the major instrument manufacturers. In 1988, Dynamic Solutions was acquired by Waters. Although Dynamic Solutions had already introduced the product Maxima, it soon became one of Waters' key software products. In a similar fashion, Perkin-Elmer acquired Nelson Analytical in 1989. The new division, known as PE-Nelson, introduced Turbochrom. These PC products became the standard for workstation-based chromatography data systems.

C. A/D Technology

The most common device used to acquire a chromatographic signal is the A/D converter. Although there were several designs commercially available, most of the analytical instrument manufacturers used one of two types: voltage to frequency (*V/F*) or successive approximation (SA). Both of these devices produce a number or value based on an analog input signal and both are capable of producing high-quality results for the chromatographer. A successive approximation A/D utilizes an automatic gain selection based on the incoming analog signal. These converters were less expensive than *V/F* converters and typically displayed a noisier signal. They often required extra software filtering.

Regardless of the type of A/D used, the most important characteristics were resolution, noise, and linearity. An A/D with *high resolution* and *low noise* enabled the scientist to integrate and quantitate small peaks of interest. This was extremely important for discovery, methods development, stability testing, and quality control. The resolution value represents how many pieces a given signal can be divided into. For example, if you have a UV detector that has a signal range of 0–2 V and you are using a 24 bit A/D, this A/D would divide the UV signal into 16777 216 parts (2 to the 24th=16777 216:1). This allows you to see a 0.0000001-V or 0.1 μV change from the detector.[2]

For those scientists who had to perform quantitation, the *linearity* of the A/D was also critical. Linearity is the condition in which the detector's response is directly proportional to the concentration or amount of a component over a specified range of component concentrations or amounts. It is imperative that the A/D not add any additional error or variability to the performance of the detector. The resulting calibration curve now becomes dependent on the combined linearity of the detector and the A/D. Accurate quantitation requires that the system is linear over the range of actual sample concentrations or amounts. Many pharmaceutical assays, like degradation and stability studies, require that the system be able to identify and quantitate very disparate levels of peaks. In many cases, this translates into a 3 to 4 order of magnitude difference between the main active component and the impurities that need to be quantitated.

Now that we have a method by which to accurately acquire a digitized representation of the chromatogram, the data system must identify the individual peaks and calculate their area or height.

D. Traditional Approach to Peak Detection

These network-based systems were terrific at providing the pharmaceutical laboratory with a centralized approach for data acquisition, data accessibility across a network, and long-term data storage on magnetic tape. Reliable peak detection and integration is based on three main parameter settings in the software:

- Sampling rate or data points per second
- Peak width or data bunching
- Liftoff threshold or slope sensitivity.

It should be noted that regardless of the CDS used, a data-acquisition rate that is too low will produce a peak that is impossible to integrate accurately. There must be enough data points to define the peak accurately. It is always safer to have too many points across a peak than too few, provided that the noise generated by the A/D is sufficiently lower than the detector noise generated by the chromatographic system.

This is where data bunching comes into the equation. As the detection algorithm tests the data for a peak, the software averages individual raw data points into discrete groups, or bunches, to produce a single point. The number of data points in a bunch is set by the peak width or data bunching parameter. Data bunching has no effect on the acquired raw data. It is an internal calculation used to enhance the process of determining peak start and peak end. All of the raw data points are used for proper peak integration. Figure 1 illustrates the effects of data bunching on a signal from a chromatogram acquired at a high data rate.

In general, the liftoff threshold defines the minimum slope of the signal at or above which the start of a peak is detected. The software would average the signal slope across two data bunch intervals and then compare it to the liftoff threshold (see Figure 2).

When the averaged slope of the signal between bunch $B1$ and $B3$ is greater than or equal to the liftoff threshold value, the software flags $B1$ as the possible peak start. The algorithm then examines the individual raw data points that make up $B1$ to determine the actual start of the

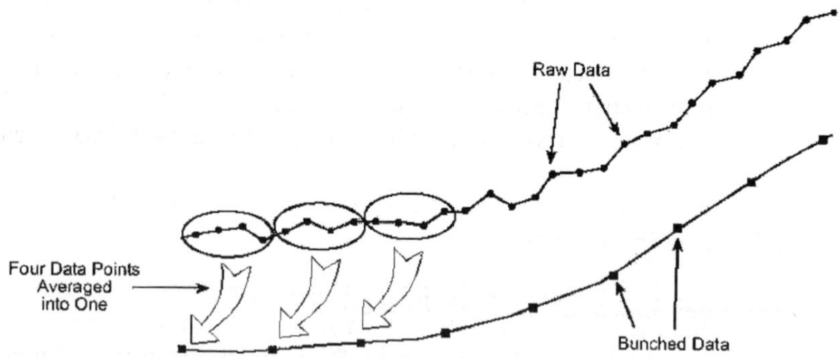

FIGURE 1 The effects of data bunching on a signal from a chromatogram acquired at a high data rate.

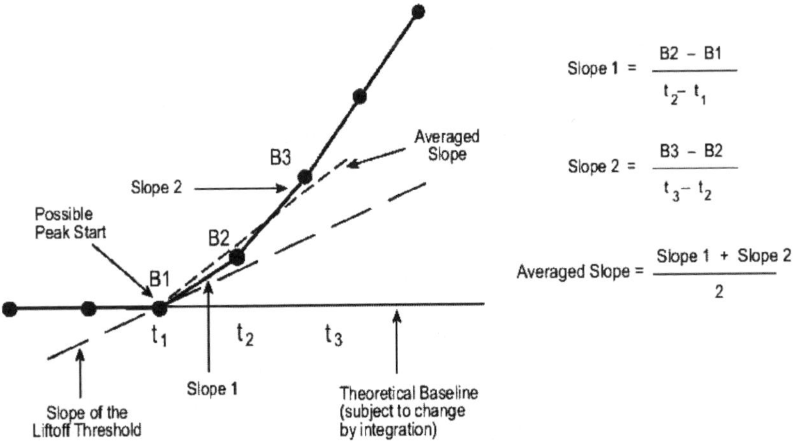

FIGURE 2 Changes in signal slope as compared by an algorithm in order to determine the start of a peak.

peak. A similar calculation is performed to determine the peak end or touchdown point.

The other key benefit to accurate peak determination is the calculation of Retention Time, often referred to as RT. For most chromatographic assays that are not using mass spectrometry as the means of compound identification, all components are identified using their retention time as compared to the standard. Although the performance of the instrument and column are the most significant factors affecting retention time, it is still important that the peak detection algorithm be capable of calculating accurate and reproducible retention times (see Figure 3).

To determine the preliminary peak apex, the software monitors the signal until the slope changes sign from positive to negative. For a negative peak, the slope changes sign from negative to positive. The software then analyzes the bunch where the slope change occurs (*B*12 in Figure 3). This is the preliminary apex of the peak. When all integration is complete and baselines are assigned, the software fits a quadratic curve to the five data points at the top of the identified peak. The results of the quadratic fit are then used to determine the actual retention time for that peak.[3]

E. Methods of Quantitation

Today, most modern data systems provide a comprehensive array of quantitation methodologies. In general, chromatographic peaks may be quantitated using *area*, *height*, *% area*, and *% height*. Within these basic techniques, the chromatographic assay may also require the use of an *Internal Standard* to help correct for injector variability or sample-prep

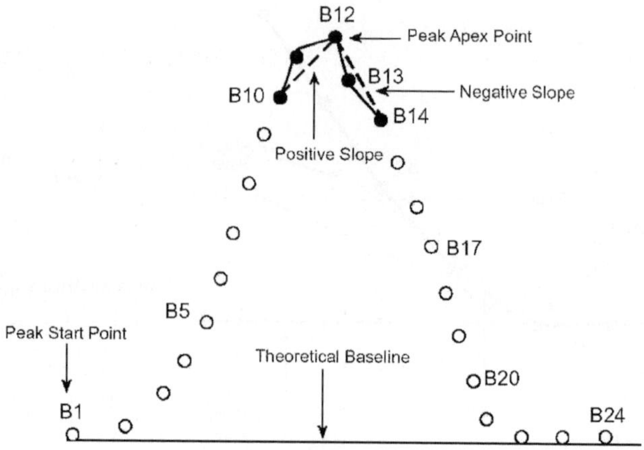

FIGURE 3 **The algorithm determines the retention time of the peak being integrated.**

errors. The final step in the quantitation method is the choice of *calibration curve fit type*. While *linear* or *linear through zero* is the most common fit type, the proper choice of fit type often depends on the behavior of the analyte and the chromatographic system being used for the assay. During the method development process, it is imperative that the detector response for each analyte of interest be properly characterized. The expected concentration range of the analyte, detector bandwidth (for UV, UV-Vis, diode array, and fluorescence, for example), column performance, and sample matrix may all contribute to a situation where the analyte response is nonlinear or is linear, but whose intercept does not pass through zero. Once the separation is fully characterized, the proper fit type may be selected. The most common choices for fit type are:

- *Single level calibration:* Linear through zero and response factor
- *Multi-level calibration:* Linear, point to point, cubic spline, inverse linear, log-log linear, quadratic, cubic, fourth-order, and fifth-order.

(For a more thorough description of the various calibration curve fit types, please see References 4–8.)

III. TECHNOLOGICAL ADVANCES

The late 1970s and early 1980s saw most of the analytical instrument manufacturers taking advantage of mainframe and minicomputer technology. Although these multi-user, time-sharing systems were well suited to perform the calculations necessary to support chromatographic assays, the user interface was typically a text-based editor with very basic graphic capability. I am sure that it is no surprise to anyone that the personal computer, which was in its infancy, would have such an effect on the entire analytical instrument industry. Although the early personal computers commercialized by Apple and IBM were not very powerful, they did accomplish two very significant goals. It was now commercially feasible to have inexpensive computing power on every desktop, and it was now possible to develop operating systems that were capable of sophisticated graphical interfaces. For analytical instrument manufacturers, this could not have come at a better time. By the late 1980s, the personal computer revolution had begun to make its mark on the laboratory. Everything from spectrophotometers to HPLCs were using Apple or PC-based computers for data acquisition, data processing, and reporting.

A. Computer Hardware

By the early 1990s, Apple and PC-based computers were capable of minicomputer performance at a fraction of the cost. Graphical user interface technology was quite advanced which made it the perfect tool for visualizing spectra, graphs, and of course, chromatograms. The laboratory workstation market exploded. It was now easier to use commercially available PC hardware to control and acquire data from most analytical instrumentation. The chromatographic market was no exception. Instrumentation was becoming more sophisticated, while the data generated was increasing in complexity and volume.

B. Network Architecture

These advancements in PC hardware and software were also having their effect on network architecture. Although multi-user, time-sharing minicomputers were still popular within the CDS marketplace, there were three significant drawbacks to the technology. First of all, they were expensive systems for most small to mid-size laboratories. These laboratories would usually stay with integrators or PC-based systems. Second, these systems were not very scalable. Since they were time-sharing computers, a single computer handled all user requests for data acquisition, calculations, and reports. As the number of users grew, the overall performance of the system would begin to degrade. Third, upgrading the system to handle more users was typically very expensive. In many cases, it would require the purchase of a new minicomputer.

As network connectivity improved within the PC operating systems, client/server computing became possible. Instead of performing all of the work on a single, multi-user computer, you could now distribute the work across many computers. In the client/server model, adding a PC to the network increases the overall processing power of the system. The client provides the graphical user interface and is responsible for some or most of the application processing. The server typically maintains the databases and processes requests from the clients to extract data from or update the database. A critical responsibility of the server is to control the application's integrity and security. Applications that could take advantage of this technology allowed even small laboratories to benefit from a client/server CDS implementation.

This type of network architecture has some key benefits for scientists in different laboratories or different locations that are using the same CDS software and need to transfer methods, review results, or even troubleshoot problematic assays. Key users may be given accounts on the CDS system. Approved methods that have been validated by development can easily be transferred to the QA/QC laboratories that need to use them. This technology also facilitates method troubleshooting. If a QA/QC laboratory is having problems with a method, it is easy for a development chemist to view the problem chromatograms in real time. The development chemist can then review the instrument and processing conditions and make any necessary suggestions or method modifications, without the use of FAXs, e-mails, telephone calls, etc.

This model for client/server has served the pharmaceutical industry for almost a decade. Although it is still a modern, scalable design, it has one drawback for today's laboratories that need to comply with current FDA regulations. As more clients or access points are added to the network, system maintenance and qualification concerns increase. PC clients in a simple, two-tier client/server are often referred to as "thick" clients. This refers to the fact that they must contain parts of the application, typically DLLs and EXE files. The concern for the corporate IT and quality organizations comes from the need to provide a computing environment that is controlled and tested. Any changes to the PC may constitute a serious change to the behavior of the applications running on it. In order to remain "in control" of this environment, most organizations have instituted strict protocols that govern any software changes to the system. In addition to the issues related to "change control" come the financial aspects of a validated system. From the regulatory perspective, any and all devices that are controlling instruments acquiring data, processing results, and generating reports should be routinely qualified as part of the corporate validation plan. For a CDS, this means the qualification of all computer hardware, acquisition hardware, and all PC clients capable of processing and reviewing data. For laboratories with hundreds of users, the qualification of these PCs can be a time-consuming and costly activity. In an attempt to control the maintenance and qualification costs associated with client/server systems, many of today's pharmaceutical companies are looking at a technology called server-based computing. Many of you may know it as "thin" client computing. Server-based computing is a model in which applications are deployed, managed, supported and executed on a dedicated application server. Server-based computing uses a multi-user operating system and a method for distributing the presentation of an application's interface to a client device. In essence, there are no software components of the application to be installed on any PC client. The application client connects to the dedicated application server and simply acts as its display. All maintenance and qualification costs are now focused on the application servers and not the PC clients. This means improved efficiency when deploying business-critical applications. In addition, server-based computing works within the current computing infrastructure and current computing standards, and with the current and future family of Windows-based offerings. This means improved returns on computing investments—desktops, networks, applications, and training. The end result: server-based computing is rapidly becoming the most reliable way to reduce the complexity and total costs associated with enterprise client/server computing. This centralized application and client management enables large computing environments to overcome the critical application deployment challenges of management, access, performance, security, and qualification.

One other significant benefit of this technology worth mentioning is its ability to provide Web access to key applications. The Citrix MetaFrame product is a thin-client software product that extends the capabilities of server-based computing. It creates a client in the form of a Microsoft Active X control, allowing its users to run their Windows applications from within a standard web browser like Internet Explorer or Netscape.

C. Database Technology

Since their introduction in the late 1970s, chromatography data systems were all designed around a directory structure with methods and raw data stored as individual files called "flat files" (a flat file is a data file that does not physically interconnect with, or point to, other files). Similar to the way you organize information on your own PC, these systems would create a directory hierarchy based on an instrument, an assay, or perhaps a user. Within each major group, there would exist subgroups or "folders" that would help to organize and separate the different file types (see Figure 4).

Initially, these simple systems only needed to manage the acquired raw data files, method files, result files, and report files. The raw data files contained the raw x–y coordinates of the detector signal, and also contained basic information about the sample called "header information". The method file contained basic A/D control information, the parameters used for peak integration, as well as calibration information if the assay performed quantitation against a standard curve. The result generated was often a separate file that contained all of the calculated or quantitated

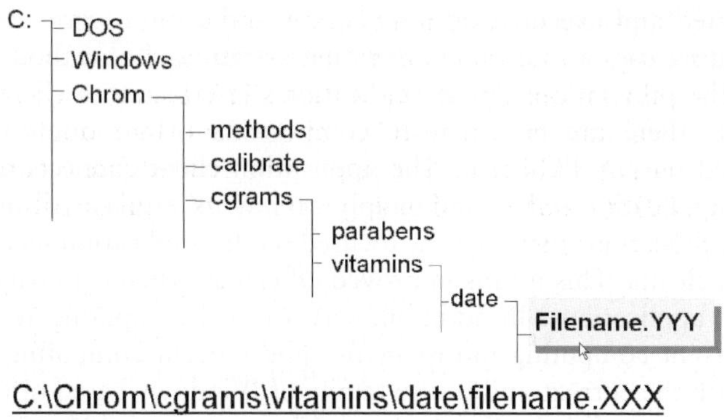

FIGURE 4 **A typical directory structure found in a "flat-file" data system.**

values while the report file held all of the scaling information and table information that was required to generate and print the final report. Although these systems worked well, there were a few drawbacks. The most significant was the potential for methods and raw data to be accidentally overwritten. As the number of system users increased, so did the potential for disaster. In many cases, the only way to prevent this was for the system administrator to protect the old "folders" and create new "folders" for their users. This would often happen on a weekly, or even daily, basis. The other drawback of the "flat-file" structure pertained to traceability. Even within a single "folder", it was difficult to deal with samples that had to be integrated more than once. This was a problem that both the pharmaceutical industry and the CDS manufacturers were struggling with long before the FDA began to enforce 21 CFR Part 11. Laboratories wanted a clear mechanism by which they could relate a specific result to the specific version of the method used for calculations, and quantitation to the original raw data file coming from the instrument. The only way to accomplish this was through the creation of yet another file that could encapsulate the raw data with each method version and its associated result. These files were called "snapshot" files because they captured a snapshot in time based on the creation of a new result. Although this technique provided traceability within each version of the created result, it had no way of relating multiple results that came from the same raw data file. Each "snapshot" file was a discrete file unto itself.

All that changed in 1992 when Waters introduced its revolutionary new CDS product "Millennium". This was the first commercially available CDS to use an embedded relational database (Oracle) to manage all aspects of the system. System security, user privileges, method versions and their respective results were all managed by the database. While this technology was new to CDS, it had been widely used by the Laboratory Information Management System (LIMS) market. All of the major LIMS products were, and continue to be, based on relational database technology. A relational database stores information in tables that have defined relationships between them. This concept makes it easy for a CDS to maintain full traceability of the data. In addition, these links form the basis by which audit trails can be created (see Figure 5).

Embedding database technology into a CDS product brings with it three significant benefits:

- Databases "date and time stamp" all information. This makes accidental overwriting of raw data and methods impossible.
- The "relational" nature of the database makes it easy to tie all of the "metadata" together. Metadata can be defined as all of the related information about a sample. This information covers all aspects of data creation, data processing, result generation, review and approval, and finally, the underlying audit trail that ties it all together. From a simple result, you can easily access the specific version of the instrument method and the analyst who ran the sample, the specific version of the processing method that integrated the sample, the chemist who processed the data and when, and what was the exact calibration curve used to generate the final result. The best part is that all of this is happening in the background.
- Finally, the relational database provides the necessary audit trail as methods are modified, samples are re-processed, and system settings and policies are changed. All aspects of the CDS are managed and audited by the embedded relational database. For today's pharmaceutical laboratories, that means complete traceability from sample introduction, to analysis, to data processing and reporting, to electronic sign-off, and finally to LIMS (see Figure 6).

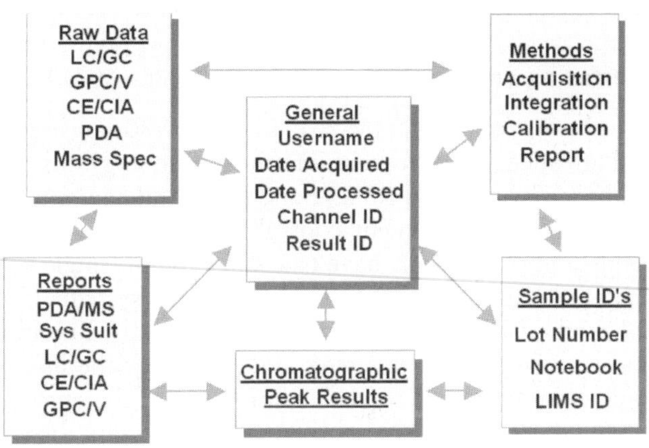

FIGURE 5 An example of the relationship that exists between the different tables of a CDS that is based on a relational database.

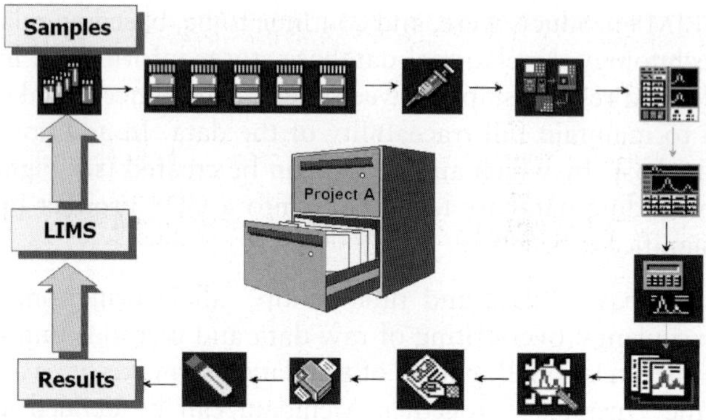

FIGURE 6 A typical sample life cycle from sample injection, through data processing, and reporting. The information finally ends up in the corporate LIMS after final review.

From a practical viewpoint, there are two key benefits to this technology. First, it makes retrieval of data very fast and simple. No one wants to search through reports and notebooks in order to find the associated data for a result that is in question. Second, during a regulatory audit it is crucial that your laboratory is able to produce the specific information with as much detail as is required by the auditor. Again, this task is much simpler when a relational database is managing the information (see Figure 7).

Figure 7 shows the relationship between a raw data channel and its associated metadata. If we were to choose the item highlighted "Instrument Method", the embedded relational database would retrieve the exact version of the instrument method that was used to acquire the raw data. All this occurs in a fraction of a second. Imagine how long it would take using a conventional flat-file system (see Figure 8).

D. Instrument Interfacing and Control

From the instrument perspective, the late 1980s and early 1990s saw some real advances in instrument interface technology. Instrument manufacturers began using serial RS-232 and IEEE-488 communication protocols to provide full instrument control capabilities to the CDS market. This enabled laboratories to have true single-point, single keyboard control of their instruments in the chromatographic system. This allows the re-use of methods that were created. This not only guaranteed that the proper instrument parameters were used, but that the instrument metadata could now be linked to each sample. From the chromatographer's perspective, there was also another very important benefit to direct instrument control. In many cases, the direct digital connection used to download instrument control parameters was also capable of collecting the digital signal from the detector. This eliminated the need for an A/D

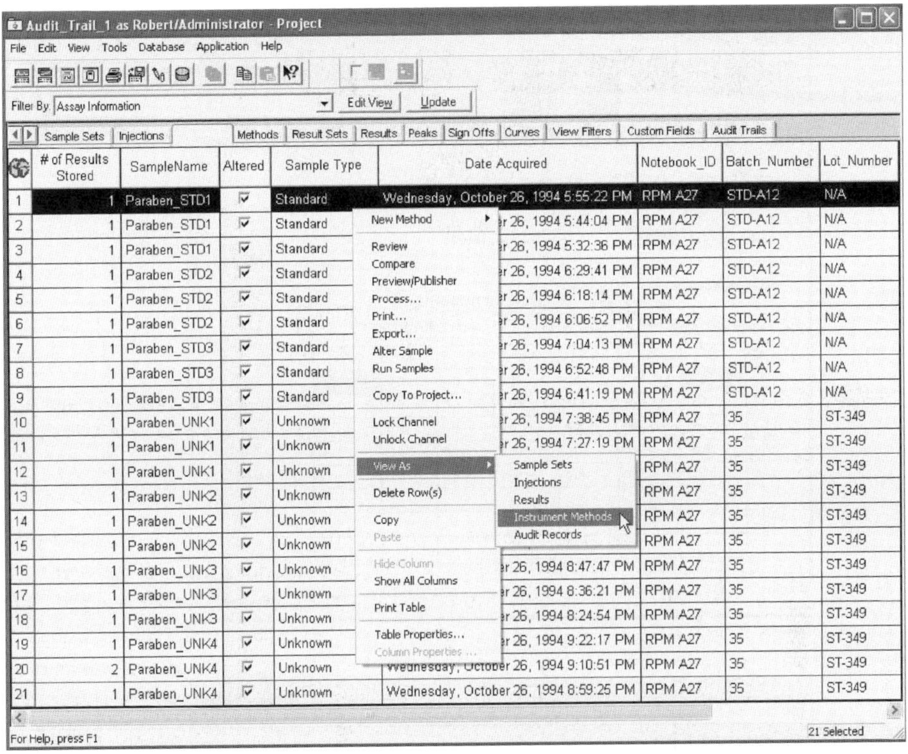

FIGURE 7 **Related information (metadata) is accessed using a CDS with an embedded database.**

converter, and resulted in a chromatographic signal with less noise. All detectors internally produce digital signals that must be converted to an analog (voltage) signal in order for them to be connected to strip-chart recorders, integrators, and chromatography data systems that use A/D converters. Once the analog signal is captured by the A/D, it must then be converted back to a digital signal in order for the CDS to be able to use it. Both of these conversion steps have some level of noise associated with them. Systems that can acquire the direct digital signal from the detector will often see a 20% to 30% reduction in signal noise. For laboratories performing low-level analyses, this can significantly improve the limit of detection (LOD) and the limit of quantitation (LOQ). This improvement in signal noise has one other key benefit: more reproducible integration. The amount of detector noise directly impacts the way in which the CDS establishes baselines and determines peak start and peak stop.

Although A/D interface technology is still widely used, there is a continued push by the pharmaceutical industry towards digital instrument control. Much of this is being driven by two factors: information-rich detectors which are more commonplace in today's laboratories, and the regulatory pressure that is associated with electronic records (21 CFR

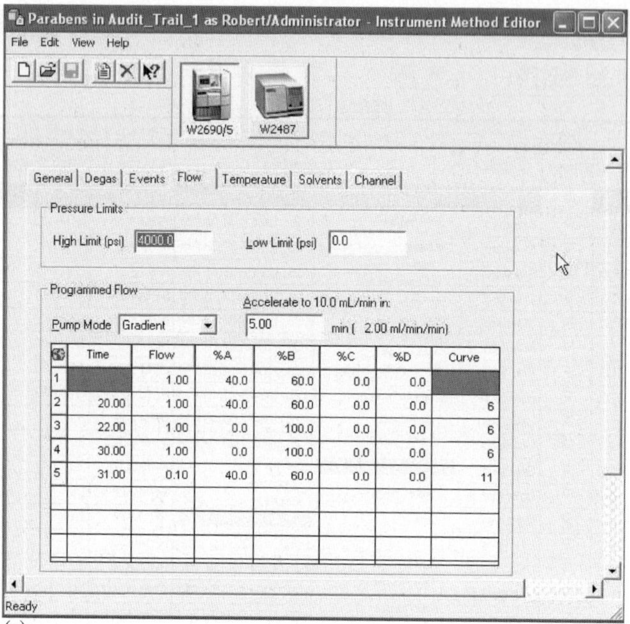

(a)

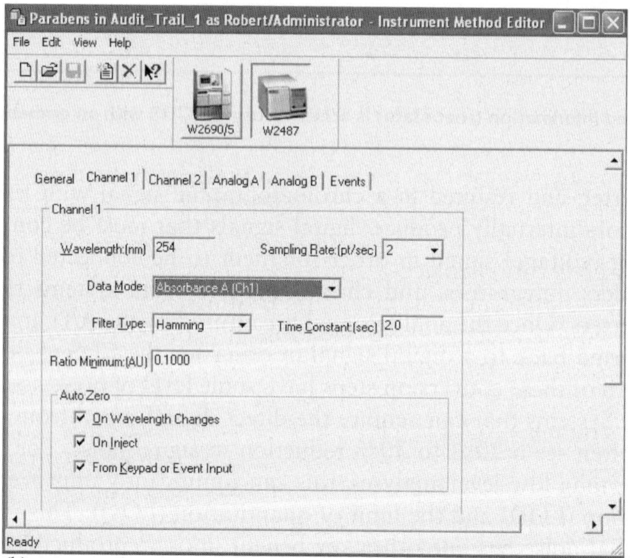

(b)

FIGURE 8 The retrieved instrument method showing the exact conditions used to acquire the raw data. The first screen shows the pumping conditions while the second screen shows the detector conditions.

Part 11). Over the past 10 years, there has been a significant increase in the number of laboratories that are using photodiode array, mass spectrometry, and to a lesser degree, scanning fluorescence as part of their routine HPLC assays. As instrument manufacturers reduce the size, complexity, and cost of these technologies, we see them becoming the detectors of choice for many laboratories. Today, photodiode array detectors are as common as single wavelength detectors, and we are beginning to see LC-MS used in routine QA/QC applications. The reasons for this are simple: these technologies provide more information with better specificity, which enables us to make better decisions. The issue that arises, however, revolves around regulatory compliance. Most of these detector technologies were originally designed to be workstation-based products with no real strategy for data security or networkability. Many CDS manufacturers have had to redesign their products to better handle the needs of the pharmaceutical industry.

As more and more laboratories rely on centralized network solutions, system "up time" becomes critical. One of the most important aspects of today's CDS is its ability to keep the laboratory running during network failures or system maintenance. Much of this burden has fallen on the interface technology used. Data "buffering" is a technology whereby all required elements of the assay that are necessary for instrument control and data acquisition are temporarily stored within the acquisition system. In the event of a network failure or scheduled server maintenance, the acquisition system continues to control the instrument while it temporarily stores the raw data for the analyses. Once the network connection is re-established, the acquisition system uploads the raw data that were "buffered". For simple, two-dimensional raw data (UV, RI, etc.), these data files are quite small. The real challenge comes when your detector of choice is photodiode array or mass spectrometry. These information-rich detectors can easily generate data files in the range from 500 kilobytes to hundreds of megabytes (see Figure 9).

Although most systems are capable of "buffering" simple single channel data, today's laboratories are generating significantly larger volumes of data using these detector technologies. In order to deal effectively with these new "buffering" requirements, some CDS manufacturers have moved away from static RAM technology to standard hard disk technology. It is now very common to find CDS acquisition systems capable of buffering gigabytes of data. These detectors have also driven CDS manufacturers to look for faster, more reliable interface protocols. Although serial RS-232 and IEEE-488 are still commonly used, ethernet connectivity is emerging as the new standard. With current networking technology capable of 100 megabit to gigabit data transfer speeds, this protocol is very well suited for instrument control applications.

From an instrument control perspective, one of the most significant advances is related to the type of graphical user interface (GUI) that we

Average Data File Size based on Detector Technology

Detector	Sampling Rate	Run Time	Data File Size
2-Dimensional UV, RI, Fluor, etc. General purpose detectors	1 point/second	20 minutes	5 KiloBytes
3-Dimensional UV-Vis Photodiode Array 220 to 400 nm. 1.2 nm. Resolution	1 scan/second	20 minutes	720 KiloBytes
3-Dimensional Single Quadrupole MS Small molecules 100 to 1200 AMU 1 AMU Resolution	1 scan/second	20 minutes	6 MegaBytes
3-Dimensional Single Quadrupole MS Large molecules/proteins 500 to 2500 AMU 1 AMU Resolution	1 scan/second	20 minutes	60 MegaBytes

FIGURE 9 The impact of detector technology on data file size.

have all come to expect. As the Microsoft Windows graphical environment improved, so did the way we as scientists interacted with our instrumentation. Early chromatography data systems provided instrument control through the use of command lines. These were nothing more than text entries that helped to define the instrument conditions. As instrumentation became more sophisticated, this type of interface became more cumbersome. The early 1990s saw the stabilization of the Microsoft Windows operating system and the introduction of the Pentium computer. These technologies allowed software developers to experiment with new ways for us to interact with our laboratory instruments. The ChemStation, originally developed by Hewlett-Packard, was one of the first systems to use a graphical depiction of the instrument for the purpose of monitoring its status and operating conditions (see Figure 10).

Today we see more and more instrument manufacturers taking advantage of this type of user interface. Most would agree that it adds a level of simplicity to the instrument and, in many cases, is a more intuitive approach for the laboratory.

E. Intelligent Systems

Within the last 5 years, there has also been a big push towards intelligent systems. The current trend of "doing more with less" has put a real burden on laboratory productivity. Pharmaceutical companies are looking

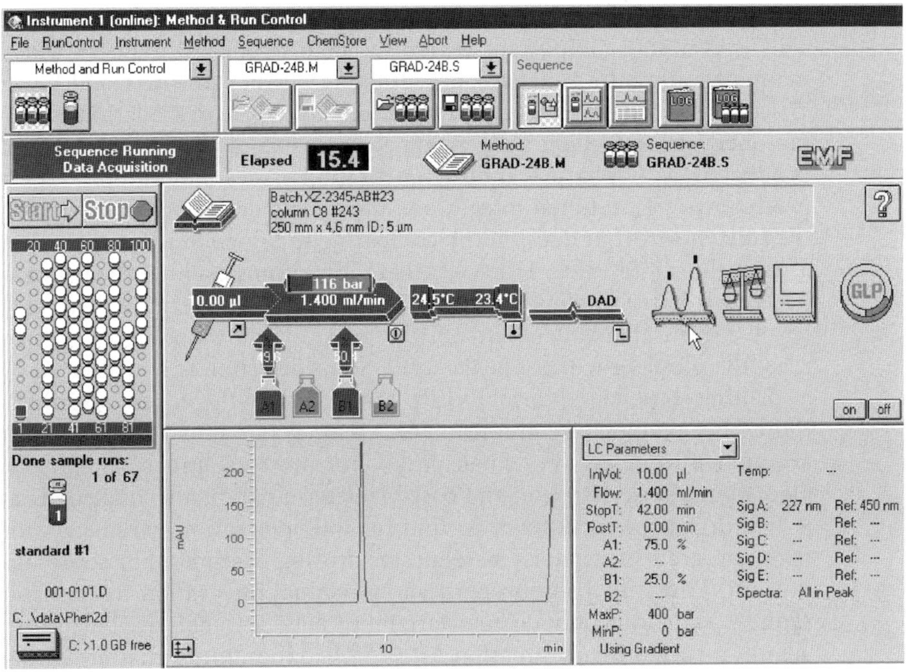

FIGURE 10 The graphical instrument control panel of the ChemStation.

at systems that can improve productivity while also enhancing the decision-making process. If we take a closer look at current CDS technology, we see four areas where the focus is on intelligent systems:

- CDS communication with instruments
- Reliable peak integration
- Automated limit testing with some decision-making capability
- Bi-directional interfacing to existing business systems (LIMS, SAP, etc.).

More and more of today's analytical instruments are being designed with sophisticated on-board diagnostics. These diagnostics are capable of identifying problems with the instrument, and may even be capable of monitoring sub-systems that are prone to failure like detector lamps, pump check valves, and autosampler valves. Data systems with direct digital control of the instrument are now capable of verifying system performance during the analyses. If problems are detected, the CDS can be set to interrupt the sequence and even run a shutdown method if possible. With many of today's pharmaceutical laboratories falling under GMP/GLP regulations, assays that do not run to completion will often require investigative action. Having the CDS provide an instrument error log detailing the shutdown can be a real benefit.

Quality, reliable integration is an absolute requirement for any CDS. This is even more important within the pharmaceutical industry where many of today's formulations are attaining efficacy at lower and lower doses. For laboratories performing stability studies, degradation profiling, metabolic profiling and, in some cases, manufacturing, if the final product is a biopharmaceutical, integration issues can severely affect productivity. Traditional integration algorithms detect peaks by finding the start or liftoff. It does this by comparing the slope of the data against a fixed threshold value. However, these algorithms do have their limitations. It is hard for slope criteria to detect shoulders because there may be no measurable valley between the peaks. If the slope never goes to zero, the shoulder will not be detected. Slope detection is also affected by a drifting baseline. Normal drift associated with solvent gradients (HPLC) or temperature gradients (GC) may dictate peak thresholds that are unable to detect very small peaks. For most traditional integration algorithms, peak detection and baseline determination are coupled. That is why it can be so difficult to find reliable integration parameters for chromatograms with noisy baselines or drifting baselines. For some laboratories, the only solution is to allow manual integration for difficult samples. This not only creates variability between analysts, but it also creates a very difficult situation for QA/QC laboratories that are trying to follow validated methods.

In 2002, Waters Corporation introduced ApexTrack™, a new integration algorithm aimed at dealing with these problems. Within the Empower software, ApexTrack™ integration can be used in place of traditional integration with any type of data; however, the algorithm was specifically designed to replace traditional integration when analyzing data where time-consuming manual manipulations are minimizing productivity or with data where traditional integration is not detecting the required peaks. ApexTrack™ integration implements a new approach to the basic peak detection and baseline determination algorithms. With ApexTrack™, the processing parameters that control peak detection and baseline placement are independent of each other. This greatly reduces the need to adjust integration parameters and the need to manually integrate difficult chromatograms (see Figure 11).

For laboratories that are running validated methods, it is common for the standard operating procedure (SOP) to specify assay limits. There may also be limit testing for the "check standards" that are commonly run before the actual assay so that the analyst may verify the suitability of their HPLC or GC system. Either way, the review of this data is often time-consuming, and in some cases requires the use of external programs or spreadsheets. This is an area where many of today's CDS are focused. It is now possible to migrate many, if not all of your custom calculations to be performed by the CDS. This not only has a significant impact on laboratory productivity, but also deals with the regulatory concerns

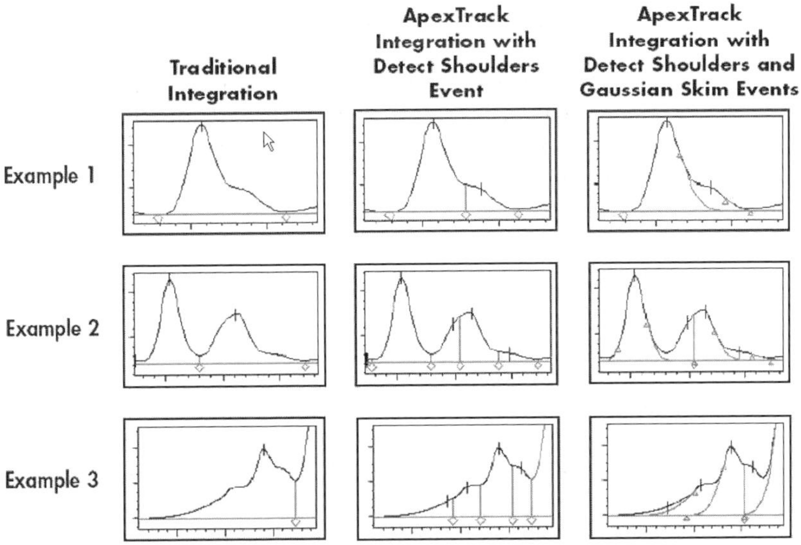

FIGURE 11 The ApexTrack™ integration algorithm and its ability to easily identify and quantitate shoulders. These shoulder peaks may be quantitated using simple dropped perpendicular lines or through Gaussian peak fitting.

regarding the use of external programs. Another key benefit that can be derived from direct instrument control is "interactive system suitability". Systems with this capability can perform real-time calculations and compare the results against your assay limits. Within a single method you can specify upper and lower limits for your system suitability calculations, as well as the assay limits that are specified in your SOP. The real benefit to this technology occurs when there is a limit failure or fault. The analyst can specify the immediate action that is appropriate for the specific assay. If the CDS detects a failure, the running samples may be stopped, the sample where the failure occurred may be reinjected, the system can move to the next sample, or the system can abort the current sequence and move on to the next sequence that may be in the queue. This capability may be able to give your laboratory the type of real-time information that is necessary to prevent running an entire sequence, only to find that it had a failure.

Another area that has seen significant attention by CDS manufacturers is the ability for today's systems to interface more easily with other business systems. In the past, LIMS was the predominant system that was linked to the CDS. Most of the interfacing was file based, and usually required that the CDS send out a result file that could be read and parsed by the LIMS. It was also possible for the LIMS to send a sample list or sequence to the CDS. Most laboratories, however, were more concerned about moving sample result information from CDS to LIMS.

Today, pharmaceutical companies are linking CDS to business systems such as SAP, chemical structure databases, electronic laboratory notebooks, and corporate data warehousing solutions. Many commonly used software products utilize COM Automation (Microsoft's Component Object Model) to provide a powerful array of programmable objects that allow seamless connections between applications.

IV. SPECIFIC NEEDS OF THE PHARMACEUTICAL INDUSTRY

The pharmaceutical industry has had an enormous impact on the evolution of existing technologies and the development of new ones. This is especially true for the CDS manufacturers. Software requirements were somewhat stagnant through the early 1990s. Most laboratories were satisfied with network-based systems that could acquire data through A/D hardware. These systems also did a reasonable job at dealing with the needs of Research and Development (R&D), quality control (QC), and manufacturing. For the most part, CDS provided adequate data handling for most of the chromatographic assays that were based on simple, two-dimensional detectors such as UV-Vis, RI, fluorescence, and electrochemical. For laboratories that needed mass detection or photodiode arrays, these were often treated as very specialized systems connected to dedicated workstations and, in some cases, were run by dedicated scientists.

All of this was to change quite dramatically. Advances in detector technology, the impact of regulatory compliance, biotechnology and bioengineered drugs, combinatorial chemistry, and high throughput screening techniques would all play major roles in the evolution of the CDS market.

A. Departmental Requirements

Looking at the industry from a CDS viewpoint, we see some very significant changes in the various laboratories within R&D, QC, and manufacturing. The corporate demands for bringing products to market faster have forced laboratories to work together. If this is to be accomplished, the entire process must become more efficient and the information that is generated must become more accessible to the various parts of the organization that need it. Figure 12 shows a model of research-based pharmaceutical laboratories that perform chromatographic analyses. If we now take a closer look at the laboratories that are labeled as research, development, and QC, we will begin to see how this affects CDS development.

Research and development: When we look within drug discovery, we typically find laboratories performing combinatorial chemistry, targeted

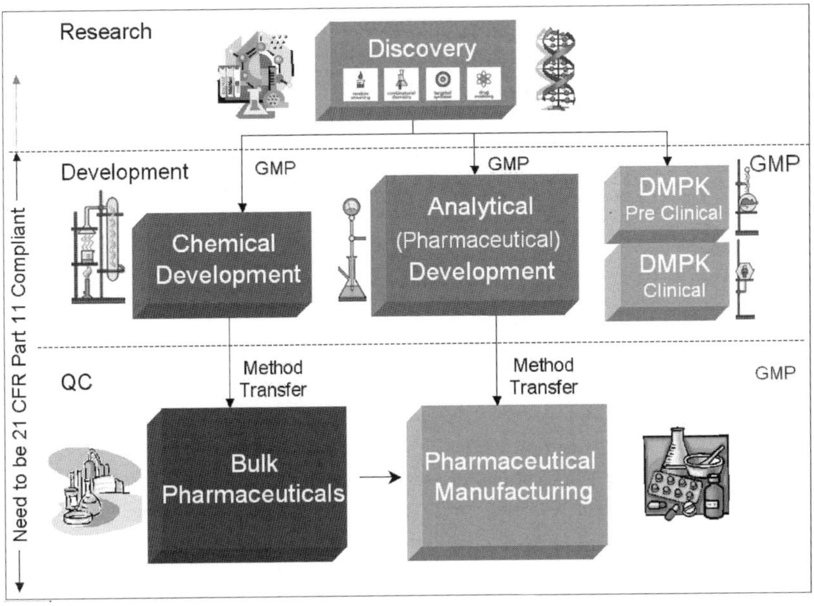

FIGURE 12 An illustration of research pharmaceutical laboratories where chromatographic analyses are performed.

synthesis, drug modeling, and high throughput screening. These laboratories perform millions of analyses a year in an attempt to discover a new chemical entity (NCE) that will prove to have therapeutic activity. These laboratories rely heavily on mass spectrometry linked to chromatographic systems. Data output for these laboratories is enormous and can easily reach the terabyte range per year. From a patent protection/intellectual property viewpoint, it is critical that the organization be able to trace any new compound from the date and time of its identification through its development and clinical testing phase. As the compound moves its way through chemical development, analytical development, and drug metabolism and pharmacokinetics (DMPK), pharmaceutical companies need to "weed out" the ones that are not capable of becoming a viable product. With the introduction of a new drug in the 300–700 million dollar range, early determination is critical. These are some of the reasons why pharmaceutical companies are linking chemical libraries, CDS, LIMS, and data warehousing systems together. Information access and management across departments leads to faster, high-quality decisions. This is a critical element in pharmaceutical product development. Figure 13 shows a chromatographic report where the scientist uses a mass detector and associates chemical structures to the results. This gives all scientists who have access to the data the ability to search the chromatographic database using mass, chemical structure, or chemical sub-structure as the search

Empower LC-MS Report

Reported by User: Robert P. Mazzarese Project Name: Sales Project

SAMPLE INFORMATION			
Sample Name:	7BenzoMix 50ng 050401 1	Acquired By:	System
Sample Type:	Standard	Sample Set Name:	
Vial:	61	Acq. Method Set:	30to40 in20 35TRT
Injection #:	2	Date Acquired:	Saturday, May 05, 2001 8:56:31 AM
Run Time:	25.0 Minutes	Injection Volume:	5.00 ul

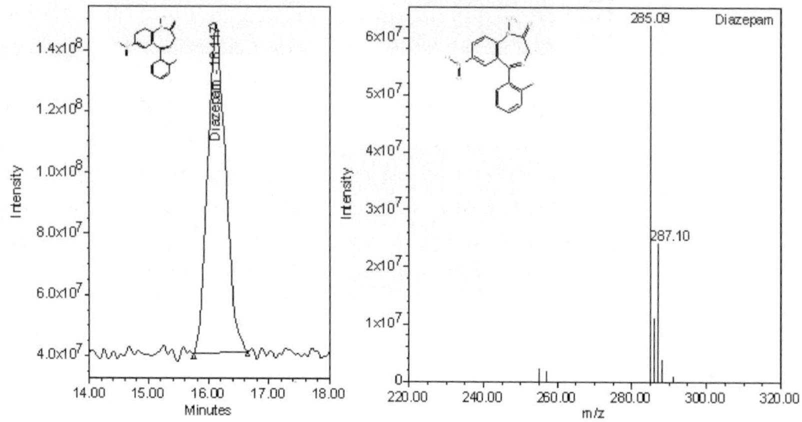

	Channel Name	Retention Time (min)	Name	Area (µV*sec)	Structure 1 Mol Wt	% Purity	USP Tailing
1	MS TIC	16.117	Diazepam	2235852444	313.280	100.00	1.135

FIGURE 13 An HPLC report using mass detection and chemical structure libraries.

criteria. This is a powerful capability for laboratories that are trying to streamline their discovery activities.

Quality control: Although many QC laboratories are performing routine analyses using standard methods, we are beginning to see their needs change as well. The methods that are used in QC were originally created by the chemical development or analytical development organizations. Since method troubleshooting is often necessary, we are beginning to see R&D and QC groups standardizing on a common CDS platform. This not only simplifies the method transfer process, but it also makes it easy for the development laboratory to view assays in real time so that issues may be quickly resolved. This too is critical to the business, especially if the development laboratories and manufacturing sites are in different cities or countries. We are also seeing QC laboratories beginning to use more sophisticated detectors like photodiode array and mass spectrometry. In some cases it is being used for the purpose of method troubleshooting. For some QC laboratories, there is the potential advantage for mass spectrometry to provide positive compound identification with the added ability to confirm peak purity. Either way, these information-rich detectors will significantly impact the data footprint for the laboratory.

Another area of concern involves the actual results that are generated. Many of today's laboratories are struggling with customized, external applications that are needed for specialty calculations and custom reports. The use of commercially available spreadsheets, internally developed software, and custom reporting tools are often a source for transcription errors and an inefficient use of time in the laboratory. This has driven many CDS suppliers to incorporate these capabilities into their products. A list of CDS Suppliers and their products is given in Table 1. Specialty calculations for dissolution, % label claim, and % recovery can now be performed as a part of the processing method. Customized reports that show only the required information are now simple to produce, and do not require the use of external software packages. This can be a significant improvement in productivity for the laboratory when you consider the countless hours spent to perform these functions outside of the chromatography data system. see Figures 14 and 15 for example reports.

B. Regulatory Requirements

Pharmaceutical laboratories have been following GLP/GMP guidelines for many years. The most significant change occurred in 1997 with the FDA issuing a formal response to the industry's request for guidelines surrounding the use of paperless record systems in a current Good Manufacturing Practice (cGMP) environment. On August 20, 1997 the FDA issued its official ruling known as 21 CFR Part 11. Its purpose was to provide general guidelines for the proper use of electronic records, electronic signatures, and handwritten signatures

TABLE 1 CDS Suppliers and Their Latest Products

Company	CDS	Website
Agilent Technologies, Inc.	Cerity	www.agilent.com
Dionex Corporation	Chromeleon	www.dionex.com
Perkin-Elmer, Inc.	TotalChrom	www.perkinelmer.com
Scientific Software, Inc.	EZ Chrom Elite	www.scisw.com
Shimadzu Scientific Inst.	Class VP	www.ssi.shimadzu.com
Thermo Electron Corp.	Atlas	www.thermolabsystems.com
Varian Inc.	Galaxie	www.varianinc.com
Waters Corporation	Empower	www.waters.com

Empower software

Inset Chrom and Sign off Book

Reported by User: Robert P. Mazzarese (Robert)

Project Name: Audit_Trail_1

SampleName	Paraben STD1	Sample Type	Standard
Vial	1	Date Acquired	Wednesday, October 26,
Injection	1	Acq Method Set	Parabens
Injection Volume	10.00 ul	Processing Method	Paraben_ProcessingMth1
Channel	486M1 Ch1	Date Processed	Friday, February 06, 1998
Run Time	10.0 Minutes		

Auto-Scaled Chromatogram

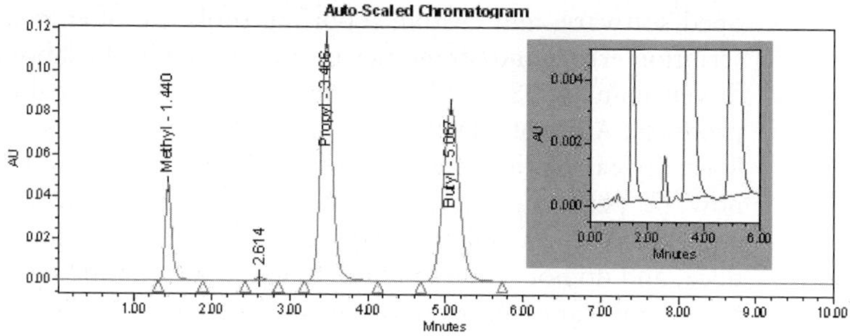

Component Results with System Suitability

	Component	RT	Area	Height	Amount	Units	USP Resolution	USP Tailing	Plate Count
1	Methyl	1.440	269034	46397	25.000	ug/ml		1.31	1563
2	Propyl	3.466	1188245	114492	25.000	ug/ml	3.49	1.11	2602
3	Butyl	5.067	1187618	82402	25.000	ug/ml	4.86	1.07	2866

Result Sign Off

	Sign Off Full Name	Sign Off Date	Sign Off Reason
1	Robert Paul Mazzarese (Robert)	Wednesday, January 19, 2000 1:56:47 PM	Sign Off Level 1, Reason: Approved by Laboratory Manager.
2	Robert Paul Mazzarese (Robert)	Wednesday, January 19, 2000 2:29:20 PM	Sign Off Level 1, Reason: Reviewed by Laboratory Manager.

FIGURE 14 A custom report with system suitability and electronic sign off.

executed to electronic records as being equivalent to paper records with handwritten signatures. Although this ruling affected many different types of electronic systems used by the pharmaceutical industry, it was highly applicable for CDS. To this day, CDS manufacturers are implementing capabilities in their products to assist the industry in achieving this new level of compliance. One other outcome of this regulation pertains to the areas of the company that fall under its jurisdiction. For years, laboratories that were performing drug discovery type assays often fell outside of the scope of their companies' GLP regulations as their work related to data archiving and instrument qualification (see Figure 12). If an experiment was not successful, it was OK to delete the data. Today, that attitude has changed. First and foremost, this is valuable information that is the property of the company. Even failed experiments need to be part of the corporate database of information. Second, the scope of 21 CFR Part 11 requires that all electronic records

Empower software

Custom Dissolution Report

Reported by User: Robert P. Mazzarese (Robert) Project Name: Sales_Project

Dissolution Plot : Prednisone

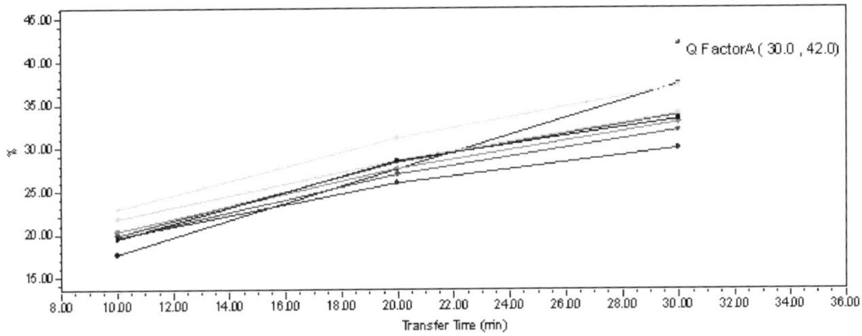

Dissolved % for Component: Prednisone Channel: 2487Channel1

	Bath	Vessel	Injection	10.0 min	20.0 min	30.0 min
1	A	1	1	19.47	28.51	33.28
2	A	2	1	19.61	26.95	32.01
3	A	3	1	21.79	28.55	33.90
4	A	4	1	22.92	31.15	37.11
5	A	5	1	20.36	27.64	32.91
6	A	6	1	19.66	25.95	29.90
7	A	7	1	17.70	27.58	37.33
8	A	8	1	19.89	28.35	33.76
Mean				20.2	28.1	33.8
Max				22.9	31.1	37.3
Min				17.7	25.9	29.9
% RSD				7.8	5.4	7.3

FIGURE 15 A custom dissolution report.

adhere to the ruling. What better place to start than with the discovery assays that first identified the NCE.

Automated qualification software is yet another area of CDS development that has been affected by regulatory compliance. When you look at today's CDS solutions, it is typical to see the system providing direct digital control of the instrumentation. Many CDS manufacturers offer software tools that are capable of automating the installation qualification (IQ) and operational qualification (OQ) process. In addition, some manufacturers also offer software that can automate the performance qualification (PQ) for the various forms of instrumentation that it is capable of controlling. These automated software tools are not only time savers for the laboratory, but they also help to properly document the system qualification effort.

One final aspect of CDS development that has been affected by regulatory compliance is multi-vendor instrument control. In an ideal world, you would choose a CDS and it would control any instrument that you needed to connect. Although technically feasible, the regulatory impact can be significant. If the CDS that your laboratory is using is capable of controlling instrumentation manufactured by another supplier, you may want to discuss this with your internal quality organization. The reason for this is quite simple; for any instrument that is controlled, you should determine whether the communication protocols were legally obtained with the support of the instrument manufacturer. For any instruments where the protocols were developed through reverse engineering, the burden of test falls on your company. You must plan for additional qualification and acceptance testing in order to validate the accuracy and reliability of the control provided.

V. FUTURE DIRECTION

Although changes in instrumentation and laboratory compliance will continue to drive the evolution of CDS products, the real challenge is related to globalization and its impact on *information accessibility* and *information management*.

Laboratories are already investigating the use of wireless devices that can be used to more efficiently access their software applications and data. Tablet PCs and personal digital assistants (PDA) are now capable of providing secure, wireless access to many of the applications that are used in the laboratory.

We are also very close to a time when electronic laboratory notebooks will capture text, tables, spreadsheets, images, drawings, spectra, video, chemical structures, and chemical reactions, and place the information in a searchable corporate-wide database. These electronic notebooks will be seamlessly linked to CDS, LIMS, structure databases, enterprise resource planning (ERP) systems, production planning systems, and corporate-wide data warehousing systems. Pertinent information about a compound, an assay, a project, or study will be automatically linked regardless of where the data originated. Scientists will be able to submit searches based on chemical structures and substructures in the hope of identifying work previously performed on similar compounds. Laboratories at different sites that are working on the same project will have a common location where all the data resides. It is likely that the web will play a significant role in the evolution of this type of system. As multiple laboratories from multiple sites around the world feed information into this system, it is easy to imagine access to this immense base of knowledge through a web portal. For the pharmaceutical industry, this will be the beginning of a true corporate information management system.

ACKNOWLEDGMENTS

The author acknowledges the historical information regarding Hewlett-Packard products from Michael Kraft and Ludwig Huber of Agilent Technologies.

I would also like to thank Todd Miller, John Van Antwerp, and Jeff Jalinski of Waters Corporation, and Michael Dong of Purdue Pharma for their helpful ideas and review of the material included in this chapter.

REFERENCES

1. Freedman, A., *The Computer Glossary,* 6th ed., Amacom, New York, 1993.
2. Types of Analog to Digital Converters, Waters Corporation, Milford, MA, Reference Documentation, 1992.
3. Data Acquisition and Processing Theory Guide, Waters Corporation, Milford, MA, 2002, 71500031209 Revision A.
4. Flannery, B. P., *Numeric Recipes in C,* Cambridge University Press, Cambridge, UK, 1988.
5. Massart, D. L., et al., *Chemometrics: A Textbook,* Elsevier Science Publishers, Amsterdam, 1988.
6. Papoulis, A., *Signal Analysis,* McGraw-Hill, New York, 1977.
7. Snyder, L. R. and Kirkland, J. J., *Introduction to Modern Liquid Chromatography,* 2nd ed., Wiley-Interscience, New York, 1979.
8. Strang, G., *Linear Algebra and its Applications,* Harcourt Brace Jovanovich, Inc., New York, 1988.

22 NEW DEVELOPMENTS IN HPLC

SATINDER AHUJA[a] AND DIANE M. DIEHL[b]

[a]*Ahuja Consulting, Calabash, NC*
[b]*Waters Corporation, Chemistry Operations, Milford, MA*

ABSTRACT

This chapter is intended to serve as a general overview of new and emerging HPLC technologies and is divided into four sections: simplifying sample preparation, new column technologies, improvements in detectors, and improvements in HPLC throughput.

I. INTRODUCTION

The earlier chapters in this book focus on the current state-of-the-art in HPLC and provide useful practical advice for analyzing pharmaceuticals by HPLC. In this chapter, we plan to offer a glimpse into the new developments in HPLC and discuss how these advances can help improve the performance of HPLC. The progress in HPLC technology has been driven by the need to improve sample throughput because project time lines are short, and the pressure to bring the next blockbuster drug to

Handbook of Pharmaceutical Analysis by HPLC,
S. Ahuja and M.W. Dong, editors.

market in the shortest time possible has increased. Many of these advances revolve around reducing HPLC analysis time, which entails improving the columns and the instrument performance. Additionally, research has centered on reducing sample preparation time.

This chapter is divided into four major subsections:

Simplifying Sample Preparation
New Column Technologies
Improvements in Detectors
Improvements in HPLC Throughput

Section II covers the latest trends in reducing sample preparation time, including direct sample infusion/injection and on-line solid phase extraction (SPE). In Section III, we focus on newer trends in stationary phases and how these phases hope to offer different selectivities compared to current C18-based phases. Section IV briefly provides a few observations on how new detectors are increasing the versatility of HPLC. Finally, in Section V we examine monolithic columns, small particles packed in short columns, high-temperature LC, ultra high-pressure LC, and parallel injection techniques.

II. SIMPLIFYING SAMPLE PREPARATION

Analysts are often under pressure to analyze and quantify an increasing number of samples every day. These samples arise from combinatorial chemistry libraries in our search of new drugs, stability studies, or clinical studies. Sample preparation is often the bottleneck in these studies, and much research has focused on eliminating most sample preparation steps. The direct infusion of samples into an electrospray ionization–mass spectrometry (ESI-MS) source can provide this option. One such commercial system is the NanoMate and ESI chip (Advion BioSciences). This automated system serves as both the autosampler and the nanoelectrospray ionization source. The potential upsides of this technique are automation, the elimination of sample carryover, low sample consumption, and high sensitivity. Prior to analysis, simple and quick protein precipitation of samples is recommended. The chip itself is a device that consists of a 10×10 array of nozzles etched from the surface of a silicon wafer. The autosampler picks up a new pipette tip, aspirates 3 μL of sample, and delivers the sample to the inlet side of the ESI chip. A channel extends from the nozzle through the chip to an inlet etched on the opposite planar surface. The pipette tip seals against the inlet channel of the chip, a voltage is applied, and nitrogen is allowed to flow to infuse the sample into the source. After infusion the pipette tip is ejected.[1]

While the idea of limited sample preparation and direct infusion into the MS are promising, this technique has not yet caught on in many laboratories. Ion suppression continues to be a limitation of this technique. It is also a limitation in other fast sample preparation methods and fast analyses.[2] Protein precipitation is a quick and dirty technique, and many components from the plasma matrix remain. With this direct infusion into the MS, these matrix components are also sprayed into the source with the analytes of interest and can cause ion suppression. Additionally, many samples can contain several analytes, often including internal standards. In many instances, these analytes cause self-suppression when infused together into the source. With a proper sample cleanup such as SPE, either off-line or on-line, this technique may prove to be more useful in the future.

If direct infusion does not provide the desired faster sample preparation, then the automation of SPE may be tried. The research is continuing on the development of on-line SPE instruments. There are several ways of approaching the design of such systems. For example, Spark Holland has developed systems that utilize individual SPE cartridges (10 mm length), called Prospekt II devices. In their system, they have an automated cartridge exchanger (ACE) that automatically changes the cartridge for each sample. Samples can be loaded into 96-well plates and then placed into the automated system for on-line sample preparation. After extraction, the samples can be injected onto an HPLC column for separation and quantification. The advantages of this system are minimal sample handling and a new cartridge for each sample. However, some of the disadvantages include potential sample carryover because of the system hardware itself and overall cycle time for analysis. Nevertheless many researchers in the pharmaceutical industry have found this system to be robust and useful in their assays.[3–6]

A second approach to on-line SPE is to use an SPE extraction column that can be used for hundreds of samples. In the simplest of systems, two pumps (either HPLC systems or stand-alone pumps) are connected to an extraction column and an analytical column via 6 or 10 ports, and these are further linked to an MS system.[7] The pump that is connected in-line with the autosampler loads the sample under high flow rate (3 to 5 mL/min). The large molecules from the matrix are not retained by the SPE sorbent and are diverted to waste. The analytes of interest are retained by the sorbent. The valve then switches so that the second pump with the elution solvent is now in-line with the SPE column and elutes the analytes onto the analytical column for HPLC/MS analysis. This type of system has proved useful for the analysis of small molecules in a variety of sample matrices such as plasma and urine.[8,9] While it is relatively straightforward to plumb this type of system with components already in the laboratory, commercial systems are available from such companies as Cohesive Technologies, who have put together a variety of other vendors' components into one comprehensive system.

One of the drawbacks of these simple systems is the inability to fully utilize the power of SPE. Because of the limitation of having only a load and elute pump, selective wash steps cannot be utilized to eliminate interfering components that can lead to ion suppression. This can prevent achieving the low limits of quantification (LOQ) that can be routinely achieved using an off-line SPE system. To address these issues, a novel system utilizing four pumps, one HPLC and 4-, 6- and 10-port switching valves, was constructed as a proof-of-concept example.[10] In this system, an Oasis® HLB extraction column was used for SPE extraction, and an analytical column was also utilized for further separation. The sample is loaded at 4 mL/min onto the SPE column. The other pumps are used to wash the SPE column with various solvents to selectively wash off the interfering, unwanted components from the sample matrix. The last pump is then brought on-line to elute and focus the analytes on the analytical column. By utilizing a mixed-mode SPE sorbent (reversed-phase and ion-exchange retention), a more selective SPE cleanup can be achieved. Detection limits are of the order of 0.1 ng/mL for the simple cleanup method and can be pushed even lower with further optimization of the wash steps. As of this writing, this system is not available as one comprehensive commercial system, but can be plumbed by following the example in Reference 10. Future systems can be designed that would include multiple extraction and analytical columns in line with rapid HPLC systems to maximize sample throughput.

III. NEW COLUMN TECHNOLOGIES

Although HPLC column technology is considered to be a mature field now, improvements and new developments are being made continuously in the stationary phases. One of the improvements has been the reduction in particle sizes. Smaller particles help to improve mass transfer and provide better efficiency. Manufacturers are producing particles down to 1.5 μm in diameter, although 3- and 5-μm particles are still the most popular. Because of the smaller particle sizes, the backpressure increases proportionally to the inverse of the square of the particle size. Most commercially available HPLC systems cannot accommodate the pressures required to operate these columns at optimum flow rates. This has led to the introduction of systems that run at high pressures.

Chapter 4 provides a discussion of zirconia and hybrid columns. We will only lightly cover here the hybrid columns and expand on the zirconia discussion. Hybrid columns were developed to reduce the amount of silanol activity as well as to impart better hydrolytic stability, especially at high pH. By incorporating different functional groups into the matrix, the hardness and pressure stability are preserved and often enhanced. The high pH stability allows the use of buffers and additives with pHs above 9—which suppresses the ionization of basic analytes and leads to better retention and strong differences in the selectivity of the separation. We look forward to continued improvements in hybrid particle technology that will allow for even better stability at high and low pH, as well as novel selectivities compared to current HPLC columns.

A. Metal Oxide Phases

Two recent papers[11,12] offer excellent reviews of the research into using ultra-stable metal oxide-based stationary phases for HPLC. The reviews cover zirconia-, alumina-, and titania-based phases, from the surface properties and preparation to the use of these columns. Metal oxides allow HPLC to be run at both ends of the pH scale where silica-based columns cannot withstand high-pH mobile phases, allowing more flexibility in method development. Additionally, as mentioned earlier, these phases permit the use of high temperatures well above traditional silica-based columns, enabling faster HPLC methods to be run. The selectivity is tunable by the choice of buffer, and the selectivity of acids and bases on these phases is very different compared to silica-based particles. However, to use these columns effectively, the end user has to put aside all silica particle-based knowledge and understand the underlying Lewis acid–base chemistry that is inherent in these phases. Other disadvantages are limited surface areas and pore sizes and a lack of commercially available titania phases and manufacturers to date.

Titania is being examined for use as a base material for bonded stationary phases because of its stability and its ability to separate bases under normal-phase conditions.[13] Other researchers are investigating silica that has been surface modified with titanium oxide followed by the immobilization of poly(methyloctylsiloxane).[14,15] These columns show promise for routine chromatographic analyses, but more research needs to be performed to evaluate the best manufacturing methods.

Future directions in these phases include metal oxide monoliths, phases with smaller pores and larger surface areas for preparative separations, and exploration of MS-compatible phases.[12]

IV. IMPROVEMENTS IN DETECTORS

Detectors have been discussed at some length in Chapter 3. Here we will highlight some of the improvements that are being made in HPLC detectors to help improve detectability. Examples of such detectors are a UV detector that allows simultaneous recording of absorbance at many wavelengths and a refractive index detector that can perform IQ, OQ, and PQ validation with onboard noise, drift relating to lamp age, and temperature readouts.

The light-scattering detector has made inroads into size exclusion chromatographic research, quality control, and process monitoring. The evaporative light scattering detector (ELSD) described in Chapter 3 is finding greater use in pharmaceutical analysis because it provides better detection limits than the refractive index detector and is compatible with gradient elution. In fluorescence detection, laser-stimulated fluorescence detectors can help improve detection up to 10 000 times over conventional detectors.

LC-MS has revolutionized the field of detection in HPLC so much so that MS has become a required detector for pharmaceutical analysis. Improvements are being made constantly even with this detector. For example, with ion-trap MS a 10-fold improvement in detectability has been reported. This improvement is achieved by increasing the trap capacity, trap geometry, and new scanning algorithms. The use of LC-MS for monitoring pharmaceutical impurities is continuously increasing. After evaluating time-of-flight, single quadrupole, and triple quadrupole mass analyzers for quantifying drug product impurities, it was concluded that LC-MS offers substantial improvements in selectivity and enables more rapid method development than LC-UV methods commonly used for this purpose. Furthermore, it should be noted that LC-MS has become the method of choice for monitoring the types of compounds that form the bulk of the drug discovery process. Further discussion on this information-rich method of analysis can be found in the next section.

V. IMPROVEMENTS IN HPLC THROUGHPUT

A. Monoliths

Monolithic columns have been available for many years and continue to be a part of the high-throughput discussion. Chapter 4 provides a brief description of the structure of monolithic columns as well as a discussion of some of their advantages and disadvantages. In this section we will briefly discuss some of the applications developed as related to fast analyses and what chemists are hoping to achieve by using monoliths.

The allure of running fast analyses have driven many researchers to the monoliths. For example, Chromolith by Merck KgA has been investigated by many researchers who are looking for faster separations of small molecules. There have been reports of successful direct injections of plasma onto the column[16,17] resulting in sub-1 minute run times that are reproducible, sensitive, and applicable to high-throughput PK screening. Other researchers have had similar results doing a small amount of sample pretreatment (i.e., using protein precipitation) for analytes in human plasma samples.[18] Still other researchers have had success in fast analyses by coupling on-line SPE with monoliths for 1.2-min total cycle times.[19] However, many limitations exist and have prevented monoliths from taking on a larger role in the laboratory. One of these limitations is that only one "silica-based chemistry" is available. As a group of researchers discovered, this current phase does not adequately retain polar analytes.[20] Instead of the monolith, they were able to use an HILIC column over a monolith with high flow rates and low backpressures, and separate nicotinic acid and six metabolites in under 1 min, as shown in Figure 1.

Because of the current limitations in the manufacture of monoliths, internal diameters of less than 4.6 mm (preferably less than 2 mm) are not commercially available. Therefore, the high flow rates (i.e., often greater than 5 mL/min) prohibit their use because of solvent consumption and waste generation, as well as the limitations of some HPLC systems to accurately and consistently deliver the high flow of solvents. If mass spectrometry is used, the flow needs to be split. The disadvantage of the combination of the large column diameter and the need to split before the detector results in a significant loss in sensitivity. Additionally, the reproducibility of monolithic columns can be poor because of the cladding process; as a

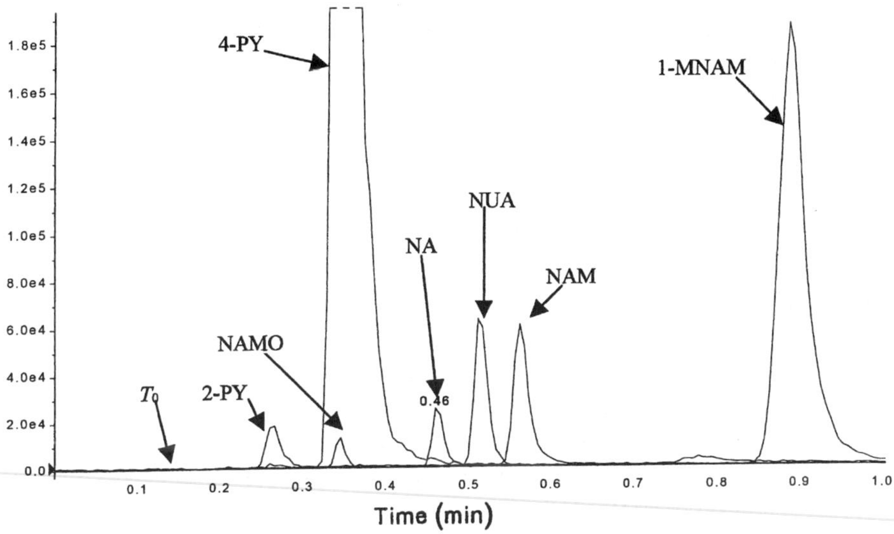

FIGURE 1 A reconstructed extracted ion chromatogram of nicotinic acid and its six metabolites under HILIC conditions. Column: Hypersil silica (4.6 × 50 mm) at a flow rate of 4 mL/min. Mobile phase A is water, mobile phase B is acetonitrile, both containing 1% formic acid. Gradient is 0.01–0.25 min 90% B to 65% B; 0.25–0.90 min 65% B to 50% B. NA: nicotinic acid; NAM: nicotinamide; NUA: nicotinuric acid; 2-PY: 1-methyl-2-pyridone-5-carboxamide; 1-MNAM: 1-methylnicotinamide; NAMO: nicotinamide-*N*-oxide; 4-PY: 1-methyl-4-pyridone-5-carboxamide. (Reprinted with permission from Reference 20.)

result, each column is a unique batch. Until a variety of stationary phases and preparative columns are available, monoliths will not find a strong place in small-molecule research. An alternative to monoliths is to use shorter columns packed with smaller particles.

B. Smaller Particles Packed in Shorter-Length Columns*

Using current HPLC technology, we know that by optimizing the column length, particle size, flow rate, and temperature of the separation, we can achieve faster separations. We can reduce the length of the column to shorten our run time and also give us lower backpressures. However, the efficiency of the column is directly proportional to the length of the column. Therefore, we might lose efficiency by reducing the length of the column. We can regain some efficiency by using a smaller diameter particle. However, with smaller particles, we can also observe higher backpressures. We can also increase the flow rate to make our run time shorter, but we might lose some of our resolution, and our backpressures might be outside the limits of the HPLC system. By increasing the temperature, we can also speed up a separation and reduce the viscosity of the mobile phase, thereby reducing the backpressure. However, many silica-based particles cannot be run above 45°C without suffering from short lifetimes. Taking all of this into consideration, we must find the right balance of particle size, column length, flow, and temperature that gives us the fastest separation with the desired resolution.

In order to talk about fast gradient separations, we need to review the principles involved in measuring the performance of a gradient separation. We can use the concept of peak capacity to measure the separation power of a particular gradient on a given column.[21,22] The peak capacity (P) is defined as follows:

$$P = 1 + \frac{t_g}{w} \tag{1}$$

where t_g is the gradient run time and w is the peak width. By making various substitutions of chromatographic relationships, we can obtain a mathematical relationship that may be used to assess the gradient performance as a function of the gradient duration, column length, particle size, linear velocity, and diffusivity of the analyte.[21,22]

Several manufacturers such as Agilent and Waters have developed short columns (less than 20 mm in length) packed with 2.5- and 1.8-μm particles. In fact, a decade ago MICRA selected 1.5-μm as the desired particle size for nonporous particles. These columns offer faster analyses

*For further extensive discussion on the theoretical basis of fast HPLC, see Chapter 4.

while maintaining similar efficiency and resolution as larger particles in longer columns. However, we need to discuss how to transfer and develop methods on these shorter columns. A separation was developed for six analytes on a 4.6×150 mm, 5 µm C_{18} column. A gradient separation in 20 min with a 25-min total cycle time is shown in Figure 2a. In order to scale the separation to a 20-mm length column, Eq. (2) was used to scale the gradient time:

$$\frac{L_2}{L_1} \times t_{g1} = t_{g2} \tag{2}$$

where L_1 is the length of the longer column, L_2 is the length of the shorter column, t_{g1} is the gradient time on the long column, and t_{g2} is the new gradient time on the short column. To run the same separation on a 4.6×20 mm 3.5-µm column, the new gradient time is 2.7 min. The separation was run at 3 mL/min and the gradient time was reduced to 2 min

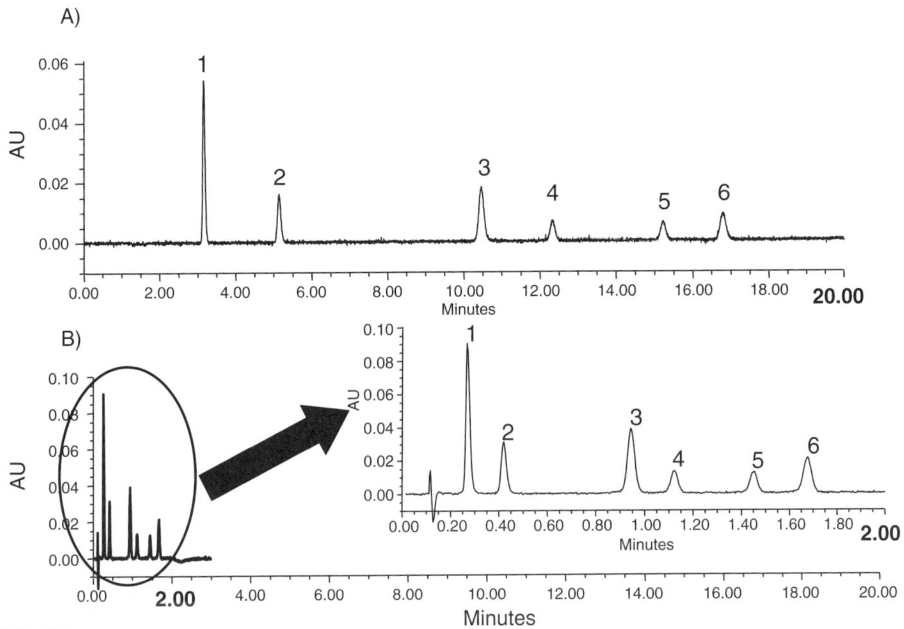

FIGURE 2 **Example of a scaled-down separation. (a) Separation of six analytes on a 4.6×150 mm, 5 µm, XTerra® MS C_{18} column. Total cycle time of 25 min. (b) Scaled-down separation on a 4.6×20 mm*IS*™, 3.5 µm, XTerra® MS C_{18} column. Total cycle time of 3 min. Sample: 1, caffeine; 2, aniline; 3, *N*-methylaniline; 4, 2-ethylaniline; 5, 4-nitroanisole; and 6, *N-N*-dimethylaniline. Mobile phase consisted of A, water; B, acetonitrile; and C, 100 mM ammonium bicarbonate buffer, pH 10. Gradient separation conditions were 80% A, 10% B, and 10% C to 50% A, 40% B, and 10% C over 20 min, with 5 min re-equilibration time, at a flow rate of 1.4 mL/min, and over 2 min, with a 1 min re-equilibration time, at a flow rate of 3 mL/min, on the 150-mm and 20-mm length columns, respectively. (Courtesy of Waters Corp.)**

for a total cycle time of 3 min. This separation is shown in Figure 2b. This is an 8.3-fold reduction in the cycle time from the separation on the longer column. In terms of quantifying cost reductions, this means 8 times as many samples can be analyzed in a given time period, freeing up systems and analyst time for other projects. Additionally, even at a flow rate of 3 mL/min, the short cycle times also allow for less solvent to be consumed and disposed for a given laboratory.

While smaller particles in shorter columns can reduce run times, resolution and efficiency are sometimes compromised. We would like to pack smaller particles in longer columns for better efficiency, but the backpressures generated are outside the capability of current commercial HPLC instrumentation. One way to alleviate the pressure issue is to use high-temperature HPLC.

C. High-Temperature HPLC

The main benefit of operating at high temperatures is that it is possible to obtain ultrafast separations. At high temperatures, the diffusivity of the analytes increases, thereby reducing retention and decreasing run times. Additionally, pure water can often be used as the eluent, eliminating toxicity issues and solvent costs. A full discussion of the heat transfer calculations and the effect of extra column tubing on efficiency is covered in Reference 23. The results from this study indicate that the practical maximum flow rate at 150°C is 15 mL/min. Increasing the column temperature at high flow rates significantly improves the efficiency. A separation of alkylphenones was run 50 times faster than at room temperature (from 20 min to 20 s). By operating at these high temperatures, pure water was used to separate a group of phenolic compounds, as shown in Figure 3.

However, the user must be aware that there are three main drawbacks of using high temperatures in HPLC that must be overcome: the stationary phase must be stable, the temperature of the eluent must match the temperature of the column, and the analytes must be thermally stable on the time scale of the chromatographic run. Recent publications cover these issues in high-temperature ultrafast liquid chromatography (HTUFLC)[23–26] and offer solutions that allow HTUFLC to be utilized.

From both theory and experimental evidence, raising the temperature by 10°C decreases the retention time by about 20% in isocratic chromatography and decreases the backpressure by 10% to 20% because of a reduction in the viscosity of the mobile phase. This can help to overcome the instrument limitations associated with running shorter columns packed with smaller particles, i.e., the pressure limitations of current HPLC systems. However, since the majority of reversed-phase columns available are silica-based, operating at temperatures above

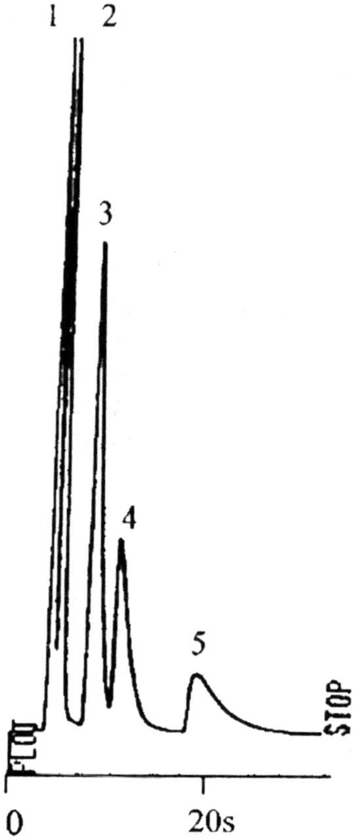

FIGURE 3 **Chromatogram showing the separation of phenols on a PS-ZrO_2 column at 120°C. Mobile phase is 100% water at a flow rate of 12 mL/min. 1, phenol; 2, 2-methylphenol; 3, 4-ethylphenol; 4, 2,4,6-trimethylphenol; 5, 4-*tert*-butylphenol. (Reprinted with permission from Reference 23. Copyright 2000, American Chemical Society.)**

45–50°C is not recommended and high-temperature LC has not been rigorously explored. With the advent of thermally stable phases such as zirconia-based stationary phases, temperatures in excess of 150°C can now be utilized. Many zirconia-based phases are available, so the stability of stationary phases is less of an issue.

By simultaneous optimization of the percent organic modifier in the eluent and the column temperature to keep the retention factors fixed, very efficient, ultrafast separation can be achieved.[24] The researchers conclude that for fast separations, the relationship between retention, temperature, and volume fraction of organic modifier needs to be taken into account. As the temperature increases, a lower volume of organic modifier is needed to speed up HPLC. Therefore, a highly retentive column should be used to counteract the loss of retention at higher temperatures. One of the potential drawbacks of using high temperature is the stability of analytes. To address that issue, Thompson and Carr[2] proposed criteria to reject or accept analytes for high-temperature analyses. They demonstrated that complex molecules can be quantified even at temperatures above 100°C in aqueous mobile phases. Also, as the time on the column decreases, the extent of on-column reaction decreases for those analytes that show some evidence of degradation.

These researchers established a series of six questions to determine whether a compound is thermally stable:[25]

1. Does it decrease the reliability of the analytical calibration curve?
2. Does it induce a significant intercept in the calibration curve?
3. Does the use of elevated temperature significantly diminish the sensitivity?
4. Does an on-column reaction distort the peak shape?
5. Does it introduce new peaks that interfere with the resolution/quantitation of the analyte or impurities?
6. Does a high-temperature eluent cause chemical reactions that alter the concentration or produce products that would interfere with quantification of the impurities?

Based on experimental data, if the answers to these questions are no, then the method can be run at high temperature.

Until recently, commercial systems to run high-temperature HPLC have not been available. Selerity Technologies has developed a forced-air column heater with mobile phase pre-heating and post-effluent cooling capable of heating to temperatures as high as 200°C. Utilizing this new column heater, the researchers examined six columns that reportedly can withstand these high temperatures.[26] The Selerity Blaze C8, Hamilton PRP-1, and Thermo Hypersil-Keystone HyperCarb columns can be used at maximum temperatures between 100°C and 200°C. The Zirchrom PBD, CARB, and DiamondBond column all exhibited some evidence of column bleed, which could lead to interferences in quantification. A new column from Zirchrom, the Zirchrom-MS column, is designed to be both chemically and thermally stable for use with mass spectrometry,[27] but data are not yet available on the thermal stability of this column.

D. Ultra High-Pressure LC

To fully utilize small particles (i.e., less than 2 μm) packed in columns greater than 50 mm in length, an HPLC system must be developed that can operate at high pressures. For the best results using columns with 1–2-μm particles, extra-column effects, dwell volumes, and detectors must be optimized for optimum performance. Additionally, commercial instrumentation capable of handling the high pressures required (i.e., 10 000 to 50 000 psi) must be designed, rigorously tested, and manufactured reproducibly and reliably.

Column efficiency is usually described by the height equivalent of a theoretical plate (H). Smaller H values (and therefore higher efficiencies) can be attained if we use smaller particles. If we look at the van Deemter equation,

$$H = A + \frac{B}{u} + Cu \tag{3}$$

where A, B, and C are constants that can vary from one column to another and u is the flow rate. We can reduce the contributions of the A- and C-terms by reducing particle size. The A-term is generally found to be proportional to the particle diameter and the C-term is proportional to the square of the particle diameter. In Figure 4, we see a comparison of the van Deemter curve for particles with diameters of 10, 5, and 1.7-μm. As the particle diameter decreases, the contribution from C is flatter, allowing for higher velocities (i.e., faster separations) without loss of efficiency.

However, the pressure drop over the column is given by

$$\Delta P = \frac{\Phi \eta L u}{d_p^2} \tag{4}$$

where Φ is the flow resistance factor, η is the viscosity, L is the column length, and u is the linear velocity of the mobile phase. The optimum flow

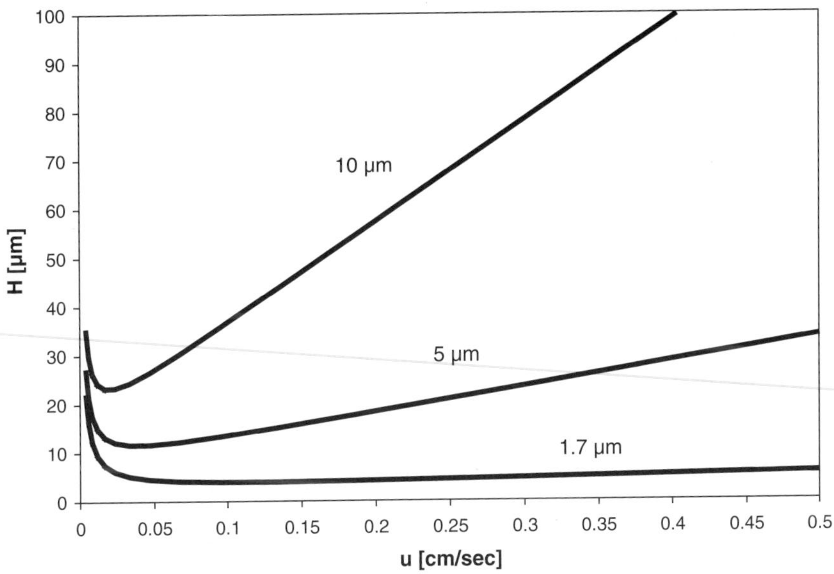

FIGURE 4 A van Deemter curve for 10-, 5-, and 1.7-μm particles. (Courtesy of Waters Corp.)

velocity is also inversely proportional to d_p. Therefore, the pressure required to run at the optimum velocity is inversely related to the cube of the particle diameter. For example, if we pack 1.7-μm particles into the same length column as 5-μm particles, the backpressure on the column would be 25 times greater.

A few research institutions have developed pumping systems that can create pressures of up to 50 000 psi (3500 bar),[28–31] but creating commercial systems has been a challenge. The engineering challenges involved with developing the pumps, seals, injectors, and detectors to operate at these pressures are enormous, but not insurmountable. Additionally, special attention must be paid to the frictional heating of the mobile phase in the column at these pressures. There can be uneven temperature profiles across the column diameter and length. This uneven radial temperature profile may lead to peak broadening. Therefore, reducing the column diameter is important.

A commercial HPLC system and columns capable of performing ultra high-pressure LC were recently introduced at PITTCON 2004 (ACQUITY Ultra Performance LC™ System by Waters). This HPLC system was designed to take full advantage of the potential of novel, sub-2-micron particles to give scientists chromatographic run times that are up to 9 times shorter than current fast HPLC systems, up to 2 times better peak capacity or resolution, and 3 times better routine sensitivity.

There are several components that are crucial and are therefore responsible for the success of this system. The first is column-packing technology. In order to realize the potential speed, sensitivity, and resolution of ultra high-pressure LC, the particle must be pressure-tolerant. This advanced packing material is a new second-generation patented hybrid particle that is made with both inorganic (e.g., silica) and organic (e.g., organosiloxane) elements.[32] These particles possess improved attributes for efficiency, ruggedness, and peak shape. Ultra high-pressure LC separations can occur only if the particles are synthesized within a narrow particle size distribution. A new sizing technology was developed that produces production-scale quantities of particles with an extremely tight particle size distribution. Additionally, learning how to pack these particles into stable columns necessitated extensive research and development.

Designing a system for sub-2-μm particle operation was a challenging task that required looking at all aspects of system design in a holistic manner. The challenge called for new thinking in many areas; new designs for solvent management, which, when combined with new software algorithms, are capable of a constant linear flow at elevated pressures; high-speed detectors capable of capturing fast separations; and utilizing faster injection cycles to keep up with the instrument's higher throughput. All of this was designed while optimizing the flow path of the system to reduce dispersion and while adding new communications technology that tracks and reports column usage. The result is a system that accentuates chromatographic speed, sensitivity, and resolution.

The power of ultra high-pressure LC is shown in Figure 5, with examples of peptide mapping. The same enolase tryptic digest sample was used to generate both chromatograms. The top chromatogram was run on a 2.1×100-mm column packed with 4.8-μm second-generation hybrid particles. The total number of peaks is 70, and the peak capacity is 143. However, the same sample run on a 2.1×100-mm column packed with 1.7-μm second-generation hybrid particles results in 168 peaks and a peak capacity of 360. Additionally, the chromatograms are plotted on the same *y*-axis, and one can see the improved sensitivity with the 1.7-μm particles. Clearly, this system provides improvements in resolution and sensitivity.

This new development in HPLC is truly exciting and has great potential. Continuing research by various groups in academe and the instrumentation industry will continue to result in expanded horizons. Future directions include developing a high-pressure system for use with capillary columns, developing systems to go up to even higher pressures

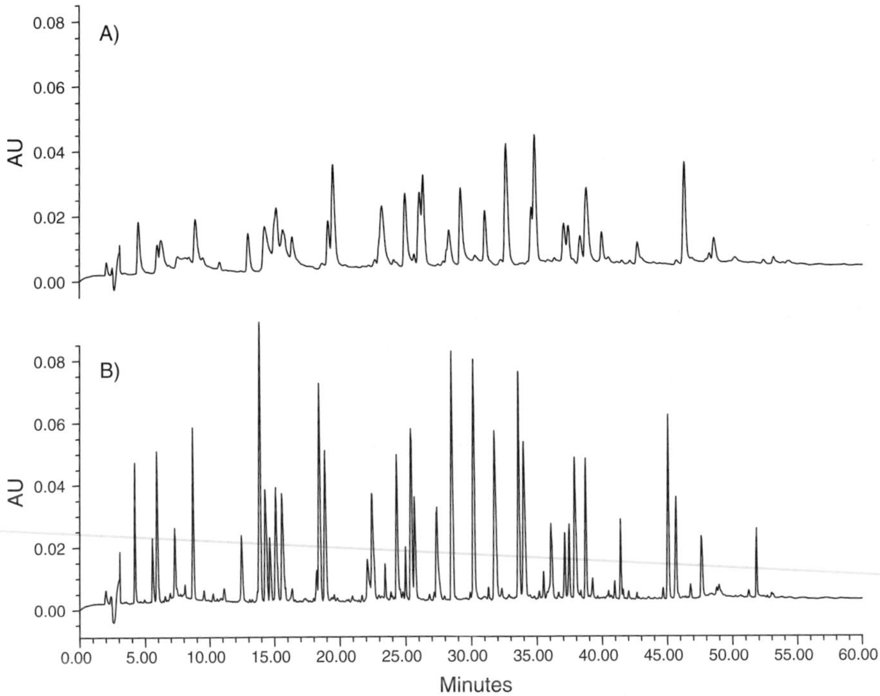

FIGURE 5 **(a) Peptide digest run on a 4.8-μm particle on a traditional HPLC system. Peak count is 70, and peak capacity is 143. (b) Peptide digest run on a 1.7-μm particle on the ACQUITY UPLC system. Peak count is 168, and peak capacity is 360. (Courtesy of Waters Corp.)**

(i.e., 50 000 psi), and creating new phases that can be used at these higher pressures.

E. Parallel Analysis

Combinatorial chemistry techniques produce hundreds, if not thousands, of samples that need to be analyzed in a relatively short period of time. While fast HPLC analyses with short columns and fast gradients have increased throughput, there are still limits on the number of samples that can be processed. Some systems have evolved that incorporate multiple injectors, pumps, and UV detectors, but are still limited by having only one mass spectrometer available for detection and analyte identification.

To fully utilize the mass spectrometer in high-throughput analyses, a multiplexed sprayer interface (MUX®) was developed that allows for parallel analysis.[33–37] In this design, the ion source consists of either four or eight electrospray needles around a rotating dish that contains an opening, allowing independent sampling from each sprayer. The four or eight samples are introduced into the mass spectrometer in parallel, in a sampling cycle that is fast relative to the LC peak width, and therefore allows four or eight samples to be analyzed simultaneously.

There are many variations on setting up the inlets to the MUX interface, depending on the needs of the laboratory. In one setup, one pump is used to deliver the flow through a flow splitter and multiple probe injector to four LC columns.[33] The flow from each column is then fed into the MUX interface. In another variation, on-line SPE is coupled with LC.[34] One pump, followed by a flow splitter, and one four-injector autosampler are used to feed samples into four extraction columns. A second pump is used (again with a flow splitter) to run gradients on the four LC columns into the MUX interface. In yet another system, an eight-channel UV MUX system is utilized.[35]

Clearly, in such a system, issues such as cross talk and reproducibility need to be investigated. Bayliss et al. concluded that the inaccuracy and imprecision of the concentrations of analytes were 15% or better, and no significant interference occurred from the adjacent sprayer (i.e., cross talk).[33] Other researchers have also found the inter-channel reproducibility to be maintained and that the cross talk was about 0.1%.[34] They mention that although the data on this system were not able to achieve the same LOQ as on a single-channel system, a comparison of the data on the MUX vs. a single-channel system showed less than a 20% difference in concentrations—leading them to conclude that the parallel system can work in drug discovery. Fang and others also conclude that even with some sacrifice in sensitivity and peak shapes on the parallel system, this system is well-suited for combinatorial library analysis.[35] Recent applications of this type of parallel system include Caco-2 permeability studies[37] as well as mass-directed purification of 10-mg on-column samples.[36]

Another potential application might be rapid method development screening—looking at four or eight columns under a certain set of conditions to investigate selectivity and retention.

A novel 24-channel HPLC by Nanostream called Veloce was introduced at PITTCON '04. The column cassette contains 24 parallel microbore columns. The eluted samples are detected by a 24-channel UV filter photometer. The advantage of such a system is that it allows one to work with multiple samples simultaneously. Other interesting systems for parallel HPLC were those introduced by Eksigent, based on microfluidic flow control, and Sepiatec GmbH, which allows the processing of 75 multiple-well plates.

F. Future Developments in Chromatography Data Systems

Chapter 21 provides discussion on chromatography data systems (CDS) and future developments in them. The real challenge in future developments in CDS arises from globalization and its impact on information accessibility and information management. Investigations are being made to utilize wireless devices that can be used to more efficiently access their software applications and data. Tablet PCs and Personal Digital Assistants (PDAs) are now capable of providing secure, wireless access to many of the applications that are used in the laboratory.

In the near future, electronic laboratory notebooks will capture text, tables, spreadsheets, images, drawings, spectra, video, chemical structures, and chemical reactions, and place the information in a searchable corporate-wide database. These electronic notebooks will be seamlessly linked to CDS, LIMS, structure databases, enterprise resource planning (ERP) systems, production planning systems, and corporate-wide data warehousing systems. Pertinent information about a compound, an assay, a project, or study will be automatically linked regardless of where the data originated. Pharmaceutical researchers will be able to submit searches based on chemical structures and substructures in hopes of identifying work previously performed on similar compounds. The Web will play a significant role in the evolution of this type of system. As various laboratories from multiple sites around the world will feed information into this system, it is easy to imagine access to this immense base of knowledge through a Web portal. For the pharmaceutical industry, this will be the desired corporate information management system.

VI. SUMMARY

In this chapter, a few glimpses have been provided into the new developments in HPLC. Many of these advances revolve around reducing not only HPLC cycle times by improving the columns and the instrumentation, but also on improving sample preparation techniques. In Section II, the latest trends in reducing sample preparation time such as direct sample infusion/injection and on-line SPE have been discussed. In Section III, newer trends in stationary phases such as metal oxide-based phases and hybrids may offer novel selectivities as compared to C18-based phases. Section IV provides some observations on how new detectors are increasing the versatility of HPLC. In Section V, monolithic columns, small particles packed in short columns, high-temperature HPLC, ultra high-pressure LC, and parallel injection techniques were covered. All of these approaches focus on developing both columns and instruments capable of achieving the types of separations that theory predicts. It is expected that new developments in lab-on-chip will eventually produce more functional systems for small molecules. All in all, the future of HPLC is exciting, and time will unravel newer advances that will make the work of the chromatographer even faster and more fruitful.

ACKNOWLEDGMENTS

Many thanks to Uwe Neue and Jeff Mazzeo for helpful comments on the chapter and to Eric Grumbach, Kim Tran, and Uwe Neue for the figures.

REFERENCES

1. Zhang, S., Van Pelt, C. K. and Wilson, D. B., *Anal. Chem.*, 75:3010–3018, 2003.
2. Tiller, P. R. and Romanyshyn, L. A., *Rapid Commun. Mass Spectrom.,* 16: 92–98, 2002.
3. Schellen, A., Ooms, B., van de Lagemaat, D., Vreeken, R. and van Dongen, W. D., *J. Chromatogr. B*, 788:251–259, 2003.
4. Kato, K., Shoda, S., Takahashi, M., Doi, N., Yoshimura, Y. and Nakazawa, H., *J. Chromatogr. B*, 788:407–411, 2003.
5. Visser, N. F. C., Lingeman, H. and Irth, H., *J. Pharm. Biomed. Anal.*, 32:295–309, 2003.
6. Raices, R. S. L., Salvadori, M. C., Estrela, R. C. E., de Aquino Neto, F. R. and Suarez-Kurtz, G., *Rapid Commun. Mass Spectrom.*, 17:1611–1618, 2003.
7. Jemal, M., Huang, M., Jiang, X., Mao, Y. and Powell, M. L., *Rapid Commun. Mass Spectrom.,* 13:2125–2132, 1999.
8. Jeanville, P. M., Estape, E. S., Torres-Negron, I. and Marti, A., *J. Anal. Toxicol.,* 25:69–75, 2001.
9. Hsieh, S., Tobien, T., Koch, K. and Dunn, J., *Rapid Commun. Mass Spectrom.,* 18:285–292, 2004.
10. Mallet, C. R., Lu, Z., Mazzeo, J. and Neue, U., *Rapid Commun. Mass Spectrom.,* 16:805–813, 2002.
11. Nawrocki, J., Dunlap, C., McCormick, A. and Carr, P. W., *J. Chromatogr. A,* 1028:1–30, 2004.
12. Nawrocki, J., Dunlap, C., Li, J., Zhao, J., McNeff, C. V., McCormick, A. and Carr, P. W., *J. Chromatogr. A*, 1028:31–62, 2004.
13. Winkler, J. and Marme, S., *J. Chromatogr. A*, 888:51–62, 2000.

14. Silva, R. B., Gushikem, Y. and Collins, C. H., *J. Sep. Sci.*, 24:49–54, 2001.
15. Fonseca, D. A., Collins, K. E. and Collins, C. H., *J. Chromatogr. A*, 1030:209–215, 2004.
15A. Kolodsick, K. J., Rossi, D. T. and Kingmill, C. A., *Pharmagenomics*, p. 46, October, 2003.
16. Hsieh, Y., Wang, G., Wang, Y., Chackalamannil, S. and Korfmacher, W. A., *Anal. Chem.*, 75:1812–1818, 2003.
17. Plumb, R., Dear, G., Mallett, D. and Ayrton, J., *Rapid Commun. Mass Spectrom.*, 15:986–993, 2001.
18. Chen, Y., Junga, H., Jiang, X. and Naidong, W., *J. Sep. Sci.*, 26:1509–1519, 2003.
19. Zeng, H., Deng, Y. and Wu, J., *J. Chromatogr. B*, 788:331–337, 2003.
20. Li, A. C., Chen, Y.-L., Junga, H., Shou, W. Z., Jiang, X. and Naidong, W., *Chromatographia*, 58:723–731, 2003.
21. Neue, U. D. and Mazzeo, J. R., *J. Sep. Sci.*, 24:921–929, 2001.
22. Cheng, Y-F., Lu, Z. and Neue, U., *Rapid Commun. Mass Spectrom.*, 15:141–151, 2001.
23. Yan, B., Zhao, J., Brown, J. S., Blackwell, J. and Carr, P. W., *Anal. Chem.*, 72: 1253–1262, 2000.
24. Thompson, J. D. and Carr, P. W., *Anal. Chem.*, 74:4150–4159, 2002.
25. Thompson, J. D. and Carr, P. W., *Anal. Chem.*, 74:1017–1023, 2002.
26. Marin, S. J., Jones, B. A., Felix, W. D. and Clark, J., *J. Chromatogr. A*, 1030:255–262, 2004.
27. McNeff, C. V. and Yan, B., *PittCon 2004*, Oral Presentation #12300–600.
28. MacNair, J. E., Lewis, K. C. and Jorgenson, J. W., *Anal. Chem.*, 69:983–989, 1997.
29. MacNair, J. E., Patel, K. D. and Jorgenson, J. W., *Anal. Chem*, 71:700–708, 1999.
30. Tolley, L., Jorgenson, J. W. and Moseley, M. A., *Anal. Chem.*, 73:2985–2991, 2001.
31. Ziang, Y., Yan, B., Yue, B., McNeff, C. V., Carr, P. W. and Lee, M. L., *J. Chromatogr. A.*, 983:83–89, 2003.
32. Wyndham, K. D., O'Gara, J. E., Walter, T. H., Glose, K. H., Lawrence, N. L., Alden, B. A., Izzo, G. S., Hudalla, C. J. and Iraneta, P. C., *Anal. Chem.*, 75:6781–6788, 2003.
33. Bayliss, M. K., Little, D., Mallett, D. N. and Plumb, R. S., *Rapid Commun. Mass Spectrom.*, 14:2039–2045, 2000.
34. Deng, Y., Zeng, H., Unger, S. E. and Wu, J-T., *Rapid Commun. Mass Spectrom.*, 15:1634–1640, 2001.
35. Fang, L., Cournoyer, J., Demee, M., Zhao, J., Tokushige, D. and Yan, B., *Rapid Commun. Mass Spectrom.*, 16:1440–1447, 2002.
36. Xu, R., Wang, T., Isbell, J., Cai, Z., Sykes, C., Brailsford, A. and Kassel, D. B., *Anal. Chem.*, 74:3055–3062, 2002.
37. Fung, E. N., Chu, I., Li, C., Liu, T., Soares, A., Morrison, R. and Nomeir, A. A., *Rapid Commun. Mass Spectrom.*, 17:2147–2152, 2003.

INDEX